THERMAL COMPUTATIONS
FOR ELECTRONICS

THERMAL COMPUTATIONS FOR ELECTRONICS

Conductive, Radiative, and Convective Air Cooling

Gordon N. Ellison

CRC Press
Taylor & Francis Group
Boca Raton London New York

CRC Press is an imprint of the
Taylor & Francis Group, an **informa** business

CRC Press
Taylor & Francis Group
6000 Broken Sound Parkway NW, Suite 300
Boca Raton, FL 33487-2742

First issued in paperback 2022

ISBN-13: 978-0-367-46531-5 (hbk)
ISBN-13: 978-1-03-233631-2 (pbk)
DOI: 10.1201/9781003029328

Publisher's Note

The publisher has gone to great lengths to ensure the quality of this reprint but points out that some imperfections in the original copies may be apparent.

Library of Congress Cataloging-in-Publication Data
Names: Ellison, Gordon N., author.
Title: Thermal computations for electronics : conductive, radiative, and convective air cooling / Gordon Ellison.
Description: Second edition. | Boca Raton, FL : CRC Press/Taylor & Francis Group, 2020. | Includes bibliographical references and index.
Identifiers: LCCN 2019060142 (print) | LCCN 2019060143 (ebook) | ISBN 9780367465315 (hardback) | ISBN 9781003029328 (ebook)
Subjects: LCSH: Electronic apparatus and appliances--Thermal properties--Mathematical models. | Electronic apparatus and appliances--Cooling--Mathematics.
Classification: LCC TK7870.25 .E43 2020 (print) | LCC TK7870.25 (ebook) | DDC 621.381/044--dc23
LC record available at https://lccn.loc.gov/2019060142
LC ebook record available at https://lccn.loc.gov/2019060143

ISBN: 9780367465315 (hbk)
ISBN: 9781003029328 (ebk)

Visit the eResources: https://www.crcpress.com/9780367465315

Dedication

I dedicate this book to my wife, Sharon, who has stood by me with patient encouragement as I labored uncountable weekends and evenings from my student years to the present in an effort to be the best that I could be. Those people who know me well understand the important role played by my family both in the United States and Norway (counties of Sogn og Fjordane and Innlandet) and I dedicate this book to them also. With regard to the latter country, genealogist Bjørn Løkken of Trondheim, Norway, deserves special recognition for his contribution of countless hours devoted to the discovery of the life and exploits of my grandfather, Peder Eliassen.

Contents

Preface to the Second Edition

It continues to be this writer's experience that many universities are doing their mechanical engineering students a great disservice by not better preparing these students for employment in the electronics industry by requiring *or at least offering*, a course in the subject of electronics cooling. It seems that the typical student's only recourse is to find access to expensive computer codes for even the most modest design scenario. Rarely does one encounter a mechanical design engineer with the most basic background in the analysis and prediction of temperatures in electronic systems. My hope is that this book or those by other authors will find its way to those individuals who could profit by adding to their individual capabilities.

Since the publication of the first edition, the author has used this book as a text for a single quarter course, *Thermal Computations for Electronics*, at Portland State University and local electronics companies. During this time several typographical and a few mathematical errors have been noticed. These have been corrected. Other improvements have been made, some as modest clarifications and others of a more significant nature. The most important changes are listed as follows.

1. Chapter 1, Section 1.5: updated *FEA* illustrated example using free software which can be readily downloaded from the Internet.
2. Chapter 4, Section 4.7: modified Lee's flow bypass theory for non-zero bypass resistance.
The Application Examples in Sections 4.8 and 4.9 illustrate the direct incorporation of units conversion factors so that the analyst need not convert air velocity and head loss back and forth between SI and the convenient mixed units used throughout this book.
3. Chapter 6, Section 6.2: added derivation of dimensionless numbers for force convection heat transfer.
4. Chapter 7, Section 7.12 (new): Lee's flow bypass theory adapted to non-zero bypass resistance and compared with the Jonsson & Moshfegh empirical correlation.
5. Chapter 8, Section 8.1: added derivation of dimensionless numbers for natural convection heat transfer.
6. Chapter 10, Section 10.2: improved clarity of derivations for radiation spacial effects and the view factor.
7. Chapter 12, Section 12.4: improved derivation of the theory for a rectangular-source, time dependent, semi-infinite media solution.
8. Chapter 12, Section 12.15: added thermal spreading resistance feature to the Section 7.9 heat sink analysis problem.
9. Chapter 13, Section 13.8: re-wrote the derivation of the finite element theory, one-dimensional Euler-Lagrange equation.
10. Chapter 13, Section 13.9: changed the example to a more relevant "pin fin" application of one-dimensional *FEA*.
11. Appendix *ix*: added proof of the reciprocity for the three-dimensional time-dependent Green's function.
13. Added Appendix *xii*: Altitude effects for fan cooled enclosures with an application example.
14. Added Appendix *xiii*: Altitude effects for buoyancy cooled enclosures with an application example.

Some first edition material has been excluded from this volume. The author considers these omissions of a minor nature and largely consist of fewer examples and exercises.

The author is especially pleased to have the opportunity to incorporate the derivation of dimensionless heat transfer numbers and the *very important* consideration of altitude effects for fan and buoyancy driven enclosures. The placement of the altitude effect considerations in appendices rather than separate chapters was largely because the use of separate and new chapters would have required a very large number of equation and graph changes in successive chapters. This could have easily led to too many typographical errors.

Readers should be aware that the publisher, CRC Press, maintains downloadable files for authors texts. In the case of this book, student readers and adopting professors have access to any errata, Mathcad© files (*.xmcd for Version 15, *.mcdx for Prime Version 3) associated with many of the text examples. In addition to Mathcad© exercise solutions, adopting professors will also have access to downloadable "lecture aides" consisting of a quite complete landscape version of the text, suitable for classroom projection.

Not to be overlooked is at least one *FEA* tutorial download. The writeup is intentionally kept as brief as possible so as not to make an unwieldy download. Thus only the major illustrations are included. It is sincerely hoped that readers will take advantage of the many hours of exploration that were required to learn the steps required to set up and solve a multi-part problem with this software. You are encouraged to explore ways to use the software that may be easier than those used in the example.

Previous and new readers of my books recognize that I reproduce most derivations and many of the numerical examples in excruciating detail. With regard to derivations, it is possible that some readers will wish to modify some of my results or hopefully only rarely, disagree with something that they may wish to modify. The detailed derivations should make desired changes straight-forward. In the event of my unavailability, the reader may re-do work to his/her own satisfaction.

Students often have difficulty in obtaining correct results when trying to reproduce a text example. By including detailed calculations, it should be easier for the student to trace just where his/her error occurred.

Finally, a scientific work as complex as the one here is bound to have errors. The author apologizes for any inconvenience that this may cause the reader.

Gordon N. Ellison

Preface to the First Edition

It has been 25 years since the publication of *Thermal Computations for Electronic Equipment* (*TCEE*). Although many readers will certainly recognize my preferences in analysis methods and topics of interest, *Thermal Computations for Electronics: Conductive, Radiative, and Convective Air Cooling* is a total rewrite and bears little resemblance to the earlier book. Thus I decided to use a slightly different title to emphasize the newness of the work.

An author of an engineering or scientific book must decide when to stop adding material appearing in journals during the manuscript preparation. When I am evaluating newly published work, I usually take a considerable amount of time, sometimes a few months, before I choose to include it in my class lectures. It should not be surprising, then, that articles of somewhat recent vintage are not addressed in this new work. Certainly if you are an author or proponent of relevant information that you would wish me to have included, I would appreciate hearing from you. Do not consider omission of your work as a personal slight as time is limited for all of us and it is easy to overlook worthwhile endeavors of our peers. Nevertheless, the reader will note numerous modern theories and correlations scattered throughout the book.

Readers familiar with *TCEE* will surely note the absence of FORTRAN source code and software instructions for the *TAMS* (Thermal Analyzer for Multilayer Structures) and *TNETFA* (Thermal NETwork Flow Analyzer) computer programs. While these programs are as useful as ever, have enhanced capability, continue to be useful in my consulting, and are a significant aspect of my formal university course (for term projects), I have chosen not to include this material. There are two reasons for this choice: (1) I believe that without the software listings and description, the text material is more time independent, and (2) the frequency of upgrades to operating systems and language compilers discourages the required software recoding. Those readers with an interest in the program codes might still find copies of the book available from the reprint edition publisher, Robert Krieger Publishing Company, Malabar, FL. Several application examples are described in *TCEE*. Unfortunately, I am not able to provide support for debugging of attempted implementations of these programs.

As with *TCEE*, I have not included topics with which I have little working experience; therefore you will find no discussion of liquid cooling, heat pipes, and thermoelectric devices. I also believe these devices are best covered in the technical literature and monographs obtainable from the manufacturers. The subtitle, *Conductive, Radiative, and Convective Air Cooling,* is intended to clarify this issue at the outset.

With regard to topic organization, I have deviated from the usual practice in heat transfer texts where the early topics consist of conduction and convection. Instead, the first five chapters are largely devoted to the problem of predicting airflow in systems and heat sinks. It is only after the reader has sufficient knowledge to predict this airflow and well-mixed air temperatures that convective heat transfer is explained. Airflow and local air temperatures are, of course, necessary as component heat transfer boundary conditions.

There are those who might wonder why they should invest time in studying a book such as this when there are perhaps a half dozen or more thermal analysis programs available. Unfortunately, these same people probably have the desire to bypass traditional methods of analysis using their calculus, vector analysis, some of the simpler numerical analysis methods, and even the dimensionless heat transfer and fluid mechanics correlations. There are all too many engineers that do not have the knowledge to make even the most basic of engineering estimates, which is most unfortunate. A thorough study of this book should do much to improve this situation. Furthermore, a not-to-be-dismissed issue is that many engineers are employed in small companies where there is a preference or necessity, due to either time or budget constraints, to use traditional analysis methods that are more than adequate, particularly where great precision is not required. The design engineer with competence in application of heat transfer theory to electronics is also better prepared to construct and interpret models using advanced computational software.

I fully realize some readers would prefer that I had not continued with my use of a mixed system of units, primarily employing watts, °C, in., in.^{2}, ft/min, ft^3/min (CFM), etc. However, a primary intention in preparing this work was that it be as error free as possible. Thus, as I incorporated my large amount of preexisting theory, application examples, and student exercises, I had no choice but to stay with the mixed units system.

Rather than include only the list of references cited in this book, I have listed many journal articles and books not specifically referenced; therefore, I am providing a Bibliography that includes cited references plus many other articles, books, and reports that may attract your interest.

Finally, I wish the reader to be aware that I specifically requested that the publisher permit this book to be "author formatted." The advantage of this method is that it eliminates typographical errors introduced by the additional intervention. Therefore, any faults in the book layout or other aspects of this product are the sole responsibility of myself and I sincerely apologize for any difficulty that I may cause the reader. Those of you who have written large, complex documents consisting of more than a couple of pages understand the impossibility of creating an error free, perfect manuscript, and I am not an exception to that failing.

To the Student

This book is written for two types of readers. The first is the engineering student studying the subject in a formal course setting for which exercises are provided at the end of each chapter. The second type of reader is the practicing engineer or scientist studying on his/her own and for whom the very detailed text examples should be helpful.

Students have the choice of three tools for obtaining numerical results for problems in this book: (1) a scientific calculator, (2) a spreadsheet program, and (3) a math scratch-pad program such as Mathcad™. The one tool that I refuse to allow my students to use in my course is the spreadsheet program. I understand that this powerful program exists on nearly every computer, but I almost always get unsatisfactory results from students that turn in assignments from spreadsheets. My recommendation is that you either invest in the Mathcad™ program (low-price academic versions are available to qualified students) or use an authorized version installed on a computer at your university or place of work. Your formulae will look almost exactly as they do in the source from which you obtained them and your work will be easy to check for errors. I have had many students who also do very well using only their scientific calculator.

Whether you are studying this material in a classroom or you are on your own, your learning experience will be maximized if you first attempt to solve the text examples. The text solutions are usually listed in excruciating detail so that you can better find any errors that you might make in your own calculations.

Most topics in this book are not particularly difficult to understand, with the exception of some of the material in Chapters 12, 13, and Appendices *vi-xi*. These latter sections may be challenging. However, as I take an overview of the entire subject as presented herein, I realize that a person new to the subject may very well find it difficult. Be patient in your studies; few worthwhile endeavors are completed without some struggle.

To the Instructor

The author intended that this text be used for a complete course on the subject of electronics cooling, but also believes that the book is useful as a reference from which selected sections may be used as application examples in a traditional heat transfer course. If you are new to teaching an electronics cooling course, you might wish to consider the following approximate schedule for a four-credit-hour course, given in a ten-week quarter system.

Week 1: Chapters 1 and 2 in entirety.
Week 2: Chapter 3 in entirety, beginning of Chapter 4.
Week 3: Remainder of Chapter 4, beginning of Chapter 6.

Week 4: Completion of Chapter 6, Chapter 7 in entirety.

Week 5: Chapter 5 in entirety, beginning of Chapter 8.

Week 6: Completion of Chapter 8, Chapter 11 basics (Fourier's law, simple 1-D conduction, fin efficiency details, 3-D conduction eq., physics of thermal conductivity, *PCB* thermal conductivity, interface resistance).

Week 7: Chapter 9 in entirety, first part of Chapter 12 (fixed angle theories, circular source semi-infinite media with Bessel functions).

Week 8: Appendices *vii, viii*, Chapter 12-Section 12.6 for rectangular source, finite media, one convecting surface [through Eq. (12.44)], Appendix *ix*.

Week 9: Chapter 12-Section 12.6 (remainder) rudiments, Chapter 12-Sections 12.7 through 12.9, Chapter 12 Section 12.14.

Week 10: Chapter 10 (as much as possible in two hours), Chapter 13-Sections 13.1 through 13.5 (thermal networks).

The schedule omits time for examinations because I prefer giving take-home exams for both the midterm and final. Both of my exams consist of a single problem resembling one of those from Appendix *i*, Supplemental Exercises. In the case of the midterm, I try to include as many of the application topics from the first twenty classroom hours. For example, a complete midterm problem would be a forced air cooled enclosure with circuit boards, a plate-fin heat sink, and a power supply; at least one circuit board requiring prediction of component temperatures using adiabatic heat transfer coefficients and temperatures. A satisfactory final exam problem is usually a natural convection, vented system with one or more vertical circuit boards, and perhaps an externally finned heat sink (with power transistors). If students have mastered the subjects of these two examinations, I am convinced that I have served them well. Testing for understanding of the basics of topics not covered in these exams is easily addressed by regular, unannounced, short quizzes.

At this point, I must admit that the above schedule is quite ambitious. In my own case, I have to address only the basics of some of the subjects. This is in part due to the fact that in the middle of the course I introduce portions of Chapters 12 (boundary conditions and differential equation for rectangular sources) and 13 (steady-state networks), and work through the solution of a single chip package on a circuit board using my *PTAMS, TAMS,* and *TNETFA* software. The students are assigned a term project consisting of an IC package mounted on an epoxy-glass circuit board and are expected to successfully solve the problem using the above-named software. This requires that I offer in-class assistance on two successive Saturdays near the end of the term. Certainly the software project allows them to successfully solve a somewhat complex conduction problem (approximately 1000 network nodes) that is far beyond the capability of a calculator or easy spreadsheet solution. I seriously doubt that both the theory and user instructions for an *FEA* program could be successfully introduced in the same amount of time. I have rarely had the time to present the calculus of variations and finite element theory in Chapter 13.

With regard to the manner in which I personally solved the text and exercise problems, I have used Mathcad™ almost exclusively. It would be a simple matter for you as an instructor to formulate variations of the text and exercise problems merely by using different dimensions, heat dissipation, etc. I remind you that Appendix *i* is a listing of exercises that require knowledge from more than one chapter and many of these are ideal for examination problems. Worked solutions using Mathcad™ in worksheet form or PDF file format will be available for college or university instructors. Some of the exercises require an iterative solution. The Mathcad™ worksheets are not programmed to automatically cycle through the problem until convergence is achieved: Each iteration requires that the final results from the previous iteration be entered at the beginning of the next iteration. The worksheet solutions provided will usually show only the final iteration.

As a direct aid in your lectures, an extensive set of PDF landscape files will also available to instructors adopting this book. The lecture material is very complete and you should have little need to add supplementary material unless you really wish to do so. If you have an interest in using the *PTAMS*, *TAMS*, and *TNETFA* software for a term project assignment, I encourage you to speak directly with me via email or telephone. At some point, I expect to post errata on the website of either or both that of the publisher or myself.

About the Author

Gordon N. Ellison has a BA in physics from the University of California at Los Angeles (UCLA) and an MA in physics from the University of Southern California (USC). His career in thermal engineering includes twelve years as a technical specialist at NCR and eighteen years at Tektronix, Inc., retiring from the latter as a Tektronix Fellow. Over the last fifteen years Ellison has been an independent consultant and has also taught the course, Thermal Analysis for Electronics, at Portland State University, Oregon. He has also designed and written several thermal analysis computer codes. His publications include journal articles, conference presentations, monographs, and a prior book, *Thermal Computations for Electronic Equipment*.

Gordon Ellison presently lives in Newberg, Oregon, with his wife to whom he has been married for approximately sixty years. His non-professional activities include spending as much time as possible with his wife, two nearby daughters, five grandchildren, and also his *very active*, *extremely intelligent*, border collie, Danny. He continues to learn about his Norwegian heritage, which includes the discovery that his surname was originally Eliassen, and that his grandfather, Peder Eliassen, is still considered to be the greatest cross-country skier of all time for the area surrounding the village of Løten, Innlandet, Norway.

Acknowledgments

It is appropriate to give credit to people that have been particularly influential in my career. The late Dr. John Nodvik, Professor of Physics at the University of Southern California (USC), taught me much about mathematical physics. I was fortunate to be able to take many of his courses, the most important of which was perhaps his Methods of Theoretical Physics, taught from the famous two-volume set of books of the same name and written by P.M. Morse and H. Feshbach (their book is back in print as of this writing). Early in my working career I enrolled in an evening extension course, also in the subject of Mathematical Physics, at the University of California at Los Angeles (UCLA) and taught by the late Professor Charles Lange. The latter part of this course was devoted to Green's functions. Following his presentation of Green's functions, Dr. Lange went far beyond the required minimum to give me personal guidance in my theoretical studies. It is not possible for me to overstate the contributions of these two teachers to my professional endeavors. A turning point in my life occurred when I was working at NCR in the late 1960s. Several managers (Don Meier, Anthony Kolk, the late Dr. Henry White, and Majid Arbab) provided me the opportunity to work in the microelectronics heat transfer field. I will be humbly grateful to them to the end of my days.

My careers at NCR and Tektronix, Inc., were rewarding in both technical challenges and interpersonal relationships with fellow employees. I must acknowledge my gratefulness to the late Bill Snell for bringing me into Tektronix and Jack Hurt for his support in the interview process and also during my early years in the company. I was privileged to work for Mr. Imants Golts for several years, many of which were fraught with challenges. Mr. Golts supported my work every step of the way and was successful in insulating me from many distractions. It is relevant to also acknowledge the short period during which I worked for Mr. Edward Hershberg who provided challenging project assignments and also supported my position within the company. Mr. Dana Patelzick provided numerous contributions to our electronics cooling work.

Since my retirement from Tektronix, I have had a rewarding career as an assistant adjunct professor at Portland State University in the Department of Mechanical and Materials Engineering. I wish to thank Professors Graig Spolek and Gerald Recktenwald for their unrelenting support of my course. As you would expect, my course notes served as the foundation of this book.

I also wish to thank others who contributed to this project. They include Mrs. Cynthia Wilson for her superb editing, my four reviewers who used their valuable time to critique my manuscript, and the staff at CRC Press/Taylor & Francis Group including Nora Konopka, Brittany Gilbert, Jessica Vakili, Theresa Delforn, Michele Dimont, and countless others.

Gordon N. Ellison

CHAPTER 1

Introduction

This chapter might be described as a very brief survey of modeling air cooled electronics, but little detail will be included to show you how to make quantitative predictions. If you have limited electronics thermal analysis experience, then hopefully the following sections will encourage you to read more of this book. If you do have some experience in analyzing electronics cooling designs, then you will at least get a preview of the author's inclinations in this field.

Not surprisingly, we will begin with the conventional basics, i.e., conduction, convection, and radiation, and include a few illustrative examples. We will consider these topics in greater detail in succeeding chapters. Finally, this book would seem incomplete without a couple of examples that require digital computer analysis, namely using the thermal network and finite element methods, for which software is sufficiently affordable for individuals desiring to conduct their own consulting activities.

1.1 PRIMARY MECHANISMS OF HEAT FLOW

Engineering thermal analyses of electronic systems are based on any or all of three methods of thermal-energy transport: conduction, convection, and radiation. Conduction takes place within a medium, but without obvious transport of the medium itself.

Convective heat transfer also requires a medium for energy flow, but in this instance there is mass transport of the medium. One of the most visible examples of material transport is water heated in an open kettle. The warm water rises in the center of the kettle and falls to the bottom as it becomes cooled upon transferring heat to the kettle walls. The liquid flow not only mixes the fluid, but actually aids in the rate of heat transfer in the vicinity of the walls. A very careful examination of the fluid immediately adjacent to the wall would show negligible fluid flow. In this very thin layer, heat is transferred by conduction.

In a manner similar to the heated water in a kettle, circulating convective air currents in the interior of sealed electronic enclosures aid the transfer of heat energy to the cabinet wall. Upon conduction through the metal or plastic enclosure, external convective heat transfer aids removal of thermal energy from the system.

Radiation heat transfer is totally unique when compared with conduction and convection in that no transport medium is required because radiation energy transport occurs via the propagation of an electromagnetic radiation field through space. Although this field typically covers the entire electromagnetic spectrum, most of the thermal radiation encountered from conventional microelectronic components and systems is located in the infrared region.

All three heat transfer mechanisms obey the second law of thermodynamics in the sense that there is a net energy flow only from a higher temperature to a lower temperature region.

1.2 CONDUCTION

Heat conduction in a one-dimensional bar is illustrated in Figure 1.1, where the heat flow Q_k at the location x is in the positive x-direction. The cross-sectional area at x is A_k, where there is a temperature gradient dT/dx. The manner in which different materials conduct heat is represented by a "constant" of proportionality, the thermal conductivity k. The conductive heat transfer is, with few exceptions, quantified by Fourier's law:

$$\boxed{Q_k = -kA_k \, dT/dx\big|_x}$$

Fourier's law (1.1)

where

$$Q_k \equiv \text{heat transferred } \left[\text{W}\right], k \equiv \text{thermal conductivity } \left[\text{W}/\left(\text{in.} \cdot {}^\circ\text{C}\right)\right], A_k \equiv \text{area } \left[\text{in.}^2\right],$$

$$x \equiv \text{location in one-dimension } \left[\text{in.}\right], T \equiv \text{temperature } \left[{}^\circ\text{C}\right]$$

The minus sign in Fourier's law is due to the fact that we use the convention of positive heat flow in the positive x-direction for a negative temperature gradient. In this example, temperature gradients in directions other than x are assumed negligible.

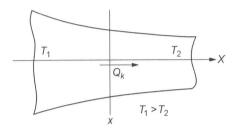

Figure 1.1. Heat conduction in a one-dimensional bar.

Electronics thermal analysis problems require that we consider a wide range of thermal conductivity values. Figure 1.2 illustrates that a single problem that includes air and solid conduction may include conductivity ratios as great as 10^5. This may require consideration when we use numerical analysis techniques. Careful consideration of property values sometimes allows us to eliminate pieces of a problem due to insignificance. Additional thermal conductivity data is listed in Appendix *iii*. Equation (1.1) is not particularly useful without additional effort. It is for this reason that we rewrite Fourier's law in a different form by rearranging the terms and integrating over the length of a one-dimensional bar of length L with a temperature T_1 at $x=0$ and T_2 $(T_1 > T_2)$ at $x=L$.

$$\frac{Qdx}{A_k} = -k\left(T\right)dT; \qquad \int_0^L \frac{Qdx}{A_k} = -\int_{T_1}^{T_2} k\left(T\right)dT = \int_{T_2}^{T_1} k\left(T\right)dT$$

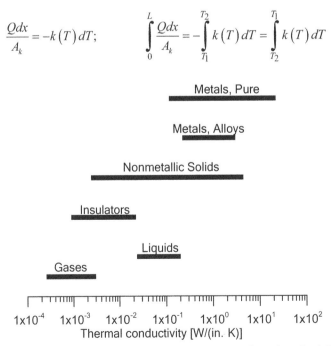

Figure 1.2. Comparison of solids, liquids, and gases. Indicated conductivity ranges are approximate.

When the thermal conductivity is not temperature-dependent, we end up with

$$\Delta T = RQ, \qquad \Delta T = T_1 - T_2$$

$$\boxed{R = \frac{1}{k}\int_0^L \frac{dx}{A_k}} \qquad \text{Thermal resistance (1.2)}$$

where R is defined as the thermal resistance of the bar. If the area is uniform over the length of the bar, we have

$$\boxed{R = \frac{L}{kA_k}} \qquad \begin{array}{l}\text{Temperature independent } k \ (1.3)\\ \text{uniform } A_k\end{array}$$

The thermal conductivity of many materials varies little over the range of temperatures encountered in electronics applications. We can usually estimate the expected operating temperature and use a mean thermal conductivity k_m. In a formal statement of this latter declaration, we return to our integration of Fourier's law.

$$\int_0^L \frac{Qdx}{A_k} = -\int_{T_1}^{T_2} k(T)\,dT = -(T_2 - T_1)\left(\frac{1}{T_2 - T_1}\right)\int_{T_1}^{T_2} k(T)\,dT = (T_1 - T_2)k_m$$

$$\Delta T = (T_1 - T_2) = Q\frac{1}{k_m}\int_0^L \frac{dx}{A_k} = RQ, R = \frac{1}{k_m}\int_0^L \frac{dx}{A_k}$$

$$\boxed{k_m = \left(\frac{1}{T_2 - T_1}\right)\int_{T_1}^{T_2} k(T)\,dT} \qquad \text{Mean conductivity } k_m \ (1.4)$$

A linear temperature dependence of k is often adequate, particularly over a small temperature range,

$$k = k_0 (1 + \alpha T)$$

in which case the mean conductivity is given by Eq. (1.5).

$$k_m = \frac{k_0}{T_2 - T_1}\int_{T_1}^{T_2} (1 + \alpha T)\,dT = \frac{k_0}{T_1 - T_2}\left(\int_{T_2}^{T_1} dT + \alpha\int_{T_2}^{T_1} T\,dT\right)$$

$$k_m = \frac{k_0}{T_1 - T_2}\left[(T_1 - T_2) + \frac{\alpha}{2}(T_1^2 - T_2^2)\right]$$

$$\boxed{k_m = k_0\left[1 + \frac{\alpha(T_1 + T_2)}{2}\right]} \qquad \text{Linear } k \text{ dependence on } T \ (1.5)$$

A power law fit is sometimes useful. For example, Lasance (1998) recommends that the conductivity for pure silicon may be approximated by

$$\boxed{k_{Si}\left[\text{W}/(\text{in.}\cdot{}^\circ\text{C})\right] = 3.81\left(\frac{T\left[{}^\circ\text{C}\right] + 273.16}{300}\right)^{-4/3}} \qquad \text{Pure silicon } k\,(1.6)$$

and highly doped silicon calculated at 80% of the pure value.

The mean thermal conductivity for pure silicon is calculated using Eqs. (1.4) and (1.6).

$$k_{m-Si} = \left(\frac{1}{T_1 - T_2}\right)\int_{T_2}^{T_1} k(T)\, dT = \left(\frac{3.81}{T_1 - T_2}\right)\int_{T_2}^{T_1}\left(\frac{T + 273.16}{300}\right)^{-4/3} dT$$

A change of independent variable T to v is made to facilitate the integration.

$$v = \frac{T + 273.16}{300}, \quad dT = 300\, dv$$

$$k_{m-Si} = \frac{(3.81)(300)}{T_1 - T_2}\int_{v_2}^{v_1} v^{-4/3}\, dv = \frac{-(3.81)(300)(3)}{T_1 - T_2}\left(v_1^{-1/3} - v_2^{-1/3}\right)$$

$$k_{m-Si} = \frac{(3.81)(300)(3)}{T_1 - T_2}\left(v_2^{-1/3} - v_1^{-1/3}\right)$$

$$\boxed{k_{m-Si}\left[\text{W}/\left(\text{in.}\cdot{}^\circ\text{C}\right)\right] = \frac{3429.00}{T_1 - T_2}\left[\left(\frac{T_2 + 273.16}{300}\right)^{-1/3} - \left(\frac{T_1 + 273.16}{300}\right)^{-1/3}\right]} \quad \text{Pure Si } k_m \quad (1.7)$$

1.3 APPLICATION EXAMPLE: SILICON CHIP RESISTANCE CALCULATION

Suppose our example is a silicon chip that measures 0.50 in.$\times$ 0.50 in. and has a thickness of 0.020 in. Furthermore assume that we have calculated that the base of the chip is at a temperature $T_2 = 100°\text{C}$ and the power dissipation at the chip surface is $Q = 50$ W. We shall use Eq. (1.7) to calculate the appropriate mean thermal conductivity. However, note that we need a T_1 to begin the calculation. Let's choose $T_1 = 150°\text{C}$. The first calculation of k_m is

$$k_m = \frac{3429.00}{150 - 100}\left[\left(\frac{100 + 273.16}{300}\right)^{-1/3} - \left(\frac{150 + 273.16}{300}\right)^{-1/3}\right] = 2.62\ \text{W}/\left(\text{in.}\cdot{}^\circ\text{C}\right)$$

We calculate the chip surface temperature T_1 to be

$$T_1 = RQ + T_2 = \left(\frac{L}{k_m A_k}\right)Q + T_2 = \left[\frac{0.02}{(2.62)(0.50)^2}\right](50) + 100 = (0.031)(50) + 100 = 101.53\,^\circ\text{C}$$

Clearly our first guess of $T_1 = 150°\text{C}$ was too large and a new calculation of R and T_1 is in order.

$$k_m = \frac{3429.00}{101.53 - 100}\left[\left(\frac{100 + 273.16}{300}\right)^{-1/3} - \left(\frac{101.53 + 273.16}{300}\right)^{-1/3}\right] = 2.84\ \text{W}/\left(\text{in.}\cdot{}^\circ\text{C}\right)$$

$$T_1 = RQ + T_2 = \left(\frac{L}{k_m A_k}\right)Q + T_2 = \left[\frac{0.02}{(2.84)(0.50)^2}\right](50) + 100 = (0.028)(50) + 100 = 101.41\,^\circ\text{C}$$

A third *iteration*, not shown here, gives the same result as the second iteration: the mean thermal conductivity, resistance, and chip surface temperature are 2.84 W/(in.·°C), 0.028 °C/W, and 101.41°C, respectively.

1.4 CONVECTION

Convective heat transfer from a surface is illustrated in Figure 1.3. The basic relation, Eq. (1.8), that describes surface convection presumes a linear dependence on the surface temperature rise $T_S - T_A$ above ambient.

$$Q_c = \bar{h}_c A_S \left(T_S - T_A \right) \qquad (1.8)$$

where we define a convection *conductance* $C_c = \bar{h}_c A_S$. Eq. (1.8) describes *Newtonian cooling*. The units used in this book are

$$Q_c \equiv Q_c \left[\mathrm{W} \right]$$
$$\bar{h}_c \equiv \bar{h}_c \left[\mathrm{W}/\left(\mathrm{in.}^2 \cdot {}^\circ\mathrm{C} \right) \right]$$
$$A_S \equiv A_S \left[\mathrm{in.}^2 \right]$$
$$C_S \equiv C_S \left[\mathrm{W}/{}^\circ\mathrm{C} \right]$$
$$T_S, T_A \equiv T_S, T_A \left[{}^\circ\mathrm{C} \right]$$

A common form of Eq. (1.8) is

$$\Delta T = \left(1/\bar{h}_c A_S \right) Q_c = R_c Q_c \qquad (1.9)$$

where $R_c = 1/\left(\bar{h}_c A_S \right)$ is the convection resistance and has units of °C/W. The reader will note that the preceding heat transfer coefficients indicate an average value. This implies that the surface temperature or temperature rise is also averaged over the surface. Much of the detailed data in the heat transfer literature is given as a surface average, and where it is given as position-dependent, it is still more convenient (and adequately accurate) to use the average value.

Several comments are in order before we proceed to a sample calculation. First, you should recognize that Eq. (1.8) is not a fundamental law of heat transfer in the sense of Fourier's law of conduction. Rather, it is a definition of the heat transfer coefficient. This definition appears rather simple and straightforward: It specifies the quantity of heat convected from a unit area and transferred through a temperature difference. The oversimplification is that h_c may depend quite significantly on the surface and ambient temperatures, and fluid velocity in the case of forced convection.

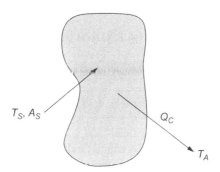

Figure 1.3. Convection Q_c to an ambient temperature T_A from a surface area A_S at temperature T_S.

Fluid properties such as viscosity and density, and surface geometry are important. Detailed theoretical studies of convective heat transfer that bring out these details may be found in undergraduate heat transfer texts. The results of some of these studies will be summarized in Chapters 6 through 9 in formulae and graphs useful for application to design problems. For the present introductory purposes, it is sufficient to indicate a range of values encountered.

Table 1.1. Approximate range of convective heat transfer coefficients for air.

Mechanism	$h_c \left[W \middle/ \left(\text{in.}^2 \cdot \,^{\circ}C \right) \right]$
Natural air convection	0.001 - 0.01
Forced air convection	up to 0.2

The values listed in Table 1.1 are approximate. They are calculated from some of the heat transfer coefficient correlations described in later chapters. The range of h_c for forced convection does not list a lower limit because it would overlap the natural convection range and would rightly have to be identified as *mixed-convection* where both phenomena would be present.

1.5 APPLICATION EXAMPLE: CHASSIS PANEL COOLED BY NATURAL CONVECTION

The geometry for this problem is illustrated in Figure 1.4. A power transistor, for which we show only the approximate footprint, dissipates $Q = 7$ W. We shall estimate the average plate temperature. In practice, there will be a *hot spot* directly beneath the heat source, but this topic shall not be addressed until a later chapter. At this point in our study, we are not prepared to calculate what the actual heat transfer coefficient really is, so we shall just use a value of $h_c = 0.004$ W/(in.$^2 \cdot \,^{\circ}$C).

We begin by using Eq. (1.9).

$$\Delta T = R_c Q_c = \left(Q_c \middle/ \overline{h}_c A_S \right) = 7 \text{ W} \middle/ \left[\left(0.004 \frac{\text{W}}{\text{in.}^2 \cdot \,^{\circ}C} \right) \left(2 \times 4.0 \text{in.} \times 9.0 \text{in.} \right) \right] = 24 \,^{\circ}C$$

The resistance is

$$R_c = 1 \middle/ \left(\overline{h}_c A_S \right) = 1 \middle/ \left[\left(0.004 \frac{\text{W}}{\text{in.}^2 \cdot \,^{\circ}C} \right) \left(2 \times 4.0 \text{in.} \times 9.0 \text{in.} \right) \right] = 3.5 \,^{\circ}C/W$$

Note that we have used both sides of the plate for the convecting surface area. Both the surface temperature rise ΔT and the resistance R_c are average values because we used an average h_c.

1.6 RADIATION

Newtonian cooling may also be used to define a radiation heat transfer coefficient, h_r.

$$\boxed{Q_r = \mathcal{F} h_r A_S \left(T_S - T_A \right)} \tag{1.10}$$

$\mathcal{F}$ is a factor that includes both surface finish and geometry effects. A common, but elementary example is that of the exterior surface of a cabinet radiating to the surrounding room walls which are very nearly the same temperature as the ambient air. In this case $\mathcal{F} = \varepsilon$ where ε is defined as the cabinet surface emissivity, a quantity that is less than 1.0 for all real surfaces ($\varepsilon \cong 0.8$ for many painted surfaces).

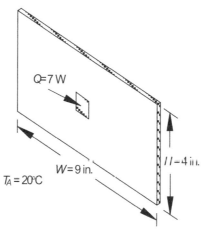

Figure 1.4. Application Example 1.5: Chassis bulkhead cooled by natural convection. Both sides of panel convect to ambient.

When a surface has a temperature T_S that is within 20°C of T_A and T_A is between 0 and 100°C, a reasonable engineering approximation for h_r is

$$h_r = 1.463 \times 10^{-10} \left(T_A + 273.16\right)^3 \qquad (1.11)$$

where $T_A = T_A$ [°C] and $h_r = h_r$[W/(in.$^2 \cdot$ °C)] .

If we perform a modest manipulation of Eq. (1.10), we see that we have a formula for the radiation resistance between a surface and ambient.

$$R_r = 1 / \left(\mathcal{F} h_r A_S\right) \qquad (1.12)$$

As a final comment on this brief introduction to radiation, it should be noted that some authors choose to incorporate the emissivity directly into formulae for the radiation heat transfer coefficient. Such a treatment is not unreasonable because the resultant h_r is then the actual W/(in.$^2 \cdot$ °C) from the surface. Such a preference will not be used in this book.

1.7 APPLICATION EXAMPLE: CHASSIS PANEL COOLED ONLY BY RADIATION

Consider the outer-top surface of a chassis panel illustrated in Figure 1.5. The panel has an average temperature of $T_S = 30$°C in an ambient of $T_A = 20$°C. Assuming that the panel emissivity is about $\varepsilon = 0.8$, we can estimate the heat transfer to ambient by radiation using Eqs. (1.10) and (1.11).

$$h_r = 1.463 \times 10^{-10} \left(T_A + 273.16\right)^3 = 1.463 \times 10^{-10} \left(20 + 273.16\right)^3 = 3.69 \times 10^{-3} \ \text{W} / \left(\text{in.}^2 \cdot \text{°C}\right)$$

$$Q_r = \varepsilon h_r A_s \Delta T = \left(0.8\right)\left(3.69 \times 10^{-3}\right)\left(10 \times 12\right)\left(10\right) = 3.54 \ \text{W}$$

We see that 3.54 W are radiated to ambient from this panel.

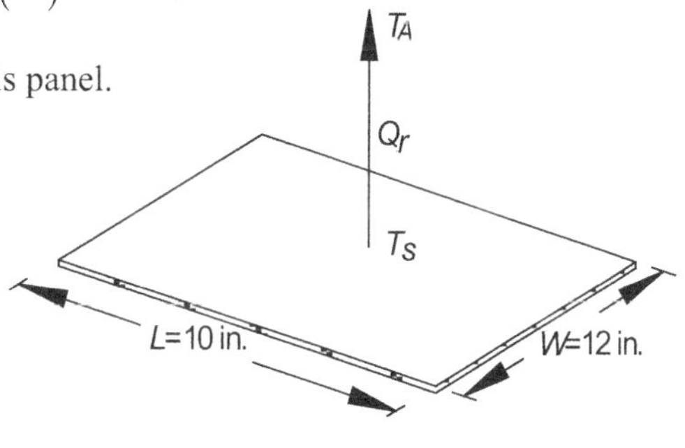

Figure 1.5. Application Example 1.7: Chassis panel top surface cooled by radiation.

1.8 ILLUSTRATIVE EXAMPLE: SIMPLE THERMAL NETWORK MODEL FOR A HEAT SINKED POWER TRANSISTOR ON A CIRCUIT BOARD

In the next several sections we shall see applications of network theory to a variety of problems. An important aspect of network methods is that there is nearly always a one-to-one correspondence between the physical and network models. Furthermore, we can solve both temperature/heat transfer and pressure/airflow problems with this methodology. Many network problems are solvable with a scientific calculator or math "scratchpad" software. When conduction problems become complicated, we usually resort to using commercial thermal network software. Very complex geometry may require a finite element program. Nevertheless, the intent of this book is to teach you how to solve many of your design problems independent of "high-end" software. Some of these examples will be quantitative, but as in previous examples, don't expect to be able to understand every detail at this stage of your study. You will learn the details in later chapters.

A power transistor with a simple, stamped heat sink is illustrated in Figure 1.6. The heat dissipating chip is bonded to the metal base of the power transistor. The chip is protected from the environment by a metal chip cover. The heat sink is bonded to the chip cover. The illustration shows a nut and bolt used to attach the metal transistor base to the printed circuit board (*PCB*). We shall assume that the circuit board is composed of mostly FR4, an epoxy-glass composite that, in its raw form, is a poor heat-conducting material. The poor heat-conducting nature of the circuit board is somewhat compensated for in this example by a nearly solid copper layer that is 0.0014 in. thick. Realistic dimensions could be 0.5 in. × 0.5 in. for the power transistor "footprint" and 1.5 in. × 1.5 in. for the *PCB*, respectively. We assume that the transistor dissipates a total of six watts.

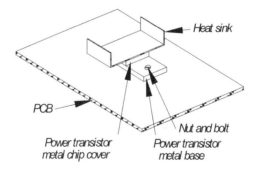

Figure 1.6. Power transistor with heat sink on a printed circuit board.

The various thermal resistances shown in Figure 1.7 can be calculated using analytical techniques and design curves provided in succeeding chapters. You may recognize portions of or all of this problem later in your reading. We shall only show the results here. Table 1.2 lists the various resistances that have been calculated.

Table 1.2. Resistance values for Figure 1.7.

Resistor No.	Value [°C/W]
1	59
2	225
3	0.124
4	0.8
5	25

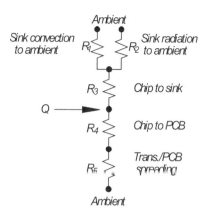

Figure 1.7. Thermal network circuit for heat sinked power transistor on printed circuit board (*PCB*).

Using the resistor values, it is a simple matter to "synthesize" the circuit, i.e., add the resistances to obtain the total resistance from the chip to ambient. If you have completed Exercise 1.4, you know that the rule for adding thermal resistances R_a and R_b in parallel is

$$(1/R) = (1/R_a) + (1/R_b) \tag{1.13}$$

which is what we use to combine the sink convection and radiation. This result is added to R_3 (using a "series addition" formula $R = R_a + R_b$) to get $R_{Chip\text{-}Top} = 46.9$ °C/W, the resistance from the chip to ambient above the assembly. You should recognize that the resistances R_4 and R_5 are added as a series sum to get the resistance from the chip, through the *PCB* to the ambient beneath the assembly to get $R_{Bot} = 25.8$ °C/W. Finally, the resistances from chip to top ambient and from chip to bottom ambient are added as a parallel sum $R = 17$ °C/W. The chip temperature rise above ambient is calculated as

$$\Delta T_{Chip-to-Amb} = QR = (6\,\text{W})(17\,°\text{C}/\text{W}) = 102\,°\text{C}$$

If the assembly were in an electronics enclosure where the local air temperature was 50°C, then the transistor chip would be at a temperature of 152°C.

1.9 ILLUSTRATIVE EXAMPLE: THERMAL NETWORK CIRCUIT FOR A PRINTED CIRCUIT BOARD

One of the reasons that the thermal network method is so flexible is that it is *geometry-independent*, thus any phenomenon for which you can calculate thermal resistance values may be represented by a network model. An example of a portion of a hypothetical *PCB* is shown in Figure 1.8. Circuit board heat conduction in in-plane directions is represented by thermal resistances R_B (through-plane resistances are possible, but not used here); *PCB*-to-fluid convection is represented by the $R_{PCB\text{-}f}$; component surface-to-fluid convection by $R_{C\text{-}f}$; component conduction from case to *PCB* via leads and an air gap or connector is shown by R_L; conduction for the chip junction-to-component case shown as R_{JC}; fluid temperature rise from the direction of inlet-to-exit is modeled by the R_f. The latter are *one-way* resistances in the sense that downstream nodes (where fluid is going) *see* only upstream nodes (where fluid is coming from). Upstream nodes do not see downstream nodes. The "eye-symbols" in Figure 1.8 are a reminder that the R_f are one-way resistors.

Temperatures are calculated at *nodal* points, which are geometrically represented in Figure 1.8 by the small circles. You should be able to recognize T_J as a chip junction temperature within the component case.

Clearly, accurate thermal analysis of a circuit board is a very complex problem and such an analysis is approximate at best. Heat conduction analysis within the *PCB* must take into account the metal content and distribution, details of the component structure must be considered, and convective heat transfer correlations for components on circuit boards are difficult to find in the literature. Finally, network modeling of the fluid flow is very approximate. Nevertheless, an analyst can obtain considerable design information if reasonable care is taken.

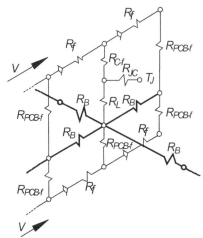

Figure 1.8. A portion of a thermal network model for a printed circuit board.

1.10 COMPACT COMPONENT MODELS

Manufacturers of transistors and integrated circuit chips have a long history of providing thermal test data for design use. The simplest set of such data consists of two thermal resistances, R_{JA} and R_{JC}, the junction-to-ambient and junction-to-case resistances. R_{JC} was defined as the chip junction temperature rise above the case temperature divided by the chip heat dissipation. R_{JA} was defined as the junction temperature rise above the ambient temperature divided by the chip power dissipation. Historically, it was not uncommon for system designers and thermal analysts alike to use these for temperature prediction.

A major change in chip junction prediction methodology was brought about by Andrews (1988), who recognized that changes were necessary to improve the accuracy of junction temperature prediction. Bar-Cohen et al. (1989) proposed a *star-model* as an improved method of characterization. Ultimately, the appearance of computational fluid dynamics (*CFD*) dedicated to electronics systems analysis led to a flurry of research in thermal network models for components where the number of resistors was sufficiently small to be computationally efficient in large *PCB* or systems analysis, but also boundary condition independent. Early work in this area was initiated largely by Lasance et al. (1997) with the intent of incorporating accurate compact component models into *CFD* system simulation.

Some of the major difficulties with the R_{JA}, R_{JC} models were (1) resistance-dependence on the selected case point, (2) boundary conditions, i.e., the value of the heat transfer coefficient, (3) *PCB* properties, and (4) test method variations among component vendors. Successful compact models address and eliminate some of these problems.

Some thermal networks considered for compact model development are illustrated in Figure 1.9. The resistance values in the circuits can be obtained by constructing very detailed finite element or finite difference models and applying an optimization method that minimizes the temperature differences between the compact and detailed models. Successful compact component models may then be incorporated into *FEM* or *FDM* circuit board models. If the board model is part of a *CFD* problem, then presumably the *CFD* program would compute the convective heat transfer coefficients for the component surfaces. The analyst could also use heat transfer coefficient correlations if so desired.

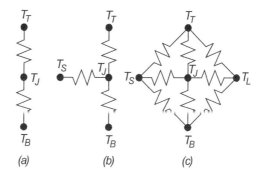

Figure 1.9. Some compact component models consisting of (a) two-resistor model, (b) three-resistor star model, and (c) eight-resistor star model. The temperatures T_J, T_S, T_L, T_T, and T_B represent junction, case side, case leads, case top, and case bottom, respectively.

1.11 ILLUSTRATIVE EXAMPLE: PRESSURE AND THERMAL CIRCUITS FOR A FORCED AIR COOLED ENCLOSURE

A cutaway drawing of a hypothetical personal computer is shown in Figure 1.10. This example is not intended to be an example of good thermal design practice, but is adequate for the purpose of illustrating the idea of a pressure/airflow circuit. There are three air inlets: the sink-fan inlet, the system inlet (labeled as "Inlet"), and air gaps around the removable and fixed drives module.

The sink-fan module is a module that contains a fan, heatsink, and microprocessor. Air enters a grill at the sink-fan inlet, flows through the fan, impinges onto the heatsink, and exits the sink-fan module, exiting the system enclosure. The fan-sink is an entity separate from the system and is presumed to be adequately designed by the manufacturer. The air from the two left-side inlets flows through a channel defined by the enclosure top panel and the uppermost expansion board, through the system fan, and finally exits the system through a protective grill following the system fan.

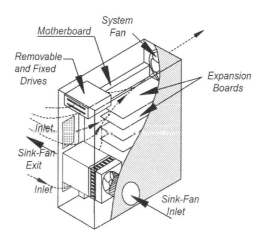

Figure 1.10. A hypothetical personal computer with expected airflow pattern.

Airflow paths are not well defined in the regions between and around the lower three expansion boards, as well as around the lower regions of the motherboard. It is to be expected that these areas are not well managed by the system fan and are therefore cooled by natural convection.

An airflow/pressure circuit for the illustrated PC is shown in Figure 1.11. Each of the resistor symbols represents an expected resistance to airflow. Airflow regions with small flow resistance do not have circuit resistances. In practice, the major system resistances are usually inlet and exit grills, louvers, slots, or similar devices. A major task that we have as analysts is to quantify these resistances. Calculation of resistances for the inlets, circuit board channels, and exits are usually straightforward. It would be preferable to obtain the resistance for the microprocessor heat sink from the vendor of either the heat sink or the microprocessor, although it would certainly be possible to estimate a heat sink resistance for initial calculations.

In order to actually compute airflow and pressures for this example, we also need pressure/airflow properties of the fan so that we know how it interacts with the system. These fan properties are always in the form of a graph of pressure across the fan vs. airflow through the fan and are readily available from the fan vendor.

We can solve many pressure/airflow circuits by using rules for adding parallel and series elements with the ultimate result of obtaining a single total resistance for the entire system. We then combine this system resistance with the fan curve to obtain the actual operating airflow. Then by working backward, we obtain the airflow in each resistor. We shall not discuss this circuit synthesizing procedure further here, but postpone it to a later chapter where we shall go into it in much greater detail.

It is important, however, to realize that some circuits are constructed in ways we cannot solve by using simple series and parallel addition of resistors. In these cases we must resort to numerical analysis methods, which are rather simple for circuits consisting of only a few elements. Occasionally the circuit may have so many elements that the solution technique is more laborious. In the latter case, you should study your problem to be certain that you do not have too much complexity for what is, in reality, an approximate solution.

You may be wondering when we get to the "thermal" part of the problem. Well, this is the place, but we could not get here without first calculating the airflow, which we now pretend we have completed. We need the airflow to calculate the air temperature local to circuit boards and the attached-components. The system airflow and card channel airflow is always first calculated as a *volumetric flow rate*, e.g., ft³/min. If we wish to estimate surface temperatures for components in an airstream,

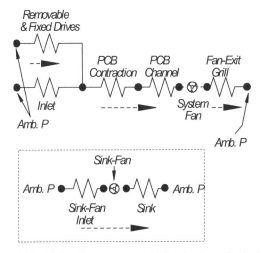

Figure 1.11. Airflow/pressure circuit for the hypothetical personal computer. Dashed arrows indicate airflow direction. Nonlabeled nodes represent internal pressures and "*Amb. P*" indicates a fixed ambient pressure, usually taken to be *gauge pressure* = 0.

we also need to calculate an *air speed*, e.g., ft/min. You probably remember that most formulae for forced air convective heat transfer coefficients have the airspeed as an independent variable.

Perhaps you are not surprised that we can construct a temperature/heat transfer circuit for our PC problem. My suggested circuit for this example is shown in Figure 1.12. There are several thermal resistances with differing subscript labels. In this case, there is only one resistance that is a convection resistance and it is labeled as $R_{PCBs\,Conv}$. This is intended to be a natural convection resistance because we shouldn't expect well-defined forced air flow in the vicinity of most of the so-called expansion boards or even in the lower regions of the motherboard.

We could have included one or more forced convection thermal resistances, but that might have been overdoing the detail for this problem. However, some thermal analysts would incorporate considerably more detail than has been included here. You need to develop your experience by keeping your models simple and gradually adding the level of detail with which you find yourself successful and comfortable.

Now back to the thermal circuit. Note that our one convection resistance, $R_{PCBs\,Conv}$, does not have a one-way resistance indicator, the symbol that looks like an eyeball looking upstream. For example, the *node* labeled T_{Air} sees an upstream node at temperature T_A, but this latter ambient air node does not see the T_{Air} node. There are several other resistors that are also one-way. As it turns out, the airflow-thermal-resistances are not really resistances at all, but it is necessary to put them into the circuit to manage the air temperatures. This is not usually a particularly important issue, but you are a little more knowledgeable if you know about this.

Hopefully you recognize that the nodes in Figure 1.12 are all thermal nodes. The nodes labeled with a T_A must be set at a fixed temperature, perhaps at 20°C, a typical office temperature. The node labeled T_{Air} is interpreted as the result of mixed, heated air flowing from (1) the drives module, and (2) fresh air from the system inlet, the T_A ambient air node just to the left of T_{Air}. The mixed system air T_{Air} then flows through the upper expansion board channel, through the system fan in the back panel, and through the fan grill. All the air nodes labeled $T_{Drives\,Exit}$, T_{Air}, $T_{PCB\,Exit}$, and T_{Exit} must be computed. Computations for the heat sink are not necessary because it is assumed that if the fan-sink vendor guidelines are followed, then the device will be satisfactory. The one and only surface temperature node, $T_{Exp\,PCBs}$, must also be computed.

Nodal temperatures for the circuit shown are very easy to calculate. We only need the airflow values G in the R_f elements. The ambient temperature T_A, and the various heat source values Q must be given. In a later chapter you will learn the derivation of a formula for calculating air temperature rise

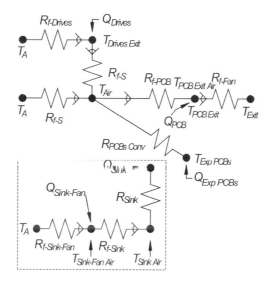

Figure 1.12. Heat transfer/temperature circuit for the hypothetical personal computer.

ΔT_{Air} $\left[{}^{\circ}C \right]$, where $Q[W] \equiv$ heat injected into airstream, G [ft^3/min] $\equiv$ volumetric airflow rate. The air temperature calculations use simple arithmetic. We use T_{Air} as a boundary condition for calculating the *lumped mass* temperature $T_{Exp\ PCBs}$ for the expansion boards.

$$\boxed{\Delta T_{Air} = 1.76Q/G} \qquad\qquad \text{Mixed air } \Delta T \ (1.14)$$

Considerable effort is devoted in Chapter 2 to the derivation of Eq. (1.14). This formula, or some variation, is frequently found in the literature. However, you must be cautious about the application of it to design problems. At this point in your studies, just remember that Eq. (1.14) is a calculation of *well-mixed air*, or is sometimes defined as the *cup-mixing temperature*.

1.12 ILLUSTRATIVE EXAMPLE: A SINGLE CHIP PACKAGE ON A PRINTED CIRCUIT BOARD - THE PROBLEM

This text is largely concerned with analysis techniques that are usually well managed using your calculator or perhaps a math analysis program. However, there are occasions when you will have a complex problem that can only be solved with some kind of commercial software, typically finite element analysis for conduction dominated problems. In these cases, you can use information from this book to calculate your boundary conditions. You may also conduct a low order analysis, i.e., not highly accurate, as an approximate check on the solution from the high-end software.

Results from a conduction analysis (with Newtonian cooling boundary conditions) are shown for a single chip package using three different analysis codes: (1) software based on a Fourier series-based analytical method, (2) software based on the thermal network method, and (3) finite element analysis software.

The problem's geometry is shown in Figures 1.13 and 1.14. Figure 1.13 illustrates the entire problem, a 32 lead, single chip package mounted on a small *PCB*. The chip is not visible because it is mounted on the underside of the ceramic substrate. Figure 1.14 shows the underside of the ceramic with the chip displayed. The details of thermal conductivities, heat transfer coefficients, dimensions, and heat dissipation are not of interest here. The purpose of this section is merely to remind you that some problems are amenable to solution by more than one method.

1.13 ILLUSTRATIVE EXAMPLE: A SINGLE CHIP PACKAGE ON A PRINTED CIRCUIT BOARD - FOURIER SERIES ANALYTICAL SOLUTION

The Fourier series method has a long history in heat transfer analysis. Researchers have found it applicable to some microelectronics problems, in particular those with a regular rectangular geometry and Newtonian cooling boundary conditions from the two largest surfaces on opposing planes

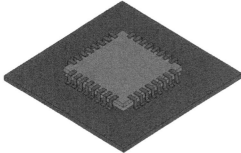 

Figure 1.13. A single chip package on a small printed circuit board. FreeCad used to create one lead per edge and ceramic substrate of solid model, Salome for remainder.

Figure 1.14. Inverted view of ceramic substrate with attached silicon chip. FreeCad used to create one lead per edge and ceramic substrate of solid model, Salome for remainder.

(Ellison, 1996). The Fourier series permits the simulation of one or more finite sources placed at specific coordinates on either one or both large surfaces. In the present example, if the *PCB* alone is considered, it is possible to model the first part of the problem, the *PCB*, as a rectangle with a specified thickness and orthotropic thermal conductivity (*PCB*), and a heat source with the dimensions of the ceramic substrate. The *PCB* with source is shown in Figure 1.15.

The use of the Fourier series method for the ceramic substrate is a little more complicated than the analysis of only a single source on a *PCB*. The extra issue is concerned with the actual heat dissipated into the board. The total silicon chip source value Q is given in the problem. However, a portion Q_L is conducted from the ceramic substrate to the *PCB* via the leads and air gap and the remainder Q_c is convected from the ceramic to local air at a temperature T_{AL}. Fortunately there is a math technique for us to determine just how the source Q is divided. You can read about this in Appendix *xiii* if you are interested.

The last issues to discuss with regard to the *PCB* are the boundary conditions. In this part of the problem we specify the heat transfer coefficients h_1 and h_2. By the time you read the chapters on natural and forced convection, you should have some confidence that you know how to calculate these values. The ambient temperature has to be set equal to zero, thus all computed temperatures are risen above ambient.

The result of the *PCB* analysis is rather simple. We only need the circuit board surface temperature within the footprint of the source. This gives us a boundary condition to specify for the ends of the leads from the ceramic package, and also a boundary temperature for conduction via the air gap between the ceramic and *PCB*. The analysis of the ceramic package using these boundary conditions is the subject of the next paragraph.

The Fourier series model of the ceramic substrate is shown in Figure 1.16. The substrate attachment points of each of the 32 leads are shown on the upper ceramic surface, but only one resistance symbol is actually shown. The calculated *PCB* surface temperature is used in the substrate model as the temperature T_0 at the end of each lead resistance. The heat transfer coefficient h_1 quantifies convection from the ceramic top. The heat transfer coefficient $h_2 = (k/t)_{\text{Air Gap}}$ can be used to account for conduction through the thin air gap between the ceramic base and *PCB* top surface. The heat source Q_S is represented by the footprint of the silicon chip. The 32 lead attachment regions are shown as filled squares on the ceramic.

1.14 ILLUSTRATIVE EXAMPLE: A SINGLE CHIP PACKAGE ON A PRINTED CIRCUIT BOARD - THERMAL NETWORK SOLUTION

The thermal network method (*TNM*) is very nearly identical to the finite difference method (*FDM*), which will be described in detail in Chapter 13. For the present purposes, it is sufficient to note that

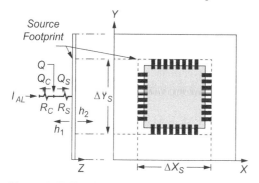

Figure 1.15. Fourier series model of *PCB* portion of single chip package example. Envelope of ceramic package and leads form footprint of source geometry.

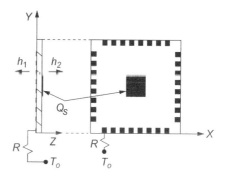

Figure 1.16. Fourier series model of ceramic substrate portion of single chip package example.

most resistances in the *TNM* can be calculated from node center to node center using Eq. (1.3). A node may be defined as the center of a cuboid mass that has uniform temperature. Each node has a conduction resistance connection to nearest neighbor solid nodes and may have a convection resistance connected to an air node. Radiation resistances are used to interconnect node surfaces that exchange significant radiation.

A very significant advantage of the *TNM* or *FEM* over a totally analytical model (as in the case of our Fourier series model) is that we can construct and solve the entire problem in a single analysis. The advantage is not due to time saving, but rather that we get more accurate results when we do the problem all at once. In the present problem, the *TNM* model results in the correct set of temperatures on the printed circuit board so that the ceramic-to-*PCB* leads are at the correct temperature on the *PCB*. We should expect that since the ceramic is a relatively good heat conductor, a lot of the heat will conduct to the *PCB* via the leads and not through the air gap. This results in peak circuit board temperatures at the lead attachment regions. The Fourier series method used a uniform source over the ceramic footprint on the *PCB* and thus would result in a peak *PCB* temperature in the center of the footprint on the *PCB*.

The thermal network model used for this problem is shown in Figures 1.17 and 1.18. Figure 1.17 reveals the network "mesh" for the *PCB* and Figure 1.18 shows greater detail for the planar region of the ceramic substrate. There are also a few resistors indicated for convection to ambient air nodes. The illustrated model has two layers of nodes for the *PCB*, one layer at the top surface and one layer at the bottom surface. The ceramic is approximated by a single layer of interior nodes. Where there are two layers of nodes, the model must also have thermal resistances interconnecting the two layers. We shall postpone showing the results until we compare with the Fourier series and *FEA* model.

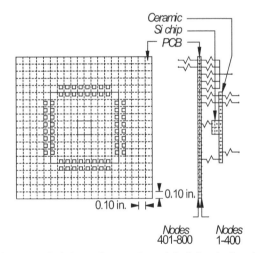

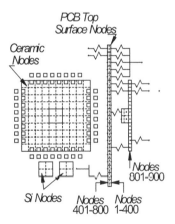

Figure 1.17. A thermal network model of the single chip package on a *PCB*. Planar detail of *PCB* emphasized. Note that the circuit board has two layers of nodes, one top layer and one bottom layer.

Figure 1.18. A thermal network model of the single chip package on a *PCB*. Planar detail of ceramic substrate and silicon chip emphasized. Note that the ceramic has only one layer of nodes.

1.15 ILLUSTRATIVE EXAMPLE: A SINGLE CHIP PACKAGE ON A PRINTED CIRCUIT BOARD - FINITE ELEMENT METHOD SOLUTION

The third and last modeling method is the finite element method (*FEM*). Some of the thermal contours are shown in Figures 1.19 to 1.21. As expected, the highest temperatures are in the center of the silicon chip source plane. The contours in Figure 1.21 are interesting in that the board temperature in the region of the lead attachments is greater than other board locations, probably by at least 5°C. As in the case of the *TNM*, an advantage of the *FEM* is that we have modeled the entire package, resulting in what should be more accurate temperatures.

1.16 ILLUSTRATIVE EXAMPLE: A SINGLE CHIP PACKAGE ON A PRINTED CIRCUIT BOARD - THREE SOLUTION METHODS COMPARED

We have outlined three analysis methods: (1) Fourier series based, (2) thermal network, and (3) finite element. The Fourier series solution is an analytical solution and the *TNM* and *FEM* are numerical analysis solutions. Although the analytical solution might be considered more exact in certain aspects, the fact that the entire problem can be modeled and analyzed as one with the numerical methods, the *TNM* and *FEM* are usually more accurate. As the mesh of the numerical solutions is refined, any concerns about the accuracy should be discounted. Of course, we always have uncertainties in the thermal conductivities and heat transfer coefficients. On the other hand, once the computer code has been created for the analytical method, it can be used repeatedly and very quickly to generate the effects of many design variations.

Figures 1.19 through 1.21 display temperature contours for the *FEM*. Computational results from the three methods are listed in Table 1.3. We see that the *TNM* and *FEM* produce similar results, whereas the temperatures from the analytical solution are nearly 10°C greater in some instances. Much of the error in the analytical solution is because the *PCB* heat source model is based on a uniformly distributed source over the ceramic footprint region. In addition, each of the three methods uses different modeling of epoxy-glass/Cu layers for the *PCB*. That of the analytical model was the most accurate.

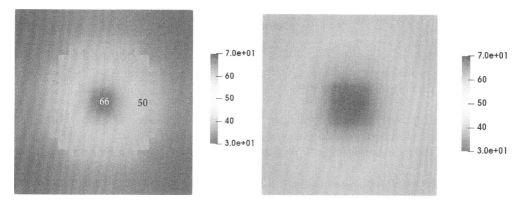

Figure 1.19. *FEA* results for single chip package on *PCB*. This view shows thermal contours for the entire exterior surface as seen from ceramic side of assembly. Temperatures are $T - T_{Amb}$. ElmerFem software was used in the analysis. Paraview software was used for visualization.

Figure 1.20. FEA results for single chip package on *PCB*. This view shows thermal contours of the ceramic and chip surfaces. Temperatures are $T - T_{Amb}$. ElmerFem software was used in the analysis.

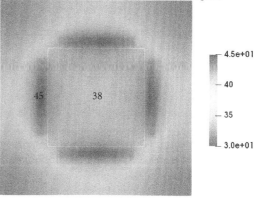

Figure 1.21. *FEA* results for single chip package on *PCB*. This view shows thermal contours for the *PCB* surface on the ceramic/Si side. Air gap, ceramic, and silicon are hidden from view. Temperatures are $T - T_{Amb}$. ElmerFem software was used in the analysis. Paraview software was used for visualization.

Table 1.3. Comparison of temperatures [°C] computed using three different methods.
Temperatures are T - T_{Amb}.

Location	Fourier	*TNM*	*FEM*
Chip surface	80	71	68
Ceramic, beneath chip	79	71	68
Ceramic edge	53	47	50
Lead on PCB	48	43	43
PCB center top	48	40	40

Reference to commercial names for the Fourier series and *TNM* based software used in the example is avoided because both software packages were written by this author. Interested readers may refer to many of the references listed under the author's name for more detail and also use the Internet for additional information.

Most readers should have interest in the *CAD* and *FEM* software used. This writer has been exploring the use of free, open-source software. Initial portions of the *FEM* model, i.e., the ceramic and one lead per edge were constructed in FreeCad, then exported as an stp file. The stp file was imported into *Salome* where the remainder of the model was built from various size cuboids. The remaining leads were translations of the original four leads. *Salome* was also used for meshing. The meshed file was solved using *ElmerFem* and visualizations produced with *Paraview*. All of the named software is freely available as binaries.

EXERCISES

1.1 Lasance (2006) recommends a curve fit $k = 1448/(T + 273.16)^{1.23}$ for the thermal conductivity of GaAs in the range of -50 to +300°C, where $T=T[°C]$ and $k=k[W/in.·°C]$. Derive the formula for k_m, the mean thermal conductivity GaAs.

1.2 Using the result from Exercise 1.1, calculate the mean thermal conductivity, thermal resistance, and chip surface temperature for a GaAs chip with dimensions of 0.5 in.×0.5 in.×0.020 in., a base temperature of 100°C, and dissipating 50 W at the chip top surface.

1.3 Use the geometry of Application Example 1.5 to determine what convective heat transfer coefficient is required to limit the average surface temperature rise to ΔT =50°C when the power transistor dissipates Q = 25 W.

1.4 Derive a law for the addition of convective and radiative thermal resistances in parallel. Hint: Add the convective heat transfer, Eq. (1.8), and the radiative heat transfer, Eq. (2.0) to get the total Q and proceed from there.

1.5 Assuming the geometry and temperatures given for Application Example 1.7, plus a convective heat transfer coefficient of h_c = 0.003 W/(in.²·°C), calculate the convective heat transfer to ambient, the total heat transfer to ambient, the radiation resistance, and the total of the convective and radiative resistances.

1.6 Use the geometry of Application Example 1.5, but consider the panel to be the rear panel of an electronics enclosure. Assume that the power transistor is attached to the side of the panel that faces the chassis interior. Furthermore, the air temperature within the enclosure is approximately the same as the chassis panel, i.e., there is negligible convection and radiation from one side of the panel. Assuming a panel radiation emissivity of 0.7, calculate: (1) the convective, radiative, and total resistances, (2) the average panel temperature, and (3) the heat transferred by convection and radiation.

1.7 Derive the law for adding two thermal resistances, R_1 and R_2, in series to obtain a single equivalent resistance R.

Exercise 1.7. Series addition of thermal resistors.

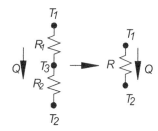

1.8 Referring to Illustrative Example 1.8, calculate the total thermal resistance, $R_{Chip\text{-}to\text{-}Amb}$, the $\Delta T_{Chip-to-Amb}$, the heat transfer from chip through the heat sink to the ambient above the sink, and the heat transfer from the chip through the *PCB* to the ambient below the *PCB*. One way to check your work is to add Q_{Top} and Q_{Bot} to see if you get 6 W.

1.9 This problem is designed to see what you can do without having a similar example already solved for you. You will need to use what knowledge you have gained from earlier examples/exercises, and the information in Section 1.11. You are given airflow and thermal data for a few portions of the personal computer in Figure 1.9 (you have not learned enough in the current chapter to totally solve the problem). You are asked to calculate: (1) the air temperature at the sink fan exit, (2) the air temperature at the heat sink exit, (3) the heat sink temperature, and (4) the air temperature at the drives exit. The following is given: $G_{Sink\,Fan} = 25$ ft³/min, $Q_{Sink\,Fan} = 4$ W, $Q_{Sink} = 50$ W, $R_{Sink} = 0.5$ °C/W, $G_{Drives} = 10$ ft³/min, $Q_{Drives} = 10$ W, $T_A = 20$°C.

CHAPTER 2

Thermodynamics of Airflow

The principal goal of this chapter is to describe the derivation of a formula for the temperature rise of steady airflow with a constant rate of heat input over a section of a duct. This duct is our representation of a circuit board or a plate-fin heat sink channel. The result is so commonly used that you should, at least once in your life, see how the equation comes about. Surprisingly, the derivation is often omitted from both heat transfer texts and reference books on electronics cooling and, although I would like to take credit for the effort, you should know that the work is taken from one of J.P. Holman's excellent books (see Holman, 1974).

You are cautioned in your application of the air temperature rise formulae derived in this chapter. In the case of ducts formed by plate-fin heat sinks, particularly those that have covering shrouds, the air temperature rise formulae are usually consistent with applicable convective heat transfer formulae. In many other situations, e.g., components mounted on circuit boards, the air temperature varies considerably across any given channel cross-section. This is more likely to be the rule than the exception in most air cooled electronics. Expect to encounter hot spots that are not quantifiable using most analytical methods and in these cases, learn to use well-mixed air temperature rise calculations as more qualitative than quantitative. You will gain some experience with this situation when you study convective heat transfer for forced air cooled circuit boards in a later chapter.

2.1 THE FIRST LAW OF THERMODYNAMICS

Before we can get directly to our topic, you must submit to some basic thermodynamics. A reversible thermodynamic process is one in which the process produces exactly the same result when taken in either of two directions when traversing around a pressure-volume curve (referred to as the *P-V* curve) for the particular gas of interest. Such a curve is illustrated in Figure 2.1. The directional convention is that work done by the gas is positive when the curve is followed in a clockwise direction.

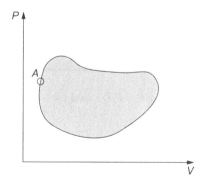

Figure 2.1. Thermodynamic *P-V* curve for an arbitrary gas.

Empirical studies have always shown that a quantity of work appears or disappears for each cycle

$$W = \oint \bar{d}W = \oint P dV$$

around the curve, beginning and ending at point *A*, for example. The barred symbol, $\bar{d}$, means that the result depends on the path taken, the direct mathematical implication being that $\bar{d}W$ is not an exact differential and cannot be obtained by differentiating a function.

If we define an infinitesimal quantity of thermal energy, i.e., heat, as $\bar{d}q$, then the conservation of energy in a cyclic, closed (thermally isolated) system is

$$\oint \left(\bar{d}q - \bar{d}W \right) = 0 \qquad (2.1)$$

which means that heat and work taken together are equivalent. Equation (2.1) cannot be proven analytically, but it has also never been shown to be false if all forms of work are included.

Now suppose we are dealing with an open system, i.e., one that is not thermally isolated. In such a system the heat minus the work in a complete cycle is not conserved so that we say

Heat into system - Work done by system $\neq 0$

This is not a concern because *the system has changed state* and a balancing amount is taken from the system. Therefore, the mathematical equivalent to Eq. (2.1) is

$$\boxed{dE = \bar{d}q - \bar{d}W}$$

Conservation of energy, (2.2)
open system, *First Law*

which is our expression for *The First Law of Thermodynamics*. The quantity E is the *total system energy*. The change in energy also means a *change of state* for the system.

We next consider a *quasi-static* system, by which we mean that any motion is very slow, so slow that each successive step is calculated using static processes. We illustrate this new system in a manner that is closer to our electronics cooling applications. Figure 2.2 shows a piston moving an infinitesimal distance L within a cylinder. Because the process is so slow, the internal pressure is very nearly equal to the external hydrostatic pressure. Within this static system, the internal force therefore equals the external force and P is constant.

The total energy E for the gas in our piston system is the sum of the internal (thermal) energy U, the kinetic energy KE, and the potential energy PE, etc. Then

$$E = U + KE + PE + \cdots$$

But since we shall consider thermodynamic systems where the only changes are in the internal (thermal) energy, i.e., $dE = dU$, we write

$$\boxed{\bar{d}q = dE + \bar{d}W = dU + \bar{d}W} \qquad (2.3)$$

We can rewrite Eq. (2.3) using $\bar{d}W = PdV$ and use

$$dU = \left(\frac{\partial U}{\partial T} \right)_V dT + \left(\frac{\partial U}{\partial V} \right)_T dV$$

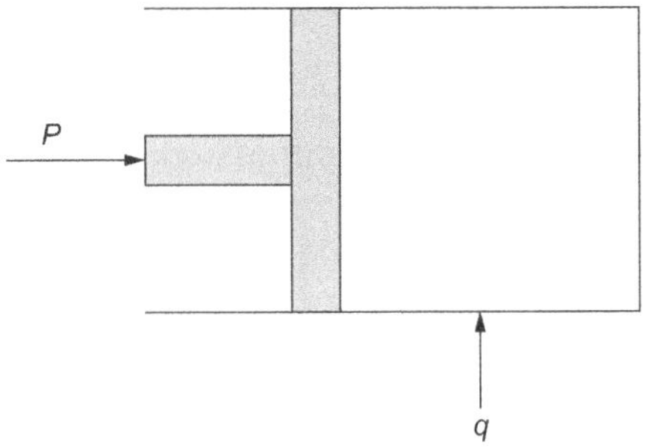

Figure 2.2. A quasi-static process with only hydrostatic processes.

Then

$$\bar{d}q = dU + PdV = \left(\frac{\partial U}{\partial T}\right)_V dT + \left(\frac{\partial U}{\partial V}\right)_T dV + PdV$$

$$\boxed{\bar{d}q = \left(\frac{\partial U}{\partial T}\right)_V dT + \left[\left(\frac{\partial U}{\partial V}\right)_T + P\right]dV} \qquad (2.4)$$

2.2 HEAT CAPACITY AT CONSTANT VOLUME

The definition of heat capacity at constant volume using Eq. (2.4) is

$$C_V \equiv \left(\frac{\bar{d}q}{dT}\right)_V = \left(\frac{\partial U}{\partial T}\right)_V$$

but it is more common practice to use $u \equiv$ internal energy per unit mass so that

$$c_V \equiv \text{Heat capacity per unit mass}$$
$$c_V \equiv \text{Specific heat capacity}$$
$$c_V = \left(\frac{\partial u}{\partial T}\right)_V$$

2.3 HEAT CAPACITY AT CONSTANT PRESSURE

The first steps are to define the quantity of enthalpy H and to temporarily remove any conditions of constant P, V. Then

$$\boxed{H = U + PV} \qquad \text{Enthalpy} \ (2.5)$$
$$U = H - PV$$

and

$$dU = dH - PdV - VdP$$

$$dU = \left(\frac{\partial H}{\partial T}\right)_P dT + \left(\frac{\partial H}{\partial P}\right)_T dP - PdV - VdP$$

Then

$$dU_P = \left(\frac{\partial H}{\partial T}\right)_P dT - PdV$$

and using

$$\bar{d}q_P = dU_P + \bar{d}W = dU_P + PdV$$

we obtain

$$\bar{d}q_P = \left(\frac{\partial H}{\partial T}\right)_P dT - PdV + PdV = \left(\frac{\partial H}{\partial T}\right)_P dT$$

We now define the heat capacity at constant pressure and use the preceding result to get

$$C_P \equiv \left(\frac{\bar{d}q}{dT}\right)_P = \left(\frac{\partial H}{\partial T}\right)_P$$

Using the usual heat capacity per unit mass, i.e., specific heat, c_P, we have

$$\boxed{c_P = \left(\frac{\partial h}{\partial T}\right)_P}$$ Specific heat at constant P (2.6)

where we have used $u \equiv$ internal energy per unit mass and $h \equiv$ enthalpy per unit mass . This ends the specific heat capacity discussion with the result that we will use in the next section.

2.4 STEADY GAS FLOW AS AN OPEN, STEADY, SINGLE STREAM

We shall make the assumption of a reversible process. Consider the stream shown in Figure 2.3 with the definitions of

$$\dot{m} \equiv \text{mass flow rate}$$
$$u \equiv \text{internal energy per unit mass}$$
$$\overline{V} \equiv \text{volume per unit mass}, \overline{V} = \overline{V_1} = \overline{V_2}$$
$$Q \equiv \text{net heat rate input}$$

We apply the conservation of energy, which is written as

Transport-rate of internal energy into control volume + net heat rate added + work-rate to push $\overline{V}$ *into control volume = Transport-rate of internal energy out of control volume + work-rate to push volume* $\overline{V}$ *out of control volume.*

The preceding statement of conservation of energy as a mathematical statement is

$$\dot{m}\,u_1 + Q + \dot{m}\,P_1\overline{V} = \dot{m}\,u_2 + \dot{m}\,P_2\overline{V}$$
$$\dot{m}\left(u_1 + P_1\overline{V}\right) + Q = \dot{m}\left(u_2 + P_2\overline{V}\right)$$

Substituting the enthalpy definition, Eq. (2.5) as a per/unit mass $h = u + P\overline{V}$ into the preceding,

$$\dot{m}\,h_1 + Q = \dot{m}\,h_2$$
$$Q = \dot{m}\left(h_2 - h_1\right) = \dot{m}\,\Delta h$$
$$\boxed{\Delta h = Q/\dot{m}}$$ Enthalpy change in stream flow (2.7)

If a gas obeys the ideal gas law, as air certainly does at modest pressures, the internal energy and enthalpy are functions of temperature only (see Holman, 1974 or Fermi, 1937). We can write

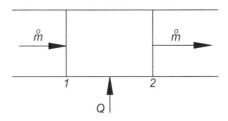

Figure 2.3. A steady, open, single stream flow.

$\left(\partial h/\partial T\right)_P = dh/dT$ and using our result, Eq. (2.6), from the previous section we immediately know that $c_P = \left(\partial h/\partial T\right)_P = dh/dT$ which we can then integrate.

$$\Delta h = h_2 - h_1 = \int_{T_1}^{T_2} c_P dT$$

If c_P is temperature-independent, $\Delta h = c_P\left(T_2 - T_1\right)$. Setting this equal to Eq. (2.7) we obtain a very important equation for single stream, ideal gas flow.

$$\boxed{Q = \dot{m}c_P\left(T_2 - T_1\right)}$$
Gas flow temperature rise, (2.8)
general formula 1

2.5 AIR TEMPERATURE RISE: TEMPERATURE DEPENDENCE

The formula for a gas flow temperature rise has been derived as Eq. (2.8). Since we will often use this result for air, it now makes sense to insert as many "constants" as possible. We first rewrite Eq. (2.8) as

$$\boxed{\Delta T = Q/\dot{m}c_P = Q/\rho G c_P}$$
Gas flow temperature rise, (2.9)
general formula 2

where

$\dot{m} \equiv$ mass flow rate

$c_P \equiv$ specific heat, constant pressure

$Q \equiv$ heat convected into air stream

$\rho \equiv$ air density

$G \equiv$ volumetric airflow

We shall use the ideal gas law $PV = \left(m/M\right)RT'$ where P and V are the gas pressure and volume, respectively, m and M are the mass and molecular weight, respectively, R is the gas constant, and $T' = T + 273.16$ for gas temperature T [°C]. Then $\rho = \left(m/V\right) = PM/RT'$ is the gas density at an absolute temperature T' and $\rho_0 = PM/RT_0'$ is the gas density at some reference temperature T_0. Assuming negligible pressure change from temperature T_0 to temperature T, we have an expression for the density at temperature T.

$$\rho = \rho_0\left(\frac{T_0 + 273.16}{T + 273.16}\right)$$

$$\boxed{\Delta T = \frac{Q\left(T + 273.16\right)}{\rho_0 c_P\left(T_0 + 273.16\right)G}}$$
Gas flow temperature rise, (2.10)
general formula

Now we are ready to insert appropriate values for ρ_0, c_P, and T_0 into Eq. (2.10). Referring to Appendix *ii*, a table of air properties, and taking $T_0 = 0°C$ we get

$$\Delta T = \frac{Q[\text{J/s}]\left(T + 273.16\right)[\text{K}]}{\left(0.021\,\text{gm/in.}^3\right)\left(1.01\,\text{J/gm}\cdot\text{K}\right)\left(273.16\right)[\text{K}]G\left[\text{in.}^3/\text{s}\right]}$$

$$\Delta T = \frac{0.17Q[\text{J/s}]\left(T + 273.16\right)[\text{K}]}{G\left[\text{in.}^3/\text{s}\right]}$$

The author's preference is to use G in the units of [ft^3/min]. Then we must write

$$G\left[\frac{\text{in.}^3}{\text{s}}\right] = G\left[\frac{\text{ft}^3}{\text{min}}\right]\left(\frac{\text{min}}{60\,\text{s}}\right)\left(\frac{12\,\text{in.}}{\text{ft}}\right)^3$$

so that

$$\Delta T = \frac{Q[\text{J/s}](T + 273.16)[\text{K}]}{\left(0.021\,\text{gm/in.}^3\right)\left(1.01\,\text{J/(gm}\cdot\text{K)}\right)(273.16)[\text{K}]\,G\left[\text{in.}^3/\text{s}\right]}$$

$$\Delta T = \frac{0.1726\,Q[\text{J/s}](T + 273.16)[\text{K}]}{G\left[\frac{\text{ft}^3}{\text{min}}\right]\left(\frac{\text{min}}{60\,\text{s}}\right)\left(\frac{12\,\text{in.}}{\text{ft}}\right)^3}$$

with the result

$$\boxed{\Delta T\left[^\circ\text{C}\right] = \frac{5.99\times10^{-3}\left(T\left[^\circ\text{C}\right] + 273.16\right)Q[\text{W}]}{G\left[\text{ft}^3/\text{min}\right]}}$$

Air ΔT, (2.11)
bulk air T

It should be clear to you that the temperature T in Eq. (2.11) is the air temperature. But what air temperature does this refer to, i.e., where in the air stream is this temperature taken? First of all, be sure that you understand that we are using a "cup mixing temperature," i.e., at any cross-section in the stream, we are assuming the air is well mixed so that the temperature is uniform at any given cross-section. Of course, we know that this is not really the situation, but making this assumption does let us make some calculations. You might wish to think of the temperature calculation at any region in the stream as an average temperature. We still have not totally resolved the problem as to what the air temperature T refers. We expect that the air progressively heats up as it flows down a card channel duct or between plate fin passages. The next two sections help us to resolve what number to insert for T in Eq. (2.11).

2.6 AIR TEMPERATURE RISE: T IDENTIFIED USING DIFFERENTIAL FORMS OF
$$\Delta T, \Delta Q$$

If we shrink the finite T, Q in Figure 2.3 down to the infinitesimals $\Delta T, \Delta Q$ in Figure 2.4 we get

$$dT = \frac{C}{G}(T + 273.16)dQ, \quad C = 5.99\times10^{-3}$$

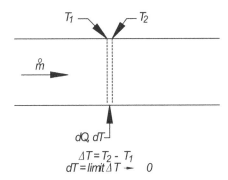

Figure 2.4. Infinitesimal heat transfer into fluid with infinitesimal temperature rise.

$$\frac{dT}{(T+273.16)} = C\frac{dQ}{G}, \quad u = T+273.16, \quad \int \frac{du}{u} = \frac{C}{G}\int dQ, \quad \ln u = \frac{CQ}{G} + B$$

Define $A = e^B$ or $B = \ln A$, then $\ln\left(\dfrac{T+273.16}{A}\right) = \dfrac{CQ}{G}, \quad T = Ae^{CQ/G} - 273.16$

where A, B = constants. At $Q = 0$, $T = T_I$, $A = T_I + 273.16$, where T_I is the duct inlet air temperature. Then we finally arrive at

$$(T+273.16) = (T_I + 273.16)e^{CQ/G}$$

which, in a more useful form, is given by

$$\boxed{\Delta T = (T_I + 273.16)\left(e^{CQ/G} - 1\right)}$$

$C = 5.99 \times 10^{-3}$ (2.12)
Air temperature rise

You may view Eq. (2.12) as an unexpected result, particularly if you have previous heat transfer experience. One way to check this result is to expand the exponential portion in a series as in

$$e^{CQ/G} = 1 + \left(\frac{CQ}{G}\right) + \frac{1}{2!}\left(\frac{CQ}{G}\right)^2 + \cdots$$

Equation (2.12) then becomes

$$\Delta T = (T_I + 273.16)\left[\left(\frac{CQ}{G}\right) + \frac{1}{2!}\left(\frac{CQ}{G}\right)^2 + \cdots\right]$$

$$\Delta T \simeq (T_I + 273.16)\left(\frac{CQ}{G}\right), \quad \frac{CQ}{G} \ll 2$$

and at $T_I = 20^\circ$C we get our useful result of

$$\boxed{\Delta T = 1.76 Q/G}$$

Air temperature rise (2.13)

where the units are ΔT [°C], Q [W], and G [ft³/min]. If you are rather particular, you might wish to replace the 1.76 with a different value when your inlet air temperature is different than 20°C.

2.7 AIR TEMPERATURE RISE: *T* IDENTIFIED AS AVERAGE BULK TEMPERATURE

Now let us look at Eq. (2.11) in a different way by identifying the bulk air temperature T, as the *average bulk temperature* $\overline{T}_B$ [°C] so that

$$\Delta T\left[^\circ\text{C}\right] = \frac{5.99 \times 10^{-3}\left(\overline{T}_B + 273.16\right)Q\left[\text{W}\right]}{G\left[\text{ft}^3/\text{min}\right]}$$

and $\overline{T}_B = (T_I + T_E)/2$ where T_I [°C] and T_E [°C] are the duct inlet and exit air temperatures, respectively. We then substitute $\overline{T}_B$ [°C] into the preceding formula for ΔT to get

$$\Delta T\left[^\circ\text{C}\right] = \frac{5.99 \times 10^{-3}\left[\left(\dfrac{T_I + T_E}{2}\right) + 273.16\right]Q\left[\text{W}\right]}{G\left[\text{ft}^3/\text{min}\right]} = \frac{5.99 \times 10^{-3}\left[\left(\dfrac{2T_I + \Delta T}{2}\right) + 273.16\right]Q\left[\text{W}\right]}{G\left[\text{ft}^3/\text{min}\right]}$$

Solving for ΔT,

$$\Delta T = \dfrac{2(T_I + 273.16)}{\left(\dfrac{2G}{CQ} - 1\right)}$$

$C = 5.99 \times 10^{-3}$ (2.14)
Air temperature rise

where T_I has units of [°C] which you may also consider another unexpected result.

We again resort to one of our check methods for these strange looking results. In the case of Eq. (2.14), we make a small change to our last result and expand the denominator.

$$\Delta T = \frac{2(T_I + 273.16)}{\left(\dfrac{2G}{CQ} - 1\right)} = \frac{2(T_I + 273.16)}{\left(1 - \dfrac{CQ}{2G}\right)}\left(\frac{CQ}{2G}\right) = (T_I + 273.16)\left(\frac{CQ}{G}\right)\left(1 - \frac{CQ}{2G}\right)^{-1}$$

$$= (T_I + 273.16)\left(\frac{CQ}{G}\right)\left[1 + \left(\frac{CQ}{2G}\right) + \left(\frac{CQ}{2G}\right)^2 + \cdots\right]$$

$$\Delta T \simeq (T_I + 273.16)\left(\frac{CQ}{G}\right), \qquad \frac{CQ}{G} \ll 2$$

If we substitute $T_I = 20°C$ and the value for the constant C, we get the same result as Eq. (2.13) and both Eqs. (2.12) and (2.14) lead to the same approximate result, Eq. (2.13), when $CQ/G \ll 2$ or $Q/G \ll 300$ or $Q/G \leq 30$ (approximately). It is left as an exercise for you to further examine this approximation.

One final comment is in order regarding Eq. (2.14). It is evident that when $2G/CQ$ is exactly equal to one, the expression "blows up." In addition, Eq. (2.14) is nonsense when Q/G is such that the denominator is negative. It is very unlikely that you will ever have such a difficulty when your problem is reasonable. The only instances when this has been a problem for myself were when I used Eq. (2.14) in a computer program where iterative calculations were such that there was one iterative step where $2G/CQ$ was exactly 1.0. This can be avoided by using Eq. (2.12) in your code.

EXERCISES

2.1 Derive the equivalent of Eq. (2.11) for helium gas, bulk gas T. Use the properties of density $\rho_0 = 3.15 \times 10^{-3}$ gm/in.^{3}, $c_p = 5.18$ J/(gm K), temperature $T_0 = -17.78°C$ (0°F).

2.2 Compare the representations of the three air temperature rise formulae, Eqs. (2.12)-(2.14), assuming an inlet air temperature of 20°C. Label the calculated air temperature rises with subscripts E, A, and B for Eqs. (2.12), (2.13), and (2.14), respectively. Use the ratio $Q/G = 1$, 10, 30, 50, 100, 200, and 300 as your independent variable, and additional values if you wish. Present your results in both tabular and graphical form. Be sure to use logarithmic axes for both Q/G and the temperature rise, otherwise your results will not be very meaningful. Assuming that Eq. (2.12) is the most exact, calculate the percent error that you get from Eq. (2.13) vs. Q/G for 1 to 100. Comment on the latter result and how it compares to the series approximations for $Q/G \leq 30$.

2.3 In those situations where you wish to use an equation like Eq. (2.13) to calculate air temperature rise, with what values would you replace the 1.76 for inlet air temperature of 30°C and 50°C?

CHAPTER 3

Airflow I: Forced Flow in Systems

During my many years of working with mechanical design engineers, I have found that forced air prediction is often the only analytical effort devoted to preliminary design activities. Certainly it must be the first step of a system thermal analysis because airflow estimates in card channels, heat sink passages, etc., are necessary for air temperature and convective heat transfer calculations. Even if you are fortunate enough to have access to computational fluid dynamics software, you most certainly should always begin with the fundamental calculations explained in this chapter to ensure that you either have hope of an adequate design or should instead revise your initial concepts.

In a later chapter we will study buoyancy driven flows, but in the present chapter we shall restrict our study to forced airflow, i.e., those systems containing fans. We shall begin with the fundamentals of duct flow followed by fan characterization. Once we have mastered some of these necessary fundamentals, we can examine enclosure flows as related to grills, circuit boards, and other important geometrical entities. Much of the discussion herein is also applicable to buoyancy driven systems.

3.1 PRELIMINARIES

Later in this chapter we shall calculate some flow quantities that are common to both heat transfer and fluid flow. In addition, in this book we tend to use mixed units where length L has the units of in., area A is in in.^{2}, velocity is in ft/min, and volumetric flow G is in ft^3/min. As you would expect, we will assume (unless you are told otherwise) that the velocity is an average over a duct cross-section. Suppose then that you wish to calculate the average velocity from a given duct cross-section area A and volumetric flow G. We will usually use

$$G\left[\text{ft}^3/\text{min}\right] = V\left[\text{ft/min}\right]A_c\left[\text{in.}^2\right]/144\left(\text{in.}^2/\text{ft}^2\right) \quad \text{Flow in mixed units} \quad (3.1)$$

The next quantity of interest to us is the *Reynolds number Re*. This dimensionless quantity is used to estimate whether a given flow is either laminar or turbulent. In addition, it will frequently appear as a variable in some of our calculations. We will discuss laminar and turbulent flow in more detail when we begin to make heat transfer calculations. For now, it is sufficient to recognize Re as a variable. Most discussions that require a Reynolds number also provide some kind of length scale pertinent to the problem. For example, if we are considering flow over a flat plate of length L, then this length is the required parameter. If we are considering flow through a duct, then we expect to see a diameter, or more correctly, a *hydraulic diameter*, as the required length. The standard formula for a Reynolds number is

$$Re_P = VP\rho/\mu = VP/\nu$$

where ρ, μ, and ν are the density, dynamic viscosity, and kinematic viscosity, respectively, and all in consistent units such that Re_P is dimensionless. In our system of mixed units we will use

$$Re_P = VP/(5\nu) \quad \text{Reynolds number, mixed units} \quad (3.2)$$
$$V\left[\text{ft/min}\right], \nu\left[\text{in.}^2/\text{s}\right], P\left[\text{in.}\right]$$

where the factor of five in the denominator is exactly the correct units conversion factor.

Unfortunately, the geometries that we encounter in electronics cooling problems are so complicated that one geometric parameter is not sufficient to accurately predict flow properties to great accuracy. Nevertheless, this is the situation as it is at present.

When we encounter problems requiring a Reynolds number, we are usually forced to use one based on a single length parameter, generally written as P in this book.

Flat plate geometry:

$P = L = $ plate length

$Re_L \cong 5 \times 10^5$ for laminar to turbulent flow transition

Duct geometry:

$P = D_H = 4A_c / p$

$A_c = $ cross-sectional area, $p = $ wetted perimeter

$Re_{D_H} < 2000$ for laminar flow

$2000 < Re_{D_H} < 10,000$ for transitional flow

$Re_{D_H} > 10,000$ for fully turbulent flow

We shall use a kinematic viscosity $\nu = 0.023 \rightarrow 0.024 \, \text{in.}^2/\text{s}$ in most of our calculations.

3.2 BERNOULLI'S EQUATION

We begin our discussion of basic airflow and pressure concepts with the introduction of Bernoulli's equation, the derivation of which is described in all fluid mechanics texts and even most introductory physics texts. We refer to Figure 3.1, an illustration of a basic duct system where we have airflow from left to right, i.e., from "test station 1" to "test station 2." The Bernoulli equation for this system describes flow and pressure conditions at the two test stations.

$$p_1 + \frac{1}{2}\rho V_1^2 + \rho g z_1 = p_2 + \frac{1}{2}\rho V_2^2 + \rho g z_2 \qquad \text{Bernoulli's equation (3.3)}$$

The quantities p_1, p_2 are the pressures at stations 1, 2, respectively; V_1, V_2 are the flow speeds at stations 1, 2, respectively; z_1, z_2 are the elevations of stations 1, 2, respectively. ρ is the air density and g is the acceleration due to gravity. Our typical electronics cooling application with forced air allows us to assume that density and elevation do not vary significantly.

It is sometimes worthwhile to be aware of the dimensions of the terms in Bernoulli's equation. This is made easy if we look at the velocity term. It is customary to refer to each term in the second

$$p + \frac{1}{2}\rho V^2 = \text{Constant}, \quad \frac{\text{Energy}}{\text{Volume}}$$

$$\frac{p_1}{\rho} + \frac{1}{2}V^2 = \text{Constant}, \quad \frac{\text{Energy}}{\text{Mass}}$$

$$\frac{p_1}{\rho g} + \frac{1}{2g}V^2 = \text{Constant, Height (length)}$$

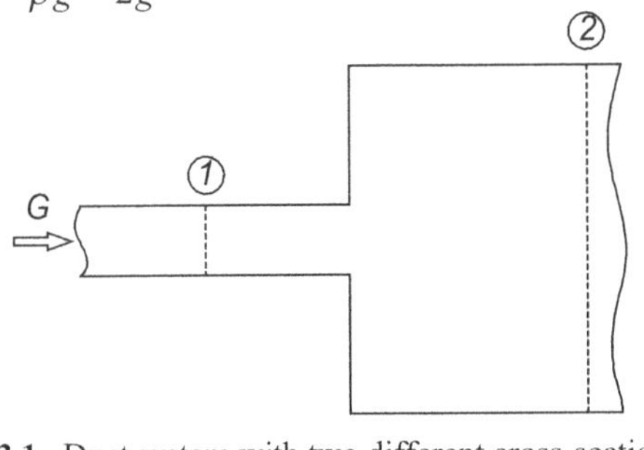

Figure 3.1. Duct system with two different cross-sectional areas.

and third equations of the three representations of Bernoulli's equation as "head" definitions, as for example, in hydraulics. In this book, we shall refer to flow losses (which we have not yet discussed) in the same manner as in hydraulics. We will use the term "head" h, not to be confused with the height h in Bernoulli's equation. Our *head h* will have the dimensions of pressure (energy/volume), but with dimensions of in. H_2O, i.e., inches of water column height. Now we are ready to discuss a more applicable form of Bernoulli's equation: Bernoulli's equation with losses.

3.3 BERNOULLI'S EQUATION WITH LOSSES

Equation (3.3) is derived on the assumption that there are no energy losses between stations 1 and 2. In reality, there are unaccounted-for energy losses as the flow proceeds from left to right, friction perhaps being the most obvious. But in Figure 3.1, the change in cross-sectional area is also the cause of a loss. For our present purposes we shall not restrict ourselves to any particular loss mechanism; we just accept that some real loss is present. Thus we can immediately write another form of Eq. (3.3) with a loss term Δp_{1-2} that accounts for all losses between locations 1 and 2.

$$p_1 + \frac{1}{2}\rho V_1^2 = p_2 + \frac{1}{2}\rho V_2^2 + \Delta p_{1-2} \qquad \text{Bernoulli's equation} \atop \text{with losses} \qquad (3.4)$$

which, rewritten with each term expressed as a "head," is

$$h_{s1} + h_{V1} = h_{s2} + h_{V2} + h_L \qquad (3.5)$$

The subscripts s and V are used according to conventional practice to denote *static* and *velocity* heads, respectively, and L to denote that h_L is the head loss between locations 1 and 2. The sums $h_{T1} = h_{s1} + h_{V1}$ and $h_{T2} = h_{s2} + h_{V2}$ are each referred to as a *total heads*, for obvious reasons.

The static, velocity, and total heads are also explained in terms of the pressures that one might typically measure in a duct system. For this description, you are referred to Figure 3.2. Three pressure measuring manometers are shown in Figure 3.2. The *total head* manometer senses flow aimed directly at the manometer entrance, the other end exposed to ambient; the *static head* manometer inlet is at a right angle to the flow direction, the other end exposed to ambient; one end of the *velocity head* manometer is exposed directly to the flow direction, the other end is exposed to the internal flow, but at a right angle to the flow direction. Because each of the two velocity head manometer ends is connected to a pressure, the result is the difference of the total head and static head. It is important to note that in the instance of the measurement system in Figure 3.2, the room ambient velocity pressure head is zero ($V = 0$). Since both the velocity and static head manometers have one end exposed to ambient pressure, these pressures read on the manometer scales are pressures *differences* from the room pressures. The pressures are often called gauge pressures.

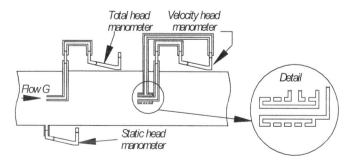

Figure 3.2. Measurement of total, velocity, and static heads.

3.4 FAN TESTING

A sketch of a typical fan test system is shown in Figure 3.3. In this particular case, the fan under test is blowing into the test system with a volumetric airflow G. "Settling screens" are used to distribute the flow throughout the initial plenum. Following the settling screens, the flow passes through a calibrated pressure/flow nozzle that is used to determine the total airflow. The variable speed exhaust blower with an adjustable gate valve is used to establish the desired steady flow value. The fan under test is set to run at its desired voltage, assuming it is controlled by direct current.

The manometer just downstream of the fan under test is connected to several static pressure taps distributed around the periphery of the chamber. A similar manometer measures static pressure across the calibrated nozzle. Charts or equations are used to convert the pressure difference across the manometer into the total system airflow.

A typical test procedure would begin with the exhaust blower/gate valve adjusted so that with the fan operating, the fan static pressure (measured with manometer just downstream of the fan) is determined to be at zero gauge pressure, i.e., a pressure equal to the room pressure. After the measured fan static pressure and fan airflow are determined, the exhaust blower and gate valve are adjusted to determine a full set of fan static pressure/airflow values, including a data pair that represents a nearly zero flow, maximum fan pressure condition.

Although the test system shown is designed to test fans under a positive pressure condition, other designs are intended for negative pressure conditions. Test system dimensions, location and design of the settling screens, nozzle configuration, etc., are available from sources such as the Air Movement and Control Association (AMCA), Bulletin 201. These systems are also useful for measuring the actual operating airflow of your system designs.

Before we advance to the next topic, we shall carefully define the various pressures involved:

$h_{fs} \equiv$ fan static pressure head

$h_{sd} \equiv$ static pressure head at the fan discharge plane

$h_{si} \equiv$ static pressure head at fan inlet

$h_{Vi} \equiv$ velocity pressure head at fan inlet

$h_{fT} \equiv$ fan total pressure

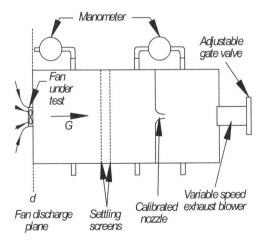

Figure 3.3. A fan test apparatus.

3.5 ESTIMATE OF FAN TEST ERROR ACCRUED BY MEASUREMENT OF DOWNSTREAM STATIC PRESSURE

We must now examine the static pressure that we are measuring for a fan with the test system and how that pressure relates to our actual electronics enclosure in which we are installing the fan. Your electronics enclosure will certainly have little resemblance to the test system in Figure 3.3. It shouldn't surprise you that it would be desirable for your cooling design to resemble the test system, but of course that is impractical. The best that you can probably hope for is that you actually have access to a single test chamber. Some of these chambers are as large as a few feet in diameter, but your electronics design may have a height not much in excess of the diameter of the fan that you would like to use, for example four or five inches. We shall now attempt to evaluate the ramifications of assuming that the fan characterization considers the fan pressure taken directly at the fan discharge, when in reality the pressure is measured somewhere in the flow downstream of the fan.

A fan is only aware of the pressure difference from the fan inlet to the fan exit, thus our fan static pressure head is defined using differences, i.e.,

$$h_{fs} = \Delta h_{fT} - \Delta h_{fV} = \left(h_{Td} - h_{Ti}\right) - \left(h_{Vd} - h_{Vi}\right) = \left(h_{Vd} + h_{sd} - h_{Vi} - h_{si}\right) - \left(h_{Vd} - h_{Vi}\right) \qquad (3.6)$$

$$\boxed{h_{fs} = h_{sd} - h_{si}} \qquad \text{Fan static pressure } (3.7)$$

and since, for the test setup in Figure 3.3, we have $h_{si} = 0$ we write

$$\boxed{h_{fs} = h_{sd}} \qquad \begin{array}{l}\text{Fan static pressure } (3.8)\\ \text{for positive pressure test}\end{array}$$

Our concern is that there is a head loss h_L between the fan discharge plane and the plane where the static head is actually measured. We therefore begin by writing the total head at the fan discharge plane as equal to the total head h_{T2} at the plane of actual static pressure measurement plus the head loss, h_L between these two locations.

$$h_{Td} = h_{T2} + h_L$$
$$h_{sd} + h_{Vd} = h_{s2} + h_{V2} + h_L$$
$$h_{sd} - h_{s2} = h_{V2} - h_{Vd} + h_L$$

We are looking for the difference or error

$$\Delta h_s = h_{sd} - h_{s2} = h_{V2} - h_{Vd} + h_L$$

For the time being, you are asked to believe that an acceptable formula for the loss term h_L, which is an expansion loss written as

$$h_L = h_{Vd}\left(1 - \frac{A_d}{A_2}\right)^2, \qquad h_{Vd} = \frac{1}{2}\rho V_d^2$$

where A_d, A_2, V_d are the cross-sectional area of the fan discharge area, the cross-sectional area at the plane of static pressure head measurement, and the air speed at the fan discharge plane, respectively. Obviously the air speed must be taken as an average value. The term h_{Vd} is called the *velocity head* at the fan discharge.

Returning to the measurement error Δh_s,

$$\Delta h_s = h_{V2} - h_{Vd} + h_L = h_{V2} - h_{Vd} + h_{Vd}\left(1 - \frac{A_d}{A_2}\right)^2$$

$$\Delta h_s = h_{V2} + h_{Vd}\left[\left(1 - \frac{A_d}{A_2}\right)^2 - 1\right]$$

We can easily draw conclusions by the consideration of two extreme cases:

Case A: Chamber diameter = Fan diameter, i.e., $A_2 = A_d$.

$$\Delta h_s = h_{V2} + h_{Vd}\left[\left(1 - \frac{A_d}{A_2}\right)^2 - 1\right] = h_{Vd} + h_{Vd}\left[\left(1 - \frac{A_d}{A_d}\right)^2 - 1\right] = 0$$

The error is zero, just as one would expect.

Case B: The chamber diameter is very large compared to the fan diameter, i.e., $A_2 \gg A_d$.

$$\Delta h_s = h_{V2} + h_{Vd}\left[\left(1 - \frac{A_d}{A_2}\right)^2 - 1\right] = h_{V2} = \frac{1.29 \times 10^{-3}}{A_2^2} G^2$$

The formula for the velocity head term h_{V2} is derived in the next section where the resulting units for Δh are in. H_2O and for G are ft^3/min (commonly written as CFM). We evaluate this last loss Δh_s for a typical large test system. Suppose then that the chamber diameter is about three feet or $r_2 = 20$ in. You should be able to use such a system to measure airflow between about 5 and 1000 CFM. Then

$$\Delta h_s = \frac{1.29 \times 10^{-3}}{A_2^2} G^2 = \frac{1.29 \times 10^{-3}}{\left[\pi (20)^2\right]^2} G^2 \sim 1 \times 10^{-9} G^2$$

Table 3.1. Static head loss error at downstream pressure taps.

G[CFM]	Δh_s [in. H_2O]
1	10^{-9}
10	10^{-7}
100	10^{-5}
1000	0.001

The results listed in Table 3.1 lead us to the conclusion that the error in measuring the static head downstream of the fan discharge plane is trivial.

3.6 DERIVATION OF THE "ONE VELOCITY" HEAD FORMULA

We are taking a short side trip to simplify the formula

$$h_V = (1/2)\rho V^2$$

for one velocity head of air (insert and combine as many constants as possible). The symbols ρ and V represent the air density and speed averaged over the appropriate flow cross-sectional area. It turns out that when we perform an analysis to determine the total system airflow and internal distribution, we often need h_V with a volumetric airflow G as the independent variable. The density of air at 20°C and standard pressure can be taken to be $\rho = 1.20$ kg/m^3. The air speed is changed to units of ft/min.:

$$V\left[\frac{m}{s}\right] = V\left[\frac{ft}{min}\right]\left(\frac{0.305\,m}{ft}\right)\left(\frac{min}{60\,s}\right) = 5.08 \times 10^{-3} V\left[\frac{ft}{min}\right]$$

Inserting $\rho = 1.20$ kg/m^3 and the preceding for V will give h_V in the units of Pa, but we wish to have units of in. H$_2$O:

$$h_V\left[\text{Pa}\right] = \frac{1}{2}\rho V^2 = \frac{1}{2}(1.20)\left(5.08 \times 10^{-3}\right)^2 V^2 \left[\frac{\text{ft}}{\text{min}}\right]^2$$

$$h_V\left[\text{Pa}\right] = 1.55 \times 10^{-5} V^2 \left[\frac{\text{ft}}{\text{min}}\right]^2$$

$$h_V\left[\frac{\text{lb}_f}{\text{ft}^2}\right] = h_V\left[\text{Pa}\right]\left[\frac{\left(\text{lb}_f/\text{ft}^2\right)}{47.88\,\text{Pa}}\right] = \frac{1.55 \times 10^{-5}}{47.88} V^2 \left[\frac{\text{ft}}{\text{min}}\right]^2$$

$$h_V\left[\text{in.H}_2\text{O}\right] = h_V\left[\frac{\text{lb}_f}{\text{ft}^2}\right]\left[\frac{\text{in.H}_2\text{O}}{5.198\left(\text{lb}_f/\text{ft}^2\right)}\right] = \frac{\left(1.55 \times 10^{-5}\right)}{(47.88)(5.198)} V^2 \left[\frac{\text{ft}}{\text{min}}\right]^2$$

$$h_V\left[\text{in.H}_2\text{O}\right] = \frac{\left(1.55 \times 10^{-5}\right)}{(47.88)(5.198)}\left(\frac{144\,\text{in.}^2}{\text{ft}^2}\right)^2 \frac{G^2\left[\text{ft}^3/\text{min}\right]^2}{A^2\left[\text{in.}^2\right]^2}$$

which gives the next result, for which we will find much use.

$$h_V\left[\text{in.H}_2\text{O}\right] = 1.29 \times 10^{-3}\frac{G^2\left[\text{ft}^3/\text{min}\right]^2}{A^2\left[\text{in.}^2\right]^2} \qquad \text{One velocity head} \quad (3.9)$$

Note: As long as we have made this diversion we might as well describe a convenient conversion factor, the conversion of pressure from $\left(\text{lb}_f/\text{ft}^2\right)$ to $\left(\text{in. H}_2\text{O}\right)$. We begin with the pressure due to a fluid of density ρ of height h, which is $p = \rho g h$. For a one-inch high water column at sea level conditions,

$$p = \rho g h = \left(1000\frac{\text{kg}}{\text{m}^3}\right)\left(9.8\frac{\text{m}}{\text{s}^2}\right)(0.0254\,\text{m}) = 2.489 \times 10^2 \frac{\text{N}}{\text{m}^2}$$

$$p = \frac{2.489 \times 10^2\left(\dfrac{\text{N}}{\text{m}^2}\right)}{6894.76\left(\dfrac{\text{N/m}^2}{\text{lb}_f/\text{in.}^2}\right)} = 3.61 \times 10^{-2}\frac{\text{lb}_f}{\text{in.}^2} = 3.61 \times 10^{-2}\left(\frac{\text{lb}_f}{\text{in.}^2}\right)144\left(\frac{\text{in.}^2}{\text{ft}^2}\right) = 5.198\frac{\text{lb}_f}{\text{ft}^2}$$

hence the result, $\qquad 5.198\dfrac{\text{lb}_f}{\text{ft}^2}\Big/\text{in.H}_2\text{O}$

3.7 FAN AND SYSTEM MATCHING

A forced air cooled system has two major flow issues that must be reconciled: (1) the pressure/flow fan curve and (2) the enclosure system response. We have somewhat addressed the former and now we shall examine the latter. We shall assume such a response exists, but will not address the details of constructing this response just yet.

We can classify the possible enclosures as one of three types: (1) negatively pressurized, (2) positively pressurized, and (3) negatively and positively pressurized. We now consider the first category, the negatively pressurized system.

The negatively pressurized system depicted in Figure 3.4 has a fan attached to what we shall describe as the rear panel. The fan draws air from the system in an exhaust fashion. The "station" 1 to the left is at room ambient pressure. The air flows through an entrance grill, through circuit board channels, a power supply, and other geometric entities that we need not describe here. We carefully define a fan inlet "i" at some distance prior to the actual geometric inlet plane of the fan. You might wish to think of the station i as an ambient local to the fan, but not yet within the inlet flow pattern. After i, the air converges in some pattern as it enters the actual fan. The air departs the fan at the defined discharge plane "d", which in our analysis is at the actual geometric discharge plane of the fan. Lastly we have "station" 2, which is the room ambient (pressure identical to that for station 1) downstream and exterior to the fan exit flow. It is important to note that the fan discharge plane at d in Figure 3.4 is to be identified with the fan discharge plane d in Figure 3.3. The most general test/application mode that we can use is this discharge plane at the geometric fan exit. Of course, we also know from Section 3.4 that the error in placing static pressure manometers downstream of the fan exit is usually trivial. All pressures in this system are less than the ambient, and thus negative. If the pressures were not negative, there would be no flow from station 1 into the enclosure and out the fan discharge plane.

Now let's consider the various pressure heads in our system beginning at station 1. We use Bernoulli's equation with losses, Eq. 3.5, defining H_L as the total system loss from 1 to i. This is the *only* loss we consider in this case.

From 1 to i:
$$h_{T1} = h_{Ti} + H_L, \qquad h_{V1} + h_{s1} = h_{Ti} + H_L$$
$$h_{V1} \equiv 0, \quad h_{s1} \equiv 0$$
$$h_{Ti} = -H_L$$

From d to 2:
$$h_{Td} = h_{T2} + H_{L,d-2} = h_{V2} + h_{s2} + H_{L,d-2}$$
$$h_{V2} \equiv 0, \, h_{s2} \equiv 0$$
$$h_{Td} = H_{L,d-2}$$

Subtracting the result of 1 to i from that for d to 2,

$$h_{Td} - h_{Ti} = H_{L,d-2} + H_L$$

and substituting this result into Eq. (3.6),

$$h_{fs} = \left(h_{Td} - h_{Ti} \right) - \left(h_{Vd} - h_{Vi} \right)$$
$$h_{fs} = H_{L,d-2} + H_L - \left(h_{Vd} - h_{Vi} \right)$$

It will be shown, or rather given later when individual loss elements are introduced, that for the infinite expansion from the fan discharge plane we have $H_{L,d-2} = h_{Vd}$ (one velocity head at fan

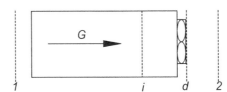

Figure 3.4. Negatively pressurized forced air cooled system.

discharge). The result is

$$\boxed{h_{fs} - h_{Vi} = H_L}$$ Exhaust fan system (3.10)

which defines the operating point of a negatively pressurized enclosure with total losses H_L. Equation (3.10) is shown graphically in Figure 3.7.

We should discuss this system a little more before we proceed to analyze the next situation. You might ask why we don't consider losses due to the flow converging into the fan and expanding from the fan. Think about the fan test system that we described in Section 3.4. In that system, we had flow convergence from the left side and into the fan and also flow divergence from the fan discharge into a plenum. This means that these two aspects of the fan flow are *built into* the fan curve that is developed during the fan characterization process, whether you perform the characterization yourself or the fan curve is provided by the fan vendor. This whole fan characterization and system analysis process is one where we must accept differences between the test system and the application. The test system that we read about is such that the fan inlet flow is from a "very large" ambient; however, the system in Figure 3.4 seems to indicate flow from a smaller inlet chamber. Certainly the application fan inlet chamber loss is different than what we might have during the fan test. So what do we do about this? You can usually just ignore this issue, but if this ever concerns you, you should have sufficient understanding to provide an adequate remedy. Some practitioners would suggest that your fan test system should resemble your application geometry. If your applications are usually small enclosures, you might consider constructing or purchasing a fan test system that is of a size that approximates that of your application problems.

The next configuration that we need to examine is the positively pressurized enclosure illustrated in Figure 3.5. The ambient pressure at i is zero and the fan builds a positive pressure at the fan discharge plane, thus the enclosure is referred to as positively pressurized and the flow is from left to right. The operating conditions are derived in a manner similar to that of the previous system, beginning with Bernoulli's equation with losses H_L that are considered from the fan discharge plane d to the ambient at 2:

$$h_{Td} = h_{T2} + H_L = h_{V2} + h_{s2} + H_L$$
$$h_{V2} \equiv 0, \quad h_{s2} \equiv 0$$
$$h_{Td} = H_L$$

Using Eq. (3.6) and solving for h_{Td}, $\quad h_{fs} = \left(h_{Td} - h_{Ti}\right) - \left(h_{Vd} - h_{Vi}\right)$
$$h_{Td} = h_{fs} + h_{Ti} + \left(h_{Vd} - h_{Vi}\right)$$

and substituting $h_{Td} = H_L$ into this result,

$$h_{fs} + h_{Vd} = H_L - h_{Ti} + h_{Vi} = H_L - \left(h_{Vi} + h_{si}\right) + h_{Vi} = H_L - h_{si}$$

and since $h_{si} = 0$,

$$\boxed{h_{fs} + h_{Vd} = H_L}$$ Blower fan system (3.11)

which defines the operating point of a positively pressurized enclosure with total losses H_L. Equation (3.11) is shown graphically in Figure 3.8.

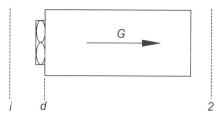

Figure 3.5. Positively pressurized forced air cooled system.

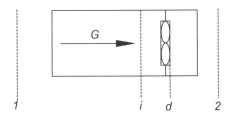

Figure 3.6. Positively/negatively pressurized forced air cooled system.

Our final system is a positively/negatively pressurized system, sometimes referred to in this book as an intermediate fan system. The configuration is shown in Figure 3.6. This system is referred to as positively/negatively pressurized because the air is at a negative pressure to the left of the fan and is at a positive pressure to the right of the fan. The pressure changes from negative to positive from the fan inlet to the fan exit. Bernoulli's equation with losses is applied between station 1 and the fan inlet i.

$$h_{T1} = h_{Ti} + H_{L,1-i}$$

$$h_{V1} + h_{s1} = h_{Ti} + H_{L,i-i}$$

$$h_{V1} \equiv 0, \; h_{s1} \equiv 0$$

$$h_{Ti} = -H_{L,1-i}$$

Bernoulli's equation with losses is next applied to the region d to 2.

$$h_{Td} = h_{T2} + H_{L,d-2}$$

$$h_{Td} = h_{V2} + h_{s2} + H_{L,d-2}$$

$$h_{V2} \equiv 0, \; h_{s2} \equiv 0$$

$$h_{Td} = H_{L,d-2}$$

Now we subtract $h_{Ti} = - H_{L,1-i}$ from $h_{Td} = H_{L,d-2}$

$$h_{Td} - h_{Ti} = H_{L,d-2} - \left(-H_{L,1-i}\right) = H_L$$

where we recognize that $H_L = H_{L,1-i} + H_{L,d-2}$, the sum of all system losses, excluding of course any loss effects at the fan inlet. Once again we use Eq. (3.6) from the fan test section.

$$h_{fs} - \left(h_{Td} - h_{Ti}\right) - \left(h_{Vd} - h_{Vi}\right)$$

$$h_{Td} - h_{Ti} = h_{fs} + \left(h_{Vd} - h_{Vi}\right)$$

We substitute $h_{Td} - h_{Ti} = H_L$ into the preceding equation to obtain the operating condition for a positively/negatively pressurized system (illustrated in Figure 3.9).

$$\boxed{h_{fs} + \left(h_{Vd} - h_{Vi}\right) = H_L}$$ Intermediate fan system (3.12)

3.8 ADDING FANS IN SERIES AND PARALLEL

It sometimes occurs that one fan does not provide adequate cooling and a more powerful fan is either not available or is undesirable. If there is adequate space you might be able to fix the problem by adding a second fan. Two methods are available: two fans in parallel or two fans in series. If your system is definitely low resistance to the air flow, parallel fans are appropriate; if the system is high resistance, fans in series should be your choice. You will see this to be true if you study Figures 3.10

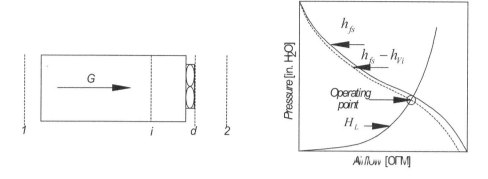

Figure 3.7. Operating point for exhaust fan system.

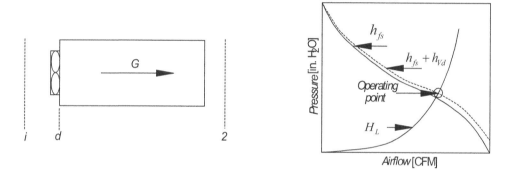

Figure 3.8. Operating point for blower fan system.

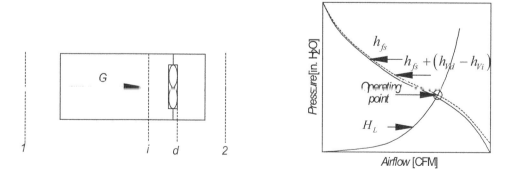

Figure 3.9. Operating point for intermediate fan system.

and 3.11. Figures 3.10 and 3.11 are appropriate for either exhaust or intermediate fans, depending on the relative contributions of the velocity heads at the fan inlet and exit.

In order to accomplish your analysis, you will need to create a new fan curve that is the equivalent of more than one fan. The procedure for constructing this new curve for fans in parallel is described next. Pick several pressure points along the fan curve varying from the pressure at zero flow to the zero pressure at maximum flow. For each pressure point that you select, add the airflow values for the multiple fans to create a new pressure-airflow data pair. If the fans are identical, you accomplish the same thing by just multiplying each airflow value of your original curve by the number of fans. Figure 3.10 was constructed by assuming two identical fans in parallel. You should probably avoid using nonidentical fans in parallel because you may find that the air prefers to exit the weaker of the two fans instead of leaving the enclosure via the intended exits.

If instead you need to use fans in series, the procedure for creating the equivalent fan curve is similar to the procedure used for parallel fans except that you add fan pressures for identical airflow values. If the fans are identical, then for each airflow value, multiply the single fan pressure by the number of fans. Figure 3.11 is intended to represent two identical fans in series.

You won't be surprised to learn that the flow expected when multiple fans are combined is not always realized in practice. You should expect something less than what your calculations lead you to believe. Another issue to be aware of is when a fan pressure curve has a dip in the curve somewhere between the minimum and maximum airflow. This is a good region to avoid as the fan operation could be unstable and "hunt" between two different airflow values.

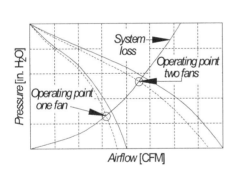

Figure 3.10. Two identical fans in parallel.

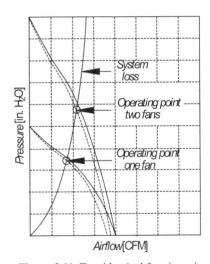

Figure 3.11. Two identical fans in series.

3.9 AIRFLOW RESISTANCE: COMMON ELEMENTS

In this section we begin the process of learning how to build a system resistance model, which we use with the fan curve to determine the system airflow. Some of these elements are based on geometric entities which are of interest in fluid applications besides electronics cooling. For example, air expansion, contraction, and flow around right angle elbows are relevant to electronics problems. In addition, we often encounter perforated, grilled, and slotted panels. There is also pressure loss information for printed circuit board channels and smooth walled channels such as those we find in extruded plate-fin heat sinks. The airflow resistance data that is included here may be all that you need in your design studies. However, you are encouraged to follow the literature because you may find data that better suits your particular needs.

The common geometric elements that you will most often use are shown in Figure 3.12. A resistance formula is given with each resistance type. Note that the resistance units for these formulae are [in. $H_2O/(CFM)^2$]. This is consistent with the basic element head loss formula, Eq. (3.13) for Δh[in. H_2O] and G[CFM].

$$\boxed{\Delta h = RG^2}$$ Basic head loss (3.13)

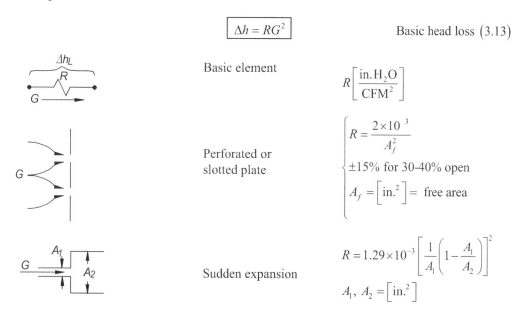

Basic element $R\left[\dfrac{in.H_2O}{CFM^2}\right]$

Perforated or
slotted plate

$\begin{cases} R = \dfrac{2\times 10^3}{A_f^2} \\ \pm 15\% \text{ for 30-40\% open} \\ A_f = \left[in.^2\right] = \text{ free area} \end{cases}$

Sudden expansion

$R = 1.29\times 10^{-3}\left[\dfrac{1}{A_1}\left(1-\dfrac{A_1}{A_2}\right)\right]^2$

$A_1, A_2 = \left[in.^2\right]$

Figure 3.12(a). Some common elements. The ratio $\sigma = A_1/A_2 < 1$ and the use of $1/A_1$ in the multiplier of ( ) indicates a "device velocity" resistance, where the "device" has the area A_1.

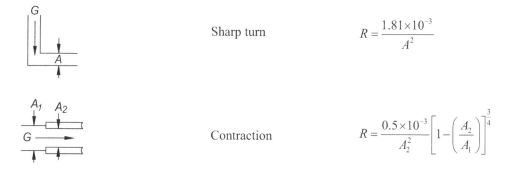

Sharp turn $R = \dfrac{1.81\times 10^{-3}}{A^2}$

Contraction $R = \dfrac{0.5\times 10^{-3}}{A_2^2}\left[1-\left(\dfrac{A_2}{A_1}\right)\right]^{\frac{3}{4}}$

Figure 3.12(b). Some common elements. The ratio $\sigma = A_2/A_1 < 1$ and the use of A_2 in the multiplier of [ ] indicates a "device velocity" resistance, where the "device" has the area A_2.

The formula for perforated plates in Figure 3.12(a) should be adequate for the very common open perforation area of 30 to 40%. It is also easy to remember for situations such as meetings where a quick evaluation of a design is needed.

A more detailed formula for the perforated plate problem is obtained from Fried and Idelchick (1988). They provide formulae, Eq. (3.14), for the loss coefficient for thin ($t/D_H < 0.015$, $Re = VD_H/v > 10^5$) perforated plates with sharp-edged orifices. Even though most problems would appear to be outside the range of the indicated Reynolds number Re, the result works surprisingly well for smaller Re. Equations (3.14) and (3.15) are combined in Figure 3.13 in the more convenient form of an airflow resistance. The Reynolds number uses a hydraulic diameter defined by $D_H = 4A/P$, for single hole area A and perimeter P. f is the fraction of the open area in the plate.

$$K_a = \left(0.707\sqrt{1-f} + 1 - f\right)^2 \Big/ f^2 \qquad \text{Perforated plate } (3.14)$$

$$K_d = \left(0.707\sqrt{1-f} + 1 - f\right)^2, \qquad t/D_H < 0.015, \ \text{Re} = VD_H/\nu > 10^5$$

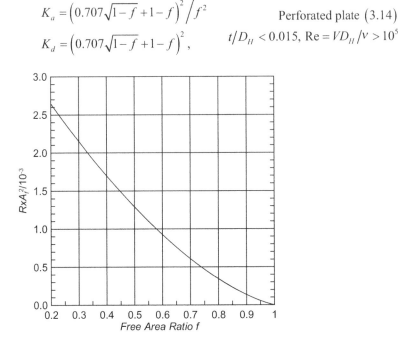

Figure 3.13. Perforated plate airflow resistance from Fried and Idelchick (1988),
$t/D_H < 0.015$, $Re = VD_H/\nu > 10^5$.

A comment is required regarding use of the terms "device velocity" and "approach velocity." You are most apt to encounter these terms when using a handbook or textbook containing head loss coefficients. In the case of Figures 3.12 and 3.13, you need not be concerned because you are given your loss data in terms of resistances to use in your flow network. However, if you have a loss coefficient, you must note whether the coefficient K is based on the device or approach velocity. You can convert between the two using $K_d = K_a f^2$ (one of the exercises asks you to derive this formula).

$$R = \frac{1.29 \times 10^{-3} K_d}{A_f^2} \qquad \text{Resistance } (3.15)$$
$$\text{for any } K_d, A_f$$

3.10 AIRFLOW RESISTANCE: TRUE CIRCUIT BOARDS

We shall consider two different ways of modeling circuit boards: (1) "true" circuit boards where the correlation formulae are based on test data from actual boards and (2) modeled circuit boards where the correlation formulae are based on "simulated" board test data, i.e., boards where the components are a uniform array of cuboid shaped elements. In this section we will consider the former. The circuit board geometries for which test-data-fitted formulae are shown in Figure 3.14. The appropriate formulae for the illustrated geometry are listed in Table 3.2.

We shall use Table 3.2 and Figure 3.14 together. The way we do this is we first look at Figure 3.14 and select from the three major choices, card (a), (b), or (c). Then we look at Table 3.2 and for our card selection of (a), (b), or (c) pick the *free passage area* and card spacing that best approximates the geometry of our particular problem. Then we just go to the column entitled R_L Formula to get the resistance formula that we need to use to actually calculate the airflow resistance. You can use these formulae for a single card channel or an entire card cage. You will note that the first three card formulae have a term that is subtracted from the area exponent. This author never represents more than one card cage at a time and always uses *n* equal to one. And even then, it is convenient to ignore this correction to the exponent. Your biggest problem will be that much of the time you won't have geometry, card spacing, and free passage area that match anything in the table. The recommendation is that you use a free passage area and card spacing that best matches your problem, and then pick the actual (a), (b), or (c) card that is available to you. This all sounds very approximate, and it is. But the method is very quick and adequate for most purposes.

An important point: The card drawings in Figure 3.14 are redrawn with a great deal of liberty taken as to the accuracy of this reproduction. This is not a serious issue as modern electronic systems will not have circuit boards and components resembling the devices shown here. The most important data for this "correlation" is the card spacing and percent open area as represented in Table 3.2.

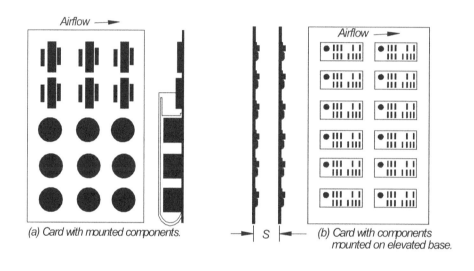

(a) Card with mounted components.

(b) Card with components mounted on elevated base.

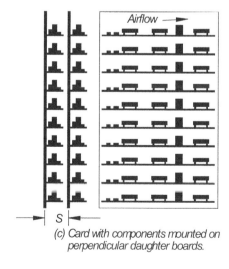

(c) Card with components mounted on perpendicular daughter boards.

Figure 3.14. Circuit board geometry for estimating resistances. See Table 3.2 (Adapted from: Hay, Donald, 1964, *Cooling Card-Mounted Solid-State Component Circuits*, McLean Engineering Division of Zero Corporation, Trenton, NJ 08691).

Table 3.2. True circuit board data referencing Figure 3.14.

Figure 3.14 Ref.	Add-ons to Basic Card Geometry	Card Spacing S [in.]	Free Passage	R_L [in. H$_2$O/CFM^2] Formula
a	--	0.50	62%	$1.35nL10^{-3}(1/A)^{(2.00-0.03n)}$
a	--	1.0	81%	$3.08nL10^{-4}(1/A)^{(2.00-0.01n)}$
a	--	0.50	70%*	$1.93nL10^{-3}(1/A)^{(2.00-0.03n)}$
b	‖ daughter	0.80	74%	$1.95nL10^{-3}(1/A)^2$
b	‖ daughter	1.60	87%	$1.43nL10^{-3}(1/A)^2$
c	⊥ daughter	0.80	58%	$5.18nL10^{-4}(1/A)^2$
c	⊥ daughter	1.60	79%	$3.24nL10^{-4}(1/A)^2$

Adapted from: Hay, Donald, 1964, *Cooling Card-Mounted Solid-State Component Circuits*, McLean Engineering Division of Zero Corporation, Trenton, NJ 08691.

n = number of card rows through which air flows.

L = card dimension [in.] parallel to flow.

A = total cross-sectional area [in.^{2}] at entrance including card edges.

Note: * This formula includes the pressure drop caused by the card holder while this is omitted in those above.

3.11 MODELED CIRCUIT BOARD ELEMENTS

Teerstra et al. (1997) presented an analytical model that predicts pressure loss for fully developed airflow in a parallel plate channel with an array of uniformly sized and spaced cuboid blocks on one wall. They used a composite solution, based on the limiting cases of laminar and turbulent smooth channels. The results are offered in the form of a friction factor that is applicable to a full range of Reynolds numbers. They quote an accuracy to within 15% of experimental data from other authors.

Although some confusion is possible, a notation nearly identical to that of Teerstra is used here to permit you an easier study of the original article, if you choose to do so. The geometry labeling is shown in Figure 3.15 and the asymptotic models are shown in Figure 3.16. From the laminar and

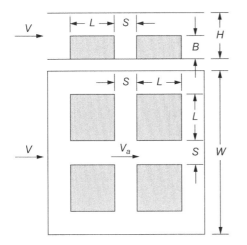

Figure 3.15. Geometry for modeled circuit board elements.

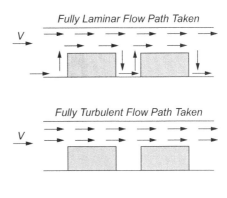

Figure 3.16. Asymptotic flow paths for modeled circuit board elements.

turbulent flow models, Teerstra et al. constructed a composite solution. The friction path length is calculated from a total "in and out" path for the fully laminar flow as shown in the top portion of Figure 3.16 and the "top-surface only" path length for fully turbulent flow as shown in the bottom portion of Figure 3.16. The investigators use a B/H biasing for flow intermediate to fully laminar and fully turbulent. The following paragraphs provide a summary of the manner in which a final correlation is constructed. At the end of this section, a formula for an airflow resistance is given and it is this latter result that you can use in your system modeling. A dimensionless friction factor f_{2H} is defined as

$$f_{2H} = -\left[\frac{dp}{dx}\right](2H) \bigg/ \left[\left(\frac{1}{2}\right)\rho V^2\right]$$

where x is the straight-through flow path. Teerstra et al. conclude their formulation with

$$f_{2H} = \left[\left(\frac{96\underline{A}}{Re_{2H}}\right)^3 + \left(\frac{0.347\underline{B}}{Re_{2H}^{1/4}}\right)^3\right]^{1/3}, \quad Re_{2H} = \frac{2VH}{\nu}$$

where it turns out that for $\underline{A} = 1$ and $\underline{B} = 0$, f_{2H} is a fit to the laminar asymptote and for $\underline{A} = 0$ and $\underline{B} = 1$, f_{2H} is a fit to the turbulent asymptote.

$$\underline{A} = \frac{\gamma^2}{\varsigma^3\chi}, \qquad \underline{B} = \frac{\gamma^{5/4}}{\varsigma^3\xi}$$

$$\gamma = 1 + \left(\frac{B}{H}\right)\left(\frac{H}{L}\right)\left(\frac{L}{L+S}\right), \varsigma = \left[1 - \left(\frac{B}{H}\right)\left(\frac{L}{L+S}\right)\right]$$

$$\chi = \left[\frac{B}{H} + \left(1 - \frac{B}{H}\right)\left(1 + \frac{2B}{H}\frac{H}{L}\frac{L}{L+S}\right)\right], \quad \xi = \frac{B}{H} + \left(1 - \frac{B}{H}\right)\left(\frac{L}{L+S}\right)$$

The pressure gradient is rewritten from the previous statement of f_{2H} to be

$$\frac{dp}{dx} = -f_{2H}\left(\frac{1}{2}\rho V^2\right)\bigg/2H$$

Then assuming that the pressure loss is uniform for the entire card length L_{Card}, the pressure gradient becomes

$$\frac{\Delta p}{L_{Card}} = -f_{2H}\left(\frac{1}{2}\rho V^2\right)\bigg/2H$$

Remembering that entrance pressure loss effects are not included in this model,

$$\left|\frac{\Delta p}{\frac{1}{2}\rho V^2}\right| = \frac{L_{Card}}{2H}f_{2H}$$

Since the $(1/2)\rho V^2$ is one velocity head (h_v) based on the approach velocity V, we can write the head loss for one card channel as

$$\Delta h_{Card} = \frac{L_{Card}}{2H}f_{2H}h_V$$

The card head loss in units of in. H_2O, based on an inlet area $A = WH$, i.e., (card width) × (card-to-card spacing) is

$$\Delta h_{Card} = \left(\frac{1.29 \times 10^{-3}}{A^2}\right)\frac{L_{Card}f_{2H}}{2H}G^2$$

The desired result is

$$R_{Card} = \left(\frac{1.29 \times 10^{-3}}{A^2} \right) \left(\frac{L_{Card} f_{2H}}{2H} \right) \quad \text{Modeled card resistance} \quad (3.16)$$

$$f_{2H} = \left[\left(\frac{96\underline{A}}{Re_{2H}} \right)^3 + \left(\frac{0.347\underline{B}}{Re_{2H}^{1/4}} \right)^3 \right]^{1/3} , \quad Re_{2H} = \frac{2HV}{v}$$

3.12 COMBINING AIRFLOW RESISTANCES

There are two different methods for calculating pressure and airflow in any given circuit. The first is amenable to solving problems with your calculator or a math scratchpad by recognizing parallel and series combinations of resistances, then adding these appropriately until you have reduced the system to a single resistance that represents the entire system. The second method uses standard numerical analysis methods for solving the circuit. We will not explore the latter method in any detail in this chapter, but some appropriate mathematics will be discussed in Chapter 13. You should be aware that some circuits are not solvable except by using numerical analysis methods.

Circuit elements requiring addition as series elements are shown in Figure 3.17 where R_1 and R_2 are added to result in the equivalent resistance R. The rule for the addition of R_1 and R_2 is easy to derive.

$$\Delta h = \Delta h_1 + \Delta h_2 = R_1 G^2 + R_2 G^2$$

We can also write the pressure loss for our resultant resistance R.

$$\Delta h = RG^2$$

Setting the right side of each of the two preceding equations equal,

$$RG^2 = R_1 G^2 + R_2 G^2$$

$$\boxed{R = R_1 + R_2} \qquad \text{Series airflow resistors} \quad (3.17)$$

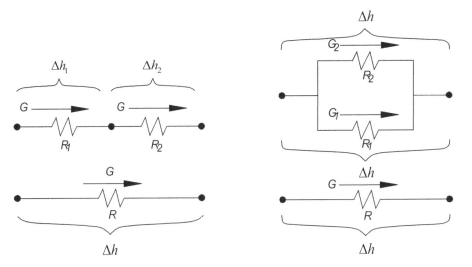

Figure 3.17. Addition of series airflow elements.   **Figure 3.18.** Addition of parallel airflow elements.

Two parallel resistances R_1 and R_2 combined to form a single resistance R are shown in Figure 3.18. Assuming conservation of airflow (see Exercise 3.3),

$$\Delta h = R_1 G_1^2, \qquad \Delta h = R_2 G_2^2, \qquad \Delta h = RG^2$$

$$G = G_1 + G_2$$

$$\sqrt{\frac{\Delta h}{R}} = \sqrt{\frac{\Delta h}{R_1}} + \sqrt{\frac{\Delta h}{R_2}}$$

$$\boxed{\frac{1}{\sqrt{R}} = \frac{1}{\sqrt{R_1}} + \frac{1}{\sqrt{R_2}}} \qquad \text{Parallel airflow resistors} \quad (3.18)$$

It is left as an exercise for you to show that

$$\boxed{\frac{G_2}{G} = \frac{1}{1 + \sqrt{R_2/R_1}}, \quad \sqrt{R/R_2}} \text{ Airflow in parallel elements} \quad (3.19)$$

We are now ready to solve a sample problem where we shall implement much of the information presented in this and the preceding chapter.

3.13 APPLICATION EXAMPLE: FORCED AIR COOLED ENCLOSURE

Our example is illustrated in Figure 3.19. Clearly this is a problem with some complexity in the airflow paths: Air enters the perforated front panel inlet and expands before entering the six card channels and the power supply. The power supply and card channels fill the enclosure from bottom to top. The air flows from front to back through the parallel channels and then contracts into the fan inlet. After flowing through the fan, the air exits through a perforated panel placed a small, but finite distance from the fan blades. The fan curve that we will use is plotted in Figure 3.21 (for the moment, ignore the system loss curve H_L).

A reasonable airflow circuit for this problem is shown in Figure 3.20. If you had to construct your own circuit without being able to look at the one drawn for you, you would probably do it somewhat differently. This is to be expected, as the procedure of constructing an airflow circuit is not an exact science. You should note that we design a circuit based on what we *think* the flow pattern is, not what it *really* is. Hopefully the differences between the model and reality are not extensive. It is reasonable to think about what the model cannot predict and use the calculated results with good engineering judgment to evaluate your design.

After constructing the flow circuit, we continue the modeling process by calculating the values of all of the resistors in the circuit. We don't have perfect formulae for all of the various flow processes. We just do the best we can. You may find studies in the literature that will expand your "library of element formulae." The following paragraphs list the details of the resistance calculations. You will probably find the arithmetic a little tedious. Some students attempt to make their calculations very quickly and often end up with erroneous results, forgetting to square an area, incorrectly calculating or writing down a power of ten, etc. The arithmetic results are important. If you are a student, forget that and put yourself in the position of a design engineer, the situation in which arithmetic results really count!

(1) Inlet perforation resistance - refer to Figure 3.13 to find $R \times A_f^2 / 10^{-3} = 1.5$ for $f = 0.45$:

$$R_{Inlet\,perf} = \frac{1.5 \times 10^{-3}}{A_f^2} = \frac{1.5 \times 10^{-3}}{\left(5.0\,\text{in.} \times 1.0\,\text{in.} \times 0.45\right)^2} = 2.96 \times 10^{-4}$$

By this point you should understand that the resistance units are (in. H_2O)/(CFM)^2.

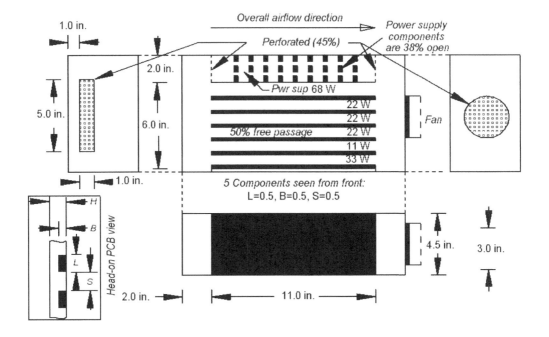

Figure 3.19. Application Example 3.13: Forced air cooled enclosure.

(2) Expansion resistance - use a formula from Figure 3.12(a):

$$R_{Inlet\,expan} = 1.29\times10^{-3}\left[\frac{1}{A_1}\left(1-\frac{A_1}{A_2}\right)\right]^2 = 1.29\times10^{-3}\left[\frac{1}{5.0\times1.0}\left(1-\frac{5.0\times1.0}{4.5\times8.0}\right)\right]^2 = 3.83\times10^{-5}$$

(3) Circuit boards taken one *channel* at a time:

We begin by calculating the open free area f_B of a circuit board channel where, looking into the channel, $N = 5$ components, each with a length and width $L = 0.5$ in., height $B = 0.5$ in., and spacing $S = 0.5$ in.

$$f_B = \frac{H_B S_B - NLB}{H_B S_B} = \frac{(4.5)(1.0)-(5)(0.5)(0.5)}{(4.5)(1.0)} = 0.444$$

We shall use $f_B = 0.5$ as an adequate value. Contraction into one channel - use the formula from Figure 3.12(b) for the contraction.

$$R_{Cont} = \frac{0.5\times10^{-3}}{A_2^2}\left[1-\left(\frac{A_2}{A_1}\right)\right]^{\frac{3}{4}} = \frac{0.5\times10^{-3}}{\left(4.5\times1.0\times0.5\right)^2}\left[1-\frac{4.5\times1.0\times0.5}{4.5\times1.0}\right]^{\frac{3}{4}} = 5.87\times10^{-5}$$

A reasonable formula choice for the circuit board is the next to last one from Table 3.2.

$$R_{Card} = \frac{5.18nL\times10^{-4}}{A^2} = \frac{5.18(1)(11.0)\times10^{-4}}{\left(4.5\times1.0\right)^2} = 2.81\times10^{-4}$$

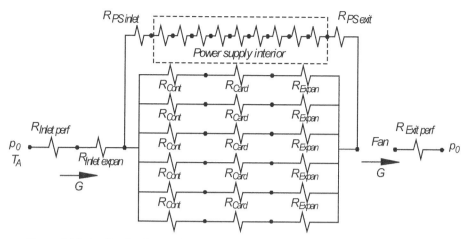

Figure 3.20. Application Example 3.13: Airflow circuit for a forced air cooled enclosure.

A small expansion for one card channel into that channel's *share* of the rear chamber uses the last formula from Figure 3.12(a).

$$R_{Expan} = 1.29 \times 10^{-3} \left[\frac{1}{A_1} \left(1 - \frac{A_1}{A_2} \right) \right]^2 = 1.29 \times 10^{-3} \left[\frac{1}{4.5 \times 1.0 \times 0.5} \left(1 - \frac{4.5 \times 1.0 \times 0.5}{4.5 \times 1.0} \right) \right]^2$$

$$R_{Expan} = 6.37 \times 10^{-5}$$

The contraction, card channel, and expansion resistances are in series so we add them according to Eq. (3.17).

$$R_{Channel} = R_{Cont} + R_{Card} + R_{Expan} = 5.87 \times 10^{-5} + 2.81 \times 10^{-4} + 6.37 \times 10^{-5} = 4.04 \times 10^{-4}$$

(4) Card cage - this is the assembly of six circuit board channels. Because these are in parallel, we add them according to Eq. (3.18).

$$\frac{1}{\sqrt{R_{Card\,cage}}} = \frac{6}{\sqrt{R_{Channel}}} = \frac{6}{\sqrt{4.04x10^{-4}}}, \qquad R_{Card\,cage} = 1.12 \times 10^{-5}$$

(5) Power supply - the power supply inlet is calculated in a fashion similar to the enclosure inlet.

$$R_{PS\,inlet} = \frac{1.5 \times 10^{-3}}{A_f^2} = \frac{1.5 \times 10^{-3}}{(2.0 \times 4.5 \times 0.45)^2} = 9.15 \times 10^{-5}$$

$$R_{PS\,exit} = R_{PS\,inlet} = 9.15 \times 10^{-5}$$

We estimate the power supply internal structure as nine contraction-expansion combinations, each 38% open.

$$R_{PS\,internal} = 9 \left\{ \frac{0.5 \times 10^{-3}}{A_2^2} \left[1 - \left(\frac{A_2}{A_1} \right) \right]^{\frac{3}{4}} + 1.29 \times 10^{-3} \left[\frac{1}{A_1} \left(1 - \frac{A_1}{A_2} \right) \right]^2 \right\}$$

$$= 9 \left\{ \frac{0.5 \times 10^{-3}}{(2.0 \times 4.5 \times 0.38)^2} \left[1 - (0.38) \right]^{\frac{3}{4}} + 1.29 \times 10^{-3} \left[\frac{1}{2.0 \times 4.5 \times 0.38} (1 - 0.38) \right]^2 \right\}$$

$$R_{PS\,internal} = 6.45 \times 10^{-4}$$

The total power supply resistance is the series sum of the inlet perforations, the internal structure, and the exit perforations.

$$R_{PS} = R_{PS\,inlet} + R_{PS\,internal} + R_{PS\,exit} = 9.15\times10^{-5} + 6.45\times10^{-4} + 9.15\times10^{-5} = 8.28\times10^{-4}$$

(6) Enclosure internal structure - we calculate the parallel sum of the power supply and card cage as

$$\frac{1}{\sqrt{R_{Enc\,internal}}} = \frac{1}{\sqrt{R_{Card\,cage}}} + \frac{1}{\sqrt{R_{PS}}} = \frac{1}{\sqrt{1.12\times10^{-5}}} + \frac{1}{\sqrt{8.28\times10^{-4}}}, \quad R_{Enc\,internal} = 9.00\times10^{-6}$$

(7) The exit perforations over the fan are 45% free.

$$R_{Exit\,perf} = 1.5\times10^{-3} \bigg/ \left[\pi \left(\frac{3.0}{2}\right)^2 \times 0.45 \right]^2 = 1.48\times10^{-4}$$

(8) System resistance - the total system resistance is the series sum of the enclosure inlet perforations, the inlet to cards/power supply expansion, the enclosure internal structure, and the fan-exit perforations.

$$R_{Sys} = R_{Inlet\,perf} + R_{Inlet\,expan} + R_{Enc\,internal} + R_{Exit\,perf}$$
$$R_{Sys} = 2.96\times10^{-4} + 3.83\times10^{-5} + 9.00\times10^{-6} + 1.48\times10^{-4} = 4.92\times10^{-4}$$

Since this is an *intermediate fan*, both h_{Vd} and h_{Vi}, the velocity heads at the fan discharge and inlet, respectively, must be calculated.

$$h_{Vd} = 1.29\times10^{-3}G^2 \big/ A_{fan-exit}^2 = 1.29\times10^{-3}G^2 \big/ \left(\pi r_{fan}^2\right)^2 = 1.29\times10^{-3}G^2 \big/ \left[\pi\left(1.5\right)^2\right]^2$$
$$h_{Vd} = 2.58\times10^{-5}G^2$$
$$h_{Vi} = 1.29\times10^{-3}G^2 \big/ A_{fan-inlet}^2 = 1.29\times10^{-3}G^2 \big/ \left(W_{Box}H_{Box}\right)^2 = 1.29\times10^{-3}G^2 \big/ \left[(8.0)(4.5)\right]^2$$
$$h_{Vi} = 9.95\times10^{-7}G^2$$

The total system head loss is $H_L = R_{Sys}G^2 = 4.92\times10^{-4}G^2$. It is instructional to arrange the data that we wish to plot in Table 3.3. The computed results in Table 3.3 as well as the fan curve, which we assume is vendor-supplied, are plotted in Figure 3.21. We see that the total system loss H_L intersects the curve $h_{fs} + (h_{Vd} - h_{Vi})$ at an airflow $G = 21$ CFM, the total airflow through the system.

The airflow in the card cage is calculated using Eq. (3.19) where we consider the enclosure internal resistance as a total resistance with the power supply in parallel. Thus we calculate G_{Card}.

Table 3.3. Application Example 3.13: Calculated heads for forced air cooled enclosure.

G (CFM)	h_{Vd}[in. H$_2$O]	h_{Vi}[in. H$_2$O]	$(h_{Vd}$-$h_{Vi})$ [in. H$_2$O]	H_L[in. H$_2$O]
0	0	0	0	0
1	2.58×10^{-5}	9.95×10^{-7}	2.48×10^{-5}	4.92×10^{-4}
5	6.46×10^{-4}	2.49×10^{-5}	6.21×10^{-4}	0.012
10	2.58×10^{-3}	9.95×10^{-5}	2.48×10^{-3}	0.049
15	5.81×10^{-3}	2.24×10^{-4}	5.59×10^{-3}	0.111
20	0.010	3.98×10^{-4}	9.93×10^{-3}	0.20
25	0.016	6.22×10^{-4}	0.0155	0.31

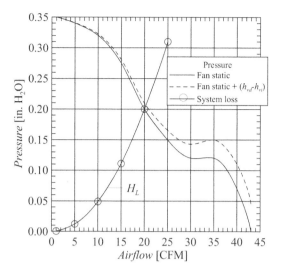

Figure 3.21. Various head curves for forced air cooled enclosure example.

$$G_{Card\ cage} = G\sqrt{R_{Enc\ internal}/R_{Card\ cage}} = 21\sqrt{9.00\times10^{-6}/1.12\times10^{-5}} = 18.8\,\text{CFM}$$

$$G_{Card} = G_{Card\ cage}/6 = 3.14\ \text{CFM}$$

We began the system analysis using a card resistance formula from Table 3.2. Now we shall recalculate the card resistance using the detailed component geometry in Figure 3.19 and the formulae provided by Teerstra et al., i.e., Eq. (3.16). There are several parameters that must be calculated to get to the R_{Card} that we want. We begin by calculating the appropriate Reynolds number. You will probably have to refer back to the Teerstra formula to follow this part.

$$Re_{2H} = \frac{2HV}{5\nu} = \frac{2(1.0\,\text{in.})\left[\left(3.14\,\text{ft}^3/\text{min}\right)/\left(1.0\,\text{in.}\times4.5\,\text{in.}/\left(144\,\text{in.}^2/\text{ft}^2\right)\right)\right]}{5(0.023\,\text{in.}^2/\text{s})} = 1745$$

Note that the numerator in the formula for the Reynolds number contains a factor of five (5), not something that you would expect. If you recall, the factor allows to use some inconsistent units, i.e., those units used in the preceding equation. You may wish to review the beginning of this chapter. Now we calculate the parameters required to obtain a card resistance.

$$\gamma = 1 + \left(\frac{B}{H}\right)\left(\frac{H}{L}\right)\left(\frac{L}{L+S}\right) = 1 + \left(\frac{0.5}{1.0}\right)\left(\frac{1.0}{0.5}\right)\left(\frac{0.5}{0.5+0.5}\right) = 1.5$$

$$\zeta = \left[1 - \left(\frac{B}{H}\right)\left(\frac{L}{L+S}\right)\right] = \left[1 - \left(\frac{0.5}{1.0}\right)\left(\frac{0.5}{0.5+0.5}\right)\right] = 0.75$$

$$\chi = \frac{B}{H} + \left(1 - \frac{B}{H}\right)\left[1 + \left(\frac{2B}{H}\right)\left(\frac{H}{L}\right)\left(\frac{L}{L+S}\right)\right] = \frac{0.5}{1.0} + \left(1 - \frac{0.5}{1.0}\right)\left[1 + \left(\frac{2\cdot0.5}{1.0}\right)\left(\frac{1.0}{0.5}\right)\left(\frac{0.5}{0.5+0.5}\right)\right]$$

$$\chi = 1.5$$

$$\xi = \frac{B}{H} + \left(1 - \frac{B}{H}\right)\left(\frac{L}{L+S}\right) = \frac{0.5}{1.0} + \left(1 - \frac{0.5}{1.0}\right)\left(\frac{0.5}{0.5+0.5}\right) = 0.75$$

$$\underline{A} = \frac{\gamma^2}{\varsigma^3\chi} = \frac{(1.5)^2}{(0.75)^3(1.5)} = 3.56, \qquad \underline{B} = \frac{\gamma^{5/4}}{\varsigma^3\xi} = \frac{(1.5)^{5/4}}{(0.75)^3(0.75)} = 5.25$$

The friction factor f_{2H} and card resistance are

$$f_{2H} = \left[\left(\frac{96\underline{A}}{Re_{2H}} \right)^3 + \left(\frac{0.347\underline{B}}{Re_{2H}^{1/4}} \right)^3 \right]^{1/3} = \left\{ \left[\frac{96(3.56)}{1712} \right]^3 + \left[\frac{0.347(5.25)}{(1745)^{1/4}} \right]^3 \right\}^{1/3} = 0.310$$

$$R_{Card} = \left(\frac{1.29 \times 10^{-3}}{A^2} \right) \left(\frac{L_{Card} f_{2H}}{2H} \right) = \frac{1.29 \times 10^{-3}}{\left[(4.5)(1.0) \right]^2} \left[\frac{(11.0)(0.310)}{2(1.0)} \right] = 1.09 \times 10^{-4} \text{ in. H}_2\text{O}/(\text{CFM})^2$$

The resistance of one card channel is a contraction plus card plus an expansion.

$$R_{Channel} = R_{Cont} + R_{Card} + R_{Expan} = 5.87 \times 10^{-5} + 1.09 \times 10^{-4} + 6.37 \times 10^{-5} = 2.31 \times 10^{-4}$$

The enclosure contains six of these channels which are in parallel with the power supply.

$$1/\sqrt{R_{Card\ cage}} = 6/\sqrt{R_{Channel}} = 6/\sqrt{2.31 \times 10^{-4}}, \quad R_{Card\ cage} = 6.42 \times 10^{-6}$$

$$\frac{1}{\sqrt{R_{Enc\ internal}}} = \frac{1}{\sqrt{R_{Card\ cage}}} + \frac{1}{\sqrt{R_{PS}}} = \frac{1}{\sqrt{6.42 \times 10^{-6}}} + \frac{1}{\sqrt{5.93 \times 10^{-4}}}, \quad R_{Enc\ internal} = 5.42 \times 10^{-6}$$

Recalculating the total system resistance,

$$R_{Sys} = R_{Inlet\ perf} + R_{Inlet\ expan} + R_{Enc\ internal} + R_{Exit\ perf} = 2.96 \times 10^{-4} + 3.83 \times 10^{-5} + 5.42 \times 10^{-6} + 1.48 \times 10^{-4}$$

$$R_{Sys} = 4.88 \times 10^{-4}$$

which is not very different than the $R_{Sys} = 4.92 \times 10^{-4}$ (in. H$_2$O)/(CFM)^2 calculated using the McLean card resistance formula, so we will keep the calculated system total airflow $G = 21$ CFM. However, the distribution between the cards and power supply requires recalculation.

$$G_{Card\ cage} = G\sqrt{R_{Enc\ internal}/R_{Card\ cage}} = 21\sqrt{5.42 \times 10^{-6}/6.42 \times 10^{-6}} = 19.3 \text{ CFM}$$

$$G_{Channel} = G_{Card\ cage}/6 = 19.3/6 = 3.2 \text{ CFM}$$

$$G_{PS} = G - G_{Card\ cage} = 21 - 19.3 = 1.7 \text{ CFM}$$

Finally we calculate the various air temperature rises *above inlet air temperature.*

22 W cards: $\Delta T_{Air\ 22W} = 1.76Q/G_{Channel} = (1.76)(22)/3.2 = 12.1°C$

11 W card: $\Delta T_{Air\ 11W} = 1.76Q/G_{Channel} = (1.76)(11)/3.2 = 6.1°C$

33 W card: $\Delta T_{Air\ 33W} = 1.76Q/G_{Channel} = (1.76)(33)/3.2 = 18.1°C$

Power supply: $\Delta T_{PS} = 1.76Q/G_{PS} = (1.76)(68)/1.7 = 70.4°C$

Comments are in order regarding the computed airflow and temperature rise values. If one presumes that desirable air temperature rises are about 10°C, the 33 W card and the power supply need work. One possibility would be to divert some air from the 22 and 11 W cards into the power supply and the 33 W card. However, this would probably result in excessive temperatures for the remaining circuit boards. Perhaps a fan with more appropriate pressure/airflow characteristics would also be a better choice.

There is another problem with regard to the enclosure design that we have not addressed. That is the fact that the front panel, perforated plate inlet does not extend from the very left side to the far right side. When you think about this, you will probably agree that the airflow in the power supply and the 33 W card will actually be considerably less than we calculated. This is because the momentum of the air after it passes through the front inlet will prevent air from turning both to the far left and right sides. The obvious option is to extend the left and right edges of the inlet grill. If this could not be accomplished, then perhaps some form of angled vane would be acceptable. A search of the literature for a formula for this kind of resistance would probably not be successful. Mock-up/airflow "smoke" visualization would probably be the best method of determining the optimum design. If you have access to computational fluid dynamics software, you could proceed with a more detailed modeling project. You will learn about airflow resistance in some common heat sinks in the next chapter.

Finally, remember that the preceding air temperatures are calculated as well-mixed. Real hot spots are not revealed in this model. At the very least, we can use well-mixed temperatures as relative indicators of various designs.

EXERCISES

3.1 Show that the conversion of an "approach" loss coefficient K_a to a "device" coefficient K_d is given by $K_d = K_a f^2$, where f = fractional free area ratio.

3.2 Derive Eq. (3.19), the law for calculating the branch airflow for parallel resistances.

3.3 Equations (3.17-3.19) are derived with the assumption of conserved airflow. Re-derive these equations with the assumption of conserved mass flow. Make provisions for an air density based on temperature for each resistance element.

3.4 An air cooled enclosure (Figure 3.4, Figure 1) with fan (Figure 3.4, Figure 2) contains eight circuit board channels and a power supply. You are asked to (1) draw an airflow circuit, (2) calculate the resistance values, (3) combine the various resistances to obtain a total system resistance, (4) match the fan curve and system losses to obtain the actual operating airflow (remember which of

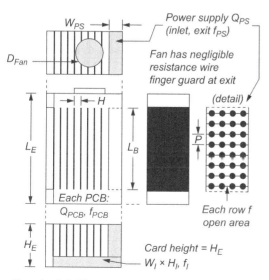

Exercise 3.4, Figure 1. Forced air cooled enclosure.

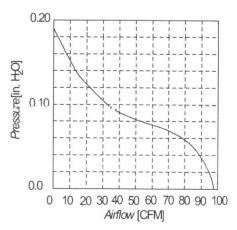

Exercise 3.4, Figure 2. Fan curve.

three enclosure types this is); calculate the airflow in each card channel and power supply, and (5) calculate the air temperature rise above ambient in the card channels, the power supply, and at the fan inlet. Refer to the following illustrations for the enclosure geometry and provided fan curve. Problem data: $H_E = 7$ in., $L_E = 18$ in., $W_I = 10$ in., $H_I = 2$ in., $f_I = 0.35$, $D_{Fan} = 4$ in., $L_B = 16$ in., $H = 1$ in., $Q_{PCB} = 25$ W, $f_{PCB} = 0.81$, $W_{PS} = 2$ in., $H_{PS} = H_E$, $Q_{PS} = 50$ W, $f_{PS} = 0.35$, $f = 0.5$, $P = 2$ in. Hints: (1) Model the entire set of eight card channels as one airflow resistance; (2) Use a McLean formula for the circuit boards.

3.5 Solve the problem detailed in Exercise 3.4, but use the Teerstra circuit board pressure loss formulae. See the detailed component illustration, Exercise 3.5, Figure 1. Problem data: $S = 0.417$ in., $L = 0.75$ in., $B = 0.3$ in., and $H = 1.0$ in. Hint: Use as much as you can of the Exercise 3.4 analysis.

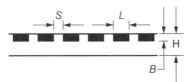

Exercise 3.5, Figure 1. Circuit board component detail.
Airflow direction into plane of page.

3.6 A small fan-cooled, high heat dissipation, component is illustrated in Exercise 3.6, Figure 1. Include the following in your analysis: (1) Draw an airflow circuit; (2) Calculate the resistance values; (3) Combine the various resistances to obtain a total system resistance; (4) Match the fan curve (Exercise 3.6, Figure 2) and system losses to obtain the actual operating airflow (remember which of three enclosure types this is), calculate the total airflow; (5) Calculate the air temperature rise above ambient at fan exit; (6) Calculate the air temperature rise above ambient at one of the assembly exits. Problem data: $W_E = 5$ in., $L_E = 5$ in., $f_{Ex} = 0.5$, $H_{Ex} = 1$ in., $W_C = 2.5$ in., $Q_C = 50$ W, $H_T = 0.75$ in., $H_C = 0.25$ in., and $Q_{Fan} = 3$ W, $D_{Fan} = 2.5$ in. Hints: (1) Ignore any expansion resistance at fan exit because of the apparent small area change; (2) Use two 90-degree turns after fan exit, one turn to the "left," and one turn to the "right"; (3) Use an expansion from fan base-component top space to exit dimensions (one for both "left" and "right" sides).

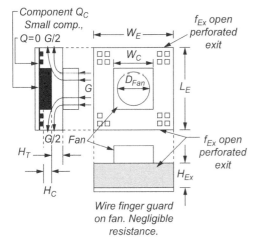

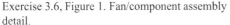

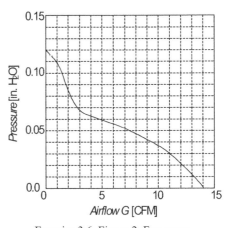

Exercise 3.6, Figure 1. Fan/component assembly detail.

Exercise 3.6, Figure 2. Fan curve.

3.7 A tower style, forced-air cooled enclosure is illustrated in Exercise 3.7, Figure 1. The fan curve is shown in Exercise 3.7, Figure 2. Use the following enclosure dimensions: $W_I = 4.0$ in., $H_I = 4.0$ in., $H_E = 4.0$ in., $f_I = 0.40$ free area ratio for the inlet. Use the following circuit board dimensions:

$H = 1.0$ in., $S = 1.667$ in., $L = 1.0$ in., $B = 0.507$ in., $L_B = 15.0$ in., $W_B = 8.0$ in., $f_B = 0.81$ free open area. The total heat dissipation for PCB_1, PCB_2, PCB_3, and PCB_4 is 25 W, 25 W, 25 W, and 50 W, respectively. The horizontal dashed line across the base of the power supply represents a perforated plate grill that has an open free area ratio $f_{PS} = 0.5$. The power supply has an airflow resistance from *after the perforated grill* to the fan inlet that has been measured to be $R_{PS} = 1 \times 10^{-4}$ (in. H_2O)/(CFM)^2. The total heat dissipation within the power supply is $Q_{PS} = 30$ W. Assume a wire finger guard of negligible resistance at the fan exit.

You are asked to (1) draw the proper airflow circuit, identifying the purpose of each element, (2) calculate the various airflow resistances where required, (3) calculate the total system airflow resistance, (4) determine the total system airflow, (5) calculate the airflow in each circuit board channel, (6) calculate the air temperature rise above the enclosure inlet for each circuit board channel, (7) calculate the air temperature rise above the enclosure inlet for the power supply inlet, and (8) calculate the air temperature rise above the enclosure inlet at the fan inlet. Hints: (1) Use the McLean circuit board head loss correlation; (2) Model the four circuit board channels as one airflow resistance; (3) Use an expansion resistance following the inlet with area $A_1 = W_I H_I$, $A_2 = W_I W_B$.

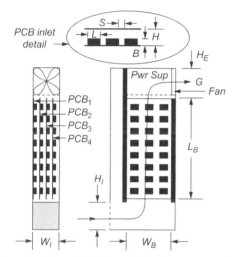

Exercise 3.7 and 3.8, Figure 1. Enclosure geometry.

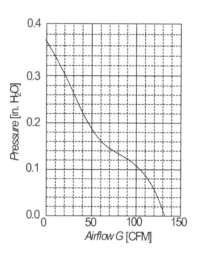

Exercise 3.7 and 3.8, Figure 2. Fan curve.

3.8 Solve the problem detailed in Exercise 3.7, but use the Teerstra circuit board pressure loss formulae. Hints: (1) Be sure to use the detailed component illustration insert labeled as "*PCB inlet detail*," Exercise 3.7 and 3.8, Figure 1 and the fan curve in Exercise 3.7 and 3.8, Figure 2; (2) Note that the illustration of the component detail suggests that the extreme left and right air channels have a spacing $S/2$, which is intended to allow use of the Teerstra model with column spacing S; (3) Use as much as you can of the Exercise 3.7 analysis.

3.9 A three-circuit board, fan-cooled enclosure (Exercise 3.9, Figure 1) and associated fan curve (Exercise 3.9, Figure 2) are illustrated. Use the following enclosure dimensions: $H_I = 1.0$ in., $W_B = 8.0$ in., $L_B = 10.0$ in., $H_E = 3.0$ in., $H = 1.0$ in., $Q_{PCB1} = 75$ W, $Q_{PCB2} = 63$ W, $Q_{PCB3} = 50$ W, $L_B = 6.0$ in., $f_1 = 0.81$, $f_2 = 0.81$, $f_3 = 0.58$ (fractional open area for $PCBs$), $f_I = 0.35$ (fractional open area at inlet), $f_D = 0.50$ (fractional open area at distribution plate), $D_{Fan} = 2.5$ in. You are asked to (1) draw an airflow circuit, (2) calculate the resistance values, (3) combine the various resistances to obtain a total system resistance, (4) match the fan curve and system losses to obtain the actual operating airflow (remember which of three enclosure types this is), calculate the airflow in each card channel, and (5) calculate the air temperature rise above ambient in the card channels, and at the fan inlet. Use the McLean circuit board correlation. Hint: Because of the very close proximity between the distribution plate and the PCB inlets, it is recommended that you do not use a PCB contraction.

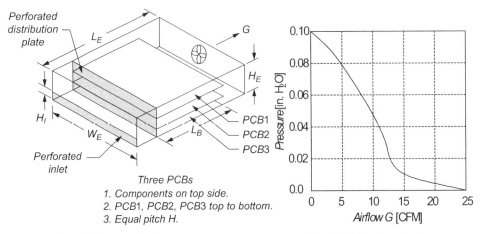

Exercise 3.9, Figure 1. Enclosure geometry.   Exercise 3.9, Figure 2. Fan curve.

3.10 Calculate and compare the airflow resistance for PCB_3 of Exercise 3.9 using both the McLean and Teerstra correlations. Assume an airflow $G = 3.5$ CFM. Recompute the Teerstra result for $G = 1.75$ CFM and 7 CFM. The PCB channel geometry is shown in Exercise 3.10, Figure 1. The dimensions are $L = 1.12$ in., $B = 0.5$ in., $N = 6$ components, and consistent with Exercise 3.9, Figure 1, $W = 8$. in., $H = 1.0$ in., and $f_3 = 0.58$.

Exercise 3.10, Figure 1. Flow into plane of page.

3.11 A forced air cooled enclosure (Exercise 3.11, Figure 1) with a fan (Exercise 3.11, Figure 2) contains five circuit board channels and a power supply. You are asked to (1) draw an airflow circuit, (2) calculate the resistance values, (3) combine the various resistances to obtain a total system resistance, (4) match the fan curve and system losses to obtain the actual operating airflow (remember which of three enclosure types this is), calculate the airflow in each card channel, power supply, and (5) calculate the air temperature rise above ambient in the card channels, at the power supply inlet, and at the fan inlet. Refer to the following illustrations for the enclosure geometry and provided fan curve. Problem data: $W_E = 4.5$ in., $L_E = 10$ in., $H_E = 4$ in., $W_{IB} = 4$ in., $H_{IB} = 1$ in., $W_{IT} = 4$ in., $H_{IT} = 0.5$ in., $f_B = 0.45$ and $f_T = 0.35$ for top and bottom inlets, respectively. Also: center to center spacing $H = 1.0$ in. for PCBs 1-4 and 0.5 in. for PCB 5, $f_{1-4} = 0.81$ fractional open area for PCBs 1-4 and $f_5 = 0.62$ fractional open area for PCB_5, $Q_{PCB1} = 20$ W, $Q_{PCB2} = 20$ W, $Q_{PCB3} = 10$ W, $Q_{PCB4} = 10$ W, $Q_{PCB5} = 5$ W, $Q_{Power\ Supply} = 12$ W, $L_B = 7$ in., $D_{Fan} = 3$ in. Each of the three major power supply entities (one cylinder with a rectangular cross-section and two circular cylinders) has an edge dimension or diameter $D = 1.05$ in. Hints: (1) Use an expansion from the top inlet region to a PCB half-height; (2) Use an expansion from the bottom inlet region to a PCB half-height; (3) Use a McLean PCB resistance formula; (4) Calculate a power supply resistance using a single, simple slot ($R = 2 \times 10^{-3}/A^2_{Open\ area}$).

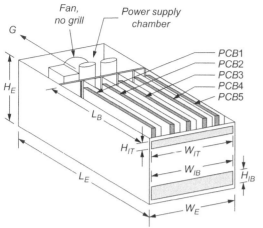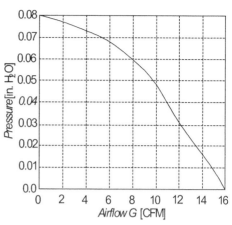

Exercise 3.11, Figure 1. Forced air cooled enclosure.   Exercise 3.11, Figure 2. Fan curve.

CHAPTER 4

Airflow II: Forced Flow
in Ducts, Extrusions, and Pin Fin Arrays

One of the most common activities in electronics thermal analysis is that of designing heat sinks. Heat sinks occur in so many different geometries that it is possible to cover only the most common. In those situations where the geometry of a heat sink does not resemble something described in this chapter, you have other options: (1) perform a literature search for a study of your particular problem, (2) contact heat sink vendors as a source of test data, (3) analyze your problem using computational fluid dynamics software, or (4) perform a test of a prototype of your design. No matter what choice you make, there is an excellent chance that you can at least make an estimate of the flow characteristics of your design using data in this chapter.

As the chapter heading indicates, we shall study the pressure loss characteristics of ducts, extrusions, and pin fin arrays. With the exception of pin fin arrays, the author has had significant experience with most of the correlations listed. The literature contains a lot of information on these problems and you should expect that new studies will continue to appear in technical journals and conference proceedings. Finally, this chapter, like others in this book, is not intended to be encyclopedic in the quantity of correlations presented, but hopefully will be sufficient to satisfy most of your daily needs in this area.

4.1 THE AIRFLOW PROBLEM FOR CHANNELS WITH A RECTANGULAR CROSS-SECTION

The geometry for this discussion is shown in Figure 4.1.* This geometry is appropriate for an extrusion with several channels or a single channel duct. Admittedly, most heat sinks of this type do not have fins that are precisely rectangular in cross-section, but instead have a taper where the fin thickness decreases from the fin root to the fin tip. We must accept this deviation and recognize that many formulae are actually approximations. Furthermore, any discrepancies in our calculations are probably due to issues other than the tapered fin.

When considering the pressure loss, we usually evaluate the effects of three major contributions to the total airflow resistance: (1) entrance effects, (2) friction effects as the flow proceeds from inlet to exit, and (3) exit effects.

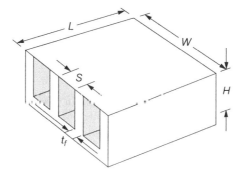

Figure 4.1. Geometry for airflow in an extruded heat sink.

---

* From this point on, we shall re-use some symbols with a meaning different than those used in earlier chapters. Learn to associate each variable name with the context of the problem.

The total airflow resistance R [in. H_2O/CFM^2] of a ducted heat sink from *prior* to the inlet past the *exit* contains the inlet, friction, and exit effects, and is written as

$$R = \frac{1.29 \times 10^{-3}}{N_c^2 A_c^2} \left[K_c + K_e + 4\bar{f}\left(\frac{L}{D_H}\right) \right] \qquad \text{Duct resistance} \quad (4.1)$$

for a head loss $h = RG^2$ where the variables are

$N_c \equiv$ Number of parallel channels

$A_c \equiv$ Cross-sectional area of one channel, $\left[\text{in.}^2\right]$

$K_c \equiv$ Inlet contraction loss coefficient

$K_e \equiv$ Exit expansion loss coefficient

$\bar{f} \equiv$ Friction coefficient averaged over duct length

$L \equiv$ Duct length, $\left[\text{in.}\right]$

$D_H \equiv$ Hydraulic diameter of one channel, $\left[\text{in.}\right]$

$D_H = 2SH/(S + H)$, closed channel

$D_H = 4SH/(S + 2H)$, open channel

The closed channel version of D_H would be appropriate for the situation illustrated in Figure 4.1 where each channel is bounded by four sides. The open channel version would be where the channels are not covered by a shroud. The more general definition of a hydraulic diameter is

$$D_H \equiv 4A_c/P \qquad \text{Hydraulic diameter} \quad (4.2)$$

for the channel with cross-sectional area A_c and *wetted* perimeter P. Clearly the closed channel has a wetted perimeter of $2(S + H)$ and the non-enshrouded channel has a perimeter wetted on only three sides so that $P = 4SH/(S + 2H)$.

On some occasions we shall be required to determine whether the channel flow is laminar or turbulent. The standard criteria using a duct Reynolds number is

Laminar duct flow $Re_{D_H} \leq 2000$

Transitional duct flow $2000 < Re_{D_H} < 10,000$

Turbulent duct flow $Re_{D_H} \geq 10,000$

You should refresh your memory by reviewing Section 3.1 where the Reynolds number is defined and also written for mixed units in the unconventional form $Re_p = VP/(5\nu)$, the form most commonly used in this book.

4.2 ENTRANCE AND EXIT EFFECTS FOR LAMINAR AND TURBULENT FLOW

The loss coefficients for the inlet and exit are plotted in Figures 4.2 and 4.3, respectively. The graphs for both K_c and K_e each indicate plots for a Reynolds number from *laminar* to *infinite*. This identification is taken directly from the well-known book by Kays and London (1964). It is not clear as to what Kays and London precisely meant by the term *laminar*. This author's best estimate is that these curves were probably calculated from theoretical considerations as no data points are shown. If this is the case, then the term *laminar* likely meant a Reynolds number of very small magnitude, which I take to be exactly $Re_{DH} = 0$. This interpretation becomes relevant when we need to interpolate for Reynolds numbers between laminar = 0 and 2000. Note that in Figure 4.3, the loss

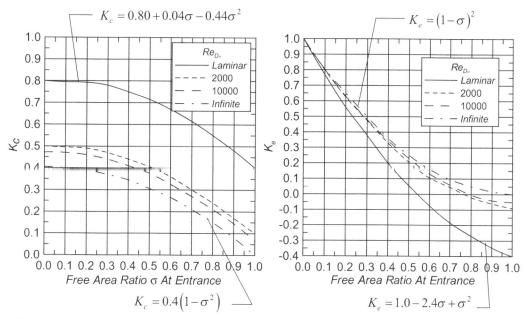

$$K_c = 0.80 + 0.04\sigma - 0.44\sigma^2 \qquad K_e = (1-\sigma)^2$$

$$K_c = 0.4(1-\sigma^2) \qquad K_e = 1.0 - 2.4\sigma + \sigma^2$$

Figure 4.2. Inlet loss coefficient, σ =smaller area/ larger area (Source: Kays, W.M. and London, A.L., *Compact Heat Exchangers*, McGraw-Hill Book Co., 1964).

Figure 4.3. Exit loss coefficient, σ =smaller area/ larger area (Source: Kays, W.M. and London, A.L., *Compact Heat Exchangers*, McGraw-Hill Book Co., 1964).

coefficient K_e for Re_{DH} infinite is, when multiplied by one velocity head at the exit, precisely our formula for a sudden expansion in Figure 3.12(a).

4.3 FRICTION COEFFICIENT FOR CHANNEL FLOW

Mean friction factors for laminar flow between parallel plates, a situation that is a reasonable match to our duct problem, are listed in tabular form in Table 4.1 and also plotted in Figure 4.4. This correlation, referred to herein as HBK-HT, is taken from the *Handbook of Heat Transfer* (HBK-HT) and may be used directly in Eq. (4.1). I have used it extensively for several years with generally satisfactory results.

Another correlation applicable to ducts and extruded heat sinks has been developed by Muzychka and Yovanovich (1998). Prior to this work, correlations for noncircular duct cross-sections used a hydraulic diameter derivation that is based on slug flow and, furthermore, does not give the correct

Table 4.1. Mean friction coefficients for laminar flow between parallel plates.

$L/(D_H Re_{D_H})$	$\overline{f}Re_{D_H}$	$L/(D_H Re_{D_H})$	$\overline{f}Re_{D_H}$	$L/(D_H Re_{D_H})$	$\overline{f}Re_{D_H}$	$L/(D_H Re_{D_H})$	$\overline{f}Re_{D_H}$
0.0_4431	168.4	0.00448	29.78	0.01067	24.77	0.01518	24.20
0.0_3209	88.89	0.00529	28.76	0.01114	24.67	0.01611	24.14
0.0_3354	73.14	0.00567	27.83	0.01165	24.57	0.01695	24.10
0.0_3686	57.60	0.00644	26.97	0.01221	24.47	0.02059	24.06
0.00159	42.63	0.00733	26.21	0.01283	24.40	∞	24.00
0.00260	35.73	0.00845	25.56	0.01352	24.32		
0.00338	32.60	0.00983	25.01	0.01427	24.25		

Source: Handbook of Heat Transfer, Rohsenow, W.H. and Hartnett, J.P., Editors, © 1973 by McGraw-Hill Book Company.

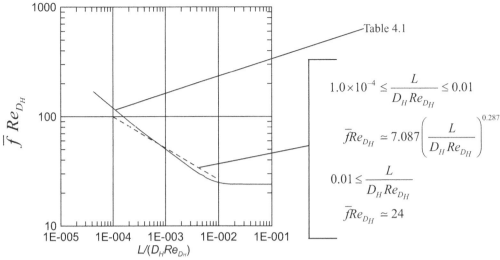

Figure 4.4. Friction factor-Reynolds number product for
laminar flow between parallel plates (Data plotted from Table 4.1).

cross-sectional area for rectangular ducts. Use of a hydraulic diameter for turbulent correlations is, however, believed to be satisfactory.

Muzychka and Yovanovich provide a correlation that combines a short duct, or entry length correlation (which is not sensitive to the cross-section shape) for developing flow. The resulting *apparent* friction factor f_{app} is determined from the following using the square root of the duct cross-sectional area A_c. This correlation is written here with nomenclature similar to the original paper.

$$Re_{\sqrt{A}} = V_f \sqrt{A}/\nu\, , \; \varepsilon = S/H\, , \qquad g = 1 \Big/ \Big[1.086957^{1-\varepsilon} \left(\sqrt{\varepsilon} - \varepsilon^{3/2} \right) + \varepsilon \Big]$$

$$f_{app} = \frac{1}{Re_{\sqrt{A}}} \left[\left(\frac{3.44}{\sqrt{\dfrac{L}{\sqrt{A}Re_{\sqrt{A}}}}} \right)^2 + \left(8\pi g \right)^2 \right]^{1/2}$$

Laminar friction (4.3)
from Muzychka and
Yovanovich

We shall use Eq. (4.4) for fully turbulent flow ($Re_{DH} > 10,000$) from Kays and London (1964).

$$\overline{f} = 0.079 \big/ Re_{DH}^{0.25}$$

Fully turbulent flow (4.4)

4.4 APPLICATION EXAMPLE: TWO-SIDED EXTRUDED HEAT SINK

A two-sided, finned, aluminum heat sink is illustrated in Figure 4.5. The heat sink was created by placing two identical heat sinks back to back. The symmetric configuration was constructed in this manner so that airflow ducted to the heat sink would split equally between the upper and lower sinks. In addition, we are assured that conduction from heat sources (not shown here) placed at the boundary between the heat sinks is evenly divided between the two devices. Convective heat

Important Note: When using f_{app} from Muzychka and Yovanovich, use $\sqrt{A}$ (A is the cross-sectional area for one channel) instead of D_H in Eq. (4.1).

transfer calculations for this device will be studied in a later chapter. Heat sink dimensions are: $H = 1.0$ in., $S = 0.23$ in., $t_f = 0.1$ in., $L = 12$ in. and 6 in. Each heat sink has 12 channels, for a total of $N_c = 24$.

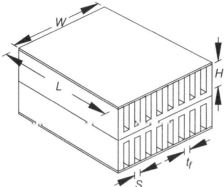

Figure 4.5. Application Example 4.4: Pressure loss calculation for two-sided, extruded heat sink.

We shall present the head loss for both lengths and see how the results compare with test data. For purposes of slightly simplifying our calculations, we assume the heat sink is not two-sided, but is single-sided and is of a width such that there are 24 fin passages ($N_c = 24$) and 25 fins ($N_f = 25$), resulting in a total, *single-sided heat sink with width* $W = 8.02$ in. Detailed calculations follow for both the HBK-HT and Muzychka-Yovanovich friction correlations for one value of total airflow $G = 25$ CFM and one length $L = 12$ in. Results for several other values of G and both values of L are tabulated and plotted.

Although all necessary geometric values are given for this problem, it is necessary to calculate the fin spacing S:

$$S = \frac{W - N_f t_f}{N_f - 1} = \frac{8.02 - (25)(0.1)}{25 - 1} = 0.23 \, \text{in.}$$

Since the heat sink has a cover shroud, our hydraulic diameter is calculated as

$$D_H = \frac{4SH}{2(S+H)} = \frac{4(0.23)(1.0)}{2(0.23+1.0)} = 0.374 \, \text{in.}$$

The free area ratio is

$$\sigma = \frac{\text{Free open area}}{\text{Total frontal area}} = \frac{N_c A_c}{WH} = \frac{(24)(0.23\,\text{in.})(1.0\,\text{in.})}{(8.02\,\text{in.})(1.0\,\text{in.})} = 0.688$$

where you will note that we have ignored the base thickness contribution to the frontal area. The thought here is that the major contraction effect is due to the fin thickness, not the heat sink base.

The airflow in one channel is used to calculate both $Re_{\sqrt{A}}$ and Re_{D_H}, the Reynolds numbers used for the Muzychka-Yovanovich and HBK-HT friction factors, respectively.

$$G_{Channel} = G/N_c = 25\,\text{CFM}/24 = 1.042\,\text{CFM}$$

$$V_f = \frac{G_{Channel}}{(A_c/144)} = \frac{1.042\,\text{CFM}}{\left[(0.23\,\text{in.})(1.0\,\text{in.})/(144\,\text{in.}^2/\text{ft}^2)\right]} = 652.17\,\text{ft/min}$$

$$Re_{\sqrt{A}} = V_f \sqrt{A_c}/5\nu = (652.17\,\text{ft/min})\sqrt{(0.23\,\text{in.})(1.0\,\text{in.})}/\left[5(0.023\,\text{in.}^2/\text{s})\right] = 2720$$

$$Re_{D_H} = V_f D_H/5\nu = (652.17\,\text{ft/min})(0.374\,\text{in.})/\left[5(0.023\,\text{in.}^2/\text{s})\right] = 2121$$

The first Reynolds number would seem to indicate flow that has some turbulence and the second is indicative of laminar flow. Usage of the plots for K_c and K_e does not require a choice of laminar or turbulent flow, but determination of the friction factor does require this consideration. We use Figures 4.2 and 4.3 to obtain

$$Re_{D_H} = 2121 \text{ and } \sigma = 0.688 \rightarrow K_c = 0.31, \ K_e = 0.05$$

where we used the Reynolds number based on hydraulic diameter to be consistent with the data for K_c and K_e in Kays and London.

We shall next calculate the Muzychka-Yovanovich friction factor f_{app}, airflow resistance R, and the total head loss H_L across the heat sink.

$$\varepsilon = \frac{S}{H} = \frac{0.23}{1.0} = 0.23, \qquad g = 1 / \left[1.086957^{1-\varepsilon} \left(\sqrt{\varepsilon} - \varepsilon^{3/2} \right) + \varepsilon \right] = 1.603$$

$$z^+ = \frac{L}{\sqrt{A_c} Re_{\sqrt{A}}} = \frac{12 \, \text{in.}}{\sqrt{(0.23)} (2720)} = 9.3 \times 10^{-3}$$

$$f_{app} = \frac{1}{Re_{\sqrt{A}}} \left[\left(\frac{3.44}{\sqrt{z^+}} \right)^2 + (8\pi g)^2 \right]^{1/2} = \frac{1}{2720} \left\{ \left(\frac{3.44}{\sqrt{9.3 \times 10^{-3}}} \right)^2 + \left[8\pi (1.603) \right]^2 \right\}^{1/2} = 0.02$$

$$R = \frac{1.29 \times 10^{-3}}{N_c^2 A_c^2} \left[K_c + K_e + 4 f_{app} \left(\frac{L}{\sqrt{A_c}} \right) \right] = \frac{1.29 \times 10^{-3}}{(24)^2 (0.23)^2} \left[0.31 + 0.05 + 4 (0.02) \left(\frac{12}{\sqrt{0.23}} \right) \right]$$

$$R = 9.93 \times 10^{-5} \, \text{in.} \, \text{H}_2\text{O} / (\text{CFM})^2$$

$$H_L = RG^2 = \left(9.93 \times 10^{-5} \right) (25)^2 = 0.062 \, \text{in.} \, \text{H}_2\text{O}$$

Now we repeat the preceding calculation for $G = 25$ CFM and $L = 12$ in., but this time we will use the HBK-HT correlation for the friction factor, Table 4.1, Figure 4.4. Parameters that do not change are not recalculated.

$$L / \left(D_H Re_{D_H} \right) = 12 / \left[(0.374)(2121) \right] = 0.015$$

We use this value of 0.015 to determine $\overline{f} Re_{D_H} = 24.2$ and

$$\overline{f} = 24.2 / Re_{D_H} = 24.2 / 2121 = 0.011$$

$$R = \frac{1.29 \times 10^{-3}}{N_c^2 A_c^2} \left[K_c + K_e + 4 \overline{f} \left(\frac{L}{D_H} \right) \right] = \frac{1.29 \times 10^{-3}}{(24)^2 (0.23)^2} \left[0.31 + 0.05 + 4 (0.011) \left(\frac{12}{0.374} \right) \right]$$

$$R = 7.724 \times 10^{-5} \, \text{in.} \, \text{H}_2\text{O} / \text{CFM}^2$$

$$H_L = RG^2 = \left(7.724 \times 10^{-5} \right) (25)^2 = 0.048 \, \text{in.} \, \text{H}_2\text{O}$$

Head losses for heat sink lengths $L = 12$ and 6 in., other airflows, and both the Muzychka-Yovanovich and HBK-HT f correlations are listed in Tables 4.2 to 4.6 and plotted in Figures 4.6 and 4.7.

Review of Figures 4.6 and 4.7 shows that the Muzychka-Yovanovich correlation provides the best fit to the test data. This correlation results in the greatest head loss and should be our choice because it will result in conservative airflow and temperature predictions. Consideration of the turbulent flow prediction in Table 4.6 indicates that just because we have exceeded a Reynolds number of 2000 does not mean that the flow is significantly, if at all, turbulent.

Table 4.2. Application Example 4.4: Calculated flow characteristics for $L = 12$ in. using the Muzychka and Yovanovich friction factor correlation, Eq. (4.3).

G [CFM]	Re_{D_H}	$Re_{\sqrt{A}}$	K_c	K_e	f_{app}	R [in.H_2O/CFM^2]	H_L [in.H_2O]
1	85	109	0.60	-0.15	0.376	0.00161	0.00161
5	424	544	0.55	-0.12	0.08	3.56×10^{-4}	0.00890
10	848	1088	0.47	-0.07	0.043	1.97×10^{-4}	0.020
15	1273	1632	0.42	-0.02	0.08	1.44×10^{-4}	0.032
20	1697	2176	0.35	0.02	0.024	1.16×10^{-4}	0.046
25	2121	2720	0.31	0.05	0.02	9.93×10^{-5}	0.062

Table 4.3. Application Example 4.4: Calculated flow characteristics for $L = 6$ in. using the Muzychka and Yovanovich friction factor correlation, Eq. (4.3).

G [CFM]	Re_{D_H}	$Re_{\sqrt{A}}$	K_c	K_e	f_{app}	R [in.H_2O/CFM^2]	H_L [in.H_2O]
1	85	109	0.60	-0.15	0.382	8.28×10^{-4}	8.28×10^{-4}
5	424	544	0.55	-0.12	0.085	1.98×10^{-4}	4.96×10^{-3}
10	848	1088	0.47	-0.07	0.047	1.17×10^{-4}	0.012
15	1273	1632	0.42	-0.02	0.034	9.00×10^{-5}	0.020

Table 4.4. Application Example 4.4: Calculated flow characteristics for $L = 12$ in. using the *Handbook of Heat Transfer* friction factor correlation, Figure 4.4.

G [CFM]	Re_{D_H}	K_c	K_e	$\dfrac{L}{D_H Re_{D_H}}$	$\bar{f}$	R [in.H_2O/CFM^2]	H_L [in.H_2O]
1	85	0.60	-0.15	0.378	0.283	1.56×10^{-3}	1.56×10^{-3}
5	424	0.55	-0.12	0.076	0.057	3.26×10^{-4}	0.00814
10	848	0.47	-0.07	0.038	0.028	1.71×10^{-4}	0.017
15	1273	0.42	-0.02	0.025	0.019	1.19×10^{-4}	0.027
20	1697	0.35	0.02	0.019	0.014	9.32×10^{-5}	0.037
25	2121	0.31	0.05	0.015	0.011	7.72×10^{-5}	0.048

Table 4.5. Application Example 4.4: Calculated flow characteristics for $L = 6$ in. using the *Handbook of Heat Transfer* friction factor correlation, Figure 4.4.

G [CFM]	Re_{D_H}	K_c	K_e	$\dfrac{L}{D_H Re_{D_H}}$	$\bar{f}$	R [in.H_2O/CFM^2]	H_L [in.H_2O]
1	85	0.60	-0.15	0.189	0.283	7.88×10^{-4}	7.88×10^{-4}
5	424	0.55	-0.12	0.038	0.057	1.72×10^{-4}	0.0043
10	848	0.47	-0.07	0.019	0.028	9.38×10^{-4}	0.00938
15	1273	0.42	-0.02	0.013	0.019	6.88×10^{-5}	0.015

Table 4.6. Application Example 4.4: Calculated flow characteristics for $L = 12$ in. using the Kays and London turbulent friction factor correlation, Eq. (4.4).

G [CFM]	Re_{D_H}	f_{Turb}	R [in.H_2O/CFM^2]	H_L [in.H_2O]
20	1697	0.012	8.23×10^{-4}	0.033
25	2121	0.012	7.85×10^{-5}	0.049

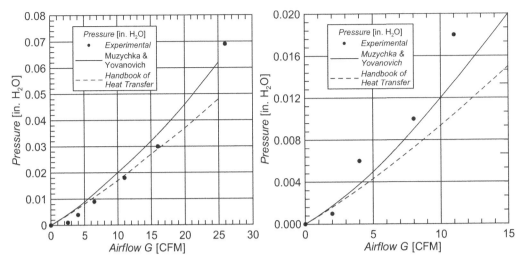

Figure 4.6. Application Example 4.4: Head loss plots for heat sink length $L = 12$ in.

Figure 4.7. Application Example 4.4: Head loss plots for heat sink length $L = 6$ in.

4.5 A PIN FIN CORRELATION

Khan et al. (2005) have published analytical correlations for the airflow and thermal problems associated with in-line (Figure 4.8) and staggered (Figure 4.9) pin fin arrays. The following paragraphs summarize their work.

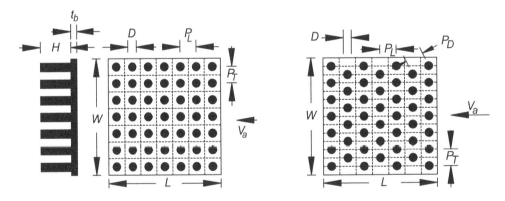

Figure 4.8. Geometry for in-line, pin fin array. Total number of fins $N = N_T N_L = 7 \times 7 = 49$.

Figure 4.9. Geometry for staggered, pin fin array. N_T is defined to give correct pitch P_T. Total number of fins $N = 46$.

Variable definitions:

$$P_L = L/N_L, \qquad\qquad p_L = P_L/D$$
$$P_T = W/N_T, \qquad\qquad p_T = P_T/D$$
$$P_D = \sqrt{P_L^2 + \left(P_T/2\right)^2}, \qquad p_D = P_D/D$$

$N_L \equiv$ Number of rows in length direction

$N_T \equiv$ Number of columns in width direction

$$V_a = \text{Approach velocity}$$

$$V_{Max} = \text{Maximum of} \left[\left(\frac{p_T}{p_T - 1} \right) V_a, \left(\frac{p_T}{p_D - 1} \right) V_a \right]$$

$$Re_D = \frac{DV_{Max}}{\nu} : \text{consistent units}, \quad Re_D = \frac{DV_{Max}}{5\nu} : \text{mixed units}$$

Khan et al. offer the following for inlet and exit losses.

$$\sigma = \frac{p_T - 1}{p_T}; \quad \boxed{\begin{array}{l} K_c = -0.0311\sigma^2 - 0.03722\sigma + 1.0676 \\[2mm] K_e = 0.9301\sigma^2 - 2.5746\sigma + 0.973 \end{array}} \qquad \text{Pin fins } (4.5)$$

and for friction coefficients,

$$\boxed{K_1 = 1.009 \left(\frac{p_T - 1}{p_L - 1} \right)^{1.09/Re_D^{0.0553}}, \; f = K_1 \left[0.233 + \frac{45.78}{(p_T - 1)^{1.1} Re_D} \right]} \quad \begin{array}{l} \text{In-line} (4.6) \\ \text{pins} \end{array}$$

$$\boxed{K_1 = 1.175 \left(\frac{p_L}{p_T Re_D^{0.3124}} \right) + 0.5 Re_D^{0.0807}, \; f = K_1 \frac{\left[378.6 / p_T^{(13.1/p_T)} \right]}{Re_D^{\left(0.68 / p_T^{0.68} \right)}}} \quad \begin{array}{l} \text{Staggered} (4.7) \\ \text{pins} \end{array}$$

The head loss is computed with

$$\boxed{\begin{array}{l} \Delta h = \left(K_c + K_e + fN_L \right) h_{V-Pins} \\[3mm] h_{V-Pins} = 1.29 \times 10^{-3} \dfrac{G^2}{(\sigma WH)^2} \end{array}} \qquad \begin{array}{l} \text{Pin fin array} \quad (4.8) \\ \text{head loss} \end{array}$$

where h_{V-Pins} is one-velocity head *within* the pin fin array and $G/(\sigma WH)$ is equivalent to V_{Max}. Remember to divide WH by 144 when you use W and H with dimensions of inches, and G in CFM. The formula for a pin fin array airflow resistance is readily deduced from Eq. (4.8).

4.6 APPLICATION EXAMPLE: PIN FIN PROBLEM FROM KHAN ET AL.

The formulae for the pin fin array problem are sufficiently complicated that it is very relevant to use precisely the same example as given in Khan's publication, thus ensuring that we have correctly interpreted the work. It is recommended that if this section becomes important to you, you should seriously consider reading the original paper for yourself (a worthwhile suggestion for other topics as well).

$$\Delta h = \left(K_c + K_e + fN_L \right) h_{V-Pins}$$

$$\Delta h = \left(K_c + K_e + fN_L \right) \left[1.29 \times 10^{-3} \frac{G^2}{(\sigma WH)^2} \right] = \frac{1.29 \times 10^{-3}}{(\sigma WH)^2} \left(K_c + K_e + fN_L \right) G^2 = RG^2$$

$$\boxed{R = \frac{1.29 \times 10^{-3}}{(\sigma WH)^2} \left(K_c + K_e + fN_L \right)} \quad \text{Pin fin resistance } (4.9)$$

We shall adhere to the geometry in Figures 4.8 and 4.9 for our problem description. The problem data is listed as follows:

$$W = L = 1.0 \text{ in.}, D = 0.07874 \text{ in.}, H = 0.394 \text{ in.}, t_b = 0.0787 \text{ in.}$$

$$N_T = 7, N_L = 7 \text{ with an in-line arrangement, } V_a = 590.55 \text{ ft/min}$$

$$k = 4.572 \text{ W}/(\text{in.} \cdot {}^\circ\text{C}), k_f = 0.0006604 \text{ W}/(\text{in.} \cdot {}^\circ\text{C}), v = 0.0245 \text{ in.}^2/\text{s}, Pr = 0.71, T_a = 27 {}^\circ\text{C}$$

noting that k and k_f are the pin and fluid (air) thermal conductivities, respectively.

We begin our calculations with determining K_c and K_e.

$$P_T = P_L = W/N_T = 1.0 \text{ in.}/7 = 0.143, \quad p_T = p_L = P_T/D = 0.143/0.07874 = 1.814$$

$$P_D = \sqrt{P_L^2 + (P_T/2)^2} = \sqrt{(1.814)^2 + (1.814/2)^2} = 0.16, \quad p_D = P_D/D = 0.16/0.07874 = 2.028$$

$$\sigma = (p_T - 1)/p_T = (1.814 - 1)/1.814 = 0.449$$

$$K_c = -0.0311\sigma^2 - 0.3722\sigma + 1.0676 = -0.0311(0.449)^2 - 0.3722(0.449) + 1.0676$$

$$K_c = 0.894$$

$$K_e = 0.9301\sigma^2 - 2.5746\sigma + 0.973 = 0.9301(0.449)^2 - 2.5746(0.449) + 0.973$$

$$K_e = 4.827 \times 10^{-3}$$

The friction factor calculation is next.

$$V_{Max} = \max\left(\frac{p_T}{p_T - 1}V_a, \frac{p_T}{p_D - 1}V_a\right) = \max(1316, 1042) = 1316 \text{ ft/min}$$

$$Re_D = \frac{DV_{Max}}{5v} = \frac{(0.07874 \text{ in.})(1316 \text{ ft/min})}{5(0.0245 \text{ in.}^2/\text{s})} = 845.74$$

$$K_1 = 1.009\left(\frac{p_T - 1}{p_L - 1}\right)^{1.09/Re_D^{0.0553}} = 1.009\left(\frac{1.814 - 1}{1.814 - 1}\right)^{1.09/(845.755)^{0.0553}} = 1.009$$

$$f = K_1\left[0.233 + \frac{45.78}{(p_T - 1)^{1.1} Re_D}\right] = (1.009)\left[0.233 + \frac{45.78}{(1.814 - 1)^{1.1}(845.75)}\right] = 0.304$$

We finish the problem by calculating the airflow resistance R and head loss.

$$G = V_a WH = (590.55 \text{ ft/min})\frac{(1.0 \text{ in.})(0.394 \text{ in.})}{(144 \text{ in.}^2/\text{ft}^2)} = 1.615 \text{ CFM}$$

$$R = \frac{1.29 \times 10^{-3}}{(\sigma WH)^2}(K_c + K_e + fN_L) = \frac{1.29 \times 10^{-3}}{[(0.449)(1.0 \text{ in.})(1.0 \text{ in.})]^2}\left[1.009 + 4.827 \times 10^{-3} + (0.304)(7)\right]$$

$$R = 0.125 \text{ in.H}_2\text{O}/\text{CFM}^2$$

$$\Delta h = RG^2 = (0.125 \text{ in.H}_2\text{O}/\text{CFM}^2)(1.615 \text{ CFM})^2 = 0.326 \text{ in.H}_2\text{O}$$

Calculation of the head loss for one airflow value is not particularly useful and we seldom have the need of calculating head loss for a greater number of airflows. More typically, we only wish to calculate the head sink resistance as part of a larger circuit.

4.7 FLOW BYPASS EFFECTS ACCORDING TO LEE

On only rare occasions do we have the need for predicting the performance of a shrouded heat sink as shown in Figure 4.1. A more typical application is that of a heat sink attached to a heat dissipating component that is mounted on a printed circuit board. Fortunately, Lee (1995) developed a nice, easy-to-understand theory for this problem. In this situation the heat sink is "open" to the air stream at the fin tips and also permits air to flow both to the left and right side of the sink. This situation is illustrated in Figure 4.10 where the flow has a velocity V_f and channel cross-sectional total area A_f within the fin channels, a bypass air velocity V_b and a total cross-sectional area A_b above and to the left and right of the heat sink, and a cross-sectional area A_{hs} of the solid, front-facing surfaces (fin and base areas).

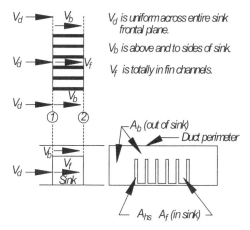

Figure 4.10. Airflow distribution for flow bypass problem (relative size of heat sink and duct not to scale).

As indicated in the preceding paragraph, the following theory is credited to Lee (1995) with the following assumptions:

1. At plane 1: The pressure is a uniform p_1 across the entire duct plane, including the heat sink and bypass; the velocity V_d is uniform across the entire duct plane. V_d is also the uniform approach velocity.
2. Between planes 1 and 2: The interfin velocity is V_f; the velocity within the bypass region (above, left, and right of the heat sink) is a uniform V_b.
3. At plane 2: The pressure is p_2 <u>outside of the heat sink envelope</u> (bypass region); the velocity is V_b <u>outside of the heat sink envelope</u> (bypass region); the interfin pressure is p_2; the interfin velocity is V_f.
4. There are head losses between planes 1 and 2.

Adhering to the preceding assumptions, we write Bernoulli's equation with losses at control surfaces 1 and 2 outside of the heat sink,

$$\frac{\rho V_d^2}{2} + p_1 = \frac{\rho V_b^2}{2} + p_2 + \Delta p_b$$

and also between control surfaces 1 and 2 in the interfin region,

$$\frac{\rho V_d^2}{2} + p_1 = \frac{\rho V_f^2}{2} + p_2 + \Delta p_f$$

where Δp_b and Δp_f are the pressure losses across planes 1 and 2 in the bypass and interfin regions, respectively. It is important to note that the above version of Bernoulli's equations is one of the few times in this book that we have used *SI* units where

$$\rho\left[kg/m^3\right], p\left[Pa, N/m^2\right], \Delta p\left[Pa, N/m^2\right], V[m/s]$$

We subtract the second from the first Bernoulli equation to obtain

$$\frac{\rho V_b^2}{2} - \frac{\rho V_f^2}{2} + \left(\Delta p_b - \Delta p_f\right) = 0$$

Next we use the conservation of mass flux for constant fluid density, i.e., conservation of airflow:

$$G_d = G_b + G_f$$

$$V_d A_d = V_b A_b + V_f A_f$$

$$V_b = \left(V_d A_d - V_f A_f\right)/A_b$$

$$V_b^2 = \left(V_d^2 A_d^2 + V_f^2 A_f^2 - 2V_d V_f A_d A_f\right)/A_b^2$$

Substituting V_b^2 into the result of subtracting the two Bernoulli equations and performing a little algebra we arrive at

$$aV_f^2 + bV_f + c = 0$$

$$a = \left[1 - \left(\frac{A_f}{A_b}\right)^2\right], b = 2\left(\frac{A_d}{A_b}\right)\left(\frac{A_f}{A_b}\right)V_d, c = \left[\frac{2\left(\Delta p_f - \Delta p_b\right)}{\rho} - \left(\frac{A_d}{A_b}\right)^2 V_d^2\right]$$

$$V_f = \frac{-b + \sqrt{b^2 - 4ac}}{2a}$$

Interfin air (4.10) velocity

where the latter is the well-known quadratic equation solution. We have rejected a second solution with a negative sign before the radical on the basis of being physically unrealistic. Strickly speaking, Lee presented his model without the presence of Δp_b, but it is included here for greater generality.

When we perform a system analysis, we can ignore the pressure/airflow effects of the heat sink if it occupies a somewhat small portion of the total circuit board cross-section. Then using Eq. (4.10) we are able to calculate the interfin air velocity. As we will learn in a later chapter, the interfin air velocity is used to calculate the heat sink thermal resistance and finally the heat sink temperature.

A different design scenario is more likely: You design a small, finned heat sink where you calculate the interfin air velocity V_f from either your analysis or vendor heat sink data. Then you would like to estimate the airflow V_d that you need in the circuit board channel that will result in the required V_f. Using a procedure nearly identical to that used in the preceding paragraphs to give us a formula for V_f, we are able to find a formula to calculate V_d. The derivation is left as an exercise, but the result is given as Eq. (4.11).

$$a = \left(\frac{A_d}{A_b}\right)^2, b = -2\left(\frac{A_d}{A_b}\right)\left(\frac{A_f}{A_b}\right)V_f, c = V_f^2\left[\left(\frac{A_f}{A_b}\right)^2 - 1\right] - \frac{2\left(\Delta p_f - \Delta p_b\right)}{\rho}$$

$$V_d = \frac{-b + \sqrt{b^2 - 4ac}}{2a}$$

Card (4.11) channel air velocity

It should be noted here that the solution of the bypass problem for V_f requires an iterative procedure because the coefficient "c" in Eq. (4.10) contains the pressure loss Δp_f across the fin channels and Δp_b across the bypass, which require us to know the fin channel air velocity V_f and the bypass air velocity V_b. In the following two examples we follow Lee (1995) by assuming that $\Delta p_f \gg \Delta p_b$. The following is offered as an outline of the iterative process:

1. Determine the card channel flow G.
2. Calculate V_d from G.
3. Guess an initial V_f, a value usually less than V_d.
4. Calculate a Reynolds Re number for a fin channel.
5. If Re indicates laminar flow, calculate a friction factor f_{Lam}.
6. If Re indicates turbulent flow, calculate a friction factor f_{Turb}.
7. Calculate K_c, K_e, R (airflow resistance for sink), and Δp_f.
8. Calculate the required quadratic constants a, b, c.
9. Solve for V_f.
10. Repeat steps 3-9 using the V_f from step 9 when you go to step 3. Note that when you return to step 3, you might wish to "dampen" or "extrapolate" your solution by using a V_f lesser or greater than what you get in step 9.

If your problem is one of solving for V_d, iteration is not required.

4.8 APPLICATION EXAMPLE: INTERFIN AIR VELOCITY CALCULATION FOR A HEAT SINK IN A CIRCUIT BOARD CHANNEL USING THE FLOW BYPASS METHOD OF LEE WITH THE MUZYCHKA AND YOVANOVICH FRICTION FACTOR CORRELATION

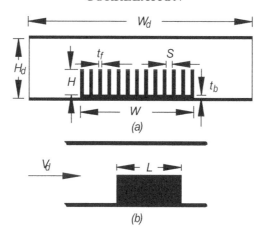

Figure 4.11. Application Example 4.8: Geometry for flow bypass problem associated with a heat sink in a circuit board channel. *(a)* Front view, flow into plane of page. *(b)* Side view, approach or duct velocity V_d.

We wish to calculate the interfin airflow for the heat sink and circuit board channel geometry illustrated in Figure 4.11. The following data apply to this problem: W_d = 10.0 in., H_d = 2.0 in., H = 0.995 in., L = 3.0 in., t_f = 0.1 in., S = 0.225 in., t_b = 0.315 in., N_f = 13 (number of fins), and W = 4.0 in., G = 50 CFM (total airflow in circuit board channel). We use the Muzychka-Yovanovich friction factor correlation for laminar flow.

Some of the variables will need to be converted to *SI* units and others occur only in ratios and may be left in their originally specified units. We will use two principal conversions:

$$\Delta p\,[\text{Pa}] = \Delta h\,[\text{in.}\,H_2O]/C_h\,[\text{in.}\,H_2O/\text{Pa}], \quad C_h = 4.019\times10^{-3}$$

and the second conversion is

$$V\,[\text{m/s}] = \left[\frac{(12\,\text{in./ft})(2.54\,\text{cm/in.})}{(100\,\text{cm/m})(60\,\text{s/min})}\right]V\,[\text{ft/min}] = C_V V\,[\text{ft/min}]$$

$$C_V = 5.08\times10^{-3}\,[\text{m/s}]/[\text{ft/min}], \quad \rho = 1.18\,\text{kg/m}^3$$

First we pre-calculate as many parts of the problem as possible in order to save effort during the required iteration cycles.

$$D_H = 4SH/(2H+S) = 4(0.225)(0.995)/[2(0.995)+0.225] = 0.404\,\text{in.}$$

$$A_d = W_d H_d = (10.0)(2.0) = 20\,\text{in.}^2$$

$$A_f = (N_f - 1)SH = (13-1)(0.225)(0.995) = 2.687\,\text{in.}^2$$

$$A_{hs} = W(H+t_b) - A_f = (4.0)(0.995+0.315) - 2.687 = 2.554\,\text{in.}^2$$

$$A_c = SH = (0.225)(0.995) = 0.224\,\text{in.}^2$$

$$A_b = A_d - A_f - A_{hs} = 20 - 2.687 - 2.554 = 14.76\,\text{in.}^2$$

$$\sigma = A_f/(A_f + A_{hs}) = 2.687/(2.687+2.554) = 0.513$$

$$V_d = \frac{G}{(A_d/144)} = \frac{50\,CFM}{\left[20\,\text{in.}^2/\left(144\,\text{in.}^2/\text{ft}^2\right)\right]} = 360\,\text{ft/min}$$

Some calculations in preparation for the friction factor evaluation are also possible.

$$\varepsilon = S/H = 0.225/0.995 = 0.226$$

$$g = \frac{1}{\left[1.086957^{1-\varepsilon}\left(\sqrt{\varepsilon}-\varepsilon^{3/2}\right)+\varepsilon\right]} = \frac{1}{\left\{1.086957^{1-0.226}\left[\sqrt{0.226}-(0.226)^{3/2}\right]+0.226\right\}} = 1.616$$

The iterative procedure previously described will be followed, and as expected, several steps are necessary. We will begin with our "guess" of the interfin air velocity. The beginning value used here is $V_f = 200$ ft/min; however, only the last iteration will be shown. The calculation method for the last step is identical to the first, but of course our "starting value" will also be the final value, i.e., our solution. This value is $V_f = 236$ ft/min.

$$Re_{D_H} = \frac{V_f D_H}{5\nu} = \frac{(236)(0.404)}{5(0.029)} = 658, \qquad Re_{\sqrt{A}} = \frac{V_f \sqrt{A_c}}{5\nu} = \frac{(236)\sqrt{0.224}}{5(0.029)} = 770$$

$$z^+ = \frac{L}{\sqrt{A_c}\,Re_{\sqrt{A}}} = \frac{3.0}{\sqrt{0.224}\,(770)} = 8.23\times10^{-3}$$

$$f_{app} = \frac{1}{Re_{\sqrt{A}}}\left\{\left(\frac{3.44}{\sqrt{z^+}}\right)^2 + (8\pi g)^2\right\}^{\frac{1}{2}} = \frac{1}{770}\left\{\left(\frac{3.44}{\sqrt{8.23\times10^{-3}}}\right)^2 + \left[8\pi(1.616)\right]^2\right\}^{\frac{1}{2}} = 0.072$$

Using the Reynolds number $Re_{D_H} = 658$, we use Figures 4.2 and 4.3 to obtain $K_c = 0.55$ and $K_e = 0.13$. We now have enough parameters calculated to obtain the airflow resistance.

$$R\left[\text{in.}\,\text{H}_2\text{O}/\text{CFM}^2\right] = \frac{1.29\times10^{-3}}{N_p^2 A_c^2}\left(K_c + K_e + 4f_{app}\frac{L}{\sqrt{A_c}}\right)$$

$$R\left[\text{in.}\,\text{H}_2\text{O}/\text{CFM}^2\right] = \frac{1.29\times10^{-3}}{(12)^2(0.224)^2}\left(0.55 + 0.13 + 4(0.072)\frac{(3.0)}{\sqrt{0.224}}\right)$$

$$R = 4.49\times10^{-4}\,\text{in.}\,\text{H}_2\text{O}/\text{CFM}^2$$

The head loss is then

$$\Delta h_f = RG_f^2, \quad G_f = V_f A_f = (236\,\text{ft/min})\left[2.687\,\text{in.}^2\left(\text{ft}^2/144\,\text{in.}^2\right)\right] = 4.40\,CFM$$

$$\Delta h_f = \left(4.49\times10^{-4}\right)(4.40)^2 = 8.70\times10^{-3}\,\text{in.}\,\text{H}_2\text{O}$$

The constant c that we use in Eq. (4.10) has units and our derivation of the formulae assumed *SI* units. It is convenient to build the velocity and head loss conversions into the quadratic equation solution constants. The constants a, b, and c are

$$a = 1 - \left(\frac{A_f}{A_b}\right)^2 = 1 - \left(\frac{2.687}{14.76}\right)^2 = 0.967$$

$$b = 2\left(\frac{A_d}{A_b}\right)\left(\frac{A_f}{A_b}\right)V_d C_V = 2\left(\frac{20}{14.76}\right)\left(\frac{2.687}{14.76}\right)(360)(5.08 \times 10^{-3}) = 0.902$$

$$c = \frac{2(\Delta h_f / C_h)}{\rho} - \left(\frac{A_d}{A_b}\right)^2 V_d^2 C_V^2 = \frac{2(8.70 \times 10^{-3} / 4.019 \times 10^{-3})}{1.18} - \left(\frac{20}{14.76}\right)^2 (360)^2 (5.08 \times 10^{-3})^2 = -2.473$$

$$V_f = \frac{-b + \sqrt{b^2 - 4ac}}{2a}\left(\frac{1}{C_V}\right) = \frac{-(0.902) + \sqrt{(0.902)^2 - 4(0.967)(-2.473)}}{2(0.967)}\left(\frac{1}{5.08 \times 10^{-3}}\right)$$

$$V_f = 236 \, \text{ft/min}$$

We began our iteration using $V_f = 236$ ft/min. Clearly we can stop here. A summary of the actual solution values is shown in Table 4.7. The entrance and exit loss coefficients are not listed in the table because each remained rather constant at $K_c = 0.55$ and $K_e = 0.13$. We shall complete the solution to this problem when we learn how to calculate thermal resistance and heat sink temperature for the final value of $V_f = 236$ ft/min.

Table 4.7. Application Example 4.8: Summary of iterative solution using the Muzychka and Yovanovich friction factor.

V_f ft/min	Re_{D_H}	$Re_{\sqrt{A}}$	f_{app}	R in.H_2O/(CFM)^2	Δh_f in. H_2O	c	V_f ft/min
200	558	653	0.082	4.94×10^{-4}	6.87×10^{-3}	-3.243	280
220	613	718	0.076	4.67×10^{-4}	7.86×10^{-3}	-2.824	257
230	641	751	0.074	4.55×10^{-4}	8.38×10^{-3}	-2.61	244
235	655	767	0.072	4.50×10^{-4}	8.64×10^{-3}	-2.54	237
236	658	770	0.072	4.49×10^{-4}	8.70×10^{-3}	-2.47	236

4.9 APPLICATION EXAMPLE: INTERFIN AIR VELOCITY CALCULATION FOR A HEAT SINK IN A CIRCUIT BOARD CHANNEL USING THE FLOW BYPASS METHOD OF LEE WITH THE *HANDBOOK OF HEAT TRANSFER* FRICTION FACTOR CORRELATION

This section is an analysis of the problem that we studied in Section 4.8, but instead of using the Muzychka and Yovanovich friction factor correlation, we shall use that from the *Handbook of Heat Transfer*, i.e., the curve plotted in Figure 4.4 or Table 4.3. We will show only the steps for the last iteration, but include more step-by-step results at the end of the problem. The final value of V_f will therefore also be equal or close to the initial value, i.e., $V_f = 266$ ft/min.

$$Re_{D_H} = \frac{V_f D_H}{5\nu} = \frac{(267 \, \text{ft/min})(0.404 \, \text{in.})}{5(0.029)} = 745$$

The variable needed for the friction factor is

$$L/\left(D_H Re_{D_H}\right) = \left(3.0\,\text{in.}\right)/\left[\left(0.404\right)\left(745\right)\right] = 9.97 \times 10^{-3}$$

and we obtain the result from Figure 4.4.

$$\overline{f}Re_{D_H} = 25.0, \qquad \overline{f} = 25.0/Re_{D_H} = 25.0/745 = 0.034$$

The inlet and exit loss coefficients for $\sigma = 0.513$ are $K_c = 0.56$ and $K_e = 0.12$. The airflow resistance is

$$R = \frac{1.29 \times 10^{-3}}{N_c^2 A_c^2}\left[K_c + K_e + 4\overline{f}\left(\frac{L}{D_H}\right)\right] = \frac{1.29 \times 10^{-3}}{\left(12\right)^2\left(0.224\right)^2}\left[0.56 + 0.12 + 4\left(0.034\right)\left(\frac{3.0}{0.404}\right)\right]$$

$$R = 3.00 \times 10^{-4}\,\text{in.}\,\text{H}_2\text{O}/\text{CFM}^2$$

for which the volumetric heat sink airflow, head loss, and pressure loss are computed next.

$$G_f = V_f A_f = \left(267\,\text{ft/min}\right)\left(2.69\,\text{in.}^2\,\frac{\text{ft}^2}{144\,\text{in.}^2}\right) = 4.98\,\text{CFM}$$

$$\Delta h_f = RG_f^2 = \left(3.00 \times 10^{-4}\right)\left(4.98\right)^2 = 7.44 \times 10^{-3}\,\text{in.}\,\text{H}_2\text{O}$$

The quadratic equation constants are

$$a = 1 - \left(\frac{A_f}{A_b}\right)^2 = 1 - \left(\frac{2.687}{14.76}\right)^2 = 0.967$$

$$b = 2\left(\frac{A_d}{A_b}\right)\left(\frac{A_f}{A_b}\right)V_d C_V = 2\left(\frac{20}{14.76}\right)\left(\frac{2.687}{14.76}\right)\left(360\right)\left(5.08 \times 10^{-3}\right) = 0.902$$

$$c = \frac{2\left(\Delta h_f/C_h\right)}{\rho} - \left(\frac{A_d}{A_b}\right)^2 V_d^2 C_V^2$$

$$c = \frac{2\left(7.44 \times 10^{-3}/4.019 \times 10^{-3}\right)}{1.18} - \left(\frac{20}{14.76}\right)^2\left(360\right)^2\left(5.08 \times 10^{-3}\right)^2 = -3.00$$

and the heat sink air velocity in mixed units is

$$V_f = \frac{-b + \sqrt{b^2 - 4ac}}{2a}\left(\frac{1}{C_V}\right) = \frac{-0.902 + \sqrt{\left(0.902\right)^2 - 4\left(0.967\right)\left(-3.00\right)}}{2\left(0.967\right)}\left(\frac{1}{5.08 \times 10^{-3}}\right) = 267\,\text{ft/min}$$

which as we see compares reasonably well with the last line in Table 4.7.

When we compare the results in Tables 4.7 and 4.8, we see that the friction coefficients differ by about a factor of two at approximately the same air flow. The final computed air velocities V_f for each of the two methods are as close as they are because the inlet and exit loss coefficients are identical at identical velocities. It is therefore worthwhile to be aware of these two methods for calculating the friction factor.

Table 4.8. Application Example 4.9: Summary of iterative solution using the *Handbook of Heat Transfer* friction factor.

V_f ft/min	Re_{D_H}	$\overline{f}$	R in.H$_2$O/(CFM)^2	Δh_f in. H$_2$O	c	V_f ft/min
200	556	0.044	3.53×10^{-4}	4.42×10^{-3}	-4.07	322
260	725	0.034	3.05×10^{-4}	7.16×10^{-3}	-3.12	273
266	742	0.034	3.00×10^{-4}	7.40×10^{-3}	-3.02	268
267	745	0.034	3.00×10^{-4}	7.44×10^{-3}	-3.00	267

4.10 FLOW BYPASS EFFECTS ACCORDING TO JONSSON AND MOSHFEGH

Jonsson and Moshfegh (2001) have reported on their extensive experimental wind tunnel study of the flow bypass problem. They placed pressure taps five cm before and after the tested heat sinks. The heat sinks had a square footprint with constant width and length of $W = 5.28$ cm, $L = 5.28$ cm (length in flow direction), $t_f = 0.15$ cm, $H = 1.0$, 1.5, and 2.0 cm. The duct and heat sink geometries were such that

$$0.33 \le W/W_d \le 0.84; \quad 0.33 \le H/H_d \le 1.0; \quad 2000 < Re_{D_H} < 16500$$

with the statement that their work is not valid for zero bypass. This latter restriction would not seem to be particularly restrictive as long as the heat sink width is slightly less than the duct width by 84 %. The geometric parameters are defined according to Figure 4.11, except that Jonsson and Moshfegh show a zero fin base thickness t_b, i.e., the fin roots are flush with the wind tunnel base. Several varieties of heat sink fins were evaluated: plate fin, in-line strip fin, staggered strip fin, in-line circular pin fin, staggered circular pin fin, in-line square pin fin, and staggered square pin fin.
The results were fit to Eq. (4.12) according to the parameters in Table 4.9.

$$\frac{\Delta p}{\left(\frac{1}{2}\rho V^2\right)} = C_2 \left(\frac{Re_{D_H}}{1000}\right)^{n_1} \left(\frac{W_d}{W}\right)^{n_2} \left(\frac{H_d}{H}\right)^{n_3} \left(\frac{S}{H}\right)^{n_4} \left(\frac{t_f, d}{H}\right)^{n_5} \quad \begin{array}{l}\text{Experimental } (4.12)\\ \text{bypass correlation}\end{array}$$

We must be careful to follow the definition of the various flow parameters used by Jonsson and Moshfegh. The velocity V is averaged over the entire bypass region as well as the open fin channel cross-sectional areas. If the bypass area is defined by A_b and the total fin channel area by A_f, then this velocity is calculated using $V = G/(A_b + A_f)$ where we must remember to include the conversion of the areas from in.² to ft² when we use V, A_b, A_f, and G with dimensions of ft/min, in.², in.², and ft³/min, respectively. The Reynolds number is based on this same velocity and also on the wind tunnel dimensions, i.e., $D_H = 2W_d H_d/(W_d + H_d)$ and $Re_{D_H} = VD_H/v$ where in this case we use consistent units in the Reynolds number, i.e., no factor of five in denominator. Thus

$$Re_{D_H} = \left(\frac{G}{A_b + A_f}\right)\left(\frac{2W_d H_d}{W_d + H_d}\right)\Big/ v$$

An important point is that Eq. (4.12) is developed for a heat sink with a length by width of 2.05 in. by 2.05 in.; therefore, this formula needs a correction factor for other dimensions. The best estimate is that the right side of Eq. (4.12) should be multiplied by the factor (L/W).

The correlation presented in this section should probably not be applied to heat sinks that are either much larger or much smaller (in width and length) than about two inches, in order to stay within the range of the study parameters.

Table 4.9. Formula parameters for flow bypass head loss according to Jonsson and Moshfegh.

Fin Style	C_2	n_1	n_2	n_3	n_4	n_5
Plate fin	4.783	-0.4778	-0.6874	-0.5979	-0.7184	0.6736
In-line strip fin	4.027	-0.3599	-0.5754	-0.6750	-0.6232	0.6268
Staggered strip fin	4.139	-0.3366	-0.6362	-0.7026	-0.6413	0.6320
In-line circular pin fin	5.375	-0.1759	-0.7161	-0.8230	-0.5401	0.5990
Staggered circular pin fin	6.967	-0.1556	-0.8533	-0.9592	-0.7740	0.7838
In-line square pin fin	3.209	-0.1551	-0.7790	-0.8169	-0.5306	0.3408
Staggered square pin fin	4.857	-0.1617	-0.8921	-0.9773	-0.6890	0.5097

4.11 APPLICATION EXAMPLE: PIN FIN PROBLEM FROM KHAN ET AL., USING THE JONSSON AND MOSHFEGH CORRELATION, NON-BYPASS

The only consideration for the "input" data for this problem that we didn't consider in Section 4.6 is the duct width and height. Since we are looking at a near zero bypass problem, we can use $W_d = W + S = 1.0$ in. $+ 0.075$ in. $= 1.075$ in., which gives us a half-interfin spacing on each side. We will use $H = H_d = 0.394$ in. Otherwise, all of our data is the same, including $G = 1.615$ CFM. The required areas are:

$$A_b = W_d H_d - HW = (1.075)(0.394) - (0.394)(1.0) = 0.029 \, \text{in.}^2$$

$$A_f = N_T HS = (7)(0.394)(0.075) = 0.206 \, \text{in.}^2$$

We find that $W/W_d = 1.0/1.075 = 0.93$ and $H/H_d = 0.394/0.394 = 1$, which are slightly out of the recommended range, but not so much as to cause us concern. The Reynolds number is within a valid range:

$$Re_{D_H} = \left(\frac{G}{A_b + A_f} \right) \left(\frac{2W_d H_d}{W_d + H_d} \right) \Big/ 5\nu = \left[\frac{1.615 \, \text{CFM}}{(0.029 + 0.206/144)} \right] \left[\frac{2(1.075 \, \text{in.})(0.394 \, \text{in.})}{1.075 \, \text{in.} + 0.394 \, \text{in.}} \right] \Big/ 5(0.023)$$

$$Re_{D_H} = 4946$$

We can now proceed with the straightforward calculation of the dimensionless pressure loss, head loss, and airflow resistance.

$$\frac{\Delta p}{\left(\frac{1}{2} \rho V^2 \right)} = C_2 \left(\frac{Re_{D_H}}{1000} \right)^{n_1} \left(\frac{W_d}{W} \right)^{n_2} \left(\frac{H_d}{H} \right)^{n_3} \left(\frac{S}{H} \right)^{n_4} \left(\frac{t_f, d}{H} \right)^{n_5} \left(\frac{L}{W} \right)$$

$$\frac{\Delta p}{\left(\frac{1}{2} \rho V^2 \right)} = (5.375) \left(\frac{4946}{1000} \right)^{-0.1759} \left(\frac{1.0}{1.075} \right)^{-0.7161} \left(\frac{0.394}{0.394} \right)^{-0.8230} \left(\frac{0.075}{0.394} \right)^{-0.5401} \left(\frac{0.07874}{0.394} \right)^{0.5990} \left(\frac{1 \text{in.}}{1 \text{in.}} \right)$$

$$\Delta h = \frac{1.29 \times 10^{-3}}{\left(A_b + A_f \right)^2} \left[\frac{\Delta p}{\left(\frac{1}{2} \rho V^2 \right)} \right] G^2$$

$$= \frac{1.29 \times 10^{-3}}{\left(A_b + A_f \right)^2} \left[\begin{array}{c} (5.375) \left(\dfrac{4946}{1000} \right)^{-0.1759} \left(\dfrac{1.0}{1.075} \right)^{-0.7161} \left(\dfrac{0.394}{0.394} \right)^{-0.8230} \\[2mm] \times \left(\dfrac{0.075}{0.394} \right)^{-0.5401} \left(\dfrac{0.07874}{0.394} \right)^{0.5990} \left(\dfrac{1 \text{in.}}{1 \text{in.}} \right) \end{array} \right] (1.615 \, \text{CFM})^2$$

$$\Delta h = 0.218 \, \text{in.} \, \text{H}_2\text{O}$$

$$R = \Delta h / G^2 = 0.218 \, \text{in.} \, \text{H}_2\text{O} / (1.615)^2 = 0.084 \, \text{in.} \, \text{H}_2\text{O} / \text{CFM}^2$$

In Section 4.6, we calculated a head loss $\Delta h = 0.326$ in. $\text{H}_2\text{O}/(\text{CFM})^2$ and $R = 0.125$ in. $\text{H}_2\text{O}/(\text{CFM})^2$. The discrepancy is quite large. Unfortunately, we don't have empirical data for this problem. It is fair to conclude, however, that either method gives us a rough estimate. It would be reasonable to select the larger of the two resistances to be conservative.

EXERCISES

4.1 Consider a shrouded, finned, extruded heat sink for airflow for G = 1, 5, 7.5 and 10 CFM. Referring to Figure 4.1, use the following geometric data: H = 0.75 in., L = 5 in., W = 4 in., t_f = 0.1 in., N_f = 10 (10 fins). Use the Muzychka and Yovanovich friction factor correlation. Summarize your calculations in a table containing G, Re_{D_H}, $Re_{\sqrt{A}}$, K_c, K_e, f_{app}, R, and H_L.

4.2 Consider a shrouded, finned, extruded heat sink for airflow for G = 1, 5, 7.5 and 10 CFM. Referring to Figure 4.1, use the following geometric data: H = 0.5 in., L = 10 in., W = 5 in., t_f = 0.1 in., N_f = 20 (20 fins). Use the Muzychka and Yovanovich friction factor correlation. Summarize your calculations in a table containing G, Re_{D_H}, $Re_{\sqrt{A}}$, K_c, K_e, f_{app}, R, and H_L.

4.3 Consider a shrouded, finned, extruded heat sink for airflow for G = 1, 5, 7.5 and 10 CFM. Referring to Figure 4.1, use the following geometric data: H = 2.0 in., L = 2 in., W = 2 in., t_f = 0.1 in., N_f = 5 (5 fins). Use the Muzychka and Yovanovich friction factor correlation. Summarize your calculations in a table containing G, Re_{D_H}, $Re_{\sqrt{A}}$, K_c, K_e, f_{app}, R, and H_L.

4.4 Recalculate the problem of Application Ex. 4.6, but use the staggered array of Figure 4.9. Hint: $N_T = N_L = 7$.

4.5 Derive Eq. (4.11), the solution for the duct air velocity, given a known interfin air velocity.

4.6 Calculate the interfin volumetric airflow G and velocity V_f for the geometry in Figure 4.11 with the following data: W_d = 10.0 in., H_d = 2.0 in., H = 1.5 in., W = 4.125 in., L = 3.0 in., t_f = 0.23 in., S = 0.203 in., t_b = 0.25 in., N_f = 10 (number of fins), and G = 25 CFM (total airflow in circuit board channel). Use the Muzychka and Yovanovich friction factor correlation if you have laminar flow. Hint: Use a starting value of V_f = 100 ft/min.

4.7 Calculate the interfin volumetric airflow G and velocity V_f for the geometry in Figure 4.11 with the following data: W_d = 10.0 in., H_d = 4.0 in., H = 2.875 in., W = 2.99 in., L = 3.0 in., t_f = 0.144 in., S = 0.425 in., t_b = 0.125 in., N_f = 6 (number of fins), and G = 15 CFM (total airflow in circuit board channel). Use the Muzychka and Yovanovich friction factor correlation if you have laminar flow. Hint: Use a starting value of V_f = 50 ft/min.

4.8 Calculate the interfin volumetric airflow G and velocity V_f for the geometry in Figure 4.11 with the following data: W_d = 1.872 in., H_d = 0.906 in., H = 0.625 in., W = 1.56 in., L = 3.0 in., t_f = 0.0891 in., S = 0.156 in., t_b = 0.125 in., N_f = 7 (number of fins), and G = 10 CFM (total airflow in circuit board channel). Use the Muzychka and Yovanovich friction factor correlation if you have laminar flow. This problem is representative of a "zero bypass" configuration. Hint: Use a starting value of V_f = 500 ft/min.

4.9 Calculate the interfin volumetric airflow G and velocity V_f for the geometry in Figure 4.11 with the following data: W_d = 10.0 in., H_d = 0.906 in., H = 2.5 in., W = 4.125 in., L = 3.0 in., t_f = 0.0736 in., S = 0.238 in., t_b = 0.312 in., N_f = 14 (number of fins), and G = 50 CFM (total airflow in circuit board channel). Use the Muzychka and Yovanovich friction factor correlation if you have laminar flow. Hint: Use a starting value of V_f = 100 ft/min.

4.10 Solve Exercise 4.8 using the *Handbook of Heat Transfer* friction factor.

CHAPTER 5

Airflow III: Buoyancy Driven Draft

This chapter is concerned with the development of a simple model that we can apply to non fan-cooled enclosures. In the first section our derived result will give us the equivalent of the fan curve, i.e., the driving "force" that results in air being sucked into the ventilation inlet and out the ventilation exit. The next step is matching the buoyancy driver to the enclosure head loss, a procedure not unlike what we learned to use with fan-cooled systems. We will use the head loss formulae that are listed in Chapter 3. This chapter ends with a short discussion on how we will later apply the buoyancy principles to an electronics enclosure that is cooled by natural convection, radiation, and of course the internal airdraft. Thus our study of air movement ends, other than for some side issues in the Appendix. We will begin our study of heat transfer in Chapter 6.

5.1 DERIVATION OF BUOYANCY DRIVEN HEAD

The first assumption in our model of the air temperature within an enclosure is that the air temperature increases linearly from the inlet air temperature T_I to a maximum air temperature T_{Max} over a vertical distance d_H. The second assumption is that this distance d_H is equal to either the vertical distance over which heat is convected into the airstream, usually a vertical circuit board height, or is equal to the distance between the inlet and exit air vents. If the board extends beyond the lower and upper vents, the recommended d_H is the inlet to exit vent distance. If the inlet and exit vents are placed above and below the bottom and top edges of the board, respectively, we presume that the air temperature varies little beyond the board.

This linear air temperature rise model is an approximation to reality. It is more likely that the warmest air stacks up toward the top cover. Ultimately, we will use our model to predict T_{Max} and we might therefore expect that this temperature will be somewhat less than the actual maximum air temperature. Referring to Figure 5.1, we see that the linear air temperature model is written as

$$\Delta T = T_{Max} - T_I$$
$$T = \left(\Delta T/d_H\right)\left(z - z_0\right) + T_I$$

where T_I is equal to or greater than the ambient air temperature T_A.

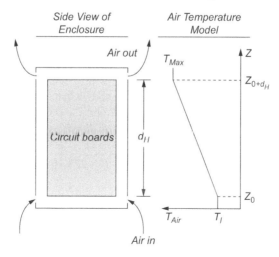

Figure 5.1. Model used for internal air temperature T_{Air} with an inlet air temperature T_I.

Now we address the buoyancy pressure for natural ventilation. This pressure is obtained from an expression for the infinitesimal pressure difference dp_B between the external pressure (external to enclosure) and internal pressure (internal to enclosure) at the same height z over an infinitesimal distance dz for which there is an internal air density decrease from that of the external air. If g is the acceleration due to gravity and ρ_0 is the external air density at temperature T_I, then we can write

$$dp_B = g(\rho_0 - \rho)dz$$

The ideal gas law provides us an expression for the air density.

$$\rho = \rho_0 \left[(273.16 + T_I)/(273.16 + T) \right]$$

Inserting the expression for air density into dp_B we have

$$dp_B = g\rho_0 \left[1 - \left(\frac{273.16 + T_I}{273.16 + T} \right) \right] dz$$

The linear air temperature model inserted into dp_B results in

$$dp_B = g\rho_0 \left[1 - \frac{T_I + 273.16}{T_I + 273.16 + \left(\dfrac{\Delta T}{d_H} \right)(z - z_0)} \right] dz$$

$$dp_B = g\rho_0 dz - g\rho_0 (T_I + 273.16) \left[\frac{dz}{T_I + 273.16 + \left(\dfrac{\Delta T}{d_H} \right)(z - z_0)} \right]$$

Integrating the buoyancy pressure from z_0 to $z_0 + d_H$,

$$\Delta p_B = g\rho_0 d_H - g\rho_0 (T_I + 273.16) \int_{z_0}^{z_0 + d_H} \frac{dz}{\left[T_I + 273.16 + \left(\dfrac{\Delta T}{d_H} \right)(z - z_0) \right]}$$

$$\Delta p_B = g\rho_0 d_H \left[1 - \left(\frac{T_I + 273.16}{\Delta T} \right) \ln \left(1 + \frac{\Delta T}{T_I + 273.16} \right) \right]$$

This result is awkward to use, but fortunately it can be shown that a simpler form of the pressure loss is accurate to within about 7% for $T_I = -40°C$ to $100°C$ and $T = 0°C$ to $100°C$.

$$\Delta p_B = g\rho_0 d_H \left[\frac{(\Delta T/2)}{(\Delta T/2) + T_I + 273.16} \right] \qquad \text{Buoyancy pressure} \quad (5.1)$$

It is left as an exercise to verify that Eq. (5.1) is a good approximation to the exact integration. Our next step is to insert as many of the constants as possible into Eq. (5.1). If we begin by using *SI* units (well, not quite, we use °C for *T*),

$$\rho_0 = 1.2 \, \text{kg/m}^3, g = 9.8 \, \text{m/s}^2, d_H \, [\text{m}], T \left[°C \right], \Delta T \left[°C \right]$$

We also use

$$d_H [\text{m}] = d_H [\text{in.}](0.0254\,\text{m/in.})$$

$$\Delta h_B [\text{in.H}_2\text{O}] = \Delta p_B \left[\text{Pa, N/m}^2\right]\left(4.019 \times 10^{-3}\right)$$

then using Eq. (5.1), we have

$$\Delta h_B [\text{in.H}_2\text{O}] = (9.8)(1.2)(0.0254)\left(4.019 \times 10^{-3}\right) d_H [\text{in.}] \left[\frac{(\Delta T/2)}{(\Delta T/2) + T_I + 273.16}\right]$$

$$\Delta h_B = 0.00120 d_H \left[\frac{(\Delta T/2)}{(\Delta T/2) + T_I + 273.16}\right] \qquad \text{Buoyancy pressure (5.2)}$$
$$\text{mixed English units}$$

where we must remember the units of $\Delta h_B [\text{in.H}_2\text{O}]$, $d_H [\text{in.}]$, $T_I [^\circ\text{C}]$, and $\Delta T [^\circ\text{C}]$.
We next substitute $\Delta T/2$ into Eq. (5.2), where ΔT is obtained from Eq. (2.14).

$$\Delta h_B = 0.00120 d_H \left[\frac{1}{1 + \dfrac{(T_I + 273.16)}{(\Delta T/2)}}\right] = 0.00120 d_H \left[\frac{1}{1 + \left(\dfrac{2G}{CQ} - 1\right)}\right]$$

$$\Delta h_B = 0.00120 d_H \left(CQ/2G\right) \qquad \text{Buoyancy head (5.3)}$$
$$C = 5.99 \times 10^{-3}$$

5.2 MATCHING BUOYANCY HEAD TO SYSTEM

There are two methods that we can employ to determine the airdraft for any given heat load. One method is to plot both the buoyancy head, Eq. (5.3), and the total head loss H_L vs. the airflow G. One can think of the buoyancy head curve as equivalent to the fan curve we used when we analyzed forced airflow systems. Thus the actual operating point of the draft-cooled system is determined by the intersection of the buoyancy and total head loss curves. Although this graphical method is certainly valid, it is too cumbersome for those problems that also include other modes of heat loss.

Fortunately, a few more algebraic calculations lead to a simple formula for the airdraft G. We shall use the "basic head loss" formula that we used for forced airflow, Eq. (3.13), which states that the total head loss H_L for an entire system with resistance R_{Sys} and total airflow G is given by

$$H_L = R_{Sys} G^2$$

and we need only to match the head loss with the buoyancy head, Eq. (5.3) as shown next.

$$H_L = \Delta h_B$$
$$R_{Sys} G^2 = 0.00120 d_H \left(CQ/2G\right)$$

Solving for G,

$$G^3 = 0.00120 d_H \left[CQ/\left(2R_{Sys}\right)\right] = 0.00120 d_H \left[\left(5.99 \times 10^{-3}\right)Q/\left(2R_{Sys}\right)\right]$$

$$G^3 = 3.59 \times 10^{-6} \left[(d_H Q)/R_{Sys}\right]$$

$$\boxed{G = 1.53 \times 10^{-2} \left(d_H Q_d / R_{Sys}\right)^{\frac{1}{3}}} \qquad \text{Buoyancy draft (5.4)}$$

The units are G[CFM], d_H[in.], Q_d [W], and R_{Sys} [in. H_2O/CFM^2]. Note that the heat dissipation is now labeled with a subscript, i.e., Q_d, as a reminder that this is the heat carried away by the airdraft. The difference between the total heat dissipated within the enclosure and Q_d is the heat that is dissipated from the system by other means, e.g., convection, radiation, and conduction. The next section illustrates the application of both the graphical and single formula methods.

5.3 APPLICATION EXAMPLE: BUOYANCY-DRAFT COOLED ENCLOSURE

It is instructive to examine Eqs. (5.3) and (5.4) by applying them to a hypothetical example. Suppose we have an enclosure that is about 12 in. high. The approximate mean distance between perforated plate inlet and exit vents is $d_H = 10$ in. The vents have free areas of 40% open and are 2.0 in. high and 9.375 in. wide. We shall suppose that the internal enclosure resistance to airflow is very small and may be neglected compared to the inlet and exit resistances. The total system airflow resistance is therefore

$$R_{Sys} = \frac{2(2\times10^{-3})}{A_f^2} = \frac{2(2\times10^{-3})}{\left[(2\,\text{in.})(9.375\,\text{in.})(0.4)\right]^2} = 7.11\times10^{-5}\,\text{in.}\,H_2O/\text{CFM}^2$$

and the total head loss for any given value of system airflow G is

$$H_L = R_{Sys}G^2 = \left(7.11\times10^{-5}\right)G^2$$

and the buoyancy pressure for this problem is

$$\Delta h_B = 0.00120 d_H\left(CQ/2G\right) = 0.00120(10)\left[\frac{\left(5.99\times10^{-3}\right)Q}{2G}\right] = 3.59\times10^{-5}\left(\frac{Q}{G}\right)$$

For the purposes of this example, we will neglect any natural convection, radiation, and conductive heat loss from the system. We can therefore assume that all of the internal heat loss from circuit boards, heat sinks, etc., is carried away by the airdraft. This is the total Q in the preceding equation. Both Δh_B and H_L are plotted in Figure 5.2 for $Q = 5$, 10, and 20 W. The system loss and buoyancy pressure curves intersect at $G = 1.36$, 1.71, and 2.16 CFM, for $Q = 5$, 10, and 20 W, respectively. One can think of the h_B curves as equivalent to the fan curves we used when we analyzed forced airflow systems. It is not surprising that the greater the heat dissipation, the greater the airflow.

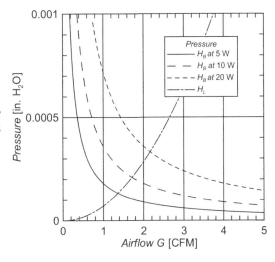

Figure 5.2. Application Example 5.3: Buoyancy airdraft characteristics for vented cabinet using a graphical method.

Next we return to the preceding example where we determined the airdrafts using a graphical method. Now we use Eq. (5.4) to calculate the airdraft for the heat dissipations of 5, 10, and 20 W.

$$G = 1.53 \times 10^{-2} \left(d_H Q_d / R_{Sys} \right)^{\frac{1}{3}} = 1.53 \times 10^{-2} \left[(10)(5\,\text{W}) / 7.11 \times 10^{-5} \right]^{\frac{1}{3}} = 1.36\,\text{CFM}$$

$$G = 1.53 \times 10^{-2} \left[(10)(10\,\text{W}) / 7.11 \times 10^{-5} \right]^{\frac{1}{3}} = 1.71\,\text{CFM}$$

$$G = 1.53 \times 10^{-2} \left[(10)(20\,\text{W}) / 7.11 \times 10^{-5} \right]^{\frac{1}{3}} = 2.16\,\text{CFM}$$

These results agree with those using the graphical method, just as they should.

We are now able to calculate the air temperature rise above the inlet air for each of the three heat dissipations. Using Eq. (2.13),

$$Q = 5\,\text{W}, \qquad \Delta T = 1.76\,Q/G = 1.76(5/1.36) = 6.5\,^\circ\text{C}$$

$$Q = 10\,\text{W}, \qquad \Delta T = 1.76\,Q/G = 1.76(10/1.71) = 10.3\,^\circ\text{C}$$

$$Q = 20\,\text{W}, \qquad \Delta T = 1.76\,Q/G = 1.76(20/2.16) = 16.3\,^\circ\text{C}$$

A final comment concerning Eq. (5.4) is in order. Note that the dependence of G on the physical variables, d_H, Q_d, and R_{Sys}^{-1} in the draft problem is that each is raised to a power of 1/3. Thus these variables do not have a particularly great effect on G. Furthermore, when we add convection and radiation losses to an enclosure, any uncertainty in d_H, Q_d, and R_{Sys} is apt to have a modest effect on the final temperature predictions for the enclosure.

5.4 SYSTEM MODELS WITH BUOYANT AIRFLOW

In this chapter we have developed a nice formula for calculating the airdraft in a vented enclosure. However, the typical application that has buoyant airflow also has losses from the external enclosure surfaces via natural convection and radiation, and natural convection from the internal air to the internal enclosure surfaces; these are subjects we have not yet discussed.

The next step is to show how we would combine the convection, radiation, and air vent heat losses into a single model. An appropriate model, which is of the thermal network type, is illustrated in Figure 5.3. The first thing to note is that all of the "elements" are represented by conductances.

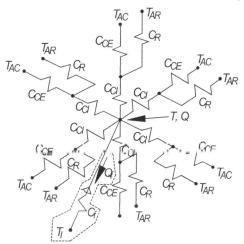

Figure 5.3. Recommended network model for a simple vented enclosure with radiation and natural convection cooling. Elements outlined with -------- are associated with draft.

The C_{CI} represent internal convection, i.e., convection from a central "lump" at temperature T (with a heat source Q) to the internal sides of the enclosure. The C_{CE} are for convection from the outer enclosure surfaces to ambient air. The C_R represent radiation from the system to ambient. The ambient temperatures T_{AR}, T_{AC} are usually at the same value, i.e., room temperature. They are separated out because if you were to set this problem up in a thermal network analyzer, you could easily identify how much heat was radiated and how much heat was convected. The reason that internal radiation is not shown is that there are often so many internal surfaces shielding the various circuit boards that a lot of the heat radiated by internal devices never makes it to the enclosure walls. Heat flow resistance through the walls for conduction is not shown because it clutters up the drawing and, furthermore, the temperature change through enclosure walls is not usually a significant effect, particularly for metal walled enclosures.

The issue of greatest relevance for this chapter is the little piece in Figure 5.3 that is outlined by a dashed line. There is a conductance labeled C_f that interconnects the main air "node" at temperature T within the vent inlet air temperature T_I. The heat that is carried away by the airdraft is just $Q_f = Q_d$. The calculation of C_f is straightforward. Using our simplest formula for air temperature rise, Eq. (2.13),

$$\Delta T = 1.76 Q/G, \; R_f = 1.76/G, \; C_f = 1/R_f$$

$$\boxed{C_f = G/1.76}$$
One-way thermal   (5.5)
fluid conductance

The term "one-way" means that the temperature T is influenced by the temperature T_I, but the reverse is not true: the temperature T_I is not influenced by T. We will provide some calculation examples of vented enclosures in a later chapter.

EXERCISES

5.1 Show that Eq. (5.1) agrees to the exact integrated result to within about 7% for $T_I = -40°C$ to $100°C$ and $T = 0°C$ to $100°C$. Hint: Compare the error in the dimensionless ratios $\Delta p_B/g\rho_0 d_H$ by using the error formula E.

(1) $E = (100) \left\{ \dfrac{\left[\Delta p_B/(g\rho d_H) \right]_{\text{Exact integration}} - \left[\Delta p_B/(g\rho d_H) \right]_{\text{Approximate}}}{\left[\Delta p_B/(g\rho d_H) \right]_{\text{Exact integration}}} \right\}$

(2) Proving: $E = (100) \left\{ 1 - \dfrac{\left[\dfrac{\left(\dfrac{T-T_I}{2} \right)}{\left(\dfrac{T-T_I}{2} \right) + T_I + 273.16} \right]}{\left[1 - \left(\dfrac{T_I + 273.16}{T - T_I} \right) \ln\left(1 + \dfrac{T - T_I}{T_I + 273.16} \right) \right]} \right\}$

(3) Using $T_I = -40°C$, calculate and plot E for $T = 0$ to $100°C$ (calculate for at least three values of T, i.e., $T = 0$, 50, and 100°C)

(4) Using $T_I = 100°C$, calculate and plot E for $T = 0$ to $100°C$ (calculate for at least three values of T, i.e., $T = 0$, 50, and 100°C).

5.2 The buoyancy draft formula, Eq. (5.4), is a nice tool. However, the derivation of this deceptively simple result required the use of Eq. (2.14), an equation for calculating air temperature rise, but is unusual in appearance and not something you are apt to encounter other than in the present or the author's first book. This exercise requires you to repeat Application Ex. 5.3, but using what may appear to be a more acceptable air temperature rise formula. Using all of the given data, including R_{Sys}, for Application Ex. 5.3, plus an inlet air temperature $T_I = 20°C$, reproduce Figure 5.2 with H_L and the buoyancy head curves Δh_B for $Q = 5$, 10, and 20 W. However, use Eq. (5.2) for the buoyancy head and Eq. (2.13) for the air temperature rise. Also calculate the approximate minimum airflow G for which your plots are valid because of Eq. (2.13).

5.3 A low power equipment rack is cooled only by a ventilation airdraft. Each bay has a heat dissipation of $Q_B = 100$ W. Many of the 15 equipment bays have adjacent equipment, which we will take advantage of to assume that each of the six surfaces to a bay is approximately adiabatic, i.e., there is negligible convection, radiation, and conduction exchange between bays. The calculations will be according to a worst case scenario wherein there is no heat loss by convection or radiation to ambient. The dimensions are $H = 10$ in., $W = 20$ in., and $L = 20$ in. The top and bottom panels of the system are perforated with a fractional open area $f = 0.35$. Each bay has vertical cards on 1-in. centers and 81% free passage area. You are asked to analyze a one-bay-wide, five-bay-high portion of the system: (1) calculate the total airflow resistance; (2) calculate the airdraft; (3) calculate the well-mixed air temperature rise for the entire height. Hints: (1) Use the McLean card cage resistance formulae in Chapter 3; (2) Use a contraction and expansion for each bay.

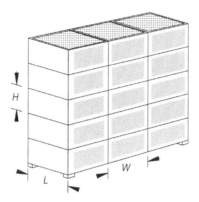

Exercise 5.3. Equipment rack.

CHAPTER 6

Forced Convective Heat Transfer I: Components

This chapter begins with an introduction to the physics of heat transfer from a forced air cooled surface. We shall first study some detail for one of the simplest geometric shapes of relevance to us, the flat plate. Our study of flat plate heat transfer is followed by convection from components on a printed circuit board. One of the most important aspects of analyzing *PCB* components is the air temperature rise. Thermal analysts understand that it is usually not satisfactory to use the well-mixed air temperature formula for components on circuit boards, thus we will encounter the concept of adiabatic temperature. The thermal analysis of heat sinks is deferred to a separate chapter.

6.1 FORCED CONVECTION FROM A SURFACE

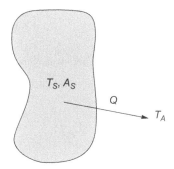

Figure 6.1. Surface area A_s at temperature T_s transfers heat Q to ambient air temperature T_A.

A phenomenological description of any type of surface heat transfer, as illustrated in Figure 6.1, is quantified by Eq. (6.1) for *Newtonian cooling*, where the symbol h is known as the *heat transfer coefficient*. This equation does not indicate the type of heat transfer mechanism. As we will learn later, the Newtonian cooling concept is also applicable to radiation. Later in this chapter, we indicate convective heat transfer by subscripting the Q and h with the character "c", but for now we shall retain generality by omitting convection-indicating subscripts.

$$Q = hA_s \left(T_s - T_A \right) \qquad \text{Newtonian cooling} \qquad (6.1)$$

The surface temperature T_s is not going to be uniform for a surface of any realistic size, and for problems other than those requiring detailed conduction analysis, we usually work with an average temperature and h. There are therefore choices to make as to what sort of symbolism we should use, e.g., we could apply an *overbar* to T and h, indicating a T_s and an average h for the surface. The author has chosen to omit overbars and will let the context of the problem indicate whether or not we are averaging. The use of average heat transfer coefficients and surface temperatures is also consistent with the fact that the former are not usually available to us as precise, position dependent formulae for the very complex airflow patterns encountered in electronic equipment.

We shall use a hypothetical, forced air example to qualitatively illustrate the convection process: A flat plate with a uniform approach velocity (from the left) is illustrated in Figure 6.2. In Figure 6.2 (a) we illustrate the aerodynamic flow situation where the entire region beneath the curve that begins at the plate leading edge is the *aerodynamic boundary layer* δ_V. A typical definition is the perpendicular-to-plate distance for which the velocity is 99% of the free stream velocity (which is also the approach V). As the illustration indicates, δ_V is near zero at the leading edge and increases with x until it remains constant. As you might expect, this description is greatly oversimplified with the flow physics being far more complicated within the boundary layer.

When the plate is a heat source with, for example, uniform heat flux or temperature, a qualitative description of a thermal boundary layer is shown in Figure 6.2 (b). The thermal boundary layer thickness is defined by the perpendicular-to-plate distance where the temperature profile is 99% of the free stream temperature T_A. In our idealized description, a downstream distance exists where the thermal boundary thickness no longer changes with x. You will find more complete descriptions of boundary layer physics in any of the numerous heat transfer texts.

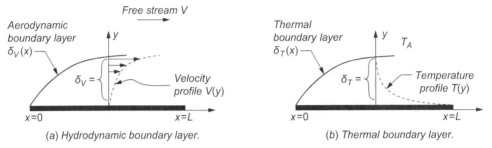

(a) Hydrodynamic boundary layer.   (b) Thermal boundary layer.

Figure 6.2. Qualitative illustration of (a) aerodynamic and (b) thermal boundary layers.

We are now at the point where we might attempt to quantify the convective heat transfer coefficient in terms of the boundary layer.

$$h_c = k_{air}/\delta_T \qquad (6.2)$$

where k_{Air} is the thermal conductivity of the air within the boundary layer. While this formula makes some sense, it is very inaccurate and we still need a formula for δ_T. Furthermore, the traditional representation of the heat transfer coefficient takes a different form and is the subject of the next section.

6.2 DIMENSIONLESS NUMBERS: NUSSELT, REYNOLDS, AND PRANDTL

The goal here is to justify the selection of certain dimensionless numbers or groups commonly used in heat transfer. There are both "Pros" and "Cons" in this method. The Pros are that (1) the number of variables for plotting experimental data is reduced and (2) evaluation of the heat transfer coefficient is possible for any flow over any size plate.

The Cons are that (1) dimensional analysis can only predict the dependency of the dimensionless numbers on particular variables, but not the actual functions, (2) dimensional analysis gives no information about the nature of the physical problem, (3) one must correctly predict all of the pertinent physical variables—a good understanding of the physics of the problem is necessary before the analysis.

There are particular cautions such as (1) an experimental determination of one dimensionless number dependence on another for various fluids is required, (2) the Buckingham Pi theorem (used next) requires that the resulting equations be independent.

Buckingham π Theorem:
        n_G is the number of dimensionless groups
        n_P is the number of physical quantities
        n_D is the number of primary dimensions

The relationship among the groups is expressed by

$$F(\pi_1, \pi_2, \pi_3, \dots \pi_{nG}) \quad \text{or} \quad \pi_1 = f_1(\pi_2, \pi_3, \dots), \quad \pi_2 = f_2(\pi_1, \pi_3, \dots), \text{ etc.}$$

Physical quantities in M, L, t, T (mass, length, time, temperature) system		
Variable	Symbol	Dimensions
Diameter or	D	$[L]$
length	L	$[L]$
Thermal conductivity	k	$\left[ML/t^3T\right]$
Velocity	V	$[L/t]$
Density	ρ	$\left[M/L^3\right]$
Viscosity	μ	$[M/Lt]$
Specific heat (const. P)	c_P	$\left[L^2/t^2T\right]$
Heat transfer coeff.	h	$\left[M/t^3T\right]$
Group	π	$[M]^0 [L]^0 [t]^0 [T]^0$

We expect n_G dimensionless groups = n_P physical quantities - n_D primary dimensions so that
$$n_G = n_P - n_D = 7 - 4 = 3$$
Each of the groups is represented by an equation of the form

$$\pi = L^a k^b V^c \rho^d \mu^e c_P^f h^g S$$

Since no group is permitted a dependence on any dimension,

$$[L]^0 [M]^0 [t]^0 [T]^0 = [L]^a \left[ML/t^3T\right]^b [L/t]^c \left[M/L^3\right]^d [M/Lt]^e \left[L^2/t^2T\right]^f \left[M/t^3T\right]^g$$

$$[L]^0 [M]^0 [t]^0 [T]^0 = [L]^{a+b+c-3d-e+2f} [M]^{b+d+e+g} [t]^{-3b-c-e-2f-3g} [T]^{-b-f-g}$$

Matching the exponents on the left and right sides of the preceding,

$$L \text{ dimension:} \quad a+b+c-3d-e+2f = 0$$
$$M \text{ dimension:} \quad b+d+e+g = 0$$
$$t \text{ dimension:} \quad 3b+c+e+2f+3g = 0$$
$$T \text{ dimension:} \quad b+f+g = 0$$

There are four equations and seven unknowns. Any set of values a - g that satisfy these four equations results in a dimensionless π.

$$(i) \quad a+b+c-3d-e+2f = 0$$
$$(ii) \quad b+d+e+g = 0$$
$$(iii) \quad 3b+c+e+2f+3g = 0$$
$$(iv) \quad b+f+g = 0$$

Since there are more unknowns than equations, three of the unknowns may be set at any value that works, i.e., results in a desired dimensionless group. One restriction: each unknown must be linearly independent. Thus

The determinant of the coefficients of the remaining unknowns must be zero.

Find the first group: $\pi = L^a k^b V^c \rho^d \mu^e c_P^f h^g$

It makes sense that to preserve the h variable, we set $g = 1$, then guess $c = 0, f = 0$:

(1) (i) $a + b + c - 3d - e + 2f = 0$ (2) (i) $a + b - 3d - e = 0$

$\quad\quad(ii)$ $b + d + e + g = 0$ $\quad\quad\quad\quad\quad(ii)$ $b + d + e = -1$

$\quad\quad(iii)\ 3b + c + e + 2f + 3g = 0$ $\quad\quad\quad\quad(iii)\ 3b + e = -3$

$\quad\quad(iv)\ b + f + g = 0$ $\quad\quad\quad\quad\quad\quad(iv)\ b = -1$

$\quad(3)$ Use $(iv)\ b = -1$ (4) Use $(iii)\ e = 0$ (5) Use $(ii)\ d = 0$

$\quad\quad(i)$ $a - 3d - e = 1$ $\quad(i)$ $a - 3d = 1$ $\quad(i)$ $a = 1$

$\quad\quad(ii)$ $d + e = 0$ $\quad\quad(ii)$ $d = 0$

$\quad\quad(iii)\ e = 0$

$$\text{Then } \pi_1 = L\frac{1}{k}h, \ Nu \equiv \pi_1, \boxed{Nu = \frac{hL}{k}}$$

Find the second group: $\pi = L^a k^b V^c \rho^d \mu^e c_P^f h^g$

Next it makes sense, that to not get the variable h again, set $g = 0$. To get something with a V in it, set $c = 1$. Then guess $f = 0$.

(1) (i) $a + b + c - 3d - e + 2f = 0$ (2) (i) $a + b - 3d - e = -1$

$\quad\quad(ii)$ $b + d + e + g = 0$ $\quad\quad\quad\quad(ii)$ $b + d + e = 0$

$\quad\quad(iii)\ 3b + c + e + 2f + 3g = 0$ $\quad\quad\quad(iii)\ 3b + e = -1$

$\quad\quad(iv)\ b + f + g = 0$ $\quad\quad\quad\quad\quad(iv)\ b = 0$

$\quad(3)$ Use $(iv)\ b = 0$ (4) Use $(iii)\ e = -1$ (5) Use $(ii)\ d = 1$

$\quad\quad(i)$ $a - 3d - e = -1$ $\quad(i)$ $a - 3d = -2$ $\quad(i)$ $a = 1$

$\quad\quad(ii)$ $d + e = 0$ $\quad\quad(ii)$ $d = 1$

$\quad\quad(iii)$ $e = -1$

$$\text{Then } \pi_2 = LV\rho\frac{1}{\mu} = \frac{LV}{v}, \ Re \equiv \pi_2, \boxed{Re = \frac{LV}{v}}$$

Find the third and last group: $\pi = L^a k^b V^c \rho^d \mu^e c_P^f h^g$

Next it makes sense that to not get the variables h and V again, set $c = g = 0$. To get something with a k in it, set $b = 1$.

(1) (i) $a + b + c - 3d - e + 2f = 0$ (2) (i) $a - 3d - e + 2f = -1$

$\quad\quad(ii)$ $b + d + e + g = 0$ $\quad\quad\quad\quad(ii)$ $d + e = -1$

$\quad\quad(iii)\ 3b + c + e + 2f + 3g = 0$ $\quad\quad\quad(iii)\ e + 2f = -3$

$\quad\quad(iv)\ b + f + g = 0$ $\quad\quad\quad\quad\quad(iv)\ f = -1$

$\quad(3)$ Use $(iv)\ f = -1$ (4) Use $(iii)e = -1$ (5) Use $(ii)\ d = 0$

$\quad\quad(i)$ $a - 3d - e = 1$ $\quad(i)$ $a - 3d = 0$ $\quad(i)\ a = 0$

$\quad\quad(ii)$ $d + e = -1$ $\quad\quad(ii)$ $d = 0$

$\quad\quad(iii)$ $e = -1$

$$\text{Then } \pi_3 = \frac{k}{\mu c_P}, \ Pr \equiv \frac{1}{\pi_3}, \boxed{Pr = \frac{\mu c_P}{k}}$$

Using the style of

$$\pi_1 = f_1(\pi_2, \pi_3) \text{ or } \boxed{Nu = f(Re, Pr)}$$

and the objective has been accomplished.

The Nusselt and Prandtl numbers are valid concepts whose definitions are independent of the convection mode, i.e., whether the heat transfer occurs via natural or forced convection. We discuss the Nusselt number first. Suppose we have a surface at temperature T_S convecting to an ambient temperature T_A as shown in Figure 6.3. The thermal boundary layer is a thin film that, very close to the wall, transfers heat from surface to air by molecular conduction. Then Fourier's law may be applied to the wall for an infinitesimal surface area dA.

$$dQ_c = -k_{Air} dA (dT/dy)\big|_{y=0} \tag{6.3}$$

By the definition of h_c and the Newtonian cooling Equation (6.1),

$$dQ_c = h_c dA (T_S - T_A) \tag{6.4}$$

Setting the right sides of Eqs. (6.3) and (6.4) equal,

$$-k_{Air} dA (dT/dy)\big|_{y=0} = h_c dA (T_S - T_A)$$

$$-k_{Air} (dT/dy)\big|_{y=0} = h_c (T_S - T_A)$$

Since T_S is a constant, $dT = d(T - T_S)$, and dropping the "$y = 0$" notation,

$$k_{Air} d(T_S - T)/dy = h_c (T_S - T_A)$$

Multiplying through by some appropriate dimensional parameter P with the dimensions of a length and rearranging,

$$h_c P/k_{Air} = [d(T_S - T)/dy]/[(T_S - T_A)/P] \tag{6.5}$$

The quantity on the left of Eq. (6.5) is the dimensionless way of specifying h_c and is defined by the *Nusselt number*.

$$\boxed{Nu_P = \frac{h_c P}{k_{Air}}} \qquad \text{Nusselt number} \tag{6.6}$$

Two different physical interpretations of the same Nusselt number are shown in Eq. (6.7). The first follows directly from Eq. (6.5) as the ratio of the temperature gradient at the wall to a reference gradient. The second is a dimensionless temperature gradient at the wall.

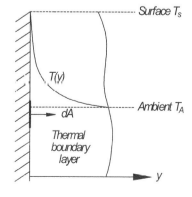

Figure 6.3. Heat convection at a solid-gas interface.

$$Nu_P = \frac{d(T_S - T)/dy}{(T_S - T_A)/P} = \frac{d\left[(T_S - T)/(T_S - T_A)\right]}{d(y/P)} \quad Nu_P \text{ interpretations } (6.7)$$

The Prandtl number (dimensionless) is defined according to Eq. (6.8),

$$\boxed{Pr = \mu c_P / k_{Air}} \qquad \text{Prandtl number } (6.8)$$

where the variables are

$\mu \equiv$ dynamic viscosity, $\left[\text{gm}/(\text{in.}\cdot\text{s})\right]$

$c_P \equiv$ specific heat at constant pressure, $\left[\text{J}/(\text{gm}\cdot{}^\circ\text{C})\right]$

$k_{Air} \equiv$ thermal conductivity of air, $\left[\text{W}/(\text{in.}\cdot{}^\circ\text{C})\right]$

Division of the numerator and denominator by the air density ρ with consistent dimensions gives us

$$Pr = \frac{\mu/\rho}{k_{Air}/c_P\rho} = \frac{v}{\alpha}$$

where $v \equiv$ kinematic viscosity [in.^{2}/s] and $\alpha \equiv$ thermal diffusivity [in.^{2}/s].

We can therefore state the ratio

$$Pr \sim \frac{\text{momentum diffusivity}}{\text{thermal diffusivity}}$$

as our physical interpretation. In the case of air at standard conditions (Pr is quite independent of temperature in the range encountered in most electronics), the Prandtl number is approximately unity ($Pr = 0.71$). We can therefore infer that the velocity and temperature profiles, i.e., the aerodynamic and thermal boundary layers, develop at about the same rate. In general,

$Pr < 1$: temperature profile develops more rapidly than velocity profile.

$Pr = 1$: temperature and velocity profiles develop at same rate.

$Pr > 1$: velocity profile develops more rapidly than temperature profile.

6.3 MORE ON THE REYNOLDS NUMBER

The Reynolds number is relevant to forced airflow. The mathematical definition of the Reynolds number in any consistent units is given by Eq. (6.9).

$$Re_P = VP\rho/\mu \qquad \begin{array}{l}\text{Reynolds number } (6.9) \\ \text{in consistent units}\end{array}$$

In this book, we use mixed units, thus our Reynolds number is given by Eq. (6.10).

$$\boxed{Re_P = (VP/v)(1/5)} \qquad \begin{array}{l}\text{Reynolds number } (6.10) \\ \text{in mixed units}\end{array}$$

where we use V [ft/min], P [in.], $\rho =$[gm/in.^{3}], v [in.^{2}/s], and $v = \mu/\rho =$ kinematic viscosity. An adequate value of the kinematic viscosity of air in approximate calculations is given by $v = 0.029$ in.^{2}/s at 50°C. The Reynolds number represents the ratio of inertial to viscous forces. The parameter P depends on the type of geometry. A P is used that is most appropriate for the geometry and data. Unfortunately, the geometry that we usually encounter has more than one dimension affecting the flow characteristics. Some problems require $P = A / L$ where A is an area and L is a length, but the result is still a P with dimensions of length. This issue of which area A and length L to use in any given investigation is an indication that dimensionless correlations are not without issues of inaccuracy and is an oversimplification of many problems. Nevertheless, correlations are useful for first-order calculations. The following section is concerned with the simplest geometry applicable to electronics cooling.

6.4 CLASSICAL FLAT PLATE FORCED CONVECTION CORRELATION: UNIFORM SURFACE TEMPERATURE, LAMINAR FLOW

A qualitative sketch of a flat plate with a forced air, aerodynamic boundary layer is illustrated in Figure 6.4. A boundary layer may be laminar, turbulent, or in an intermediate stage (called the transition region), depending on the distance from the leading edge of the plate, the free stream velocity, and the amount of upstream turbulence. A modest rate of flow somewhere in the vicinity of the leading edge tends to have the velocity profile consistent with a laminar boundary layer, i.e., the fluid particles follow streamlines. As the flow proceeds further from the leading edge in the x-direction, small disturbances start to become more significant in the "transition" region until finally inertia forces dominate viscous effects, and the boundary layer becomes turbulent with only a very thin laminar sublayer at the plate surface. The location at which the boundary becomes turbulent is called "the critical length," x_c, and is calculated from the *critical Reynolds number* Re_{x_c}. Empirical studies reported in the literature indicate Re_{x_c} may vary from 8×10^4 to 5×10^6, depending on the nature of the surface and the degree of turbulence in the flow upstream from $x = 0$. The standard treatises indicate that

$$Re_{x_c} = 5 \times 10^5 \qquad \text{Flat plate} \quad (6.11)$$

where the free stream velocity, which is assumed identical to the approach velocity, is used to estimate Re_{x_c}.

The critical Reynolds number is used to judge whether a laminar or turbulent airflow Nusselt number should be used. For laminar flat plate flow, the classical boundary layer problem was solved by Pohlhausen (see Schlichting, H., 1968), and shown to be dependent on the distance x from the plate leading edge.

$$Nu_x = 0.332 Re_x^{1/2} Pr^{1/3} \qquad \begin{array}{l} \text{Flat plate laminar} \\ \text{local, uniform } T \end{array} \quad (6.12)$$

Equation (6.12) presumes an isothermal plate temperature T_S. When the plate is not isothermal, we use an average temperature. Temperature-dependent properties are evaluated at the mean boundary layer temperature taken as $(T_S + T_A)/2$. The local heat transfer coefficient is given by

$$h_x = (k_{Air}/x) Nu_x \qquad \begin{array}{l} \text{Flat plate laminar} \\ \text{local, uniform } T \end{array} \quad (6.13)$$

The average heat transfer coefficient is readily calculated:

$$\bar{h}_L = \frac{1}{L} \int_0^L h_x dx = \frac{1}{L} \int_0^L \left(\frac{k_{Air}}{x} \right) Nu_x dx = \frac{k_{Air}}{L} \int_0^L \left[\left(\frac{1}{x} \right)(0.332) Re_x^{1/2} Pr^{1/3} \right] dx$$

$$= \left(\frac{k_{Air}}{L} \right) \int_0^L \left[\left(\frac{0.332}{x} \right) \left(\frac{Vx}{\nu} \right)^{1/2} Pr^{1/3} \right] dx = (0.332) \left(\frac{k_{Air}}{L} \right) \left(\frac{V}{\nu} \right)^{1/2} Pr^{1/3} \int_0^L \frac{dx}{\sqrt{x}}$$

$$= (0.332) \left(\frac{k_{Air}}{L} \right) \left(\frac{V}{\nu} \right)^{1/2} Pr^{1/3} \left(2\sqrt{L} \right) = 2(0.332) \left(\frac{k_{Air}}{L} \right) \left(\frac{VL}{\nu} \right)^{1/2} Pr^{1/3}$$

$$\bar{h}_L = 2(0.332) \left(\frac{k_{Air}}{L} \right) Re_L^{1/2} Pr^{1/3} = 2h_L$$

This is a curious result. The length-averaged heat transfer coefficient is identical to twice the heat transfer coefficient at the trailing edge of the plate.

We can put $\bar{h}_L$ into a more useful form by assuming a typical mean boundary layer temperature of 50°C. The following formulae are used.

$$v\left[\text{in.}^2/\text{s}\right] = 7.793\times10^{-5}\left(T\left[^\circ\text{C}\right]\right)^{1.143} + 0.022$$

$$k\left[\text{W}/\left(\text{in.}\cdot{}^\circ\text{C}\right)\right] = 2.258\times10^{-6}\left(T\left[^\circ\text{C}\right]\right)^{0.947} + 6.006\times10^{-4}$$

At $T = 50^\circ$C we obtain $v = 0.029\,\text{in.}^2/\text{s}$ and $k = 6.92\times10^{-4}\,\text{W}/\left(\text{in.}\cdot{}^\circ\text{C}\right)$. We then set up our formula for the length-averaged heat transfer coefficient.

$$\bar{h}_L = 2\left(0.332\right)\left(\frac{k_{Air}}{L}\right)Re_L^{1/2}Pr^{1/3} = 0.664\left(\frac{k_{Air}}{L}\right)\left(\frac{VL}{5v}\right)^{1/2}Pr^{1/3}$$

$$= \frac{0.664}{\sqrt{5}}Pr^{1/3}\left(\frac{k_{Air}}{\sqrt{v}}\right)\sqrt{\frac{V}{L}}$$

$$\bar{h}_L = \frac{0.664}{\sqrt{5}}\left(0.71\right)^{1/3}\left(\frac{6.92\times10^{-4}}{\sqrt{0.029}}\right)\sqrt{\frac{V}{L}}$$

where we have used the Reynolds number formula for mixed units (thus the $5v$ is in the denominator of Re_L). The factor $\sqrt{5}$ accounts for the units of V and L and we arrive at the result of

$$\boxed{\bar{h}_L\left[\text{W}/\left(\text{in.}^2\cdot{}^\circ\text{C}\right)\right] = 0.00109\sqrt{\frac{V\left[\text{ft/min}\right]}{L\left[\text{in.}\right]}}}$$

Flat plate laminar (6.14)
average, uniform T

which is plotted in Figure 6.5. We should check the largest Reynolds number that we might obtain using this figure for $L = 10$ in. and $V = 2000$ ft/min.

$$Re_L = VL/\left(5v\right) = \left(2000\,\text{ft/min}\right)\left(10\,\text{in.}\right)/\left[5\left(0.029\,\text{in.}^2/\text{s}\right)\right] = 1.4\times10^5$$

indicating that the flow could still, according to Eq. (6.11), be laminar, thus implying that all of the data in Figure 6.5 is apppropriate for laminar flow.

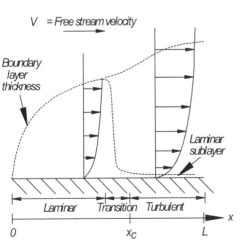

Figure 6.4. Qualitative description of flat plate airflow boundary layer phenomena.

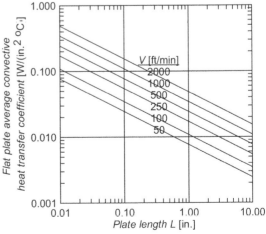

Figure 6.5. Flat plate convective heat transfer coefficient $\bar{h}_L$ from Pohlhausen plotted from Eq. (6.14). Laminar flow, uniform T.

The multiplicative constant in Eq. (6.14) has a slight temperature dependence. Constants for other temperatures are shown in Table 6.1. Clearly the temperature dependence is small. For the purpose of memorization and also for most applications, the value 0.0011 is appropriate.

Table 6.1. Multiplicative constant for Equation (6.14). Units are V [ft/min]and L [in.].

$\overline{T}\left[\,^{\circ}\mathrm{C}\right]$	$\overline{h}_L/\sqrt{V/L}\left[W/\left(\mathrm{in.}^2\cdot{}^{\circ}\mathrm{C}\right)\right]$
0	0.001084
25	0.001094
50	0.001090
75	0.001084
100	0.001078

6.5 EMPIRICAL CORRECTION TO CLASSICAL FLAT PLATE FORCED CONVECTION CORRELATION, LAMINAR FLOW

The following paragraphs summarize experiments conducted in the author's laboratory. The actual data has long been lost, but the results are reproduced here with the same accuracy as originally presented in Ellison (1984). A 0.025-in. thick, 2.0 in. × 2.0 in. ceramic substrate with several small heat-dissipating thick film resistive heat sources was suspended in a component airflow test facility, a short section of which is illustrated in Figure 6.6. A pitot tube was used to measure air speed and an "egg-crate" airflow straightener was placed upstream.

A detailed computer model of this conducting/convecting system was developed with $\overline{h}_L$ used as the variable and radiation included separately. Several values of $\overline{h}_L$ were used to compute temperature profiles on the substrate. These were then compared with profiles measured using an infrared microscope. The results of the experiments were that the theoretical, laminar flat plate flow heat transfer coefficients led to a significant overprediction of temperature unless the $\overline{h}_L$ values were increased significantly. You are advised to remember that the test was conducted for a very specific geometry and use of this data should be restricted to dimensions not greatly deviating from that which was used. However, other experiments (Ellison, 1976) validate the results in devices measuring 1.3 in., 0.6 to 3.0 in., and 0.05 in. for width, length, and thickness, respectively, where the airflow direction was parallel to the 0.6 to 3.0 in. dimension.

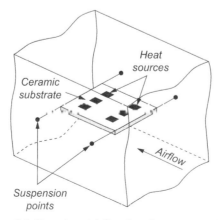

Figure 6.6. Experimental flow-bench arrangement used to determine flat plate heat transfer coefficient correction.

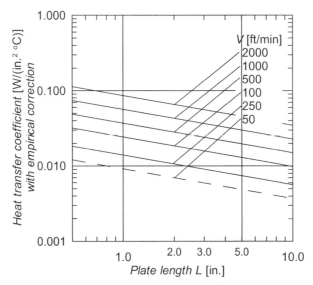

Figure 6.7. Average convective heat transfer coefficient, Eq. (6.15) with empirical correction applied. The dashed line ($V < 250$ ft/min) indicates an extrapolation of the test data.

A correction factor was therefore applied to Equation (6.14) and with the results shown in Figure 6.7. The result may also be combined into a single formula for a corrected Nusselt number or heat transfer coefficient, Eq. (6.15), that fits to within about 10% of the original data. The material properties, thermal conductivity k_{Air} and kinematic viscosity ν_{Air} at 30 °C were used.

$$\overline{Nu}_L = 0.374 Re_L^{0.607}$$

$$\boxed{\overline{h}_L = 0.374 \left(\frac{k_{Air}}{L} \right) Re_L^{0.607}}$$

Corrected flat   (6.15)
plate laminar average

6.6 APPLICATION EXAMPLE: WINGED ALUMINUM HEAT SINK

A finned, aluminum heat sink consisting of two "wings" bent up from the sink base is illustrated in Figure 6.8. We shall ignore any construction details of the component package to which the heat sink is attached. A worst case model is that all of the heat is convected to the local air ambient, i.e., conduction to the board is neglected and the surface emissivity of the heat sink is less than 0.1. The heat sink dimensions are: $W = 1.0$ in., $L = 2.0$ in., $H = 0.75$ in. and the heat dissipation is $Q = 5.0$ W. The air speed is $V = 100$ ft/min.

We wish to calculate the thermal resistance from the heat sink base to the local ambient air and also the temperature rise above the local ambient. We assume that the heat dissipated by the silicon chip spreads to the entire heat sink base. The convecting surface area includes the top of the heat sink base and both the upper and lower surfaces of the two heat sink wings.

$$A_s = 4HL + WL = 4(0.75)(2.0) + (1.0)(2.0) = 8\,\text{in.}^2$$

We can use Figure 6.7 to obtain an approximate graph value for the average heat transfer coefficient and then calculate the thermal resistance and temperature rise from the heat sink base to the local ambient.

$$h_L = 0.01\,\text{W}/\left(\text{in.}^2 \cdot {}^\circ\text{C}\right)$$

$$R_c = 1/\left(h_L A_s\right) = 1/\left[(0.01)(8)\right] = 12.5\,{}^\circ\text{C/W}, \Delta T = R_c Q = (12.5)(5) = 62.5\,{}^\circ\text{C}$$

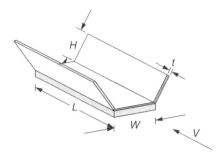

Figure 6.8. Application Example 6.6: Winged aluminum heat sink. The temperature from fin root to tip is assumed uniform, i.e., the fin efficiency (discussed in a later chapter) is unity.

An alternate method of calculating h_c is to use Eq. (6.15):

$$Re_L = VL/(5v) = (100)(2.0)/[5(0.029)] = 1.38 \times 10^3$$

$$h_L = 0.374 \left(\frac{k_{Air}}{L} \right) Re_L^{0.607} = 0.374 \left(\frac{6.92 \times 10^{-4}}{2} \right) (1.38 \times 10^3)^{0.607} = 0.010 \text{ W}/(\text{in.}^2 \cdot {}^{\circ}\text{C})$$

where we see that both methods of determining h_L agree, just as we should expect.

It is instructive to see what the resistance and temperature rise would be if the heat sink were omitted, i.e., when we have convection from only the top of the IC package case:

$$R_c = 1/(h_L LW) = 1/[(0.010)(2.0)(1.0)] = 48 \,{}^{\circ}\text{C}/\text{W}, \Delta T = R_c Q = (48)(5) = 240 \,{}^{\circ}\text{C}$$

A very simple heat sink can have significant effects!

6.7 CLASSICAL FLAT PLATE FORCED CONVECTION CORRELATION: UNIFORM HEAT RATE PER UNIT AREA, LAMINAR FLOW

Kays and Crawford (1980) provide a position-dependent correlation for uniform heat rate per unit area for laminar flow.

$$Nu_x = 0.453 Re_x^{1/2} Pr^{1/3} \qquad \text{Flat plate laminar, local } (6.16)$$
$$\text{uniform heat rate/area}$$

$$h_x = 0.453 \left(\frac{k_{Air}}{x} \right) Re_x^{1/2} Pr^{1/3}$$

If Eq. h_x is integrated from $x = 0$ to $x = L$ and $Pr = 0.71$ is used, we obtain

$$\bar{h}_L = \frac{1}{L} \int_{x=0}^{x=L} h_x dx = 0.906 \frac{k_{Air}}{\sqrt{v}} Pr^{1/3} \sqrt{\frac{V}{L}} \qquad \text{Consistent units } (6.17)$$
$$\text{uniform heat rate/area}$$

$$\bar{h}_L = 0.906 \frac{k_{Air}}{\sqrt{5v}} Pr^{1/3} \sqrt{\frac{V}{L}} \qquad \text{Mixed units}$$

Substituting $v = 0.029 \text{ in.}^2/\text{s}$, $k_{Air} = 6.92 \times 10^{-4} \text{ W}/(\text{in.} \cdot {}^{\circ}\text{C})$ into the mixed units h we arrive at

$$\boxed{\bar{h}_L = 0.00147 \sqrt{V/L}} \qquad \text{Flat plate laminar, average } (6.18)$$
$$\text{uniform heat rate/area}$$

which is adequately accurate for most engineering design purposes. $\overline{Nu}_L$ is the Nusselt number we would need to get $\bar{h}_L$. Therefore "by definition"

$$\overline{Nu}_L = (L/k_{Air}) \bar{h}_L$$

The units for h_L, V, and L are W/(in.^{2}· °C), ft/min, and in., respectively. We can quickly perform a check on Eq. (6.18) using the data provided in the example of Section 6.6.

$$\bar{h}_{L} = 0.00147\sqrt{V/L} = 0.00147\sqrt{100/2} = 0.010 \, \text{W}\Big/\!\left(\text{in.}^2\big/{}^\circ\text{C}\right)$$

which is identical to the 0.010 value obtained in Section 6.6. It is reasonable to suggest that either Eq. (6.15) or (6.18) is valid for *component-size* surfaces.

6.8 CLASSICAL FLAT PLATE (LAMINAR) FORCED CONVECTION CORRELATION EXTENDED TO SMALL REYNOLDS NUMBERS: UNIFORM SURFACE TEMPERATURE

Yovanovich and Teerstra (1998) constructed a composite model for an area-averaged Nusselt number for forced, laminar flow parallel to isothermal rectangular plates. By using a composite model based on both the thermal diffusion limited condition and also the laminar boundary layer solution ($100 < Re < 10^5$), a single range Nusselt number was derived. You may find such a correlation useful. The correlation may also be relevant for buoyancy induced flow within vertical channels.

In the present model, flow may be parallel to either a short or long direction as shown in Figure 6.9. At large Reynolds numbers, the flow follows two-dimensional boundary layer theory, whereas at lower Reynolds numbers, edge effects come into play. At very low Reynolds numbers, the thermal diffusion model is relevant, and in this regime Yovanovich and Teerstra have found that if the square root of the active area is used as the characteristic length, the Nusselt number is independent of the plate aspect ratio. *The reader is cautioned with regard to the use of the symbols L and W:* It is common, though not guaranteed practice, that in forced convective flow, the symbol L is used as the dimension in the flow direction. This should not cause confusion as only one length dimension, L, is available in most correlations, particularly for flat plates. However, in the present instance, Yovanovich and Teerstra clearly use L as the greater of the two dimensions and W as the smaller of the two.

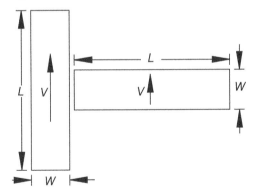

Figure 6.9. Flat plate laminar flow in either long (L) or short (W) direction following Yovanovich and Teerstra (1998).

A thermal diffusion "shape factor" $S^*_{\sqrt{A}}$ is given by Yovanovich as

$$S^*_{\sqrt{A}} = \begin{cases} \dfrac{\left(1+\sqrt{L/W}\right)^2}{\sqrt{\pi L/W}}, & 1 \le L/W \le 5 \\[4mm] \dfrac{2\sqrt{\pi L/W}}{\ln\left(4L/W\right)}, & 5 < L/W < \infty \end{cases} \qquad \begin{array}{l} \text{Diffusion} \ (6.19) \\ \text{shape factor} \end{array}$$

$$\sqrt{A} = \sqrt{LW}$$

The Nusselt number is given as

$$Nu_{\sqrt{A}} = \left[\left(S_{\sqrt{A}}^* \right)^{1.42} + \left(0.742 Re_{\sqrt{A}}^{*1/2} Pr^{1/3} \right)^{1.42} \right]^{1/1.42}$$

Yov./Teer. (6.20)
$0 \le Re_{\sqrt{A}}^* \le 5000$

The multiplicative constant of 0.742 is slightly greater than the 0.664 used by Polhausen. The final proposed model is given by Eq. (6.20) using a modified Reynolds number Re^*.

$$Re_{\sqrt{A}}^* = Re_{\sqrt{A}} \sqrt{L/W}$$ flow parallel to short dimension W

$$Re_{\sqrt{A}}^* = Re_{\sqrt{A}} \sqrt{W/L}$$ flow parallel to long dimension L

The constant of 1.42 in the exponents of Eq. (6.20) was determined by the best fit with data from a computational fluid dynamics model.

A comparison of Eqs. (6.15), (6.17), and (6.20) for a square plate with an edge dimension of 2 in. is shown in Figure 6.10. The Yovanovich and Teerstra correlation is valid for Reynolds numbers up to 10^5 and the Ellison correlation is valid for Reynolds numbers of about 3.5×10^3 to 3×10^4. In the latter range, the two correlations display the difference expected since Ellison's reflects an experimental correction to a laminar boundary layer solution and the Yovanovich and Teerstra correlation in the same region is predominantly also a laminar boundary layer solution. We see that at Reynolds numbers less than about 100, the Yovanovich and Teerstra formula shows an increasingly important dependence on the thermal diffusion limit, just as intended.

We have studied some flat plate, laminar flow correlations, but now the question is which should one use? This writer's practice is to use one of Eqs. (6.15) and (6.18) for surfaces that one would expect to extend *through* the circuit board boundary layer, e.g., for heat sinks. The experimentally corrected correlation, Eq. (6.15), can be used for air speeds from 50 to 2000 ft/min and Yovanovich's and Teerstra's work, Eqs. (6.19) and (6.20) for flow speeds less than 50 ft/min. Of course, Eq. (6.18) is also applicable to the higher flow rate problems. If the application is for a component surface on a circuit board and with a somewhat low profile, a correlation specific to circuit boards from either of the following two sections would be more appropriate.

You may be wondering why you have not seen any correlations for turbulent flat plate flow. The answer is quite simple. The author, with approximately 35 years' experience in the thermal analysis

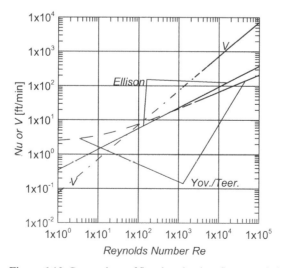

Figure 6.10. Comparison of flat plate, laminar flow correlations for $L = 2$ in., $W = 2$ in., $Pr = 0.71$, $v = 0.029$ in.$^2/$s .

of electronics, has never had a problem where the *Reynolds number indicated turbulent flow* over a flat plate. A calculation of *Re* for the greatest speed in Figure 6.7 is approximately an order of magnitude less than Re_c, e.g.,

$$Re = VL/(5v) = (2000\,\text{ft/min})(2.0\,\text{in.})/\left[5(0.029\,\text{in.}^2/\text{s})\right] = 2.8 \times 10^4 < 5 \times 10^5 \quad (\text{critical } Re)$$

6.9 CIRCUIT BOARDS: ADIABATIC HEAT TRANSFER COEFFICIENTS AND ADIABATIC TEMPERATURES

Considerable attention has been devoted to the well-mixed air temperature for channel flow. While the method may be appropriate for comparison of various cooling methods and as a figure of merit for circuit board channels, it is not recommended when trying to calculate component temperatures as accurately as possible, even though you may be attempting only a first-order analysis. An alternative and preferred computational method is to use the *adiabatic temperature and heat transfer coefficient*. The methodology used in this method must be credited to Professor R. Moffat (retired from Stanford University) and some of his students. A literature search will result in numerous references and discussions of this topic. An excellent book, *Air Cooling Technology for Electronic Equipment*, edited by Sung Jin Kim and Sang Woo Lee (1996) includes a list of some of these references as well as two articles that are used as the basis of the next two sections.

In order to discuss the adiabatic temperature and heat transfer coefficient, we may refer to Figure 6.11 where a uniform array of cuboid-shaped components is shown. The problem is made manageable by assuming a uniform velocity profile with a speed V_a which we call our approach velocity. Depending on which researcher's results we use, we may also need to reference a bypass velocity V_b within the board channel. While it is always possible to *calculate* a well-mixed air temperature, this is not the true reference temperature to which the component surface convects. There are complex temperature and velocity variations in the region between the component surface and the opposite channel wall and we should not expect to predict these with first-order analysis methods. While the methods of computational fluid dynamics have been found to be quite useful in these problems, it requires expensive software to which the analyst may not have access.

Fortunately for us, several researchers have published useful correlations based on extensive experimental data. These correlations require the assumption of component arrays where the "blocks" are uniformly distributed on the board and are uniform in size. We cannot expect that our actual

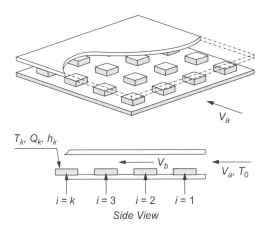

Figure 6.11. Components on printed circuit board. V_a = approach velocity, V_b = bypass velocity.

equipment designs will have this uniformity of geometry, but we are often able to select a single component size and spacing that approximates our problem. While we are not able to calculate exact temperatures, we will have more accurate results than if we used the well-mixed air temperature rise formula.

Now we shall discuss the adiabatic temperature and heat transfer coefficient in a general way by briefly describing an experimental setup that should resemble a system used by a typical researcher to obtain a correlation. Panels that form bounding channel surfaces should be heat-insulated (a guarded heater system would suffice), constructed in such a manner that the heat dissipated by each component is totally convected to the local air, i.e., the board is made of thermally insulating material. The component and board surfaces within the channel should also be of sufficiently low radiation emissivity that radiation is negligible. Calculations are necessary to verify that conductive and radiative losses are negligible. Components are simulated by high conductivity blocks and each component is instrumented with a thermocouple. The high conductivity block material will ensure that the block temperature is uniform so that any temperature measurement is also equal to the surface temperature. Most researchers' experimental models are evaluated for only a single column, i.e., flanking effects are neglected. The remainder of this section presumes that these flanking effects are indeed minor and this assumption will not diminish your understanding of the phenomena.

Two sets of measurements are made for a column of components: (1) The convective heat transfer coefficient and (2) the adiabatic temperature. The heat transfer coefficient for a given row, the kth row, is accomplished by powering only the component of interest with heat Q_k and measuring the surface temperature T_k of this component. The adiabatic heat tranfer coefficient h_k is calculated from

$$h_k = Q_k / \left[A_S \left(T_k - T_{ad} \right) \right]$$

where T_{ad} is the channel inlet temperature, and A_s is the total convecting surface area of the block (top + four edges). The adiabatic reference temperature T_{ad} is equal to T_0, the channel inlet temperature for this measurement. We can say that this reference temperature is the adiabatic temperature because if Q_k were zero, i.e., the component were an adiabatic surface, then the block temperature would be T_{ad}.

Now we consider what is referred to as the *thermal wake function*. The thermal wake function is defined as the fractional surface temperature rise (above the inlet temperature) of any component, i.e., the kth component, due to upstream heating by some other element, i.e., the ith component. Thus the thermal wake function for the kth element heated by the ith for a channel inlet air temperature T_0 is

$$\theta_{k-i} = \left(T_k - T_0 \right) / \left(T_i - T_0 \right)\big|_{\substack{Q_i \neq 0 \\ Q_k = 0}}$$

A given θ_{k-i} for the kth component is measured by powering the ith component, measuring the surface temperatures T_k and T_i, and calculating the ratio that gives θ_{k-i}. Considering the explanation of

$$h_k = \frac{Q_k}{A_k \left(T_k - T_0 \right)}\bigg|_{\substack{only \\ Q_k \neq 0}} , \quad \theta_{k-i} = \frac{T_k - T_0}{T_i - T_0}\bigg|_{\substack{Q_i \neq 0 \\ Q_k = 0}} \qquad \text{Adiabatic } h \ (6.21)$$
thermal wake func.

the preceding two equations, we can sum it with the definitions in Eq. (6.21) and the formula to calculate the component temperature, Eq. (6.22).

$$T_k - T_0 = \left(\frac{Q_k}{h_k A_k} \right) + \sum_{i=1}^{k-1} \theta_{k-i} \left(T_i - T_0 \right), \ i < k \qquad \text{Component} \ (6.22)$$
temperature

The components represented by Eqs. (6.21) and (6.22) are counted with the row nearest the leading edge as $k = 1$. The first term on the right of Eq. (6.22) is the self-heating effect. Each term within the summation of Eq. (6.22) is an upstream contribution to the total local air temperature. Note that an upstream Q is not explicitly used, but the effect of Q_i is implicitly expressed in T_i.

This author interprets the summed series in Eq. (6.22) as the local air temperature rise "seen" by a component. When you make calculations, you will find that this adiabatic temperature rise calculation is always greater than what you would obtain using the well-mixed temperature formula. The air flowing over a heated component does not mix totally into the channel.

As an example, the following equations are applicable to calculating the case temperature rise above the inlet air, $T_5 - T_0$, for a component in the fifth row.

$$T_5 - T_0 = \left(\frac{Q_5}{h_5 A_5} \right) + \sum_{i=1}^{5-1} \theta_{k-i} \left(T_i - T_0 \right)$$

$$T_5 - T_0 = \left(\frac{Q_5}{h_5 A_5} \right) + \theta_{5-1} \left(T_1 - T_0 \right) + \theta_{5-2} \left(T_2 - T_0 \right) + \theta_{5-3} \left(T_3 - T_0 \right) + \theta_{5-4} \left(T_4 - T_0 \right)$$

$$T_5 - T_0 = \Delta T_{5C-LA} + \Delta T_{5-Air}$$

where the self-heating is

$$\Delta T_{5C-LA} = Q_5 / h_5 A_5$$

and the equivalent of a local air temperature rise is

$$\Delta T_{5-Air} = \theta_{5-1} \left(T_1 - T_0 \right) + \theta_{5-2} \left(T_2 - T_0 \right) + \theta_{5-3} \left(T_3 - T_0 \right) + \theta_{5-4} \left(T_4 - T_0 \right)$$

$$\Delta T_{5-Air} = \theta_4 \left(T_1 - T_0 \right) + \theta_3 \left(T_2 - T_0 \right) + \theta_2 \left(T_3 - T_0 \right) + \theta_1 \left(T_4 - T_0 \right)$$

The air temperature rise is the sum of four terms where, beginning with the first term, we have air heating due to the first row component at T_1, the second row at T_2, the third row at T_3, and the fourth row at T_4. The subscript difference $k-i$ on the thermal wake function θ_{k-i} denote the number of rows distant from the component of interest (row 5 in this instance). Row 4 is only one row distant from row 5, thus the wake function is subscripted as "1." Row one at temperature T_1 is 4 rows distant from row 5, thus the wake function is subscripted with a "4." The wake functions are independent of any particular Q_k, thus you must calculate all of the wake functions prior to making actual temperature calculations for each row. Note that the wake functions are heat-dissipation independent. Furthermore, the method is not dependent on the components all having the same heat dissipation.

6.10 ADIABATIC HEAT TRANSFER COEFFICIENT AND TEMPERATURE ACCORDING TO FAGHRI ET AL.

Extensive work by Faghri et al. is very nicely explained in the book by Sung Jin Kim and Sang Woo Lee (1996). The geometry for their correlations is illustrated in Figure 6.12. The reported ranges of variable values are listed in Table 6.2, noting that the blocks representing the components have a square footprint and the spacing S is the same in both the flow and transverse directions.

Table 6.2. Geometry variable range of validity for correlation by Faghri et al. $Re = 400 \rightarrow 15{,}000$.

S [in.]	L [in.]	a [in.]	$H\text{-}a$ [in.]	S/L	$(H\text{-}a)/L$	a/L
$0.25 \rightarrow 0.33$	$1.0 \rightarrow 1.96$	$0.5 \rightarrow 1.0$	$0.25 \rightarrow 1.0$	$0.125 \rightarrow 0.5$	$0.125 \rightarrow 1.5$	0.5

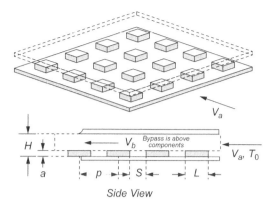

Figure 6.12. Geometry for Faghri et al. adiabatic h, T correlations.

The Nusselt number is a function of distance x from the circuit board channel leading edge.

$$Nu_L = \left[1 + 0.0786\left(\frac{x}{D_H}\right)^{-1.099}\right]\left[\frac{\left(\dfrac{H-a}{L}\right)^{-0.670}}{2.729 Re^{-0.607}\left(\dfrac{S}{L}\right)^{-0.295}}\right] \qquad \begin{matrix} Nu \;(6.23) \\ \text{Faghri et al.} \end{matrix}$$

where

$$Re = \frac{V_b(H-a)}{5\nu}, \quad D_H = \frac{2WH}{(W+H)}$$

for Re in mixed units, or any consistent units without the 5 in the denominator. W is the duct channel width. The reader is reminded that V_b is between the top component surface and the opposite wall. The Faghri thermal wake functions are

$$\theta_1 = \frac{(S/L)^{-0.540}}{2.685 Re^{0.168}}, \quad \frac{\theta_N}{\theta_1} = 0.151 + 0.849 N^{-1.314} \qquad \text{In-line } (6.24)$$

$$\frac{\theta_{fN}}{\theta_N} = 0.0575 N^{1.128} \qquad \text{Flank } (6.25)$$

6.11 ADIABATIC HEAT TRANSFER COEFFICIENT AND TEMPERATURE FOR LOW-PROFILE COMPONENTS ACCORDING TO WIRTZ

Richard Wirtz has assembled and correlated extensive data for what he calls low-profile packages on printed circuit boards, also explained in the book by Sung Jin Kim and Sang Woo Lee

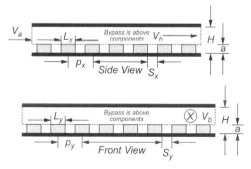

Figure 6.13. Geometry for Wirtz et al.'s adiabatic h, T correlations.

(1996). The geometry for these correlations is illustrated in Figure 6.13. Both Faghri and Wirtz use a bypass velocity V_b defined using the total volumetric airflow through the card channel and the area whose perimeter uses the duct width W and $(H - a)$ *distance between the component tops and opposing board surface*. The approach velocity is V_a.

The reported range of variable values is listed in Table 6.3, noting that in this case, the blocks representing the components are not necessarily square, nor must the spacing be equal for the flow and transverse directions. The table also includes the original source from which Wirtz took his data.

Table 6.3. Geometry variable range of validity for correlation by Wirtz.

Author	L_x [mm]	L_x [in.]	L_x/a	H/a	σ
Wirtz and Dykshoorn (1984)	25.4	1.0	4.00	$1.5 \rightarrow 4.6$	0.25
Sparrow et al. (1982)	26.7	1.05	2.67	2.7	0.64
Anderson and Moffat (1992)	37.5	1.48	3.95	$1.5 \rightarrow 4.6$	0.59
Wirtz et al. (1994)	56.0	2.21	8.75	$1.5 \rightarrow 10$	0.49
Wirtz and Mathur (1994)	69.8	2.75	6.0	2.0	0.45
Wirtz and Colban (1995)	69.8	2.75	6.0	2.0	$0.45 \rightarrow 0.69$

$$\sigma \equiv \text{Packaging density} = L_x L_y \big/ \big[\left(L_x + S_x \right) \left(L_y + S_y \right) \big]$$

Wirtz recommends Nusselt numbers for both laminar ($Re < 5000$) and turbulent ($Re > 5000$) flow.

$$
\begin{aligned}
Nu_{L_x} &= 0.6 Re_{L_x}^{0.5} Pr^{0.33}, & Re_{L_x} &\leq 5000 \\
Nu_{L_x} &= 0.082 Re_{L_x}^{0.72}, & Re_{L_x} &> 5000
\end{aligned}
$$

Very low profile packages
$$Nu_{L_x} = 0.07 Re_{L_x}^{0.718}, \qquad Re_{L_x} > 5000$$

$$Nu \ (6.26)$$
$$\text{Wirtz}$$

where $Re_{L_x} = V_b L_x / \nu$ in consistent units and $Re_{L_x} = V_b L_x / (5\nu)$ in mixed units. "Very low profile" is described as a small, but finite package height. Low profile is defined by the condition

$$L_x L_y \big/ \big[L_x L_y + 2a \left(L_x + L_y \right) \big] \geq 0.5$$

Researchers and application engineers have determined that there is usually an *entry length* effect where the heat transfer coefficient is greater for rows near the leading edge. Wirtz recommends a compensation for low row numbers.

$$
\begin{aligned}
\text{Row 1:} \quad & h_k = 1.25 \left(k_{Air} / L_x \right) Nu_{L_X} \\
\text{Row 2:} \quad & h_k = 1.10 \left(k_{Air} / L_x \right) Nu_{L_X} \\
\text{Row 3:} \quad & h_k = \left(k_{Air} / l_x \right) Nu_{L_X}
\end{aligned}
$$

Entry length effects (6.27)

We conclude from Eq. (6.27) that fully developed flow exists beginning at about the third row. The thermal wake function is just as important as the heat transfer coefficient. We will see that there are circumstances where a well-mixed air temperature rise would suggest that heating the air stream is not particularly important. However, an adiabatic temperature calculation might show that just the opposite is true. Wirtz recommends using a wake function developed by Kang (1994). Wirtz notes, however, that there is sufficient data only to substantiate a correlation for turbulent flow. Unfortunately, the dismissal of a laminar flow thermal wake function correlation does not help the design engineer. This writer has therefore chosen to offer and use correlations for both laminar and turbulent flow from Kang.

Laminar, row 1:

$$\theta_{1L} = 0.42 \left(\frac{p_x}{L_x} \right)^{-0.5}$$

(6.28)

Turbulent, row 1:

$$\theta_{1T} = 0.590 Pr^{-0.5} \left(\frac{L_x}{p_x} \right)^{0.5} \left(\frac{L_x}{H} \right)^{0.44} \left(1 - \frac{a}{H} \right)^{0.5} Re_{V_a,L_x}^{-0.22}$$

Low profile (6.29)

$$\theta_{1T} = 0.503 Pr^{-0.5} \left(\frac{L_x}{p_x} \right)^{0.5} \left(\frac{L_x}{H} \right)^{0.44} \left(1 - \frac{a}{H} \right)^{0.5} Re_{V_a,L_x}^{-0.22}$$

Very low (6.30) profile

The Reynolds number is based on the *approach velocity* V_a and of course the component length L_x. Rows beyond the first row use the following:

Laminar and turbulent:

$$\theta_{k-i} = \theta_1 \frac{1}{(k-i)^m}$$
$$m = 0.5 + 0.335 e^{(-Pe/20)} + 0.105 e^{(-Pe/100)} + 0.06 e^{(-Pe/2500)}$$
$$Pe = RePr$$

(6.31)

Using consistent units, the Peclet number Pe is given by

Laminar: $Pe_{Lam} = V_a \left[L_y^2 / (v p_x) \right] Pr$

Turbulent: $Pe_{Turb} = \left[V_b L_y^2 / (0.006 p_x) \right] \left(c_p \rho / k_{Air} \right) \left(1 / Re_{V_b,H}^{0.88} \right)$

Applications in this book utilize a convenient mixed set of units. Using dimensions of V_a, V_b [ft/min]; p, L_y [in.]; v [in.^{2}/s],

$$Pe_{Lam} = \left(V_a L_y^2 Pr \right) / (5v p_x)$$

Laminar (6.32)

Using the properties of air at 20°C, V_b [ft/min]; p, L_y [in.],

$$Pe_{Turb} = V_b L_y^2 \times 10^3 / \left(p_x Re_{V_b,H}^{0.88} \right)$$

Turbulent (6.33)

6.12 APPLICATION EXAMPLE: CIRCUIT BOARD WITH 0.82 IN. × 0.24 IN. × 0.123 IN. CONVECTING MODULES

A circuit board with eight columns and seven rows of 16-lead, dual-in-line packages is illustrated in Figure 6.14 (a) and component detail in Figure 6.14 (b). The components are displaced in the vertical direction (perpendicular to board plane) such that there exists a small space of thickness a - a_0 between the component base and mounting surface. The problem data is

$G = 14.3 \, CFM, Q_{Board} = 18.6 \, W, H = 1.0 \, in.$

$L_x = 0.24 \, in., L_y = 0.82 \, in., S_x = 0.46 \, in., S_y = 0.1 \, in., a = 0.162 \, in., a_0 = 0.123 \, in.$

The calculated parameters to test against the Faghri model criteria in Table 6.2 are (using $S = S_x, L = L_x$).

$$H - a = 1.0 - 0.162 = 0.838\,\text{in.}, S/L = 0.46/0.24 = 1.92$$
$$(H-a)/L = (1.0 - 0.162)/0.24 = 3.49, a/L = 0.162/0.24 = 0.68$$

from which we see that H - a = 0.838 in. is within the Faghri model minimum and maximum of 0.25 in. and 1.0 in., respectively; S/L = 1.92 is considerably outside the Faghri model minimum to maximum of 0.125 to 0.5, respectively; $(H$ - $a)/L$ = 3.49 is greater than twice the Faghri model maximum of 1.5; a/L = 0.68 is somewhat greater than the Faghri model value of 0.5.

The calculated parameters to test against the Wirtz model criteria in Table 6.3 are

$$L_x/a = 0.24/0.162 = 1.48, H/a = 1.0/0.162 = 6.17,$$
$$\sigma = L_x L_y \Big/\big[(L_x + S_x)(L_y + S_y)\big] = (0.24)(0.82)\Big/\big[(0.24 + 0.46)(0.82 + 0.1)\big] = 0.31$$

from which we see that L_x/a = 1.48 is nearly half of the lowest Wirtz value of 2.67, H/a = 6.17 is somewhat greater than the largest Wirtz value of 4.6, and the packaging density = 0.31 is within the Wirtz minimum and maximum of 0.25 and 0.69, respectively. The problem L_x = 0.24 in. is considerably less than the Wirtz minimum of 1.0 in.

Neither the Faghri nor the Wirtz model is precisely applicable to this problem, a situation that we often encounter in attempting to use heat transfer and fluid mechanics correlations. Perhaps the most obvious criteria is that this problem is somewhat of the low-profile type problem. The component top-to-board distance a = 0.162 in. is far less than the smallest value of 0.5 in. listed in Table 6.2. The Wirtz model will therefore be the one used here.

Only a few of the calculations are shown. The details are left as a problem for the student. The approach and bypass velocities, and the Reynolds number are

$$V_a = G/HW = 14.3\,\text{CFM}\Big/\big[(1.0\,\text{in.})(8.05\,\text{in.})\big/(144\,\text{in.}^2/\text{ft}^2)\big] = 256\,\text{ft/min}$$
$$V_b = G\Big/\big[(H-a)W\big] = 14.3\,\text{CFM}\Big/\big[(1.0-0.162\,\text{in.})(8.05\,\text{in.})\big/(144\,\text{in.}^2/\text{ft}^2)\big] = 305\,\text{ft/min}$$
$$Re_{L_x} = (V_b L_x)/(5v) = \big[(305\,\text{ft/min})(0.24\,\text{in.})\big]\Big/\big[5(0.029\,\text{in.}^2/\text{s})\big] = 505$$

The Peclet number Pe and the exponent m that are required to calculate the thermal wake functions can be shown to be

$$Pe = 1.20 \times 10^3, \qquad m = 0.537$$

The thermal wake functions are

$$\theta_1 = 0.246, \theta_2 = 0.0.169, \theta_3 = 0.0.136$$
$$\theta_4 = 0.0.117, \theta_5 = 0.0.104, \theta_6 = 0.094$$

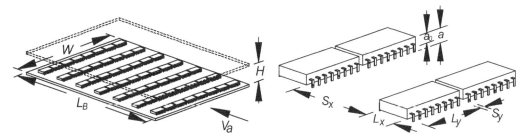

Figure 6.14 (a). Application Example 6.13: Circuit board with convecting modules.

Figure 6.14 (b). Application Example 6.13: *Details* of circuit board with convecting modules.

The calculated, fully developed heat transfer coefficient is $h_{FD} = 0.035$ W/(in.$^2 \cdot$ °C).

The final case temperatures are plotted in Figure 6.15 and are labeled as 305: *Calc, No Brd.*, the latter emphasizing that circuit board conduction effects are not included; all of the other plots include *PCB* conduction effects. The two lowest curves are plotted for measured and calculated case temperatures where board conduction effects are included. The board conduction problem is too complicated to be included at this point of our study, but the results are shown to emphasize just how important it is to include the board in an actual design analysis. Otherwise, your results will be far too much of a worst case situation to be of much value to you as a design engineer.

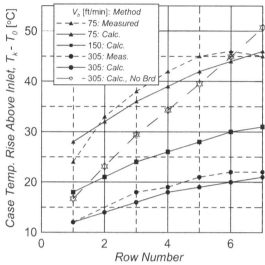

Figure 6.15. Application Example 6.12: Circuit board with 0.82 in. × 0.24 in. × 0.123 in. convecting modules. The calculations shown in this example are labeled as 305: *Calc., No Brd.*

EXERCISES

6.1 A winged heat sink is illustrated. All of the common dimensions, including V, are identical to the example in Section 6.6, except, of course, the finger fin dimensions w and S. Use $w = 0.1$ in. and $\nu = 0.029$ in.^{2}/s and $k_{Air} = 6.92 \times 10^{-4}$ W/(in. $\cdot$ °C). We assume that in order to prevent bridging of the boundary layer between the fingers, the finger spacing shall be kept equal to or greater than w. First calculate the approximate value for the maximum number of fingers along each edge in the L direction. Then, beginning with $N_f = 0$ and ending with $N_{f\text{-}Max}$, calculate the forced airflow thermal resistance. Use enough values of N_f so that you can plot the thermal resistance vs. N_f. Hints: (1) The maximum resistance should be the same as calculated for the text example in Section 6.6; (2) Be sure to separate the finger contributions from the component top contribution; (3) The fingers use a different characteristic length than does the component top.

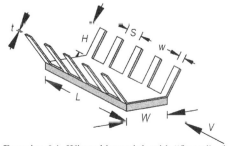

Exercise 6.1. Winged heat sink with "finger" wings.

6.2 A single aluminum fin is illustrated with dimensions: $L = 1.0$ in., $H = 2.0$ in., $t_f = 0.10$ in. and conductivity $k_{Al} = 5.0$ W/(in. °C). Air flows in a direction parallel to the fin length L with a speed V from 50 to 1000 ft/min. The surface emissivity is less than 0.1 so you may ignore radiation. Use a heat source of $Q = 5$ W entering the fin root. The fin height and air speed are great enough that you cannot assume that the fin is uniform in temperature from the fin root to tip. Unfortunately, we have not studied the method commonly used to compensate for fin temperature gradients. We shall now introduce the concept of fin efficiency, but without explanation. You can either follow the simple directions here or read the appropriate section in a later chapter. All you need is to know now is that the heat transfer coefficient has a multiplicative factor η which for the geometry of this problem is

$$\eta = \sqrt{R_s/R_k} \, \tanh\sqrt{R_k/R_s} \, ; R_s = 1/hA_s \, , R_k = H/kA_k \, ; A_s = 2LH; A_k = t_f L$$

for a fin with an insulated tip. A_s is the convecting surface area and A_k is the conduction cross-sectional area for conduction from the fin root to the tip. In this problem, use Eq. (6.15) for the Nu correlation. You are asked to calculate and plot the thermal resistance (from fin root to local ambient) and temperature rise (of fin root above the local ambient) of the convecting fin for air speeds $V = 50, 100, 250, 500,$ and 1000 ft/min. Use $\nu = 0.029$ in.^{2}/s, $k_{Air} = 6.92 \times 10^{-4}$ W/(in. · °C).

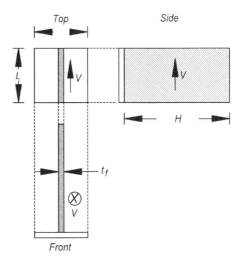

Exercise 6.2. Single fin in forced air flow.

6.3 Using the variable names for Exercise 6.2, calculate and plot the thermal resistance (from fin root to local ambient) and temperature rise (of fin root above the local ambient) of the convecting fin for air speeds $V = 50, 100, 250, 500,$ and 1000 ft/min. The data values are $L = 0.5$ in., $H = 1.0$ in., $t_f = 0.06$ in. and conductivity $k_{Al} = 5.0$ W/(in. · °C). Air flows in a direction parallel to the fin length L with a speed V from 50 to 1000 ft/min. The surface emissivity is less than 0.1 so you may ignore radiation. Use a heat source of $Q = 5$ W entering the fin root. Follow the advice given in Exercise 6.2 regarding fin efficiency. Use $\nu = 0.029$ in.^{2}/s, $k_{Air} = 6.92 \times 10^{-4}$ W/(in. · °C).

6.4 Use the geometry of Exercise 6.2 for $V = 50$ and 100 ft/min, but use Eqs. (6.16) and (6.17) for your Nu correlation. Calculate $Re, Nu, h, R,$ and ΔT. You may neglect the effects of fin efficiency in this problem. Hints: (1) Be very careful to use fin length and width dimensions that correspond to the longer and shorter of the two dimensions required by Yovanovich and Teerstra; (2) The area that you use in the Re and Nu is $L \cdot W$, but the area you use to calculate R must include both sides of the fin.

6.5 In Application Example 6.12, the details of the calculation were omitted. You are asked to calculate the complete problem results for this example using the Wirtz model. Use the example value of $G = 14.3$ CFM. Of course, your results will not include circuit board conduction.

6.6 Application Example 6.12 (and Exercise 6.5) used a channel volumetric airflow of $G = 14.3$ CFM, i.e., $V_b = 305$ ft/min. Calculate the volumetric airflow G that would result in a bypass airspeed of $V_b = 150$ ft/min. Then calculate the entire seven-row problem for $V_b = 150$ ft/min. Your results will not include circuit board conduction; therefore, you will not be able to compare your answers with the Figure 6.16 plot.

6.7 Application Example 6.12 (and Exercise 6.5) used a channel volumetric airflow of $G = 14.3$ CFM, i.e., $V_b - 305$ ft/min. Calculate the volumetric airflow G that would result in a bypass airspeed of $V_b = 75$ ft/min. Then calculate the entire seven-row problem for $V_b - 75$ ft/min. Your results will not include circuit board conduction; therefore, you will not be able to compare your answers with the Figure 6.16 plot.

6.8 Application Example 6.12 (and Exercises 6.6-6.7) used a board spacing H=1.0 in. This problem is the same except that the board spacing is H=0.5 in. Calculate the volumetric airflow G that would result in a bypass airspeed of $V_b = 37.5$ ft/min. Then calculate the entire seven-row problem for $V_b = 37.5$ ft/min. Your results will not include circuit board conduction.

6.9 Application Example 6.12 (and Exercises 6.6-6.7) used a board spacing H=1.0 in. This problem is the same except that the board spacing is H=0.5 in. Calculate the volumetric airflow G that would result in a bypass airspeed of $V_b = 75$ ft/min. Then calculate the entire seven-row problem for $V_b = 75$ ft/min. Your results will not include circuit board conduction.

6.10 Application Example 6.12 (and Exercises 6.6-6.7) used a board spacing H=1.0 in. This problem is the same except that the board spacing is H=0.5 in. Calculate the volumetric airflow G that would result in a bypass airspeed of $V_b = 150$ ft/min. Then calculate the entire seven-row problem for $V_b = 150$ ft/min. Your results will not include circuit board conduction.

6.11 Application Example 6.12 (and Exercises 6.6-6.7) used a board spacing H=1.0 in. This problem is the same except that the board spacing is H=0.5 in. Calculate the volumetric airflow G that would result in a bypass airspeed of $V_b = 300$ ft/min. Then calculate the entire seven-row problem for $V_b = 300$ ft/min. Your results will not include circuit board conduction.

6.12 A circuit board with 25 components (5 columns, 5 rows) is illustrated as Exercise Figure 6.13. You wish to know the component case temperatures of every row when the total dissipation of $Q = 50$ W is uniformly distributed over the circuit board and the total channel airflow is $G = 10$ CFM. The problem data is: $H = 0.75$ in., $L_x = 1.5$ in., $L_y = 1.5$ in., $S_x = 1.5$ in., $S_y = 1.5$ in., $a = 0.25$ in., $a_0 = a$, L_B 15 in., $W = 15$ in. Test for the applicability of both the Wirtz and Faghri models, but use only Wirtz for this problem.

6.13 A circuit board with 25 components (5 columns, 5 rows) is illustrated as Exercise Figure 6.13. You wish to know the component case temperatures of every row when the total dissipation of $Q = 50$ W is uniformly distributed over the circuit board and the total channel airflow is $G = 10$ CFM. The problem data is $H = 0.55$ in., $L_x = 1.5$ in., $L_y = 1.5$ in., $S_x = 1.5$ in., $S_y = 1.5$ in., $a = 0.25$ in., $a_0 = a$, $L_B = 15$ in., $W = 15$ in. Test for the applicability of both the Wirtz and Faghri models, but use only Wirtz.

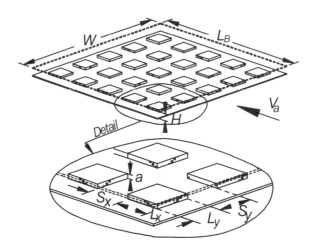

Exercise 6.13. Low-profile components on printed circuit board.

CHAPTER 7

Forced Convective Heat Transfer II:
Ducts, Extrusions, and Pin Fin Arrays

In previous chapters, we discussed forced air convective heat transfer from simple planar surfaces and modules on printed circuit boards. The present chapter will show you how to calculate forced convective heat transfer from extended surfaces, i.e., plate fin heat sinks (usually in the form of an extrusion) and pin fin arrays. You will learn about some correlations that the author has successfully used to make temperature predictions. Where available, you will also see comparison of calculated results with test data.

The literature is filled with heat transfer correlations for various fin arrays and researchers will continue to publish their work on this subject. You should not expect that all useful correlations from various journals and books are included in this chapter, but you will find that those in the following pages should meet most of your everyday needs.

7.1 BOUNDARY LAYER CONSIDERATIONS

In electronics cooling applications, nearly all airflow is contained within a space that may be defined by boundaries, thus forming a duct of some sort. Two parallel flat plates with closed spaces on either side form a duct. Aerodynamically and thermally however, boundary layer effects must be considered. For example, air with a uniform velocity profile at the entrance V_a (approach velocity) to a duct, as illustrated in Figure 7.1, begins to develop an aerodynamic boundary layer of thickness $\delta(x)$, which varies from a very small value at the duct inlet to a fully developed flow value downstream. The fully developed boundary layer thickness is obtained when the boundary layers from the two opposing walls come into contact with one another. From this point on, the boundary layer thickness remains unchanged. Within the boundary layer, the air speed V_f varies from zero at the wall to a maximum at the duct center line. Of course, this is a highly simplified explanation. Such details are beyond the scope of what we need to consider when making first-order calculations.

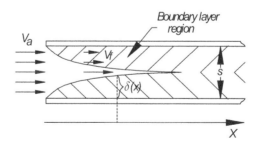

Figure 7.1. Aerodynamic boundary layer development in a duct.

If the aerodynamic boundary layer thickness for the entrance region is defined as that distance from the wall for which the air speed is 99% of the free-stream value, flat plate boundary layer theory predicts

$$\delta_{Aero-L} = 5x\big/Re_x^{1/2} \qquad \text{Laminar boundary layer thickness} \qquad (7.1)$$

$$\delta_{Aero-Th} = 0.376x\big/Re_x^{1/5} \qquad \text{Turbulent boundary layer thickness} \qquad (7.2)$$

The Reynolds number is based on V_a. We can use flat plate theory in the duct entrance region because the boundary layers don't show the effects of the duct walls until the two opposing boundary layers make contact. One might believe then that the distance at which the flow becomes fully developed is at a value of x for which $S = 2\delta$ (see Figure 7.1). However, since little air speed change occurs in the outer region of the layers, a *more appropriate value* is $S \cong \delta$. It is then easy to show that an estimate of the distance from the inlet for which the flow changes from laminar to fully developed is approximately given by

$$x_{Aero-FD} = S^2 V_f / (25\nu) \qquad \text{Fully developed} \quad (7.3)$$
$$\text{laminar duct flow}$$

in consistent units. If you are using mixed units, the "25" is replaced by "125" and the units are S[in.], x_{FD} [in.], V_f [ft/min], and ν [in.^{2}/s].

We have been discussing aerodynamic boundary layer thickness, but you might wonder how this relates to the thermal boundary layer, i.e., that distance from the wall at which the temperature is 99% of the free stream value. Laminar boundary layer theory indicates

$$\delta_{Th-L} = \delta_{Aero-L} / Pr^{1/3} \qquad \text{Conversion to thermal} \quad (7.4)$$

which, if we use $Pr = 0.71$ for air, shows us that the thermal boundary layer should be about 12% thicker than that of the aerodynamic boundary layer.

7.2 A CONVECTION/CONDUCTION MODEL FOR DUCTS AND HEAT SINKS

An illustration of a finned heat sink extrusion (identical to the one in Chapter 4, Figure 4.1) is shown in Figure 7.2. The top-covering shroud is intended to confine the front-entering air to the interfin passages and is not included as part of any heat conduction or convection. The heat sink base has both a substantial thermal conductivity, i.e., is a metal, and also has a significant thickness with regard to the direction of planar heat conduction within the base plane. We shall assume that heat-dissipating devices such as power transistors are distributed in some fashion over the exterior surface of the base. The heat sink base properties will be shown to have particular significance when we examine thermal spreading in Chapter 12.

The "temperature model" is shown in Figure 7.3. The heat sink base is assumed to have a nearly uniform temperature from the leading to trailing edge. We are depending on the thick base and high conductivity to uniformly *flatten* the heat sink temperature profile. The interfin air is assumed to

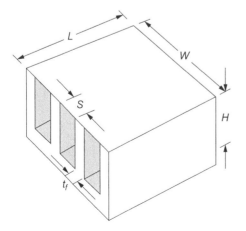

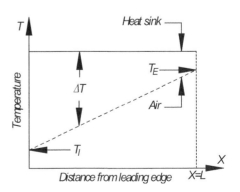

Figure 7.2. Shrouded plate-fin heat sink extrusion. Airflow is in direction parallel to heat sink length L.

Figure 7.3. Heat sink extrusion temperature distribution model.

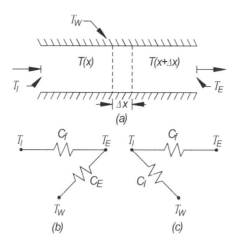

Figure 7.4. Representation of duct flow heat transfer model. (a) Model of duct section. (b) Two-element model referenced to exit air temperature. (c) Two-element model referenced to inlet air temperature.

have a uniformly increasing temperature from the heat sink inlet to exit. Perhaps the most important aspect of this model is that we must account for the decreasing temperature difference (through which heat is convected) as the air moves down the channels.

Now we are ready to build our math model. Referring to Figure 7.4 (a), the heat transferred by convection from a small section of wall length Δx, at temperature T_W to local air at (mixed mean) temperature $T(x)$ is given by

$$\Delta Q_{Wf} = hP\Delta x\left[T_W - T(x)\right] \tag{7.5}$$

for a duct perimeter P and a convective heat transfer coefficient h. The heat absorbed by an air element of length Δx with a mass flow rate $\dot{m}$ is given by

$$\Delta Q_a = \dot{m}c_p\left[T(x+\Delta x) - T(x)\right] \tag{7.6}$$

In the limit $\Delta x \to 0$, Eqs. (7.5) and (7.6) become

$$
\begin{aligned}
dQ_{Wf} &= hP\left(T_W - T\right)dx \\
dQ_a &= \dot{m}c_p dT
\end{aligned}
\tag{7.7}
$$

where the x-dependence of T is understood. Conservation of energy applied to Eq. (7.7) requires that we set

$$dQ_{Wf} = dQ_a$$

and substitute Eq. (7.7), then integrate from an inlet air temperature T_I to an exit air temperature T_E over a total heat sink length L.

$$hP\left(T_W - T\right)dx = \dot{m}c_p dT$$

$$\int_{T_I}^{T_E} \frac{dT}{T - T_W} = -\int_0^L \frac{hP}{\dot{m}c_P}dx = -\frac{hP}{\dot{m}c_P}\int_0^L dx$$

$$\ln\left(\frac{T_E - T_W}{T_I - T_W}\right) = -\frac{hA_S}{\dot{m}c_P}$$

$$T_W - T_E = \left(T_W - T_I\right)e^{-\beta}; \quad \beta = hA_S\Big/\left(\dot{m}c_P\right)$$

Adding T_I to both sides of the preceding we have

$$T_I + \left(T_W - T_E\right) = T_I + \left(T_W - T_I\right)e^{-\beta}$$

$$T_I - T_E = \left(T_I - T_W\right) + \left(T_W - T_I\right)e^{-\beta} = \left(T_W - T_I\right)\left(e^{-\beta} - 1\right)$$

$$T_E - T_I = \left(T_W - T_I\right)\left(1 - e^{-\beta}\right)$$

and since

$$T_E - T_I = Q\Big/\left(\dot{m}\,c_P\right)$$

then

$$\frac{Q}{\dot{m}\,c_P} = \left(T_W - T_I\right)\left(1 - e^{-\beta}\right), \quad \frac{Q}{\left(T_W - T_I\right)} = \dot{m}\,c_P\left(1 - e^{-\beta}\right)$$

If we *define* a conductance $C_I = Q/(T_W - T_I)$ referenced to the inlet air, we obtain

$$C_I = Q\Big/\left(T_W - T_I\right) = \dot{m}\,c_P\left(1 - e^{-\beta}\right) = \frac{hA_S}{\left[hA_S\Big/\left(\dot{m}\,c_P\right)\right]}\left(1 - e^{-\beta}\right)$$

$$\boxed{\frac{C_I}{hA_S} = \frac{\left(1 - e^{-\beta}\right)}{\beta}, \quad \beta = hA_S\Big/\left(\dot{m}\,c_P\right)}$$

Conductance ref. (7.8)
to inlet

The derivation of a formula for a conductance C_E referenced to the exit air is accomplished in only a few steps. We previously obtained

$$T_W - T_E = \left(T_W - T_I\right)e^{-\beta}$$

Then

$$C_E = \frac{Q}{\left(T_W - T_E\right)} = \frac{Q}{\left(T_W - T_I\right)}e^{\beta} = C_I e^{\beta} = hA_S\frac{\left(1 - e^{-\beta}\right)}{\beta}e^{\beta}$$

and

$$\boxed{\frac{C_E}{hA_S} = \frac{\left(e^{\beta} - 1\right)}{\beta}, \quad \beta = hA_S\Big/\left(\dot{m}\,c_P\right)}$$

Conductance ref. (7.9)
to exit

The circuit representations of C_E and C_I are shown in Figures 7.4 (b) and 7.4 (c), respectively. It should be pointed out that β is the thermal-fluid resistance (which accounts for air temperature rise in any duct) divided by the surface convection resistance, i.e.,

$$\beta = hA_S\Big/\left(\dot{m}\,c_P\right) = R_f\big/R_c = C_c\big/C_f$$

where the conductance C_f is shown in Figure 7.4.

You might ask yourself why we need two representations of the conductance in this heat sink model. The answer is that there are two different situations. In the first case, suppose we are just calculating a heat sink conductance or resistance for one or more different air flow rates. Then it makes sense to use Eq. (7.8) for C_I. As an example, let's assume that for a heat sink dissipating $Q = 30$ W, we have calculated a conductance $C_I = 0.5$ W/°C or a resistance $R_I = 1/C_I = 2.0$ °C/W. Then the temperature rise ΔT above the inlet air is 60°C. We merely add this temperature rise to some estimated T_I to get the heat sink temperature. If for example, $T_I = 50$°C, then we expect our heat sink to be at about 110°C.

Now suppose that we have completed the design of our heat sink and we wish to incorporate a heat sink thermal resistance or conductance into a large system circuit model containing other thermal elements, including thermal-fluid elements that will show various air temperatures throughout the model. If we used a conductance C_I, and inserted a heat sink dissipation at the circuit node representing the heat sink temperature T_W, as in Figure 7.4 (c), our final circuit solution would indicate a *preheating* of the inlet air node T_I. This would be incorrect. However, if instead, we used a C_E as shown in Figure 7.4 (b), we would have heating of air node T_E, but this is just as we would expect and would be correct.

Remember that although C_E and C_I reference a single temperature, the heat transfer coefficient references a *local air*, because this is how we constructed our derivation. Of course, by the nature of the approximation, this h is actually an average over the length of the heat sink. If you use a Nusselt number (i.e., h) that references the inlet temperature, then you will need to convert your inlet referenced h to a local referenced h. We will study this in the next section.

We need to address one more issue before moving on to the next section. This issue is that of computing C_f. When we analyze heat sinks, we use this quantity C_f so often that it should be simplified as much as possible. We will use nominal material properties. If you wish to change some of these so that the result is more appropriate for your applications, you can easily do so. Material properties are evaluated here at a temperature of about 30°C.

$$C_f = \dot{m} c_P \left[\frac{\text{W}}{\text{°C}} \right] = \rho G c_P = \rho V A c_P = 0.0189 \left[\frac{\text{gm}}{\text{in.}^3} \right] (12)^3 \left[\frac{\text{in.}}{\text{ft}} \right]^3 \times$$

$$V \left[\frac{\text{ft}}{\text{min}} \right] \left(\frac{1}{60} \right) \left[\frac{\text{min}}{\text{s}} \right] A \left[\text{in.}^2 \right] \left(\frac{1}{12} \right)^2 \left[\frac{\text{ft}}{\text{in.}} \right]^2 1.01 \left[\frac{\text{W} \cdot \text{s}}{\text{gm} \cdot \text{°C}} \right]$$

$$\boxed{C_f = \dot{m} c_P \left[\frac{\text{W}}{\text{°C}} \right] = \frac{V \left[\frac{\text{ft}}{\text{min}} \right] A \left[\text{in.}^2 \right]}{262}} \tag{7.10}$$

Our heat sink baseplate-convection assembly is actually a simple heat exchanger. The general heat exchanger effectiveness is defined as

$$\varepsilon = \frac{Heat\,actually\,convected}{Maximum\,possible\,heat\,convected}$$

where the *maximum possible heat convected* Q_{Max} is the heat transferred when all the fluid is at the heat sink surface temperature. This is accomplished by letting the heat transfer coefficient become infinite. The actual heat convected is found using the conductance C_I from Eq. (7.8),

$$Q_{Actual} = \dot{m} c_P \left(1 - e^{-\beta} \right) \left(T_W - T_I \right)$$

and we get Q_{Max} by

$$Q_{Max} = \lim_{h \to \infty} Q_{Actual} = \lim_{\beta \to \infty} \dot{m} c_P \left(1 - e^{-\beta} \right) \left(T_W - T_I \right) = \dot{m} c_P \left(T_W - T_I \right)$$

so that

$$\varepsilon = Q_{Actual} / Q_{Max} = \dot{m} c_P \left(1 - e^{-\beta} \right) \left(T_W - T_I \right) / \left[\dot{m} c_P \left(T_W - T_I \right) \right]$$

$$\boxed{\varepsilon = \left(1 - e^{-\beta} \right)}$$

Heat exchanger (7.11)
effectiveness

7.3 CONVERSION OF AN ISOTHERMAL HEAT TRANSFER COEFFICIENT FROM REFERENCED-TO-INLET AIR TO REFERENCED-TO-LOCAL AIR

Equations (7.5) to (7.9) implicitly assume a heat transfer coefficient $h = h_{Local}$, meaning it is referenced to a local air temperature. The thermal circuit on the right-hand side of Figure 7.5 would be appropriate in this instance. The remainder of this section shows you the derivation and resulting formula that will allow you to convert from $h = h_{Inlet}$ to h_{Local}.

We will carefully distinguish between the two heat transfer coefficients for the remainder of this section. In succeeding sections and chapters, we will understand which h we are using from the discussion context. We begin by writing Eq. (7.8) with our more detailed nomenclature.

$$C_I = \dot{m}c_P\left(1 - e^{-\beta}\right); \quad C_I\big/\dot{m}c_P = 1 - e^{-\beta}$$

which becomes $\quad e^{\beta} = 1\Big/\left[1 - \left(C_I\big/\dot{m}c_P\right)\right], \quad \ln\left\{1\Big/\left[1 - \left(C_I\big/\dot{m}c_P\right)\right]\right\} = \beta$

But by our earlier definition, C_I is referenced to the inlet air and we can write $C_I = h_{Inlet}A_S$ and the relation $\beta = h_{Local}A_S/(\dot{m}c_P)$

$$\ln\left[1\Big/\left(1 - \frac{h_{Inlet}A_S}{\dot{m}c_P}\right)\right] = \frac{h_{Local}A_S}{\dot{m}c_P}$$

$$\boxed{h_{Local} = \frac{\dot{m}c_P}{A_S}\ln\left[1\Big/\left(1 - \frac{h_{Inlet}A_S}{\dot{m}c_P}\right)\right]} \qquad \begin{array}{l}\text{Convert } h_{Inlet} \quad (7.12)\\ \text{to } h_{Local}\end{array}$$

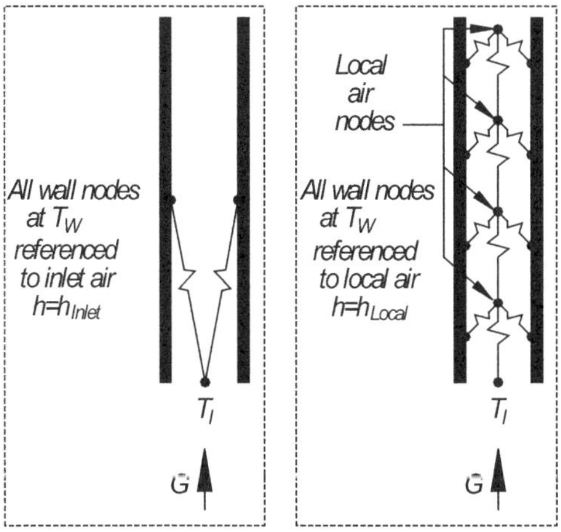

Figure 7.5. Conversion of h_{Inlet} to h_{Local}.

7.4 NUSSELT NUMBER FOR FULLY DEVELOPED LAMINAR DUCT FLOW CORRECTED FOR ENTRY LENGTH EFFECTS

Two different correlations are offered in this chapter for use in duct flow thermal analysis. Both work quite well. The correlation discussed in this section was used by Ellison (1984a), which many readers may wish to continue to use. The next section is concerned with a newer study by Y.S. Muzychka and M.M. Yovanovich (1998).

Nusselt numbers for fully developed laminar flow, constant temperature Nu_T and heat flux Nu_H, are listed in Table 7.1 for ducts of several different cross-section shapes. The shapes that are of greatest interest to us are the circular and rectangular cross-sections. The Nusselt numbers for rectangular cross-section ducts are plotted in Figure 7.6. The two Nusselt numbers for the infinite aspect ratio are plotted for $b/a = 40$. The latter value is a guess on the part of this author, but is plotted in this fashion so that the reader might find it easier to make small extrapolations for $b/a > 8$.

The problem with the Nusselt numbers shown in Table 7.1 and Figure 7.6 is that they are valid for only fully developed flow. We can use Eq. (7.3) to estimate the length of a plate-fin heat sink where fully developed laminar flow is expected. Suppose we use an example of a heat sink that has fin passages with an interfin spacing $S = 0.2$ in., a fin height $H = 1.0$ in., and an interfin air speed $V_f = 100$ ft/min. We should use a hydraulic diameter as a characteristic duct dimension. We then calculate

$$D_H = \frac{4SH}{2(S+H)} = \frac{2SH}{(S+H)} = \frac{2(0.2\,\text{in.})(1.0\,\text{in.})}{(0.2\,\text{in.}+1.0\,\text{in.})} = 0.33\,\text{in.}$$

$$x_{FD} = s^2 V_f / (125v) = D_H^2 V_f / (125v) = (0.33)^2 (100) / \left[125(0.029) \right] = 3\,\text{in.}$$

This calculation suggests that after about a 3-in. distance into the heat sink passages, we might assume fully developed flow.

Kays (1955) reported the results of a numerical simulation concerned with entry length effects for laminar flow in a circular cross-section duct. He used a velocity profile (with uniform inlet velocity and temperature) calculated by Langhaar (1942). Some of these results are plotted for average Nu for constant wall temperature (Figure. 7.7) and local Nu for constant heat input (Figure 7.8). Equations (7.13) and (7.14) were used for the graphs. The plots illustrate the rate at which Nu asymptotically approaches the fully developed limits.

$$\overline{Nu}_{D_H} = 3.66 + \frac{0.104 \left(\dfrac{Re_{D_H} Pr}{L/D_H} \right)}{1 + 0.016 \left(\dfrac{Re_{D_H} Pr}{L/D_H} \right)^{0.8}} \qquad \text{Constant wall temp. (7.13)}$$

Table 7.1. Nusselt numbers for fully developed laminar flow.

Cross-Section Shape	b/a	Nu_H	Nu_T
◯	1.0	4.364	3.66
$a\square b$	1.0	3.61	2.98
$a\square b$	1.43	3.73	3.08
$a\square b$	2.0	4.12	3.39
$a\square b$	3.0	4.79	3.96
$a\square b$	4.0	5.33	4.44
$a\square b$	8.0	6.49	5.60
—	∞	8.235	7.54
—		5.385	4.86
△		3.00	2.35

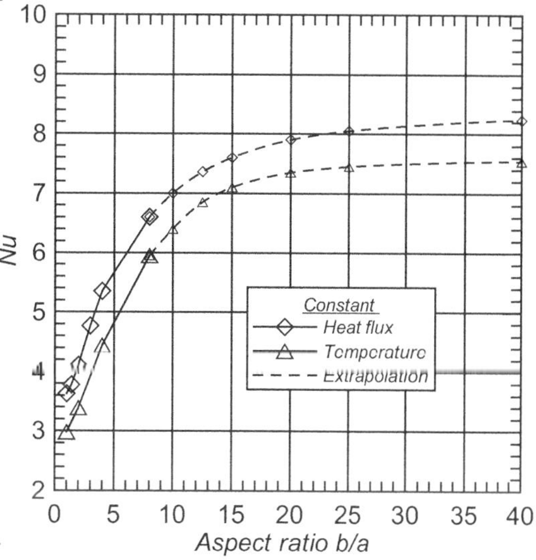

Figure 7.6. Nusselt number for fully developed laminar flow in rectangular ducts.

$$Nu_x = 4.36 + \cfrac{0.036\left(\cfrac{Re_{D_H} Pr}{x/D_H}\right)}{1+0.0011\left(\cfrac{Re_{D_H} Pr}{x/D_H}\right)}$$
Constant heat input (7.14)

It is interesting to see what Figure 7.7 suggests for a heat sink length that we can consider to have mostly fully developed flow. A look at Figure 7.7 suggests a criteria of $Re_D Pr/(L/D_H) < 4$, which can be written as

$$L > Re_{D_H} Pr D_H / 4$$
Entry length (7.15)

Our example results in a value of

$$L > \left[\frac{V_f D_H}{5v}\right] Pr D_H \Big/ 4 = \left[\frac{(100)(0.33)}{5(0.029)}\right](0.71)(0.33)\Big/ 4 = 13\,\text{in}.$$

This L is significantly greater than the result we obtained from boundary layer theory. Now we have to make a choice of using either Eq. (7.3) or (7.15) for our criteria. My choice is the latter of the two possibilities. Using the result from this example indicates that since most plate-fin heat sinks will be significantly shorter than 13 in. in length, it is important to not assume fully developed flow.

Equations (7.13) and (7.14) were developed for ducts with a circular cross-section, but most of our duct flow applications in electronics thermal analysis will require consideration of ducts with a rectangular cross-section. Ellison (1984a) used Eqs. (7.13) and (7.14), but applied what seemed to be a reasonable correction using the fully developed Nusselt numbers listed in Table 7.1 or Figure 7.6. The following outlines the procedure for laminar duct flow.

1. Calculate Nu from Eq. (7.13) or (7.14), i.e., laminar flow with entry length effects.
2. Calculate the ratio $r_{Nu} = Nu$ - rectangular / Nu - circular using the fully developed flow Nu values from Table 7.1 or Figure 7.6.
3. Multiply the results from Step 1 by results from Step 2 to obtain the desired Nu,
i.e., Nu_{D_H} (rect. duct w/ entry length) $= r_{Nu} Nu_{D_H}$ (circular duct w/ entry length)

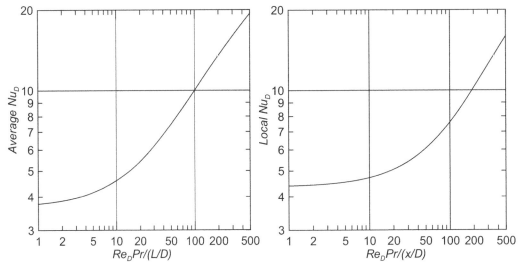

Figure 7.7. Average Nu_D for constant wall temperature on a circular duct with entry length effects. Empirical validation for $7 < Re_D Pr/(L/D) < 70$.

Figure 7.8. Local Nu_D for constant heat flux on a circular duct with entry length effects. Empirical validation for $20 < Re_D Pr/(L/D) < 120$.

7.5 A NEWER NUSSELT NUMBER FOR LAMINAR FLOW IN RECTANGULAR (CROSS-SECTION) DUCTS WITH ENTRY LENGTH EFFECTS

Many fluid and thermal flow correlations have historically been based on ducts with a circular cross-section. The standard practice in applying these correlations to ducts with a noncircular cross-section is to use a hydraulic diameter in place of the circle-based diameter. When we have no other choice but to use a circular cross-section duct, we shall continue this practice, even though this method is suspect.

Muzychka and Yovanovich (1998) have published their study of this problem. The result is

$$Nu_{\sqrt{A}} = \left\{ \left[C_1 C_2 \left(\frac{f_{app} Re_{\sqrt{A}}}{\left(\frac{L}{\sqrt{A} Re_{\sqrt{A}} Pr} \right)} \right)^{1/3} \right]^5 + \left[C_3 \left(\frac{f_{app} Re_{\sqrt{A}}}{8 \sqrt{\pi} \varepsilon^{\gamma}} \right) \right]^5 \right\}^{1/5} \quad \begin{array}{l} \text{Rectang-} \\ \text{ular duct} \\ \text{w/entry length} \end{array} \quad (7.16)$$

We see that Eq. (7.16) uses a Reynolds number based on the square root of the duct cross-sectional area. The parameters f_{app}, and ε are the *apparent friction factor* and duct aspect-ratio, respectively, as given by Muzychka and Yovanovich in their duct friction result, Eq. (4.3). Muzychka and Yovanovich use a Nusselt number that references the local mean air temperature within the duct channel. The constants C_1, C_2, C_3, and γ are listed in Table 7.2. The upper bound and lower bound are selected by symmetry considerations: (1) Upper Bound - geometries which possess two or more planes of symmetry (such as rectangular ducts); (2) Lower Bound - geometries which do not possess two or more planes of symmetry.

Table 7.2. Constant values for *Nu* Muzychka and Yovanovich correlation.

	Local		Average
C_1	1		1.5
	Isothermal	Isoflux	
C_2	0.409		0.501
C_3	3.01		3.66
	Upper Bound	Lower Bound	
γ	1/10	$-3/10$	

7.6 NUSSELT NUMBER FOR TURBULENT DUCT FLOW

My experience has been that most plate-fin heat sinks have laminar forced airflow, but we must be prepared to manage the greater airflow situations. We learned in Chapter 4 that the conventional criteria for determining the laminar or turbulent nature of a duct airflow is

$$\text{Laminar duct flow } Re_{D_H} \leq 2000$$

$$\text{Transitional duct flow } 2000 < Re_{D_H} < 10{,}000$$

$$\text{Turbulent duct flow } Re_{D_H} \geq 10{,}000$$

A correlation for fully developed turbulent flow by Dittus and Boelter (1930) and also found in Kreith and Bohn (2001) has served several generations of heat transfer practitioners.

$$\boxed{\overline{Nu}_{fD} = 0.023 Re_D^{0.8} Pr^{0.4}}$$ Fully developed turbulent (7.17)
$$T_s > T_f$$

$$\boxed{\overline{Nu}_{fD} = 0.023 Re_D^{0.8} Pr^{0.3}}$$ Fully developed turbulent (7.18)
$$T_s < T_f$$

where all properties are evaluated at the mean fluid temperature T_f and

$$0.5 < Pr < 120; \quad 6000 < Re_D < 10^7 : L/D > 60$$

Ducts shorter than $L/D < 60$ require a correction to the fully developed Nu_D (Kreith and Bohn, 2001):

$$\boxed{\begin{array}{c} \dfrac{\overline{Nu}_D}{\overline{Nu}_{fD}} = 1 + \dfrac{24}{Re_D^{0.23}}\left(\dfrac{L}{D}\right)^b \\[2ex] b = 2.08 x 10^{-6} Re_D - 0.815 \end{array}} \qquad 2 < \left(\dfrac{L}{D}\right) < 20 \quad (7.19)$$

$$\boxed{\dfrac{\overline{Nu}_D}{\overline{Nu}_{fD}} = 1 + \left(\dfrac{6D}{L}\right)} \qquad 20 < \left(\dfrac{L}{D}\right) < 60 \quad (7.20)$$

and of course Eqs. (7.17) to (7.20) use a hydraulic diameter when the duct is noncircular in cross-section.

7.7 APPLICATION EXAMPLE: TWO-SIDED EXTRUDED HEAT SINK

We shall continue the analysis of the heat sink that we studied in Section 4.4 for airflow head loss, except that the current topic is the temperature rise of the heat sink above the inlet air temperature. The heat sink is illustrated again in Figure 7.9.

The symmetric configuration was constructed so that airflow ducted to the heat sink is equally split between the upper and lower sinks. In addition, we should be assured that conduction from heat sources (not shown here) placed at the boundary between the heat sinks is equally divided between

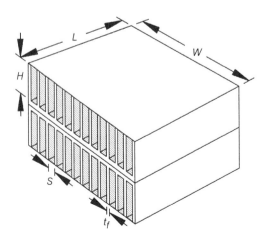

Figure 7.9. Application example: temperature rise calculation for two-sided, extruded heat sink.

the two heat sinks. Heat sink dimensions are: $W = 8.0$ in., $H = 1.0$ in., $S = 0.229$ in., $t_f = 0.1$ in., $L = 12$ in. and 6 in. Each heat sink has 12 channels, for a total of $N_c = 24$. The longer heat sink, $L = 12$ in., has a total heat dissipation $Q = 62.8$ W, and the shorter heat sink, $L = 6$ in., has a total heat dissipation $Q = 30$ W.

We shall present the head loss for both length configurations and see how the results compare with test data from the Nusselt number correlations: two for laminar flow and one for turbulent flow, outlined in Sections 7.4 to 7.6. Detailed calculations follow for the value of the total airflow $G = 26$ CFM and length $L = 12$ in. Results are tabulated and plotted for several other values of G and both lengths.

Since our heat sink is shrouded, the hydraulic diameter uses a four-sided perimeter.

$$D_H = \frac{4SH}{2(S+H)} = \frac{2SH}{(S+H)} = \frac{2(0.229\,\text{in.})(1.0\,\text{in.})}{(0.229\,\text{in.}+1.0\,\text{in.})} = 0.373\,\text{in.}$$

$$A_{c-Total} = N_c HS = (24)(1.0\,\text{in.})(0.229\,\text{in.}) = 5.55$$

$$Re_{D_H} = \frac{V_f D_H}{5\nu} = \frac{(G/A_{c-Total})D_H}{5\nu} = \frac{\left[26\,\text{CFM}/(5.55\,\text{in.}/144)\right](0.373\,\text{in.})}{5(0.029)}$$

$$Re_{D_H} = \frac{(680.8\,\text{ft/min})(0.373\,\text{in.})}{5(0.029)} = 1751$$

The adjusted Kays and Crawford Nusselt number from Eq. (7.13), Figure 7.6 for a channel aspect ratio $1.0/0.229 = 4.37$ is

$$\overline{Nu}_{D_H-Adj.} = \left(\frac{\overline{Nu}_{fD-Rect.}}{\overline{Nu}_{fD-Circ.}}\right)\left\{3.66 + \frac{0.104\left[\left(Re_{D_H}Pr\right)/(L/D_H)\right]}{1+0.016\left[\left(Re_{D_H}Pr\right)/(L/D_H)\right]^{0.8}}\right\}$$

$$\overline{Nu}_{D_H-Adj.} = \left(\frac{4.7}{3.66}\right)\left\{3.66 + \frac{0.104\left[(1751)(0.71)/(12.0/0.373)\right]}{1+0.016\left[(1751)(0.71)/(12.0/0.373)\right]^{0.8}}\right\} = 8.68$$

$$h_{Lam} = (k_{Air}/D_H)\overline{Nu}_{fD-Adj.} = (6.6\times10^{-4}/0.373)(8.68) = 0.0154$$

It would be a mistake not to consider fin efficiency so we shall calculate it here, beginning with a calculation of the single fin *primitives* R_k and R_c.

$$R_k = H/(k_{fin}Lt_f) = 1.0\,\text{in.}/\left[(5.0\,\text{W/in.}\cdot\,^\circ\text{C})(12.0\,\text{in.})(0.1\,\text{in.})\right] = 0.167\,^\circ\text{C/W}$$

$$R_c = 1/(2h_{Lam}LH) = 1/\left[2(0.0154)(12.0)(1.0)\right] = 2.71\,^\circ\text{C/W}$$

$$\eta = \sqrt{\frac{R_c}{R_k}}\,\tanh\sqrt{\frac{R_k}{R_c}} = \sqrt{\frac{2.71}{0.167}}\,\tanh\sqrt{\frac{0.167}{2.71}} = 0.98$$

which indicates a high efficiency and will be even closer to a unity value at smaller airflow values. The convection-only resistance for laminar flow is

$$A_S = (N_f -1)(2H + S)L = (25-1)(2.0\,\text{in.}+0.229\,\text{in.})(12.0\,\text{in.}) = 642.0\,\text{in.}^2$$

$$R_{Laminar} = 1/(\eta h_{Lam}A_S) = 1/\left[(0.982)(0.0154)(642.0)\right] = 0.104\,^\circ\text{C/W}$$

Now we use Eqs. (7.8) and (7.10) to calculate a thermal resistance referenced to the inlet air

$$\dot{m}c_P\left[\text{W}/^\circ\text{C}\right] = \frac{V\left[\text{ft/min}\right]A_{c-Total}\left[\text{in.}^2\right]}{262} = \frac{(680.0)(5.55)}{262} = 14.3\,\text{W}/^\circ\text{C}$$

$$\beta = 1 \Big/ \Big[\big(R_{Lam} \big) \Big(\dot{m} c_P \Big) \Big] = 1/(0.104)(14.3) = 0.676$$

$$R_I = R_{Lam} \, \beta \big/ \big(1 - e^{-\beta} \big) = (0.104)(0.676) \big/ \big(1 - e^{-0.676} \big) = 0.142 \, ^\circ\text{C/W}$$

The average heat sink temperature rise above the inlet air temperature is

$$\Delta T_{Sink-InletAir} = R_I Q = \big(0.142 \, ^\circ\text{C/W} \big)\big(62.8 \, \text{W} \big) = 8.94 \, ^\circ\text{C}$$

Now we solve the same problem using the Muzychka and Yovanovich correlation from Section 7.5. We calculate the Reynolds number differently.

$$A_{c-Total} = N_c HS = (24)(1.0\,\text{in.})(0.229\,\text{in.}) = 5.55\,\text{in.}$$

$$Re_{\sqrt{A}} = \frac{V_f \sqrt{A}}{5\nu} = \frac{\big(G/A_{c-Total} \big) \sqrt{HS}}{5\nu} = \frac{\Big[26\,\text{CFM} \big/ \big(5.55\,\text{in.}^2 /144 \big) \Big] \sqrt{(1.0\,\text{in.})(0.229\,\text{in.})}}{5(0.029)} = 2247$$

Muzychka and Yovanovich require the apparent friction factor from Eq. (4.3).

$$\varepsilon = S/H = 0.229/1.0 = 0.229, \qquad g = 1 \Big/ \Big[1.086957^{1-\varepsilon} \Big(\sqrt{\varepsilon} - \varepsilon^{\frac{3}{2}} \Big) + \varepsilon \Big] = 1.606$$

$$zPlus = L \big/ \big(\sqrt{A} Re_{\sqrt{A}} \big) = L \big/ \big(\sqrt{HS} Re_{\sqrt{A}} \big) = 12.0 \Big/ \Big[\sqrt{(1.0)(0.229)}(2247) \Big] = 0.011$$

$$f_{app} = \frac{1}{Re_{\sqrt{A}}} \Bigg[\bigg(\frac{3.44}{\sqrt{zPlus}} \bigg)^2 + \big(8\pi g \big)^2 \Bigg]^{\frac{1}{2}} = \frac{1}{2247} \Bigg[\bigg(\frac{3.44}{\sqrt{0.011}} \bigg)^2 + \big(8\pi \cdot 1.606 \big)^2 \Bigg]^{\frac{1}{2}} = 0.023$$

The total convecting surface area, fin efficiency, and flow conductance were previously calculated as $A_S = 354.0\,\text{in.}^2$, $\eta = 0.98$, and $\dot{m} c_p = 18.72\,\text{W/}^\circ\text{C}$.

We use Eq. (7.16) to calculate the Nusselt number, requiring the constants $C_1 = 1.5$, $C_2 = 0.409$, $C_3 = 3.01$, and $\gamma = 0.1$ from Table 6.2.

$$Nu_{\sqrt{A}} = \Bigg\{ \Bigg[C_1 C_2 \bigg(\frac{f_{app} Re_{\sqrt{A}}}{\big(L \big/ \big(\sqrt{A} Re_{\sqrt{A}} Pr \big) \big)} \bigg)^{1/3} \Bigg]^5 + \Big[C_3 \big(f_{app} Re_{\sqrt{A}} \big/ \big(8\sqrt{\pi} \varepsilon^\gamma \big) \big) \Big]^5 \Bigg\}^{1/5}$$

$$Nu_{\sqrt{A}} = \Bigg\{ \Bigg[C_1 C_2 \bigg(f_{app} Re_{\sqrt{A}} \Big/ \Big(\frac{zPlus}{Pr} \Big) \bigg)^{1/3} \Bigg]^5 + \Big[C_3 \big(f_{app} Re_{\sqrt{A}} \big/ 8\sqrt{\pi} \varepsilon^\gamma \big) \Big]^5 \Bigg\}^{1/5}$$

$$Nu_{\sqrt{A}} = \left\{ \begin{array}{l} \Bigg[(1.5)(0.409) \bigg((0.023)(2247) \Big/ \Big(\frac{0.011}{0.71} \Big) \bigg)^{1/3} \Bigg]^5 \\[4mm] + \Big[(3.01) \big((0.023)(2247) \big/ 8\sqrt{\pi}\,(0.229)^{0.1} \big) \Big]^5 \end{array} \right\}^{1/5} = 13.21$$

$$h_{Lam} = \bigg(\frac{k_{Air}}{\sqrt{A}} \bigg) \overline{Nu}_{\sqrt{A}} = \bigg(\frac{6.6 \times 10^{-4}}{\sqrt{(1.0\,\text{in.})(0.229\,\text{in.})}} \bigg)(13.21) = 0.0180 \, \text{W} \big/ \big(\text{in.}^2 \cdot {}^\circ\text{C} \big)$$

The average heat sink temperature rise above the inlet air temperature is calculated as follows:

$$R_{Lam} = 1/\left(\eta h_{Lam} A_s\right) = 1/\left[(0.98)(0.018)(642.0)\right] = 0.088\,^{\circ}\mathrm{C/W}$$

$$\beta = 1\Big/\left[\left(R_{Lam}\right)\left(\dot{m} c_P\right)\right] = 1/(0.088)(14.29) = 0.799$$

$$R_I = R_{Lam}\,\beta\Big/\left(1 - e^{-\beta}\right) = (0.088)(0.799)\Big/\left(1 - e^{-0.799}\right) = 0.127\,^{\circ}\mathrm{C/W}$$

$$\Delta T_{Sink-InletAir} = R_I Q = \left(0.127\,^{\circ}\mathrm{C/W}\right)(62.8\,\mathrm{W}) = 7.99\,^{\circ}\mathrm{C}$$

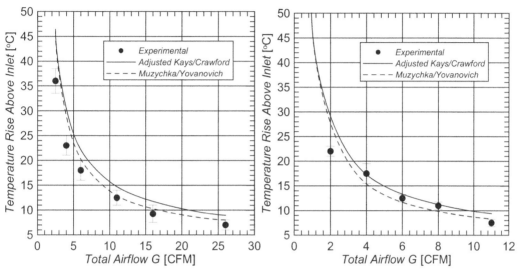

Figure 7.10 (a). Heat sink temperature rise (above inlet) for $L = 12$ in. Vertical bars represent heat sink minimum-to-maximum temperature rise.

Figure 7.10 (b). Heat sink temperature rise (above inlet) for $L = 6$ in. Vertical bars represent heat sink minimum-to-maximum temperature rise.

We see an 11% discrepancy between the 7.99°C that we calculated with the Muzychka and Yovanovich correlation and the previous result of 8.95°C using the adjusted Kays and Crawford correlation. Tables 7.3 and 7.4 list some of the intermediate and final temperature calculations for heat sink lengths of $L = 12$ and 6 in. using both correlations. The temperature rises are plotted in Figure 7.10.

Tables 7.3 and 7.4 serve two purposes: (1) they display data that is not plotted, but is nevertheless of interest in discussing the plots; (2) if you have difficulty obtaining sensible answers to problems of this type, you can check your work by calculating some of the problems yourself and comparing with the listed results. One significant point is that interfin air speeds much less than about $V_f = 50$ ft/min are perhaps in a mixed convection mode, i.e., both forced and natural convection effects may exist. This suggests that in the case of both the longer and shorter heat sinks, calculated temperatures at volumetric flow rates much less than about $G = 4$ CFM may not be reliable. It seems quite clear that the Muzychka and Yovanovich correlation results in better accuracy.

Another point of interest is that when we look at the β column of Tables 7.3 and 7.4, we see that β decreases with increasing flow rate, which is consistent with the definition. At large values of β, much of the total temperature rise is due to the low mass flow rate.

7.8 FLOW BYPASS EFFECTS ACCORDING TO JONSSON AND MOSHFEGH

In Chapter 4 we became aware of an empirical correlation for head loss analysis of heat sinks in a flow bypass situation (Jonsson and Moshfegh, 2001). These same authors also offer a bypass correlation for the Nusselt number of several different fin styles. The geometry is illustrated in

Table 7.3. Summary of intermediate and final calculations for Application Example 7.7. Correlation used: Adjusted Kays and Crawford.

$L = 12.0$ in., $Q = 62.8$ W

G [ft^3/min]	V_f	Re_D [ft/min]	Nu_D	h_{Lam} [W/(in.^{2}·°C)]	η	$\dot{m}c_p$ [W/°C]	β	R_l [°C/W]	ΔT [°C]
1.0	26.2	67.3	4.89	0.00866	0.989	0.55	10.0	1.82	114.3
2.5	65.4	168.3	5.17	0.00915	0.988	1.37	4.23	0.740	41.40
4.0	104.7	269.3	5.44	0.00964	0.987	2.20	2.78	0.485	30.46
6.0	157.1	404.0	5.79	0.00987	0.982	3.30	1.97	0.352	22.14
11.0	288.0	740.6	6.60	0.01168	0.985	6.05	1.22	0.235	14.73
16.0	419.0	1077.0	7.34	0.0130	0.983	8.79	0.933	0.188	11.78
26.0	680.8	1751.0	8.68	0.0154	0.980	14.29	0.676	0.142	8.94

$L = 6.0$ in., $Q = 30.0$ W

G [ft^3/min]	V_f	Re_D [ft/min]	Nu_D	h_{Lam} [W/(in.^{2}·°C)]	η	$\dot{m}c_p$ [W/°C]	β	R_l [°C/W]	ΔT [°C]
1.0	26.2	67.3	5.08	0.00899	0.988	0.55	5.19	1.83	54.89
2.0	52.4	134.7	5.44	0.00964	0.987	1.10	2.78	0.970	29.1
4.0	104.7	269.3	6.12	0.01084	0.986	2.20	1.56	0.576	17.28
6.0	157.1	404.0	6.75	0.01195	0.984	3.30	1.15	0.445	13.34
8.0	209.5	538.6	7.34	0.0130	0.983	4.40	0.933	0.375	11.25
11.0	288.0	740.6	8.16	0.01445	0.981	6.05	0.753	0.313	9.38

Table 7.4. Summary of intermediate and final calculations for Application Example 7.7. Correlation used: Muzychka and Yovanovich.

$L = 12.0$ in., $Q = 62.8$ W

G [ft³/min]	V_f [ft/min]	$Re_{\sqrt{A}}$	$Nu_{\sqrt{A}}$	h_{Lam} [W/(in.²·°C)]	f_{app}	η	$\dot{m}c_p$ [W/°C]	β	F_l [°C/W]	ΔT [°C]
1.0	26.2	86.4	10.06	0.0140	0.473	0.982	0.55	15.9	1.19	114.0
2.5	65.4	216.0	10.25	0.0140	0.193	0.982	1.37	6.48	0.729	45.78
4.0	104.7	345.81	10.45	0.0140	0.122	0.981	2.20	4.13	0.462	29.00
6.0	157.1	518.6	10.70	0.0150	0.083	0.981	3.30	2.82	0.322	20.25
11.0	288.0	950.8	11.36	0.0160	0.048	0.980	6.05	1.63	0.206	12.92
16.0	419.0	1383.0	12.00	0.0170	0.035	0.980	8.79	1.18	0.64	10.30
26.0	680.8	2247.0	13.21	0.0180	0.023	0.979	14.29	0.799	0.27	7.99

$L = 6.0$ in., $Q = 30.0$ W

G [ft³/min]	V_f [ft/min]	$Re_{\sqrt{A}}$	$Nu_{\sqrt{A}}$	h_{Lam} [W/(in.²·°C)]	f_{app}	η	$\dot{m}c_p$ [W/°C]	β	λ_l [°C/W]	ΔT [°C]
1.0	26.2	86.4	10.19	0.0140	0.479	0.982	0.55	8.05	1.2	54.60
2.0	52.4	172.9	10.45	0.0140	0.245	0.981	1.10	4.13	0.925	27.74
4.0	104.7	345.8	10.79	0.0150	0.128	0.981	2.20	2.17	0.514	15.41
6.0	157.1	518.6	11.49	0.0160	0.089	0.979	3.30	1.51	0.389	11.68
8.0	209.5	691.5	12.00	0.0170	0.069	0.979	4.40	1.18	0.328	9.84
11.0	288.0	950.8	12.73	0.0180	0.053	0.977	6.05	0.911	0.277	8.30

Figure 7.11. You might wish to review the limitations of this model as outlined in Chapter 4. We will, however, restate the geometric limitations.

$$0.33 \le W/W_d \le 0.84, \qquad 0.33 \le H/H_d \le 1.0; \qquad 2000 < Re_{D_H} < 16500$$

The Reynolds number is defined in consistent units for a duct total volumetric airflow rate G,

$$Re_{D_H} = \left(\frac{G}{A_b + A_f} \right)\left(\frac{2W_d H_d}{W_d + H_d} \right)\Big/ \nu$$

The Nusselt number is

$$Nu_L = C_1 \left(\frac{Re_{D_H}}{1000} \right)^{m_1} \left(\frac{W_d}{W} \right)^{m_2} \left(\frac{H_d}{H} \right)^{m_3} \left(\frac{S}{H} \right)^{m_4} \left(\frac{t_f, d}{H} \right)^{m_5} \quad \begin{array}{l}\text{Experimental} \quad (7.21) \\ \text{bypass correlation}\end{array}$$

with C_1 and the five different ms are listed in Table 7.5. It should be mentioned that the correlation presented in this section should not be applied to heat sinks that are either much larger or much smaller (in width and length) than about 2 in., in order to stay within the range of the study parameters.

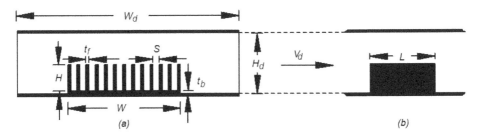

Figure 7.11. Geometry for flow bypass Nusselt number according to Jonsson and Moshfegh. (a) Front view. (b) Side view.

Table 7.5. Formula parameters for flow bypass according to Jonsson and Moshfegh.

Fin Style	C_1	m_1	m_2	m_3	m_4	m_5
Plate fin	88.28	0.6029	0.1098	0.5632	0.08713	0.4139
In-line strip fin	90.88	0.7065	-0.07122	-0.6485	0.04164	0.4700
Staggered strip fin	105.4	0.7210	-0.08695	-0.6558	0.03624	0.5327
In-line circular pin fin	169.3	0.6422	-0.1528	-0.6382	0.2626	0.2772
Staggered circular pin fin	219.6	0.6432	-0.1793	-0.6410	0.1119	0.4500
In-line square pin fin	108.3	0.6642	-0.1713	-0.6434	0.3537	0.1045
Staggered square pin fin	153.1	0.6736	-0.1943	-0.6677	0.2075	0.3102

7.9 APPLICATION EXAMPLE: HEAT SINK IN A CIRCUIT BOARD CHANNEL USING THE FLOW BYPASS METHOD OF LEE

The flow bypass problem that we studied in Chapter 4 resulted in a predicted fin channel air speed $V_f = 267$ ft/min using the head loss prediction from the *Handbook of Heat Transfer*. Now we will apply the adjusted Kays and Crawford Nusselt number correlation to predict the thermal resistance.

We begin by calculating the hydraulic diameter, but using a wetted perimeter that includes two fin sides and the interfin space. Then we calculate the Reynolds number for this D_H.

$$D_H = \frac{4SH}{(S+2H)} = \frac{4SH}{(S+2H)} = \frac{2(0.225\,\text{in.})(0.995\,\text{in.})}{[0.225\,\text{in.}+2(0.995\,\text{in.})]} = 0.404\,\text{in.}$$

$$A_{c-Total} = N_c HS = (12)(0.995\,\text{in.})(0.225\,\text{in.}) = 2.68\,\text{in.}^2$$

$$Re_{D_H} = \frac{V_f D_H}{5\nu} = \frac{[267\,\text{ft/min}](0.404\,\text{in.})}{5(0.029)} = 745$$

The adjusted Kays and Crawford Nusselt number from Eq. (7.13) for a channel aspect ratio $0.995/0.225 = 4.42$ is

$$\bar{N}u_{D_H-Adj.} = \left(\frac{\bar{N}u_{fD-Rect.}}{\bar{N}u_{fD-Circ.}}\right)\left[3.66 + \frac{0.104\left(\dfrac{Re_{D_H}Pr}{L/D_H}\right)}{1+0.016\left(\dfrac{Re_{D_H}Pr}{L/D_H}\right)^{0.8}}\right]$$

$$\bar{N}u_{D_H-Adj.} = \left(\frac{4.70}{3.66}\right)\left\{3.66 + \frac{0.104\left[\dfrac{(745)(0.71)}{3.0/0.404}\right]}{1+0.016\left[\dfrac{(745)(0.71)}{3.0/0.404}\right]^{0.8}}\right\} = 11.10$$

$$h_{Lam} = (k_{Air}/D_H)\bar{N}u_{fD-Adj.} = (6.6\times10^{-4}/0.404)(11.10) = 0.0181\,\text{W}/(\text{in.}^2\cdot{}^\circ\text{C})$$

Now we consider the fin efficiency, beginning with a calculation of the single fin *primitives* R_k and R_c.

$$R_k = H/(k_{fin}Lt_f) = 0.995\,\text{in.}/\left[(5.0\,\text{W}/(\text{in.}\cdot{}^\circ\text{C}))(3.0\,\text{in.})(0.1\,\text{in.})\right] = 0.667\,{}^\circ\text{C/W}$$

$$R_c = 1/(2h_{Lam}LH) = 1/[2(0.0181)(3.0)(0.995)] = 9.24\,{}^\circ\text{C/W}$$

$$\eta = \sqrt{\frac{R_c}{R_k}}\tanh\sqrt{\frac{R_k}{R_c}} = \sqrt{\frac{9.24}{0.667}}\tanh\sqrt{\frac{0.667}{9.24}} = 0.977$$

which is very close to unity.

The convection-only resistance for laminar flow is

$$A_s = 2N_f LH + (N_f-1)SL = 2(13)(3.0\,\text{in.})(0.995\,\text{in.}) + (13-1)(0.225\,\text{in.})(3.0\,\text{in.}) = 85.7\,\text{in.}^2$$

$$R_{Lam} = 1/(\eta h_{Lam}A_s) = 1/[(0.977)(0.0181)(85.7)] = 0.659\,{}^\circ\text{C/W}$$

Now we use Eqs. (7.8) and (7.10) to calculate a thermal resistance referenced to the inlet air.

$$A_{c-Total} = (N_f-1)SH = (13-1)(0.225\,\text{in.})(0.995\,\text{in.}) = 2.69\,\text{in.}^2$$

$$\dot{m}c_P\left[\frac{\text{W}}{{}^\circ\text{C}}\right] = \frac{V_f\left[\dfrac{\text{ft}}{\text{min}}\right]A_{c-Total}\left[\text{in.}^2\right]}{262} = \frac{(267)(2.69)}{262} = 2.74\,\text{W}/{}^\circ\text{C}$$

$$\beta = 1/\left[(R_{Lam})(\dot{m}c_P)\right] = 1/(0.659)(2.74) = 0.554$$

$$R_I = R_{Lam} \, \beta \big/ \left(1 - e^{-\beta}\right) = (0.659)(0.554) \big/ \left(1 - e^{-0.554}\right) = 0.858 \,^\circ\text{C/W}$$

Next we solve the same problem using the Muzychka and Yovanovich from Section 7.5. The Reynolds number is calculated differently. If you look back at this example in Chapter 4, you will note that we had a predicted $V_f = 236$ ft/min.

$$Re_{\sqrt{A}} = \frac{V_f \sqrt{A}}{5\nu} = \frac{V_f \sqrt{HS}}{5\nu} = \frac{236\sqrt{(0.995\,\text{in.})(0.225\,\text{in.})}}{5(0.029)} = 770$$

Muzychka and Yovanovich require the apparent friction factor from Eq. (4.3).

$$\varepsilon = S/H = 0.225/0.995 = 0.225, \qquad g = 1 \Big/ \left[1.086957^{1-\varepsilon}\left(\sqrt{\varepsilon} - \varepsilon^{\frac{3}{2}}\right) + \varepsilon\right] = 1.616$$

$$zPlus = L \big/ \left(\sqrt{A}Re_{\sqrt{A}}\right) = L \big/ \left(\sqrt{HS}\,Re_{\sqrt{A}}\right) = 3.0 \Big/ \left[\sqrt{(0.995)(0.225)}\,(770)\right] = 0.0082$$

$$f_{app} = \frac{1}{Re_{\sqrt{A}}}\left[\left(\frac{3.44}{\sqrt{zPlus}}\right)^2 + (8\pi g)^2\right]^{1/2} = \frac{1}{770}\left[\left(\frac{3.44}{\sqrt{0.0082}}\right)^2 + (8\pi \times 1.616)^2\right]^{1/2} = 0.072$$

We have already calculated the total convecting surface area $A_S = 85.79$ in.^{2}; the fin efficiency is easily calculated as $\eta = 0.977$, and $\dot{m}c_p = 2.74$ W/$^\circ$C. Equation (7.16) is used to calculate the Nusselt number, requiring the constants $C_1 = 1.5$, $C_2 = 0.409$, $C_3 = 3.01$, and $\gamma = 0.1$ from Table 6.2.

$$Nu_{\sqrt{A}} = \left\{\left[C_1 C_2 \left(\frac{f_{app}Re_{\sqrt{A}}}{\left(\dfrac{L}{\sqrt{A}Re_{\sqrt{A}}Pr}\right)}\right)^{1/3}\right]^5 + \left[C_3\left(\frac{f_{app}Re_{\sqrt{A}}}{8\sqrt{\pi}\varepsilon^\gamma}\right)\right]^5\right\}^{1/5}$$

$$= \left\{\left[C_1 C_2 \left(\frac{f_{app}Re_{\sqrt{A}}}{\left(\dfrac{zPlus}{Pr}\right)}\right)^{1/3}\right]^5 + \left[C_3\left(\frac{f_{app}Re_{\sqrt{A}}}{8\sqrt{\pi}\varepsilon^\gamma}\right)\right]^5\right\}^{1/5}$$

$$= \left\{\left[(1.5)(0.409)\left(\frac{(0.072)(770)}{\left(\dfrac{0.0082}{0.71}\right)}\right)^{1/3}\right]^5 + \left[(3.01)\left(\frac{(0.072)(770)}{8\sqrt{\pi}\,(1.616)^{0.1}}\right)\right]^5\right\}^{1/5}$$

$$Nu_{\sqrt{A}} = 14.30$$

$$h_{Lam} = \left(\frac{k_{Air}}{\sqrt{A}}\right)\overline{Nu}_{\sqrt{A}} = \left(\frac{6.6\times10^{-4}}{\sqrt{(0.995\,\text{in.})(0.225\,\text{in.})}}\right)(14.30) = 0.020 \,\text{W}\big/\left(\text{in.}^2 \cdot {}^\circ\text{C}\right)$$

The heat sink thermal resistance referenced to the inlet air temperature is calculated:

$$R_{Lam} = 1/(\eta h_{Lam} A_s) = 1/[(0.977)(0.020)(85.71)] = 0.600 \,^\circ C/W$$

$$\beta = 1/\left[(R_{Lam})\left(\overset{\circ}{m} c_P\right)\right] = 1/(0.600)(2.42) = 0.689$$

$$R_I = R_{Lam} \,\beta/(1 - e^{-\beta}) = (0.600)(0.689)/(1 - e^{-0.689}) = 0.83 \,^\circ C/W$$

Our last step in this flow bypass problem is to calculate the heat sink thermal resistance using the method of Jonsson and Moshfegh. We begin by calculating the Reynolds number according to their definition in Section 7.8.

$$Re_{D_H} = \left(\frac{G}{A_b + A_{c-Total}}\right)\left(\frac{2W_d H_d}{W_d + H_d}\right)\Big/\nu$$

$$A_b = W_d H_d - WH_T = (10.0\,\text{in.})(2.0\,\text{in.}) - (4.0\,\text{in.})(1.31\,\text{in.}) = 14.76\,\text{in.}^2$$

In mixed units we obtain

$$Re_{D_H} = \left(\frac{G}{A_b + A_{c-Total}}\right)\left(\frac{2W_d H_d}{W_d + H_d}\right)\Big/(5\nu)$$

$$Re_{D_H} = \left[\frac{50\,\text{ft}^3/\text{min}}{[(14.76 + 2.69)/144]}\right]\left[\frac{2(10.0\,\text{in.})(2.0\,\text{in.})}{10.0\,\text{in.} + 2.0\,\text{in.}}\right]\Big/[5(0.029)] = 9487$$

which along with the ratios, $W/W_d = 4.0/10.0 = 0.4$, $H/H_d = 0.995/2.0 = 0.50$, clearly meets the conditions $0.33 \le W/W_d \le 0.84$, $0.33 \le H/H_d \le 1.0$, $2000 < Re_{D_H} < 16500$. We calculate the Nusselt number according to Eq. (7.21) and Table 7.5.

$$Nu_L = C_1 \left(\frac{Re_{D_H}}{1000}\right)^{m_1}\left(\frac{W_d}{W}\right)^{m_2}\left(\frac{H_d}{H}\right)^{m_3}\left(\frac{S}{H}\right)^{m_4}\left(\frac{t_f, d}{H}\right)^{m_5}$$

$$Nu_L = 88.28\left(\frac{9487}{1000}\right)^{0.6029}\left(\frac{10.0}{4.0}\right)^{-0.1098}\left(\frac{2.0}{1.0}\right)^{-0.5623}\left(\frac{0.225}{0.995}\right)^{0.08731}\left(\frac{0.1}{0.995}\right)^{0.4139} = 71.045$$

$$h = (k_{Air}/L) Nu_L = (6.6 \times 10^{-4}/3.0)(71.045) = 0.016$$

$$R_{Sink} = 1/(hA_s) = 1/[(0.016)(85.71)] = 0.71 \,^\circ C/W$$

We can summarize our results from the three correlations.

Adjusted Kays and Crawford	0.86 °C/W
Muzychka and Yovanovich	0.83 °C/W
Jonsson and Moshfegh	0.71 °C/W

so that, at least for this problem, the three methods give results that are in excellent agreement.

A typical vendor test setup is a wind tunnel where all of the tunnel air is ducted into the heat sink using minimal, if any, bypass area. The heat sink studied in this section bears a strong resemblance to an older device in a heat sink vendor's catalog. Although we don't really know the vendor test setup, the ducting method just mentioned is likely. It is also typical for a heat sink vendor to rate the heat sink thermal resistance vs. the average approach velocity. Table 7.6 is a summary of additional calculations for this heat sink using the Adjusted Kays and Crawford and also the Muzychka and

Yovanovich correlations. The resistance R_I is also plotted in Figure 7.12 for both correlations and the vendor data. The agreement between the two correlations is very good, but the agreement with the vendor data is not quite so good. In any case, don't take each data point as precise. The approach velocity calculated from the interfin velocity can be calculated in slightly different ways, e.g., $V_a = V_f (A_{c\text{-}Total}/WH)$. The subject of thermal spreading due to a discrete heat source is studied in Chapter 12 where in this same problem we shall see that the analytical and experimental results compare a bit more favorably when the effects of a heat producing power transistor are considered.

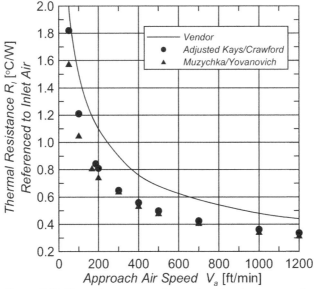

Figure 7.12. Thermal resistance referenced to inlet air vs. approach air speed for heat sink analyzed for Application Example 7.9.

7.10 IN-LINE AND STAGGERED PIN FIN HEAT SINKS

Khan, Culham, and Yovanovich (2005) published a correlation for in-line and staggered, cylindrical, pin fin heat sinks. The variables are defined for the geometry in Figure 7.13 as follows.

$$P_L = L/N_L, \; p_L = P_L/D, \; P_T = W/N_T, \; p_T = P_T/D, \; P_D = \sqrt{P_L^2 + (P_T/2)^2}, \; p_D = P_D/D$$

N_L = Number of rows in length direction, N_T = Number of columns in width direction

$$V_{Max} = \text{Maximum of } \left[p_T V_a / (p_T - 1), p_T / (p_D - 1) V_a \right]$$

$$Re_D = D V_{Max}/v : \text{consistent units}; Re_D = D V_{Max}/5v : \text{mixed English units}$$

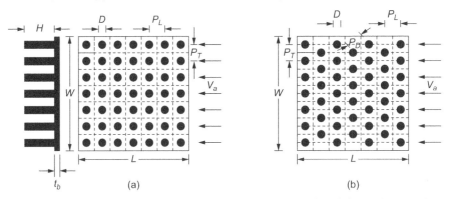

Figure 7.13. Pin fin arrays: (a) in-line and, (b) staggered. H = pin height (not illustrated).

Table 7.6. Summary of intermediate and final calculations for heat sink studied in Application Example 7.9.

Adjusted Kays and Crawford Nu Correlation

V_a [ft/min]	V_f [ft/min]	Re_D	Nu_D	h_{Lam} [W/(in.²·°C)]	η	$\dot{m}c_p$ [W/°C]	β	R_c [°C/W]	R_I [°C/W]
50.6	75	209	6.97	0.01138	0.985	0.769	1.25	1.04	1.82
100	148	412.7	8.75	0.01428	0.982	1.518	0.792	0.832	1.21
186	275	766.8	11.24	0.01836	0.976	2.82	0.545	0.651	0.844
200	296	825.3	11.61	0.01895	0.976	3.04	0.522	0.631	0.810
300	445	1241	13.86	0.02263	0.971	4.56	0.413	0.531	0.648
400	592	1651	15.70	0.02562	0.967	6.07	0.350	0.471	0.558
500	740	2063	17.27	0.02819	0.964	7.59	0.307	0.429	0.498
700	1035	2886	18.84	0.03239	0.959	10.16	0.251	0.375	0.425
1000	1490	4154	22.87	0.03734	0.953	15.28	0.200	0.328	0.362
1200	1780	4963	24.43	0.0399	0.95	18.25	0.178	0.308	0.336

Muzychka and Yovanovich Nu Correlation

V_a [ft/min]	V_f [ft/min]	$Re_{\sqrt{A}}$	f_{app}	$Nu_{\sqrt{A}}$	h_{Lam} [W/(in.²·°C)]	η	$\dot{m}c_p$ [W/°C]	β	R_c [°C/W]	R_I [°C/W]
50.6	75	245	0.188	11.46	0.01598	0.979	0.769	1.745	0.745	1.576
100	148	482.9	0.105	12.81	0.01786	0.977	1.518	0.986	0.668	1.051
167	248	809	0.069	14.50	0.02022	0.974	2.543	0.664	0.592	0.811
200	296	966	0.061	15.25	0.02130	0.973	3.035	0.548	0.564	0.745
300	445	1452	0.045	17.39	0.0243	0.976	4.563	0.442	0.496	0.641
400	592	1932	0.038	19.29	0.0269	0.966	6.07	0.367	0.449	0.536
500	740	2415	0.032	21.03	0.0293	0.963	7.59	0.319	0.413	0.482
700	1035	3377	0.026	24.13	0.0337	0.958	10.61	0.260	0.362	0.411
1000	1490	4862	0.021	28.25	0.0394	0.951	15.28	0.210	0.311	0.345
1200	1780	5844	0.019	30.60	0.0427	0.947	18.25	0.190	0.289	0.317

Khan et al. (2005) correctly distinguish between heat transfer coefficients h_b and h_{fin} for the base and pin fins, respectively, in Eqs. (7.22).

$$\boxed{\begin{aligned} h_b &= \frac{0.75k_f}{D}\sqrt{\frac{p_T-1}{N_L p_L p_T}}Re_D^{1/2}Pr^{1/3} \\ h_{fin} &= \frac{C_1 k_f}{D}Re_D^{1/2}Pr^{1/3} \end{aligned}}$$

Pin fins and base (7.22)

using the constants

$$C_1 = \left[0.2 + e^{(-0.55p_L)}\right]p_T^{0.285}p_L^{0.212} : \quad \text{In-line pins}$$

$$C_1 = \frac{0.61p_T^{0.091}p_L^{0.053}}{1 - 2e^{(-1.09p_L)}} : \quad \text{Staggered pins}$$

The areas and fin efficiency are given by

$$A_b = LW - N_T N_L \frac{\pi D^2}{4}, \quad A_{fin} = \pi DH; \quad m = \sqrt{\frac{4h_{fin}}{kD}}, \quad \eta_{fin} = \frac{\tanh(mH)}{mH}$$

and the total pins/base areas thermal resistances R_{th} are

$$R_{fin} = 1/h_{fin}A_{fin}\eta_{fin}, R_b = 1 \bigg/ \left[h_b\left(LW - N_T N_L \frac{\pi D^2}{4}\right)\right], R_{th} = 1 \bigg/ \left[\left(\frac{N_T N_L}{R_{fin}}\right) + \frac{1}{R_b}\right]$$

7.11 APPLICATION EXAMPLE: THERMAL RESISTANCE OF A PIN FIN HEAT SINK USING THE CORRELATION AND PROBLEM EXAMPLE FROM KHAN ET AL.

Khan et al. (2005) did their readers an important service by providing a numerical example using their formulae. I suggest that both prospective and active authors emulate this practice in their own work. Most journal publications are purposely abridged and we are not always positive that we know how the authors actually intend us to use their work. The following paragraphs follow the example of Khan et al., but we shall continue the practice of using mixed English units in this book. This is a continuation of the head loss problem in Section 4.6.

The in-line pin fin problem data expressed consistent with Figure 7.13 is: $W = L = 1.0$ in., $D = 0.07874$ in., $H = 0.394$ in., $t_b = 0.0787$ in., $N_T = 7$, $N_L = 7$, $V_a = 590.55$ ft/min (approach air speed), $k = 4.572$ W/(in. · °C) for the pins and base, $v = 0.0245$ in.^{2}/s, $Pr = 0.71$, and $T_A = 27$°C (ambient air temperature). The following calculations are required to arrive at the combined thermal resistance of the pins and base:

$$P_T = P_L = \frac{W}{N_T} = \frac{1.0}{7} = 0.143; p_T = p_L = \frac{P_T}{D} = \frac{0.143}{0.07874} = 1.814; \sigma = \frac{p_T-1}{p_T} = \frac{1.814-1}{1.814} = 0.449$$

$$V_{Max} = \max(\frac{p_T}{p_T-1}V_a, \frac{p_T}{p_D-1}V_a) = \max\left(1.316\times10^3, 1.042\times10^3\right) = 1.316\times10^3 \text{ ft/min}$$

The heat transfer coefficients are calculated next.

$$Re_D = \frac{DV_{Max}}{5v} = \frac{(0.07874)(1.316\times10^3)}{5(0.0245)} = 845.75$$

$$C_1 = \left[0.2 + e^{(-0.55p_L)}\right]p_T^{0.285}p_L^{0.212} = \left[0.2 + e^{(-0.55)(1.814)}\right](1.814)^{0.285}(1.814)^{0.212} = 0.765$$

$$h_b = \frac{0.75k_f}{D}\sqrt{\frac{p_T-1}{N_L p_L p_T}}Re_D^{1/2}Pr^{1/3} = \frac{0.75\left(6.604\times10^{-4}\right)}{0.07874}\sqrt{\frac{1.814-1}{(7)(1.814)(1.814)}} = 0.031\,\text{W}/\left(\text{in.}^2 \cdot {}^\circ\text{C}\right)$$

$$h_{fin} = \frac{C_1 k_f}{D} Re_D^{1/2} Pr^{1/3} = \frac{(0.765)(6.604 \times 10^{-4})}{0.07874}(845.76)^{1/2}(0.71)^{1/3} = 0.166 \, \text{W}/(\text{in.}^2 \cdot {}^\circ\text{C})$$

The areas and fin efficiency are

$$A_b = LW - N_T N_L \frac{\pi D^2}{4} = (1.0)(1.0) - (7)(7)\frac{\pi(0.07874)^2}{4} = 0.761 \, \text{in.}^2$$

$$A_{fin} = \pi DH = \pi(0.07874)(0.394) = 0.097 \, \text{in.}^2$$

$$m = \sqrt{\frac{4h_{fin}}{kD}} = \sqrt{\frac{4(0.166)}{(.572)(0.07874)}} = 1.36, \, \eta_{fin} = \frac{\tanh(mH)}{mH} = \frac{\tanh[(1.36)(0.394)]}{(1.36)(0.394)} = 0.914$$

The fin, base, and total thermal resistances, R_{fin}, R_b, and R_{th}, respectively, are

$$R_{fin} = 1/h_{fin}A_{fin}\eta_{fin}, R_b = 1\bigg/\left[h_b\left(LW - N_T N_L \frac{\pi D^2}{4}\right)\right], R_{th} = 1\bigg/\left[\left(\frac{N_T N_L}{R_{fin}}\right) + \frac{1}{R_b}\right]$$

$$R_{fin} = 1/h_{fin}A_{fin}\eta_{fin} = 1/[(0.166)(0.097)(0.914)] = 67.495 \, {}^\circ\text{C/W}$$

$$R_b = 1\bigg/\left[h_b\left(LW - N_T N_L \frac{\pi D^2}{4}\right)\right] = 1\bigg/\left\{(0.031)\left[(1.0)(1.0)-(7)(7)\frac{\pi(0.07874)^2}{4}\right]\right\} = 42.81 \, {}^\circ\text{C/W}$$

$$R_{th} = 1\bigg/\left[\left(\frac{N_T N_L}{R_{fin}}\right) + \frac{1}{R_b}\right] = 1\bigg/\left\{\left[\frac{(7)(7)}{67.495}\right] + \frac{1}{42.81}\right\} = 1.352 \, {}^\circ\text{C/W}$$

in agreement with Khan et al. (2005). There seems to be a dearth of empirical data for cylindrical pin fin heat sinks for us to check out the correlation. The reader is therefore advised to use the correlation with care.

7.12 LEE'S FLOW BYPASS ADAPTED TO NON-ZERO BYPASS RESISTANCE

The airflow analyses in Sections 4.8, 4.9, and thermal analysis in Section 7.9 were analyzed according to Lee's original theory, i.e., the head loss in the bypass region was negligible. In this section the zero bypass resistance is not ignored and the resultant adaptation of the theory is tested against the empirical correlation of Jonsson and Moshfegh (referred to as *J&M* from here on) for thermal resistance as outlined in Section 7.8. The work by J&M is readily used with a minimum of confusion and is considered here as a "laboratory test" which is best used with the exact device dimensions used by these two researchers. *J&M* used a heat sink with a specific L = 2.079 in., W = 2.079 in., t_f = 0.059 in., and H = 0.394 in. to 0.787 in. J&M constrained their experiments to the range of $0.33 \leq W/W_d \leq 0.84$; $0.33 \leq H/H_d \leq 1.0$; $2000 \leq Re_{DH} \leq 16500$ where Re_{DH} is defined in Sections 4.10 and 7.8. Eq. (7.21) states the flow bypass equations with built-in conversion constants required. Although all flow bypass calculations for the airflow portion were considered in Chapter 4, the "validation criteria" for the current problem is thermal resistance, thus both the airflow and thermal considerations are examined here.

$$a = \left[1 - \left(\frac{A_f}{A_b}\right)^2\right], b = 2\left(\frac{A_d}{A_b}\right)\left(\frac{A_f}{A_b}\right)V_d C_V, c = \left[\frac{2(\Delta h_f - \Delta h_b)/C_h}{\rho} - \left(\frac{A_d}{A_b}\right)^2 V_d^2 C_V^2\right]$$

$$V_f = \frac{-b + \sqrt{b^2 - 4ac}}{2a}\left(\frac{1}{C_V}\right) \qquad \text{Interfin air velocity} \quad (7.23)$$

$$V_b = \sqrt{V_d^2(A_d/A_b)^2 + V_f^2(A_f/A_b)^2 - 2V_d V_f(A_d A_f/A_b^2)}$$

$C_V = 5.08 \times 10^{-3}$, $C_h = 4.019 \times 10^{-3}$, Units: V [ft/min], Δh [in. H_2O], use $\rho = 1.18 \, \text{kg/m}^3$ for air

In this problem, dimensions of the device are as identical as possible to the J&M study: $W = L = 2.079$ in., $t_f = 0.059$ in., and $H = 0.787$ in., $H_d = 2.0$ in. Material constant values used are $Pr = 0.71$, $k_{Air} = 6.6\times10^{-4}$ W/(in. °C), and $v = 0.029$ in.^{2}/s. The duct width is varied from $W_d = 2.5$ in. to 20 in. to provide some variation in the bypass resistance. The number of fins on the heat sink are $N_f = 3, 5,$ and 10. Three different total channel volumetric air flow rates of $G = 10, 25,$ and 50 CFM are used. Results for all calculations are listed in Tables 7.7, 7.8, and 7.9.

The arithmetic is a bit tedious so a single problem is summarized here where $G = 25$ CFM, $N_f = 5$, and $W_d = 4$ in. The J&M method is illustrated first.

$$S = (W - N_f t_f)/(N_f - 1) = [2.079 - (5)(0.059)]/(5-1) = 0.446 \text{ in.}$$

$$W/W_d = 2.079/4.0 = 0.52, H/H_d = 0.787/2.0 = 0.39$$

$$A_f = (N_f - 1)SH = (5-1)(0.446)(0.787) = 1.404 \text{ in.}^2$$

$$A_d = H_d W_d = (2.0)(4.0) = 8.0 \text{ in.}^2, A_{Sink} = WH = (2.079)(0.787) = 1.636 \text{ in.}^2$$

$$A_b = A_d - A_{Sink} = 8.0 - 1.636 = 6.364 \text{ in.}^2$$

Using Eq. (7.21) and a Reynold's number in mixed units,

$$Re_{DH} = \frac{\left[\dfrac{G}{(A_b + A_f)/144}\right]\left(\dfrac{2W_d H_d}{W_d + H_d}\right)}{5v} = \frac{\left[\dfrac{25}{(6.364+1.404)/144}\right]\left[\dfrac{2(4.0)(2.0)}{4.0+2.0}\right]}{5(0.029)} = 8,523$$

$$Nu_L = C_1\left(\frac{Re_{DH}}{1000}\right)^{m_1}\left(\frac{W_d}{W}\right)^{m_2}\left(\frac{H_d}{H}\right)^{m_3}\left(\frac{S}{H}\right)^{m_4}\left(\frac{t_f}{H}\right)^{m_5}$$

$$Nu_L = (88.28)\left(\frac{8523}{1000}\right)^{0.6029}\left(\frac{4.0}{2.079}\right)^{-0.1098}\left(\frac{2.0}{0.787}\right)^{-0.5632}\left(\frac{0.446}{0.787}\right)^{0.08713}\left(\frac{0.059}{0.787}\right)^{0.4139} = 57.60$$

$$A_S = 2N_f LH + WL = 2(5)(2.079)(0.787) + (2.079)(2.079) = 20.684 \text{ in.}^2$$

$$h = \frac{k_{Air}}{L} Nu_L = \left(\frac{6.5\times10^{-4}}{2.079}\right)(57.60) = 0.018 \text{ W}/(\text{in.}^2 \cdot °\text{C})$$

$$R_{Sink} = \frac{1}{hA_S} = \frac{1}{(0.018)(20.684)} = 2.7 \text{ °C/W}$$

The modified Lee's method uses Eq. (7.23).where we can use units of V [ft/min], Δh [in. H$_2$O. Many of the "constants" are combined as much as possible.

$$a = \left[1 - \left(\frac{A_f}{A_b}\right)^2\right] = \left[1 - \left(\frac{1.404}{6.364}\right)^2\right] = 0.951$$

$$b = 2\left(\frac{A_d}{A_b}\right)\left(\frac{A_f}{A_b}\right)C_V V_d = 2\left(\frac{8}{6.364}\right)\left(\frac{1.404}{6.364}\right)(5.08\times10^{-3})V_d = 2.818\times10^{-3} V_d$$

$$c = \left[\frac{2(\Delta h_f - \Delta h_b)/C_h}{\rho} - \left(\frac{A_d}{A_b}\right)^2 C_V^2 V_d^2\right]$$

$$c = \left[\frac{2(\Delta h_f - \Delta h_b)/(4.019\times10^{-3})}{(1.18)} - \left(\frac{8}{6.364}\right)^2 (5.08\times10^{-3})^2 V_d^2\right]$$

$$c = \left[6.812 \times 10^{-3} \left(\Delta h_f - \Delta h_b \right) - 4.078 \times 10^{-5} V_d^2 \right]$$

$$V_f = \frac{-b + \sqrt{b^2 - 4ac}}{2a} \left(\frac{1}{5.08 \times 10^{-3}} \right) = 98.425 \left(\frac{-b + \sqrt{b^2 - 4ac}}{a} \right)$$

$$V_b = \sqrt{V_d^2 \left(A_d / A_b \right)^2 + V_f^2 \left(A_f / A_b \right)^2 - 2V_d V_f \left[\left(A_d \right) \left(A_f \right) / \left(A_b \right)^2 \right]}$$

$$V_b = \sqrt{V_d^2 \left(8 / 6.364 \right)^2 + V_f^2 \left(1.404 / 6.634 \right)^2 - 2V_d V_f \left[\left(8 \right) \left(1.404 \right) / \left(6.364 \right)^2 \right]}$$

$$V_b = \sqrt{1.58 V_d^2 + 0.045 V_f^2 - 0.510 V_d V_f}$$

Airflow resistances are calculated for both the heat sink channels and also the bypass region. The details are quite straight-forward, but it should be mentioned that the bypass calculation uses only friction term in the resistance whereas the heat sink resistance has the added contraction and exit resistances. Also remember that consistent with the model, the bypass length is the same as the heat sink length in the flow direction. Thus,

$$R_{AF-Sink} = \frac{1.29 \times 10^{-3}}{\left(N_f - 1 \right)^2 \left(HS \right)^2} \left[K_c + K_e + 4 f_{Sink} \frac{L}{\sqrt{HS}} \right], \quad \Delta h_f = R_{AF-Sink} G_{Sink}^2$$

$$R_{AF-Bypass} = \frac{1.29 \times 10^{-3}}{A_b^2} \left[4 f_b \frac{L_{Bypass}}{\sqrt{A_b}} \right], \quad \Delta h_b = R_{AF-Bypass} G_b^2$$

$$G_{Total} = G_{Sink} + G_b$$

As was the case for other flow bypass calculations using Lee's method, the interfin air speed must be iterated. The details of each iteration for this problem need not be listed here as the method has been well documented earlier in this book. The laminar friction factor f was calculated using Eq. (4.3). Calculations were completed for channel total airflows of $G = 10$, 25, and 50 CFM, and also for $N_f = 3$ ($S = 0.951$ in.), 5 ($S = 0.446$ in.), and 10 ($S = 0.165$ in.) fins. The results are tabulated in Tables 7.7, 7.8, and 7.9, respectively. Plots of the tabulated results are given in Figures 7.14a, 7.14b, and 7.14c.

A study of the tables and plots reveals that although some of the $J\&M$ calculations are out of the valid range, the heat sink thermal resistances R_{Sink} are still in good agreement with the modified Lee results. The two methods also show a tendency toward improved agreement as G becomes larger. In no case is there significant discrepancy when V_f is greater than about 100 ft/min. There are at least two areas for which one could suspect the modified Lee method: (1) the flow bypass theory and (2) the convective heat transfer coefficients. The thermal analysis in Section 7.9 would seem to indicate that even at low values of heat sink airflow, the thermal correlations used seem adequate. Overall, the modified Lee method appears useful. However, there is still some concern with the thermal community as to how much trust can be placed in this kind of analysis. A last comment concerns comparison of the Lee method for non-zero and zero bypass resistance. Tables 7.7 - 7.9 show litte difference for the problem studied.

Perhaps the most important or useful aspect of the modified Lee method is that the bypass resistance can be easily adapted to a circuit board resistance rather than just a smooth walled channel.

Table 7.7. Summary of intermediate and final calculations for heat sink analysis example adapted to non-zero bypass resistance model; $G = 10$ [CFM]. Italicized J&M values are out of the "guaranteed correct" range.

N_f	W_d [in.]	V_d [ft/min]	V_d/σ	Modified Lee $\Delta p_b \neq 0$					Lee $\Delta p_b = 0$		Jonsson & Moshfegh		
				V_f [ft/min]	G_{Sink} [CFM]	G_b [CFM]	Re_{Sink}	R_{Sink} [°C/W]	V_f [ft/min]	R_{Sink} [°C/W]	W/W_D	Re_{DH}	R_{Sink} [°C/W]
3	2.5	288	314	250	2.6	7.4	2,044	4.6	246	4.6	0.83	4,539	5.2
3	4	180	197	143	1.5	8.5	1,169	6.1	139	6.2	0.52	3,361	6.5
3	6	120	131	88	0.9	9.1	720	7.8	86	7.9	0.35	2,512	8.1
3	8	90	98	62	0.6	9.4	507	9.3	60	9.5	0.26	2,004	9.6
3	10	72	79	47	0.5	9.5	380	10.8	45	11.0	0.21	1,668	11.0
3	20	36	39	19	0.2	9.8	151	17.4	18	17.8	0.10	906	17.1
5	2.5	288	336	235	2.3	7.7	1,127	3.2	229	3.2	0.83	4,626	3.7
5	4	180	210	128	1.3	8.7	614	4.3	125	4.4	0.52	3,408	4.7
5	6	120	140	75	0.7	9.3	357	5.7	72	5.8	0.35	2,531	5.9
5	8	90	105	49	0.5	9.5	235	7.0	48	7.1	0.26	2,015	7.0
5	10	72	84	36	0.3	9.7	173	8.4	35	8.5	0.21	1,674	8.0
5	20	36	42	12	0.1	9.9	57	16.8	12	17.8	0.10	908	12.5
10	2.5	288	402	185	1.5	8.5	382	1.9	180	1.9	0.83	4,860	2.2
10	4	180	251	84	0.7	9.3	172	3.1	81	3.1	0.52	3,512	2.8
10	6	120	168	40	0.3	9.7	83	5.7	38	6.0	0.35	2,582	3.6
10	8	90	126	23	0.2	9.8	48	9.6	22	10.1	0.26	2,044	4.2
10	10	72	101	15	0.1	9.9	31	14.8	14	16.4	0.21	1,694	4.8
10	20	36	50	4	0.03	9.97	8	60	3.5	61	0.10	913	7.6

Table 7.8. Summary of intermediate and final calculations for heat sink analysis example adapted to non-zero bypass resistance model, $G = 25$ [CFM]. Italicized J&M values are out of the "guaranteed correct" range.

N_f	W_d [in.]	V_d [ft/min]	V_d/σ	Modified Lee $\Delta p_b \neq 0$					Lee $\Delta p_b = 0$		Jonsson & Moshfegh		
				V_f [ft/min]	G_{Sink} [CFM]	G_b [CFM]	Re_{Sink}	R_{Sink} [°C/W]	V_f [ft/min]	R_{Sink} [°C/W]	W/W_D	Re_{DH}	R_{Sink} [°C/W]
3	2.5	720	787	661	6.9	18.1	5,497	2.8	652	2.8	0.83	11,347	3.0
3	4	450	492	388	4.0	21	3,173	3.7	382	3.7	0.52	8,421	3.7
3	6	300	328	245	2.5	22.5	2,003	4.6	241	4.7	0.35	6,279	4.7
3	8	225	246	174	1.8	23.2	1,429	5.5	171	5.6	0.26	5,009	5.5
3	10	180	197	135	1.4	23.6	1104	6.3	133	6.3	0.21	4,167	6.3
3	20	90	98	58	0.6	24.4	482	9.6	58	9.7	0.10	2,265	9.8
5	2.5	720	839	630	6.1	18.9	3,020	1.9	621	2.0	0.83	11,566	2.1
5	4	450	524	360	3.5	22.5	1,726	2.6	354	2.6	0.52	8,523	2.7
5	6	300	350	221	2.2	22.9	1,057	3.3	217	3.3	0.35	6,328	3.4
5	8	225	262	156	1.5	23.5	750	3.9	154	4.0	0.26	5,038	4.0
5	10	180	210	118	1.1	23.9	561	4.5	115	4.6	0.21	4,186	4.6
5	20	90	105	48	0.5	24.5	225	7.2	46	7.3	0.10	2,270	7.2
10	2.5	720	1005	560	4.6	20.4	1,156	1.1	551	1.1	0.83	12,151	1.3
10	4	450	628	290	2.4	22.6	599	1.5	284	1.5	0.52	8,780	1.6
10	6	300	419	159	1.3	23.7	329	2.0	156	2.1	0.35	6,454	2.0
10	8	225	314	102	0.8	24.2	211	2.7	99	2.7	*0.26*	*5,112*	*2.4*
10	10	180	251	71	0.6	24.4	147	3.5	70	3.5	*0.21*	*4,235*	*2.8*
10	20	90	126	21	0.2	24.8	42	10.7	20	11.5	*0.10*	*2,283*	*4.4*

Table 7.9. Summary of intermediate and final calculations for heat sink analysis example adapted to non-zero bypass resistance model, $G = 50$ [CFM]. Italicized J&M values are out of the "guaranteed correct" range.

N_f	W_d [in.]	V_d [ft/min]	V_d/σ	Modified Lee $\Delta p_b \neq 0$					Lee $\Delta p_b = 0$		Jonsson & Moshfegh		
				V_f [ft/min]	G_{Sink} [CFM]	G_b [CFM]	$Re_{,Sink}$	R_{Sink} [°C/W]	V_f [ft/min]	R_{Sink} [°C/W]	W/W_D	Re_{DH}	R_{Sink} [°C/W]
3	2.5	1440	1574	1347	14	36	10,998	2.0	1333	2.0	0.83	22,695	2.0
3	4	900	984	801	8.3	41.7	6,550	2.6	793	2.6	0.52	16,842	2.5
3	6	600	56	513	5.3	44.7	4,187	3.2	507	3.2	0.35	12,558	3.1
3	8	450	492	373	3.9	46.1	3,050	3.7	369	3.8	0.26	10,017	3.6
3	10	360	394	290	3.0	47.0	2,371	4.3	287	4.3	0.21	8,333	4.2
3	20	180	197	133	1.4	48.6	1,083	6.3	131	6.4	0.10	4,530	6.5
5	2.5	1440	1678	1303	12.7	37.3	6,246	1.4	1290	1.4	0.83	23,132	1.4
5	4	900	1049	758	7.4	42.1	3,634	1.8	749	1.8	0.52	17,041	1.8
5	6	600	699	478	4.7	45.3	2,291	2.2	472	2.2	0.35	12,656	2.2
5	8	450	524	341	3.3	46.7	1,632	2.6	336	2.7	0.26	10,076	2.6
5	10	360	420	260	2.6	47.5	1,246	3.0	257	3.1	0.21	8,372	3.0
5	20	180	210	114	1.1	48.9	547	4.6	112	4.6	0.10	4,546	4.7
10	2.5	1440	2011	1245	10.1	39.9	2,571	0.7	1231	0.75	0.83	24,302	0.83
10	4	900	1257	664	5.4	44.6	1,371	1.0	656	1.0	0.52	17,560	1.07
10	6	600	838	387	3.2	46.2	799	1.3	382	1.3	0.35	12,908	1.34
10	8	450	628	264	2.1	47.9	545	1.6	260	1.6	*0.26*	10,225	*1.60*
10	10	360	503	193	1.6	48.4	398	1.8	190	1.9	*0.21*	8,470	*1.83*
10	20	180	251	67	0.5	49.5	138	3.7	66	3.7	*0.10*	4,567	*2.87*

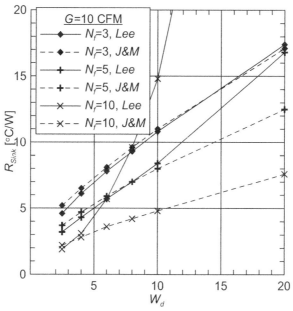

Figure 7-14a. Calculated heat sink thermal resistances comparing modified Lee with Jonsson & Moshfegh flow bypass models. Channel airflow is $G = 10$ CFM.

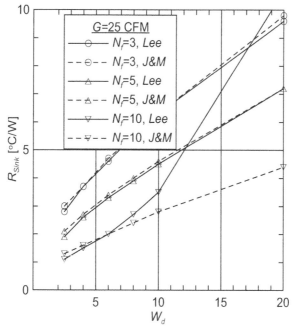

Figure 7-14b. Calculated heat sink thermal resistances comparing modified Lee with Jonsson & Moshfegh flow bypass models. Channel airflow is $G = 25$ CFM.

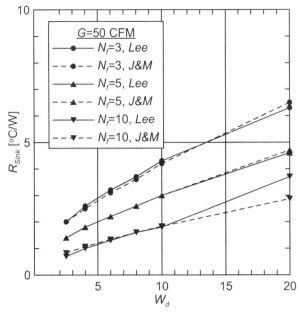

Figure 7-14c. Calculated heat sink thermal resistances comparing modified Lee with Jonsson & Moshfegh flow bypass models. Channel airflow is $G = 50$ CFM.

EXERCISES

7.1 Show that the conductances C_E and C_f, when added in series, are equivalent to the conductance C_I. Remember that $C_f = \dot{m} c_P$.

7.2 Using the nomenclature shown in the following two "system" (airflow G) illustrations, show that, in general, the circuit with C_I will give erroneous results. In what condition will the $T_W - T_0$ be approximately equal for (a) and (b)? Admittedly the "system" is very simple, but it is adequate for our purposes. Hints: (1) Derive formulae for the temperature rise of each system as $T_W - T_0$ and compare the results; (2) Use Eqs. (7.8) and (7.9) to get a relationship between C_E and C_I.

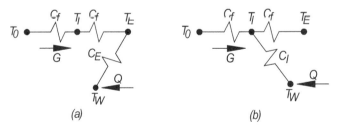

Exercise 7.2. (a) Heat sink conductance ref. to air exit. (b) Heat sink conductance ref. to air inlet.

7.3 An aluminum heat sink similar to that shown in Figure 7.2 has an average local heat transfer coefficient $h = 0.20$ W/(in.$^2 \cdot$ °C), a volumetric total airflow of $G = 25$ CFM, an inlet air temperature $T_I = 30$°C, and a uniformly distributed total heat dissipation $Q = 100$ W. The remaining data is: $W = 4$ in., $L = 10$ in., $H = 0.75$ in., $t_f = 0.10$ in., and $S = 0.25$ in. Calculate C_I, C_E, and C_f. Also calculate $T_W - T_I$ using the circuit representation for both Figures 7.4 (b) and (c).

7.4 Calculate the heat exchanger effectiveness of the hardware and flow conditions described for Exercise 7.3.

7.5 Use the variables h as the average heat transfer coefficient and A_S as the total convecting surface area for a shrouded, plate-fin heat sink. How much greater must the thermal-fluid conductance $C_f = \dot{m} c_p$ be compared to $C_c = hA_S$ for the heat sink total resistance R_I (referenced to inlet air) to be dominated by convection R_c? Hint: Use Eq. (7.8), rewriting and plotting it as R_I / R_C as a function of $\alpha = C_f/C_c = \dot{m} c_p/hA_S$.

7.6 In this problem you will consider the Application Example 7.7 for both heat sink lengths. Calculate and plot the ratio $(1/\dot{m} c_p)/R_I$ vs. volumetric flow rate G for each heat sink. What can you say about the relative effect of the volumetric airflow rate G on the total heat sink thermal resistance R_I referenced to the inlet air temperature? Select the results for either the adjusted Kays and Crawford correlation or the Muzychka and Yovanovich correlation.

7.7 Calculate the thermal resistance R_I, heat sink referenced to inlet air, following the geometry in Figure 7.11, but assume no bypass, i.e., all the air is ducted into the heat sink. The heat sink dimensions are: $W = 4.125$ in., $H = 1.5$ in., $L = 3.0$ in., $N_f = 10$, $t_f = 0.15$ in., and $t_b = 0.25$ in. Using approach velocities of $V_a = 50, 100, 250, 500,$ and 1000 ft/min, list some of your intermediate results as well as the final ones in the same manner as Table 7.6. Assume laminar flow using the Muzychka and Yovanovich correlation. For purposes of calculating R_I, allow heat transfer from the outer surfaces of the left and right fins.

7.8 Calculate the thermal resistance R_I, heat sink referenced to inlet air, following the geometry in Figure 7.11, but assume no bypass, i.e., all the air is ducted into the heat sink. The heat sink dimensions are: $W = 4.125$ in., $H = 1.5$ in., $L = 3.0$ in., $N_f = 10$, $t_f = 0.15$ in., and $t_b = 0.25$ in. Use approach velocities of $V_a = 50, 100, 250, 500,$ and 1000 ft/min. List some of your intermediate results as well as the final ones in the same manner as Table 7.6. Assume laminar flow using the adjusted Kays and Crawford correlation. This problem is identical to Exercise 7.8 except for the different correlation. For purposes of calculating R_I, allow heat transfer from the outer surfaces of the left and right fins.

7.9 Calculate the thermal resistance R_I, heat sink referenced to inlet air, following the geometry in Figure 7.11, but assume no bypass, i.e., all the air is ducted into the heat sink. The heat sink dimensions are: $W = 2.990$ in., $H = 2.875$ in., $L = 3.0$ in., $N_f = 6$, $t_f = 0.144$ in., and $t_b = 0.125$ in. Use approach velocities of $V_a = 50, 100, 250, 500,$ and 1000 ft/min. List some of your intermediate results as well as the final ones in the same manner as Table 7.6. Assume laminar flow using the Muzychka and Yovanovich correlation. For purposes of calculating R_I, allow heat transfer from the outer surfaces of the left and right fins.

7.10 Calculate the thermal resistance R_I, heat sink referenced to inlet air, following the geometry in Figure 7.11, but assume no bypass, i.e., all the air is ducted into the heat sink. The heat sink dimensions are: $W = 2.990$ in., $H = 2.875$ in., $L = 3.0$ in., $N_f = 6$, $t_f = 0.144$ in., and $t_b = 0.125$ in. Use approach velocities of $V_a = 50, 100, 250, 500,$ and 1000 ft/min. List some of your intermediate results as well as the final ones in the same manner as Table 7.6. Assume laminar flow using the adjusted Kays and Crawford correlation. This problem is identical to Exercise 7.9 except for the different correlation. For purposes of calculating R_I, allow heat transfer from the outer surfaces of the left and right fins.

7.11 Calculate the thermal resistance R_I, heat sink referenced to inlet air, following the geometry in Figure 7.11, but assume no bypass, i.e., all the air is ducted into the heat sink. The heat sink dimensions are: $W = 1.56$ in., $H = 0.625$ in., $L = 3.0$ in., $N_f = 7$, $t_f = 0.0891$ in., and $t_b = 0.125$ in. Use approach velocities of $V_a = 50, 100, 250, 500,$ and 1000 ft/min. List some of your intermediate results as well as the final ones in the same manner as Table 7.6. Assume laminar flow using the Muzychka and Yovanovich correlation. For purposes of calculating R_I, allow heat transfer from the outer surfaces of the left and right fins.

7.12 Calculate the thermal resistance R_I, heat sink referenced to inlet air, following the geometry in Figure 7.11, but assume no bypass, i.e., all the air is ducted into the heat sink. The heat sink dimensions are: $W = 1.56$ in., $H = 0.625$ in., $L = 3.0$ in., $N_f = 7$, $t_f = 0.0891$ in., and $t_b = 0.125$ in. Use approach velocities of $V_a = 50$, 100, 250, 500, and 1000 ft/min. List some of your intermediate results as well as the final ones in the same manner as Table 7.6. Assume laminar flow using the adjusted Kays and Crawford correlation. This problem is identical to Exercise 7.11 except for the different correlation. For purposes of calculating R_I, allow heat transfer from the outer surfaces of the left and right fins.

7.13 Calculate the thermal resistance R_I, heat sink referenced to inlet air, following the geometry in Figure 7.11, but assume no bypass, i.e., all the air is ducted into the heat sink. The heat sink dimensions are: $W = 4.125$ in., $H = 1.0$ in., $L = 3.0$ in., $N_f = 7$, $t_f = 0.0736$ in., and $t_b = 0.312$ in. Use approach velocities of $V_a = 50$, 100, 250, 500, and 1000 ft/min. List some of your intermediate results as well as the final ones in the same manner as Table 7.6. Assume laminar flow using the Muzychka and Yovanovich correlation. For purposes of calculating R_I, allow heat transfer from the outer surfaces of the left and right fins.

7.14 Calculate the thermal resistance R_I, heat sink referenced to inlet air, following the geometry in Figure 7.11, but assume no bypass, i.e., all the air is ducted into the heat sink. The heat sink dimensions are: $W = 4.125$ in., $H = 1.0$ in., $L = 3.0$ in., $N_f = 7$, $t_f = 0.0736$ in., and $t_b = 0.312$ in. Use approach velocities of $V_a = 50$, 100, 250, 500, and 1000 ft/min. List some of your intermediate results as well as the final ones in the same manner as Table 7.6. Assume laminar flow using the adjusted Kays and Crawford correlation. This problem is identical to Exercise 7.13 except for the different correlation. For purposes of calculating R_I, allow heat transfer from the outer surfaces of the left and right fins.

CHAPTER 8

Natural Convection Heat Transfer I: Plates

The high power dissipation encountered in the current era of electronic equipment usually requires forced air cooling. However, the ultra-simplistic nature of a natural convection cooling system without failure-prone electric fans or even the plumbing associated with liquid cooling sometimes makes natural convection a viable choice, but of course only if the heat dissipation is of modest value.

Natural convection air currents occur when a surface and nearby ambient air are at different temperatures. Heat flows to or from the air and results in an air density decrease or increase, respectively. Heated air rises and cooled air falls, the driving force being gravity in both instances. In this chapter you will learn about natural convection from the most basic of shapes encountered in electronic equipment, beginning with flat plate surfaces.

8.1 DIMENSIONLESS NUMBERS: NUSSELT AND GRASHOF

Just as we began Chapter 6 for Forced Convection, the goal here is to justify the selection of certain dimensionless groups commonly used in natural convection heat transfer. The Pros and Cons of the method are identical to those discussed in Chapter 6 and shall not be repeated here.

Buckingham - π Theorem

n_G is the number of dimensionless groups.
n_P is the number of physical quantities.
n_D is the number of primary dimensions.

Then $n_G = n_P - n_D$ and the relationship among the groups is expressed by

$$F\left(\pi_1, \pi_2, \pi_3, \dots \pi_{n_G}, \right) = 0 \text{ or } \pi_1 = f_1\left(\pi_2, \pi_3, \dots\right), \ \pi_2 = f_1\left(\pi_1, \pi_3, \dots\right), \text{etc.}$$

The natural convection problem

Physical quantities in M, L, t, T (mass, length, time, temperature) system.

Variable	Symbol	Dimensions
Temperature difference	ΔT	$[T]$
Diameter, length or	D, L	$[L]$
Thermal conductivity	k	$\left[ML/t^3 T \right]$
Kinematic viscosity	v	$\left[L^2/t \right]$
Thermal diffusivity	α	$\left[L^2/t \right]$
Heat transfer coeff.	h	$\left[M/t^3 T \right]$
Thermal expansion	$g\beta$	$\left[L/t^2 T \right]$
Group	π	$[M]^0 [L]^0 [t]^0 [T]^0$

Each of the groups is represented by an equation of the form

$$\pi = h^a \Delta T^b k^c v^d \alpha^e L^f (\beta g)^g$$

$$[L]^0 [M]^0 [t]^0 [T]^0 = \left[\frac{M}{t^3 T}\right]^a [T]^b \left[ML/t^3 T\right]^c \left[L^2/t\right]^d \left[L^2/t\right]^e [L]^f \left[L/t^2 T\right]^g$$

$$[M]^0 [L]^0 [t]^0 [T]^0 = [M]^{a+c} [L]^{c+2d+2e+f+g} [t]^{-3a-3c-d-e-2g} [T]^{-a+b-c-g}$$

Matching the exponents on the left and right sides of the preceding,

$$M \text{ dimension:} \quad a+c=0$$
$$L \text{ dimension:} \quad c+2d+2e+f+g=0$$
$$t \text{ dimension:} \quad 3a+3c+d+e+2g=0$$
$$T \text{ dimension:} \quad a-b+c+g=0$$

There are four equations and seven unknowns. Any set of values a - g that satisfy these four equations results in a dimensionless π.

$$(i) \quad a+c=0$$
$$(ii) \quad c+2d+2e+f+g=0$$
$$(iii) \quad 3a+3c+d+e+2g=0$$
$$(iv) \quad a-b+c+g=0$$

Since there are more unknowns than equations, three of the unknowns may be set at any value that works, i.e., results in a desired dimensionless group. One restriction: each unknown must be linearly independent. Thus *the determinant of the coefficients of the remaining unknowns must be zero.*

$$\pi = h^a \Delta T^b k^c v^d \alpha^e L^f (\beta g)^g$$

It makes sense that to preserve the h variable, we set $a = 1$, then guess $e = 0$, $g = 0$:

$(1)(i) \quad a+c=0$ $(2)(i)$ Use $c=-1$ (3) Use $(i) c=-1$ (4) Use $(iii) d=0$

$(ii) \quad c+2d+2e+f+g=0$ $(ii) c+2d+f=0$ $(ii) \quad 2d+f=1$ $(ii) f=1$

$(iii) \quad 3a+3c+d+e+2g=0$ $(iii) 3c+d=-3$ $(iii) \quad d=0$

$(iv) \quad a-b+c+g=0$ $(iv) -b+c=-1$ $(iv) \quad b-0$

$$\pi_1 = h \frac{1}{k} L, \ Nu \equiv \pi_1, \ \boxed{Nu = \frac{hL}{k}} \quad \text{which is defined as the Nusselt Number.}$$

Next it makes sense, that to not get the variable h again, set $a = 0$: To get something with an α in it, set $e = 1$. Guess $g = 0$.

$(1)(i) \ a+c=0$ $(2)(i) \ c=0$ (3) Use $(i) \ c=0$ (4) Use $(iii) \ d=-1$

$(ii) \ c+2d+2e+f+g=0$ $(ii) \ c+2d+f=-2$ $(ii) \quad 2d+f=-2$ $(ii) \ f=0$

$(iii) 3a+3c+d+e+2g=0$ $(iii) \ 3c+d=-1$ $(iii) \quad d=-1$

$(iv) \ a-b+c+g=0$ $(iv) -b+c=0$ $(iv) \quad b=0$

$$\pi_2 = \frac{1}{v} \alpha = \frac{\alpha}{v}, \ Pr \equiv \frac{1}{\pi_2}, \ Pr = \frac{v}{\alpha} \quad \text{which is defined as the Prandtl Number.}$$

$$\pi = h^a \Delta T^b k^c v^d \alpha^e L^f (\beta g)^g$$

Now it makes sense to avoid getting the variables h and α. Then set $a = e = 0$. Set $g = 1$:

$(1)(i)\ a+c=0$ $(2)(i)\ c=0$ (3) Use $(i)\ c=0$ (3) Use $(iii)\ d=-2$

$(ii)\ c+2d+2e+f+g=0$ $(ii)\ c+2d+f=-1$ $(ii)\ 2d+f=-1$ $(ii)\ f=3$

$(iii)\ 3a+3c+d+e+2g=0$ $(iii)\ 3c+d=-2$ $(iii)\ d=-2$ $(ii)\ b=1$

$(iv)\ a-b+c+g=0$ $(iv)\ -b+c=-1$ $(iv)\ b=1$

The result is
Grashof Number.
$$\pi_3 = \Delta T \frac{1}{v^2} L^3 (\beta g),\ Gr \equiv \pi_3,\ \boxed{Gr = \frac{g\beta\Delta T L^3}{v^2}}$$ which is defined as the

Use the style of
$$\pi_1 = f_1(\pi_2, \pi_3)\ or\ Nu = f(Gr, Pr)$$

the objective has been accomplished.

Now it is appropriate to discuss the preceding natural convection problem from a more physical point of view, including the dimensionless numbers where appropriate.

The expected air currents for a heated plate ($T_S > T_A$) in an enclosure are depicted in Figure 8.1. The curves in Figure 8.2 illustrate the concepts of thermal and aerodynamic boundary layers. Outside the thermal boundary layer δ_T, the air temperature $T(x,y)$ is approximately equal to the ambient temperature T_A. The air speed is of course approximately zero outside of the aerodynamic boundary layer δ_A.

Most of the literature concerning free convection heat transfer coefficients is written in terms of a correlation for the Nusselt number Nu_p vs. the Grashof and Prandtl numbers, Gr_p and Pr, respectively. The product $Ra_p = Gr_p Pr$ is known as the *Rayleigh* number. Many of these correlations occur in the form

$$Nu_p = C(Gr_p Pr)^n = CRa_p^n \qquad \text{Nusselt number} \quad (8.1)$$

The Grashof number is a combination of the physical variables ρ, μ, β for air density, dynamic viscosity, and coefficient of thermal expansion, respectively, and the constant $g = 386.0$ in./s². Gr_p is dimensionless with a dependence on a length (dimensional parameter) P [in.], surface temperature T_S [°C], and ambient air temperature T_A [°C]:

$$Gr_p = \frac{g\rho^2}{\mu^2} \beta(T_S - T_A) P^3 \qquad \text{Grashof number} \quad (8.2)$$

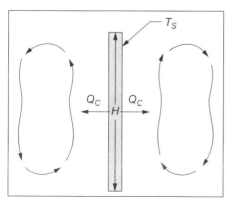

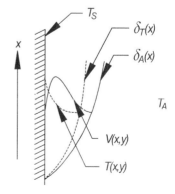

Figure 8.1. Natural convection heat transfer and convection currents for a vertical plate. Plate and ambient air are at temperatures T_S and T_A, respectively.

Figure 8.2. Natural convection velocity V and temperature T profiles for a heated, vertical flat plate. Plate and ambient air are at temperatures T_S and T_A, respectively.

The Nusselt and Prandtl numbers were discussed in Chapter 6. The air density and viscosity are usually evaluated at the mean film temperature, i.e., $(T_S + T_A)/2$, but it is more convenient to use the kinematic viscosity $\nu = \mu/\rho$. The coefficient of thermal expansion is usually evaluated at the ambient air temperature.

The various physical variables in Eq. (8.2) are tabulated in Appendix *ii*. The Rayleigh number is used so frequently in our natural convection work that it is worthwhile to construct a plot of $\gamma = Ra/(T_S - T_A)P^3$ and γ/β, which is illustrated in Figure 8.3. The two different plots are necessary because the variables ρ, μ, and Pr are evaluated at the mean film temperature and β is evaluated at the local ambient air temperature. If the surface temperature rise is modest, you may wish to make approximate calculations and use a mean film temperature for all of the variables to obtain γ, and for more precise work you will want to obtain β separately. Formulae for γ/β and β are included as Eq. (8.3), where the formula for β is based on the ideal gas law.

$$\gamma = \left(g\beta/\nu^2\right)Pr = 1.952 \times 10^3 \exp\left(-1.2000 \times 10^{-2}T\right)$$
$$\gamma/\beta = \left(g/\nu^2\right)Pr = 5.454 \times 10^5 \exp\left(-9.254 \times 10^{-3}T\right)$$
$$\beta = 1/\left(T_A + 273.16\right)$$

Convection (8.3) parameters

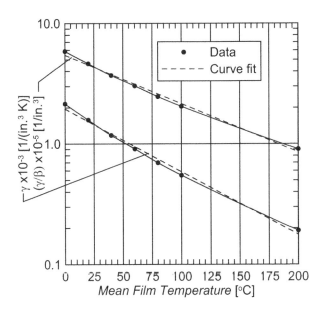

Figure 8.3. The natural convection parameters $\gamma = \left(g\beta/\nu^2\right)Pr$ and $\gamma/\beta = \left(g/\nu^2\right)Pr$ as a function of mean film temperature $(T_S + T_A)/2$.

8.2 CLASSICAL FLAT PLATE CORRELATIONS

The term *classical* is used here to indicate Nusselt number formulae that are so basic that they may be found in nearly any heat transfer text, e.g., Holman (1990). The correlations and the recommended Rayleigh number range for vertical and flat plates are listed in Table 8.1.

The correlations in Table 8.1 are too complicated for many approximate calculations so the next step is an attempt to obtain simplified formulae for heat transfer coefficients. We shall accomplish this by factoring out and grouping together all of the temperature-dependent material properties. Then we will examine the result to see how much we might really simplify the problem.

We write the heat transfer coefficient as

$$h = \left(\frac{k_{Air}}{P}\right)Nu_P = \left(\frac{k_{Air}}{P}\right)CRa_P^n = \left(\frac{k_{Air}}{P}\right)C\left(Gr_P Pr\right)^n = \left(\frac{k_{Air}}{P}\right)C\left[\left(\frac{g\beta\Delta TP^3}{v^2}\right)Pr\right]^n$$

$$h = \left(\frac{k_{Air}}{P}\right)C\left[\left(\frac{gPr}{v^2}\right)\beta\Delta TP^3\right]^n = C\left(\frac{k_{Air}}{P}\right)\left[\left(\frac{\gamma}{\beta}\right)\beta\Delta TP^3\right]^n$$

We arrive at a general formula for our flat plate heat transfer coefficients.

$$h = Ck_{Air}\left[\left(\frac{\gamma}{\beta}\right)\beta\right]^n\left(\frac{P^{3n}}{P}\right)\Delta T^n = \alpha\frac{\left(\Delta T\right)^n}{P^{1-3n}} \quad \text{Generalized } h \; (8.4)$$

$$\alpha = Ck_{Air}\left(\frac{\gamma}{\beta}\right)^n\beta^n$$

where we must now substitute values of C and n for the vertical and horizontal plates, as well as γ/β, β, and k_{Air} for several values of temperature. Since we are trying to assemble formulae that we may very well apply on a regular basis, we should consider temperature dependencies as carefully as possible. Thus we will use a mean film temperature for $k_{Air}, \gamma/\beta$ and an ambient temperature for β using Eq. (8.3) for the latter two quantities and a formula from Appendix ii for k_{Air}. Referring to Table 8.1 for a vertical flat plate in laminar flow, we obtain Eq. (8.5) for h and α. As you will see, we go to considerable effort to evaluate α. Then Table 8.2 is constructed, listing numerical values of α for several different temperature rises and mean film temperatures. Also note that β, the expansion coefficient for air is carefully managed.

Vertical Plate, Laminar Flow:

$$h = \alpha\frac{\left(\Delta T\right)^n}{P^{1-3n}} = \alpha\frac{\left(\Delta T\right)^{0.25}}{H^{1-3(0.25)}} = \alpha\left(\frac{\Delta T}{H}\right)^{0.25}$$

Table 8.1. Classical natural convection correlations for flat plates where $Nu_P = CRa_P^n$.

Geometry	C	n	Ra_P	Flow	P
Vertical	0.59	1/4	$10^4 < Ra_P < 10^9$	Laminar	H
Vertical	0.13	1/3	$10^9 < Ra_P < 10^{12}$	Turbulent	H
Horizontal ↑	0.54	1/4	$2.2\times 10^4 < Ra_P < 8\times 10^6$	Laminar	$\dfrac{WH}{2(W+H)}$
Horizontal ↑	0.15	1/3	$8\times 10^6 < Ra_P < 1.6\times 10^9$	Turbulent	$\dfrac{WH}{2(W+H)}$
Horizontal ↓	0.27	1/4	$3\times 10^5 < Ra_P < 3\times 10^{10}$	Laminar	$\dfrac{WH}{2(W+H)}$
Horizontal ↓			*Nothing offered*	Turbulent	

Note: ↑ direction of heat transfer is upward. ↓ direction of heat transfer is downward.

$$\alpha = Ck_{Air}\left(\frac{\gamma}{\beta}\right)^n \beta^n = 0.59k_{Air}\left(\frac{\gamma}{\beta}\right)^{0.25} \beta^{0.25} = 0.59\left[6.0583\times10^{-4} + 1.6906\times10^{-6}\left(\frac{\Delta T}{2} + T_A\right)\right]$$

$$\times\left\{5.460\times10^5\exp\left[-9.2817\times10^{-3}\left(\frac{\Delta T}{2} + T_A\right)\right]\right\}^{0.25} \frac{1}{\left(T_A + 273.16\right)^{0.25}} \qquad (8.5)$$

Table 8.2. α and Ra_H corresponding to various values of ΔT for a
vertical flat plate, laminar flow. $T_A = 20°C$.

ΔT [°C]	H [in.]	T_{Film} [°C]	α [W/(in.$^{1.75}\cdot$ °C$^{1.25}$)]	Ra_H
1	1	20.5	0.00237	908 [1]
5	2	22.5	0.00237	3.6×10^4
10	5	25	0.00237	1.1×10^6
50	10	45	0.00238	3.6×10^7
100	10	70	0.00238	5.7×10^7
100	50	70	0.00238	7.2×10^9 [1]

Note: [1] Ra_H indicates out of range.

We complete the process of simplifying the heat transfer coefficients by continuing for vertical plate, turbulent flow; horizontal plate with heat flow upward, laminar flow; horizontal plate with heat flow upward, turbulent flow; and horizontal plate with heat flow downward, laminar flow.

Vertical Plate, Turbulent Flow:

$$h = \alpha\frac{\left(\Delta T\right)^n}{P^{1-3n}} = \alpha\frac{\left(\Delta T\right)^{1/3}}{H^{1-3(1/3)}} = \alpha\left(\Delta T\right)^{1/3}$$

$$\alpha = Ck_{Air}\left(\frac{\gamma}{\beta}\right)^n \beta^n = 0.13k_{Air}\left(\frac{\gamma}{\beta}\right)^{1/3} \beta^{1/3} = 0.13\left[6.0583\times10^{-4} + 1.6906\times10^{-6}\left(\frac{\Delta T}{2} + T_A\right)\right]$$

$$\times\left\{5.454\times10^5\exp\left[-9.254\times10^{-3}\left(\frac{\Delta T}{2} + T_A\right)\right]\right\}^{1/3} \frac{1}{\left(T_A + 273.16\right)^{1/3}} \qquad (8.6)$$

Table 8.3. α and Ra_H corresponding to various values of ΔT for a vertical
flat plate, turbulent flow. $T_A = 20°C$.

ΔT [°C]	H [in.]	T_{Film} [°C]	α [W/(in.$^{1.75}\cdot$ °C$^{1.25}$)]	Ra_H
1	1	20.5	0.000962	200 [1]
5	2	22.5	0.000961	7.7×10^3 [1]
10	5	25	0.000959	2.4×10^5 [1]
50	10	45	0.000949	8.0×10^6 [1]
100	10	70	0.000933	1.3×10^7 [1]
100	50	70	0.000933	1.6×10^9

Note: [1] Ra_H indicates out of range.

Horizontal Plate, Upward Heat Transfer, Laminar Flow:

$$h = \alpha \frac{(\Delta T)^n}{P^{1-3n}} = \alpha \frac{(\Delta T)^{0.25}}{P^{1-3(0.25)}} = \alpha \left(\frac{\Delta T}{P}\right)^{0.25} ; \quad P = \frac{WL}{2(W+L)}$$

$$\alpha = Ck_{Air}\left(\frac{\gamma}{\beta}\right)^n \beta^n = 0.54k_{Air}\left(\frac{\gamma}{\beta}\right)^{0.25}\beta^{0.25} = 0.54\left[6.0583\times10^{-4} + 1.6906\times10^{-6}\left(\frac{\Delta T}{2}+T_A\right)\right]$$

$$\times \left\{5.454\times10^5 \exp\left[-9.254\times10^{-3}\left(\frac{\Delta T}{2}+T_A\right)\right]\right\}^{0.25} \frac{1}{(T_A+273.16)^{0.25}} \tag{8.7}$$

Table 8.4. α and Ra_P corresponding to various values of ΔT for a horizontal flat plate, upward direction heat transfer, laminar flow. $T_A = 20°C$.

ΔT [°C]	P [in.]	T_{Film} [°C]	α [W/(in.$^{1.75}$ · °C$^{1.25}$)]	Ra_P
1	1	20.5	0.00217	831 [1]
5	2	22.5	0.00217	3.3×10^4
10	5	25	0.00217	1.0×10^6
50	10	45	0.00218	3.3×10^7
100	10	70	0.00218	5.3×10^7
100	50	70	0.00218	6.6×10^9

Note: [1] Ra_P indicates out of range.

Horizontal Plate, Upward Heat Transfer, Turbulent Flow:

$$h = \alpha \frac{(\Delta T)^n}{P^{1-3n}} = \alpha \frac{(\Delta T)^{1/3}}{P^{1-3(1/3)}} = \alpha (\Delta T)^{1/3}$$

$$\alpha = Ck_{Air}\left(\frac{\gamma}{\beta}\right)^n \beta^n = 0.15k_{Air}\left(\frac{\gamma}{\beta}\right)^{1/3}\beta^{1/3} = 0.15\left[6.0583\times10^{-4} + 1.6906\times10^{-6}\left(\frac{\Delta T}{2}+T_A\right)\right]$$

$$\times \left\{5.454\times10^5 \exp\left[-9.254\times10^{-3}\left(\frac{\Delta T}{2}+T_A\right)\right]\right\}^{1/3} \frac{1}{(T_A+273.16)^{1/3}} \tag{8.8}$$

Table 8.5. α and Ra_P corresponding to various values of ΔT for a horizontal flat plate, upward direction heat transfer, turbulent flow. $T_A = 20°C$.

ΔT [°C]	P [in.]	T_{Film} [°C]	α [W/(in.$^{1.75}$ · °C$^{1.25}$]	Ra_P
1	1	20.5	0.00111	231 [1]
5	2	22.5	0.00111	9.1×10^{5} [1]
10	5	25	0.00111	2.8×10^{5} [1]
50	10	45	0.00110	9.2×10^6
100	10	70	0.00108	1.5×10^7
100	50	70	0.00108	1.8×10^9

Note: [1] Ra_P indicates out of range.

Horizontal Plate, Downward Heat Transfer, Laminar Flow:

$$h = \alpha \frac{\left(\Delta T\right)^n}{P^{1-3n}} = \alpha \frac{\left(\Delta T\right)^{0.25}}{P^{1-3(0.25)}} = \alpha \left(\frac{\Delta T}{P}\right)^{0.25} \; ; \; P = \frac{WL}{2\left(W+L\right)}$$

$$\alpha = Ck_{Air}\left(\frac{\gamma}{\beta}\right)^n \quad \beta^n = 0.27 k_{Air}\left(\frac{\gamma}{\beta}\right)^{0.25} \quad \beta^{0.25} = 0.27\left[6.0583\times10^{-4} + 1.6906\times10^{-6}\left(\frac{\Delta T}{2}+T_A\right)\right]$$

$$\times\left\{5.454\times10^5 \exp\left[-9.254\times10^{-3}\left(\frac{\Delta T}{2}+T_A\right)\right]\right\}^{0.25} \frac{1}{\left(T_A+273.16\right)^{0.25}} \tag{8.9}$$

Table 8.6. α and Ra_p corresponding to various values of ΔT for a horizontal flat plate, downward direction heat transfer, laminar flow. $T_A = 20°C$.

ΔT [°C]	P [in.]	T_{Film} [°C]	α [W/(in.$^{1.75}$ · °C$^{1.25}$)]	Ra_p
1	1	20.5	0.00108	416 [1]
5	2	22.5	0.00108	1.6×10^4 [1]
10	5	25	0.00109	5.0×10^5
50	10	45	0.00109	1.7×10^7
100	10	70	0.00109	2.6×10^7
100	50	70	0.00109	3.3×10^9

Note: [1] Ra_p indicates out of range.

We will review the results in Tables 8.2 to 8.6 for two reasons: (1) If possible, select a single α for each plate/orientation/flow regime and (2) examine the applicability of the correlations as to the Rayleigh number criteria as specified in Table 8.1. With regard to the value of α, the values are nearly invariable with temperature. The issue of the Rayleigh number seems to be that other than in the instance of the smallest temperature rise and plate dimension, all of the laminar flow correlations are valid. The two turbulent flow regimes, those for a vertical plate and horizontal plate, upward heat transfer, seem to be necessary only for a very high temperature rise *and* a large plate. Temperature rises and plate dimensions were selected for those values that one might have in most electronic cooling applications and it appears that natural convection turbulent flow is not likely to occur. Finally, note that an ambient temperature $T_A = 20°C$ was used. Calculations of α and Ra_H for $T_A = 50°C$ show that neither change by more than about 5%. Table 8.7 lists a useful summary.

The observant reader will note that there is a correlation in Table 8.7 for a horizontal plate, downward heat transfer, turbulent flow. This is the geometry/flow regime for which a correlation does not appear to exist in the literature. This author has chosen to "guess" at a correlation by using one identical to turbulent flow, upward heat transfer, but with an α exactly half of that calculated for upward flow.

8.3 SMALL DEVICE FLAT PLATE CORRELATIONS

We use this section to describe some experimental results. The impetus for the work is the poor accuracy of classical convection formulae when used with small devices such as a bracket with a 0.5-in. height or a small component. The test apparatus was simple. A flat plate was used as the convecting surface with one TO-3 power transistor centrally attached. The plate was thick enough to

Table 8.7. Summary of simplified classical heat transfer coefficient formulae for flat plates.

Laminar flow $\left(10^4 < Ra_P < 10^9\right)$

$$h = 0.0024\left(\frac{\Delta T}{P}\right)^{1/4}$$

Turbulent flow $\left(10^9 < Ra_P < 10^{12}\right)$

$$h = 0.00095\left(\Delta T\right)^{1/3}$$

Laminar flow $\left(2.2\times10^4 < Ra_P < 8\times10^6\right)$

$$h = 0.0022\left(\frac{\Delta T}{P}\right)^{1/4}$$

Turbulent flow $\left(8\times10^6 < Ra_P < 1.6\times10^9\right)$

$$h = 0.0011\left(\Delta T\right)^{1/3}$$

Laminar flow $\left(3\times10^5 < Ra_P < 3\times10^{10}\right)$

$$h = 0.0011\left(\frac{\Delta T}{P}\right)^{1/4}$$

Turbulent flow $\left(3\times10^{10} < Ra_P\right)$

$$h = 0.00055\left(\Delta T\right)^{1/3}$$

Vertical plate: $P = H\,[\text{in.}]$; horizontal plate: $P = WL/\left[2\left(W+L\right)\right]$.

cause only a modest temperature gradient from the transistor case to the plate edge. Thermocouples were used to determine a temperature profile, from which an average surface temperature was determined. One vertical and two horizontal plates were used. The dimensions and orientations are illustrated in Figure 8.4, nine configurations in all. Each was powered at three different values to provide the effects of temperature rise. The emissivity of each surface was preestablished by spraying the plates with a flat black paint, thus permitting subtraction of radiation effects from the final results. The convective conductance is the difference between the total and radiative conductances:

$$C_c = C - C_r, C_r = \varepsilon h_r A_S,\ C = \left(\text{power dissipated}\right)/\left(\overline{T}_S - T_A\right)$$

It was assumed that the h_c for the horizontal heated upper and lower surfaces differed by a factor of two, consistent with Table 8.1. The test results are plotted in Figure 8.5 (horizontal surface test results for the heated surface facing up). The parameter P is H and $WL/[2(W+L)]$ for the vertical and horizontal cases, respectively. It is tempting to speculate that the exponent $n = 0.35$ and 0.33 is due to turbulent flow, but the Rayleigh numbers in Figure 8.5 do not match the turbulent flow Ra_p in Table 8.1. The *small device* heat transfer coefficients are summarized in Table 8.8.

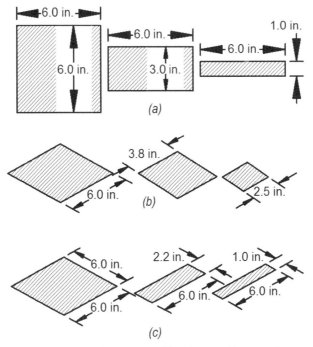

Figure 8.4. Orientation and geometry of flat plate used in natural convection, small-device correlation. (*a*) Vertical-plate fixed width, variable height. (*b*) Horizontal plate-square. (*c*) Horizontal plate-fixed length, variable width.

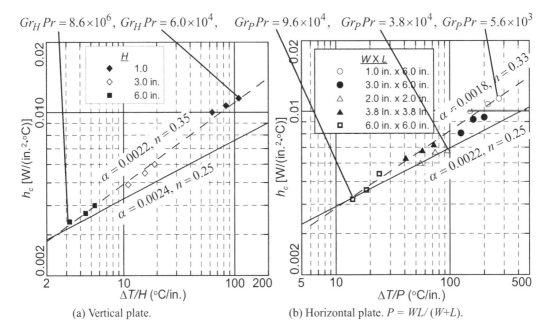

(a) Vertical plate.   (b) Horizontal plate. $P = WL/(W+L)$.

Figure 8.5. Experimental *small device*, flat plate heat transfer coefficients. $Ra_p = Gr_p Pr$ corresponds to appropriate experimental values. Line plots are fits to $h = \alpha \left(\Delta T / P \right)^n$.

Table 8.8. Summary of simplified *small device*, flat plate heat transfer coefficient formulae.

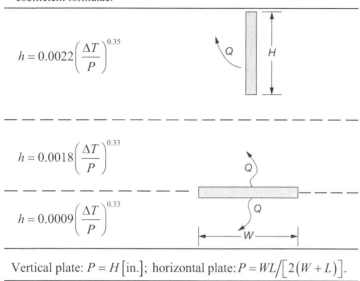

$$h = 0.0022\left(\frac{\Delta T}{P}\right)^{0.35}$$

$$h = 0.0018\left(\frac{\Delta T}{P}\right)^{0.33}$$

$$h = 0.0009\left(\frac{\Delta T}{P}\right)^{0.33}$$

Vertical plate: $P = H\,[\text{in.}]$; horizontal plate: $P = WL\big/\big[2(W + L)\big]$.

8.4 APPLICATION EXAMPLE: VERTICAL CONVECTING PLATE

A vertical flat plate is illustrated in Figure 8.6. We are not yet in a position to consider conduction effects; therefore, even though we note the discrete power source with a heat dissipation Q, we will calculate an average surface temperature, neglecting the plate thickness. The problem is typical of many conduction problems with convection boundary conditions where it is usually adequate to (1) calculate an average heat transfer coefficient and then (2) apply the value to a detailed conduction analysis program, *FEM*, *FDM*, or an analytical based program. The reader should think of this problem as step (1). We will use plate dimensions of $W = 9.0$ in., $H = 6.0$ in., a heat dissipation of $Q = 8$ W, and an ambient temperature $T_A = 20°C$. We also assume the emissivity ε is sufficiently small, i.e., less than 0.1, so that we neglect radiation.

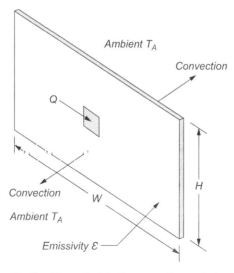

Figure 8.6. Application Example 8.4: Geometry for vertical convecting plate.

The surface area A_S includes both sides of the plate:

$$A_S = 2WH = 2(9.0\,\text{in.})(6.0\,\text{in.}) = 108\,\text{in.}^2$$

and our basic heat transfer formula is

$$Q = h_c A_S \Delta T = 0.0024\left(\frac{\Delta T}{H}\right)^{0.25} A_S \Delta T = \frac{0.0024 A_S}{H^{0.25}}(\Delta T)^{1.25}$$

which we solve for ΔT to obtain

$$\Delta T = \left[\frac{Q}{\left(0.0024 A_S / H^{0.25}\right)}\right]^{\frac{1}{1.25}} = \left\{\frac{8.0\,\text{W}}{\left[\left(0.0024 \times 108\,\text{in.}^2\right)/(6.0\,\text{in.})^{.25}\right]}\right\}^{\frac{1}{1.25}} = 22.24\,^{\circ}\text{C}$$

from which we next calculate the average heat transfer coefficient

$$h = 0.0024(22.24/6.0)^{0.25} = 0.0033\,\text{W}/\left(\text{in.}^2 \cdot {}^{\circ}\text{C}\right)$$

8.5 APPLICATION EXAMPLE: VERTICAL CONVECTING AND RADIATING PLATE

We shall reanalyze the previous problem, but now we include radiation using an emissivity $\varepsilon = 0.8$ and an ambient $T_A = 30°C$. We employ the vertical plate equation from Table 8.7 and the approximate radiation heat transfer coefficient from Chapter 1:

$$h_c = 0.0024(\Delta T/H)^{0.25}\,, \; h_r = 1.463 \times 10^{-10}(T_A + 273.16)^3$$

We may immediately calculate h_r:

$$h_r = 1.463 \times 10^{-10}(T_A + 273.16)^3 = 1.463 \times 10^{-10}(30 + 273.16)^3 = 4.076 \times 10^{-3}\,\text{W}/\left(\text{in.}^2 \cdot {}^{\circ}\text{C}\right)$$

Now we set up the appropriate heat transfer equation as

$$Q = h_c A_S \Delta T + \varepsilon h_r A_S \Delta T = \left(h_c + \varepsilon h_r\right) A_S \Delta T$$

and solving for the temperature rise we obtain

$$\Delta T = \frac{Q}{\left(h_c + \varepsilon h_r\right) A_S} = \frac{Q}{\left[0.0024(\Delta T/H)^{0.25} + \varepsilon h_r\right] A_S}$$

which must be solved iteratively. The arithmetic is shown for the first iteration using an initial guess of $\Delta T = 40°C$.

$$\Delta T = \frac{Q}{\left[0.0024(\Delta T/H)^{0.25} + \varepsilon h_r\right] A_S} = \frac{8.0\,\text{W}}{\left[0.0024\left(40\,^{\circ}\text{C}/6.0\,\text{in.}\right)^{0.25} + 0.8(0.00408)\right]\left(108\,\text{in.}^2\right)}$$

$$\Delta T = \frac{8.0}{\left[(0.00386) + 0.00326\right](108)} = 10.41\,^{\circ}\text{C}$$

We use this 10.41°C as an indication that our 40°C was too large. A new calculation of ΔT is attempted using a new guess of $\Delta T = 20°C$.

$$\Delta T = \frac{Q}{\left[0.0024(\Delta T/H)^{0.25} + \varepsilon h_r\right] A_S} = \frac{8.0\,\text{W}}{\left[0.0024\left(20^{\circ}\text{C}/6.0\,\text{in.}\right)^{0.25} + 0.8(0.00408)\right]\left(108\,\text{in.}^2\right)}$$

$$\Delta T = \frac{8.0}{\left[(0.00324) + 0.00326\right](108)} = 11.39\,^{\circ}\text{C}$$

The iteration process is continued in this fashion until we are satisfied that the solution has *converged* to a numerically accurate result. The results for each iteration are shown in Table 8.9. Our final answer is an average plate temperature rise of 12.1°C above ambient.

Table 8.9. Iteration results for Application Example 8.5.

Iteration	$\Delta T\left[°C\right]$	$h_c\left[W/\left(in.^2 \cdot °C\right)\right]$	$\Delta T\left[°C\right]$
1	40	0.00386	10.41
2	20	0.00324	11.39
3	11	0.00279	12.24
4	12	0.00285	12.11
5	12.10	0.00285	12.10

8.6 VERTICAL PARALLEL PLATE CORRELATIONS APPLICABLE TO CIRCUIT BOARD CHANNELS

Teerstra et al. (1996) published a very useful study of vertical parallel plate correlations that may be used to simulate vertical circuit board channels in natural convection (geometry illustrated in Figures 8.7 and 8.8). It is also suggested that the interested reader consult Moffat and Ortega (1988). The following summarizes both works. The first correlation was developed by Bar-Cohen and Rohsenow (1984) for isothermal surfaces and is illustrated in Figure 8.7. The left- and right-hand boards each convect Q watts into the air channel, the latter with an inlet air temperature T_I. Note that we are now using the symbol L to denote the channel height. The Nusselt number is

$$\overline{Nu}_b = \left[\left(\frac{C}{Ra_b}\right)^2 + \left(\frac{1}{0.59Ra_b^{1/4}}\right)^2\right]^{-1/2} \begin{cases} C=24: \text{Symmetrically heated walls.} \\ C=12: \text{One heated wall, one} \\ \qquad\quad\text{unheated adiabatic wall.} \end{cases} \quad \text{Iso- (8.10)}$$
$$\text{thermal}$$

where $Ra_b \equiv$ Modified Channel Rayleigh number for uniform wall T_W. Ra_b for the Bar-Cohen and and Rohsenow correlation is given by Eq. (8.11) where γ is the same combination of variables that we defined in Section 8.1 for flat plates.

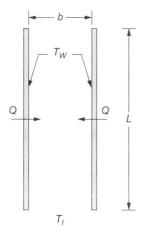

Figure 8.7. Isothermal vertical circuit board channel of height L and inlet air temperature T_I.

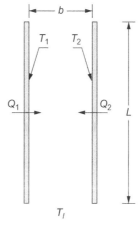

Figure 8.8. Unequal temperature, wall flux, vertical circuit board channel of height L and inlet air temperature T_I.

$$Ra_b = \frac{g\beta(T_W - T_I)b^4}{v^2 L} \cdot Pr = \left(\frac{g\beta}{v^2}\right)Pr\left(\frac{b^4}{L}\right)(T_W - T_I) = \left(\frac{\gamma}{\beta}\right)\beta\left(b^4/L\right)(T_W - T_I) \quad (8.11)$$

Bar-Cohen and Rohsenow (1984) considered an assembly of cards for which they developed a formula for the optimum spacing b_{opt} based on the total heat dissipation per unit wall temperature rise above the inlet temperature, $d[Q_{Total}/(T_W - T_I)]/db = 0$.

$$b_{opt} = 2.714\zeta^{-1/4}, \ Nu_{opt} = 1.31, \ Ra_{opt} = 54.3, \zeta = \gamma\Delta T/L \quad (8.12)$$

A second correlation, by Raithby and Hollands (1985), applies to unequal, isothermal temperature and unequal heat flux walls, where the geometry and nomenclature are illustrated in Figure 8.8. The Nusselt number in this instance is

$$\overline{Nu}_b = \left[\left(\frac{90(1+r_T)^2}{4r_T^2 + 7r_T + 4}\cdot\frac{1}{\overline{Ra}_b}\right)^{1.9} + \left(\frac{1}{0.62\overline{Ra}_b^{1/4}}\right)^{1.9}\right]^{-1/1.9} \quad \begin{array}{l}\text{Unequal} \ (8.13)\\ \text{isothermal}\end{array}$$

and $\overline{Ra}_b \equiv$ Channel Rayleigh number for $T_1 \neq T_2$ where

$$\overline{Ra}_b = \frac{g\beta(1+r_T)(T_1 - T_I)b^4}{2v^2 L}\cdot Pr = \frac{\gamma(1+r_T)(T_1 - T_I)b^4}{2L} \quad r_T = \frac{T_2 - T_I}{T_1 - T_I} \quad (8.14)$$

In both of the isothermal wall correlations, the air properties are evaluated at the average film temperature $\overline{T}_f = (T_W + T_I)/2$ except in cases of small $\overline{Ra}_b$ where the average wall temperature should be used. In both cases the thermal expansion coefficient β is based on the inlet temperature, i.e., $\beta = 1/(T_I + 273.16)$.

The third and final vertical channel correlation for this section is also by Bar-Cohen and Rohsenow (1984) for symmetric, isoflux walls where we use a nomenclature that indicates the locations for which the Nusselt number applies. Figure 8.7 best suits this situation.

$$Nu_b(L): \quad \text{wall temperature at channel exit}$$
$$Nu_b(L/2): \quad \text{wall temperature at channel mid-point}$$

The Nusselt numbers are given by

$$Nu_b(L) = \left[\frac{24C}{Ra_b^*} + \frac{2.51}{\left(Ra_b^*\right)^{0.4}}\right]^{-1/2}, \ Nu_b\left(\frac{L}{2}\right) = \left[\frac{6C}{Ra_b^*} + \frac{1.85}{\left(Ra_b^*\right)^{0.4}}\right]^{-1/2} \quad \text{Isoflux} \ (8.15)$$

$C = 2$, Symmetrically heated walls; $C = 1$, Asymmetrically heated walls (one heated, the other unheated, i.e., adiabatic)

The modified channel Rayleigh number Ra_a^* for uniform heat flux is

$$Ra_b^* \equiv \left(\frac{g\beta}{v^2}\right)Pr\left(\frac{b^5}{L}\right)\left(\frac{q}{k}\right) = \left(\frac{\gamma}{\beta}\right)\beta\left(\frac{b^5}{L}\right)\left(\frac{q}{k}\right) \quad (8.16)$$

$$q = \text{Wall heat flux} = \frac{Q_{One\ Wall\ Side}}{LW}; L, W = \text{Wall height, depth}$$

Bar-Cohen and Rohsenow (1984) developed optimum spacing b_{opt} criteria as follows.

$$\text{L: Sym. heating: } b_{opt} = 1.472\zeta^{-1/5}, \ Nu_{opt} = 0.351, \ Ra_{opt}^* = 6.9$$

$$\text{Asym. heating: } b_{opt} = 1.169\zeta^{-1/5}, \ Nu_{opt} = 0.464, \ Ra_{opt}^* = 6.9 \qquad (8.17)$$

$$\text{L/2: Sym. heating: } b_{opt} = 1.472\zeta^{-1/5}, \ Nu_{opt} = 0.62, \ Ra_{opt}^* = 6.9$$

$$\text{Asym. heating: } b_{opt} = 1.169\zeta^{-1/5}, \ Nu_{opt} = 0.49, \ Ra_{opt}^* = 2.2$$

$$\zeta = \frac{Ra_b^*}{b^5} = \left(\frac{g\beta}{\nu^2}\right) Pr \left(\frac{q}{k_{Air}L}\right) = \left(\frac{\gamma}{\beta}\right)\beta\left(\frac{q}{k_{Air}L}\right)$$

Figure 8.9 is a plot of Eqs. (8.10), (8.13), and (8.15). Note the locations of (1) $Ra_b = 54$ for optimum spacing in the Bar-Cohen & Rohsenow symmetric isothermal channel and (2) $Ra_b = 6.9$ for optimum spacing in the Bar-Cohen & Rohsenow symmetric isoflux channel. In the case of the former, it is clear that the optimum spacing occurs where the slope begins to increase dramatically as Ra_b decreases. However, in the second case, the value of Ra_b for optimum spacing is not particularly intuitive.

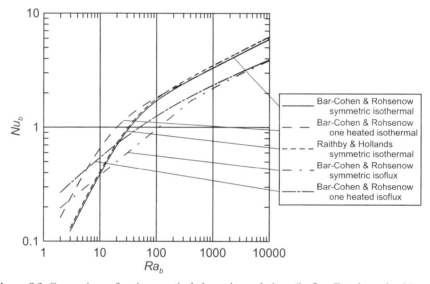

Figure 8.9. Comparison of various vertical channel correlations (isoflux T at channel exit).

Our formula for the optimum circuit board spacing, Eq. (8.12), is not convenient for quick estimates. Therefore, we carry this topic a little further with Eq. (8.12) which shows that the optimum spacing is $b_{opt} = 2.714\zeta^{-1/4}$. We shall work out our own formula for b_{opt} and end up with an easier-to-use result. We begin with the Rayleigh number

$$Ra_b = (b/L)Gr_b Pr = (b/L)\left[\left(g\beta/\nu^2\right)b^3\Delta T\right]Pr = (b/L)\left[\left(g\beta/\nu^2\right)Pr\right]b^3\Delta T = \left(b^4/L\right)\gamma\Delta T$$

where we obtain γ from either Figure 8.3 or Eq. (8.3). The preceding is rewritten as

$$Ra_{b-opt} = \left(b_{opt}^4/L\right)\gamma\Delta T, \ \ b_{opt}^4 = \frac{LRa_{b-opt}}{\gamma\Delta T}$$

Next we denote an optimum spacing b_{opt} consistent with the value of $Ra_{b-opt} = 54.3$ to obtain

$$b_{opt}^4 = \frac{LRa_{b-opt}}{(\gamma/\beta)\beta\Delta T} = \frac{54.3L(T_I + 273.16)}{(\gamma/\beta)\Delta T} \qquad (8.18)$$

$$\frac{b_{opt}}{L^{1/4}} = 2.7146 \left[\frac{\left(T_I + 273.16\right)}{\left(\gamma/\beta\right)\Delta T} \right]^{1/4} \qquad \text{Isothermal} \quad (8.19)$$
$$\text{symmetric}$$

so that γ/β is evaluated at the mean film temperature and is readily obtained from Figure 8.3 or Eq. 8.3. Equation (8.19) is plotted in Figure 8.10. Note that both b_{opt} and L must be in units of inches here.

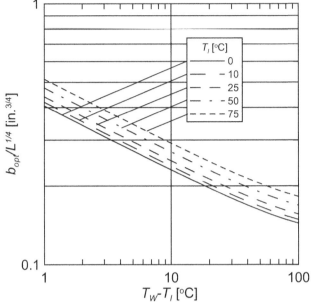

Figure 8.10. Optimum board spacing for symmetric, isothermal walls.

As was the case with the isothermal boards, Eqs. 8.17 are not convenient for quick estimates. Equation (8.17) shows that $b_{opt} = 1/472\zeta^{-1/5}$. The Rayleigh number is

$$Ra_b^* = \left(\frac{b^5}{L}\right)\left[\left(g\beta/v^2\right)\left(\frac{q}{k_{Air}}\right) \right]Pr = \gamma \left(\frac{b^5}{L}\right)\left(\frac{q}{k_{Air}}\right)$$

where we obtain γ from either Figure 8.3 or Eq. (8.3). Next we denote an optimum spacing b_{opt} consistent with the value of $Ra_{b-opt}^* = 6.9$ to obtain

$$Ra_{b-opt}^* = \gamma\left(b_{opt}^5/L\right)\left(q/k_{Air}\right), \quad b_{opt}^5 = \frac{LRa_{b-opt}^*}{\gamma\left(q/k_{Air}\right)}$$

All properties of air are evaluated at the film temperature, except the thermal expansion coefficient, $\beta = 1/\left(T_I + 273.16\right)$ so that we should instead use

$$b_{opt}^5 = \frac{LRa_{b-opt}^*}{\left(\gamma/\beta\right)\beta\left(q/k_{Air}\right)} = \frac{6.9L\left(T_I + 273.16\right)}{\left(\gamma/\beta\right)\left(q/k_{Air}\right)} \qquad (8.20)$$

$$\frac{b_{opt}}{L^{1/5}} = 1.4715 \left[\frac{\left(T_I + 273.16\right)}{\left(\gamma/\beta\right)\left(q/k_{Air}\right)} \right]^{1/5} \qquad \text{Isoflux} \quad (8.21)$$
$$\text{symmetric}$$
$$Nu(L), Nu(L/2)$$

γ/β is obtained from Figure 8.3 or Eq. 8.3. Eq. (8.21) is plotted in Figure 8.11. Both b_{opt} and L are in units of inches. $Nu(L)$ and $Nu(L/2)$ are obtained at the channel exit-height and half-height, respectively.

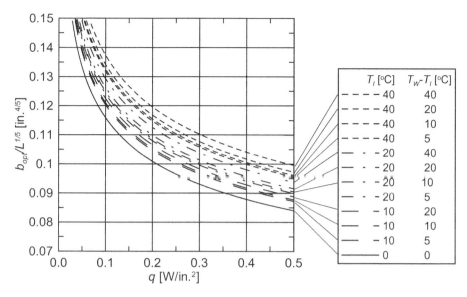

T_I [°C]	T_W-T_I [°C]
40	40
40	20
40	10
40	5
20	40
20	20
20	10
20	5
10	20
10	10
10	5
0	0

Figure 8.11. Optimum board spacing for symmetric, isoflux walls, $Nu(L)$, $Nu(L/2)$.

8.7 APPLICATION EXAMPLE: VERTICAL CARD ASSEMBLY

The geometry for this problem is shown in Figure 8.12. We have an array of equally spaced circuit boards with dimensions $L = 10$ in., $W = 10$ in., heat dissipation $Q = 25$ W per board, and an inlet air temperature $T_I = 30$°C. Assume that each Q convects equally from both the component side and backside of a board (thus a single convection resistance for one board side uses $Q/2$). The various optimization formulae offered in this chapter assume an infinitesimal board/component thickness. The practical thing to do is to use the b_{opt} as the *component top-surface to opposing board surface* distance. We then see what we get for an optimum card spacing and also calculate the temperature rise above inlet for each *PCB* and apply both the isothermal and the isoflux models to determine the differences.

First we do this the easy way by using Figure 8.10 for the isothermal model and Figure 8.11 for the isoflux model. The former requires a wall temperature rise. Let's guess T_W - T_I =25°C. Then using Figure 8.10:

$$\text{Isothermal model: } b_{opt} = \left(b_{opt}/L^{1/4}\right)_{\text{Figure 8.10}} L^{1/4} = (0.2)(10)^{1/4} = 0.356\,\text{in.}$$

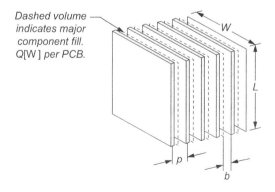

Dashed volume indicates major component fill. Q[W] per PCB.

Figure 8.12. Application Example 8.7.

Performing an approximate visual interpolation of curves in Figure 8.11 for the isoflux model, we use the same wall temperature rise, a calculated wall flux q, arrive at

$$q = (Q/2)/(WL) = (25\,\text{W}/2)/(100\,\text{in.}^2) = 0.125\,\text{W}/\text{in.}^2$$

$$\text{Isoflux model: } b_{opt} = (b_{opt}/L^{1/5})_{\text{Figure 8.11}} L^{1/5} = (0.125)(10)^{1/5} = 0.198\,\text{in.}$$

The isothermal and isoflux models give us quite different values for b_{opt}. We need to examine the two models in more detailed precision to see if we can make some judgment as to which result to use for *our design*. This means that we should use the more exact formulae for our models and perform all of the necessary iterations that we need to get the correct values for the temperature dependencies within the Rayleigh and Nusselt numbers.

When we combine Eqs. (8.3) and (8.19) for the isothermal model we get

$$\frac{b_{opt}}{L^{1/4}} = 2.7146\left[\frac{(T_I + 273.16)}{(\gamma/\beta)\Delta T}\right]^{1/4} = 2.7146\left[\frac{(T_I + 273.16)}{5.454\times10^5 \exp(-9.254\times10^{-3}T)}\right]^{1/4}\frac{1}{\Delta T^{1/4}}$$

$$= 2.7146\left[\frac{(T_I + 273.16)}{5.454\times10^5 \exp(-9.254\times10^{-3}(T_I + \Delta T/2))}\right]^{1/4}\frac{1}{\Delta T^{1/4}}$$

$$\frac{b_{opt}}{L^{1/4}} = 2.7146\left\{\frac{(T_I + 273.16)}{5.454\times10^5 \exp(-9.254\times10^{-3}[T_I + (T_W - T_I)/2])}\right\}^{1/4}\frac{1}{(T_W - T_I)^{1/4}}$$

and to actually calculate b_{opt} we see that we also need the temperature rise $T_W - T_I$. We must begin with an estimate of T_W, calculate b_{opt}, and also recalculate T_W. In order to calculate T_W, we need to combine Eq. (8.3) for $\gamma/\beta, \beta$ with (8.10) for $\overline{Nu}_b$ and (8.11) for Ra_b.

$$Ra_b = (\gamma/\beta)\beta(b^4/L)(T_W - T_I)$$

$$Ra_b = \frac{5.454\times10^5 \exp\left[-9.254\times10^{-3}\left(\frac{T_W - T_I}{2} + T_I\right)\right]}{(T_I + 273.16)}(b^4/L)(T_W - T_I)$$

$$k_{Air} = 6.0583\times10^{-4} + 1.6906\times10^{-6}T = 6.0583\times10^{-4} + 1.6906\times10^{-6}\left(\frac{T_W - T_I}{2} + T_I\right)$$

$$\overline{Nu}_b = \left[\left(\frac{24}{Ra_b}\right)^2 + \left(\frac{1}{0.59Ra_b^{1/4}}\right)^2\right]^{-1/2}, \qquad h = \left(\frac{k_{Air}}{b}\right)\overline{Nu}_b$$

$$T_W = \frac{(Q/2)}{(k_{Air}/b)\overline{Nu}_b WL} + T_I$$

The number of iterations required is dependent on the starting guess of T_W and also some modification of each calculated T_W for each subsequent iteration. In this example an initial value of T_W was used with three iterations used as shown in Table 8.10. The details of the last step follow. Thus, here we calculate b_{opt} using the final value of T_W.

$$b_{opt} = 2.7146\left\{\frac{(T_I + 273.16)}{5.454\times10^5 \exp(-9.254\times10^{-3}[T_I + (T_W - T_I)/2])}\right\}^{1/4}\left(\frac{L}{T_W - T_I}\right)^{1/4}$$

$$b_{opt} = 2.7146\left\{\frac{(30 + 273.16)}{5.454\times10^5 \exp(-9.254\times10^{-3}[30 + (74.7 - 30)/2])}\right\}^{1/4}\left(\frac{10}{74.7 - 30}\right)^{1/4} = 0.323\,\text{in.}$$

$$Ra_b = \frac{5.454 \times 10^5 \exp\left[-9.254 \times 10^{-3}\left(\dfrac{T_W - T_I}{2} + T_I\right)\right]}{(T_I + 273.16)} \left(b_{opt}^4 / L\right)(T_W - T_I)$$

$$Ra_b = \frac{5.454 \times 10^5 \exp\left[-9.254 \times 10^{-3}\left(\dfrac{74.7 - 30}{2} + 30\right)\right]}{(30 + 273.16)} \left[\frac{(0.323)^4}{10}\right](74.7 - 30) = 53.903$$

$$k_{Air} = 6.0583 \times 10^{-4} + 1.6906 \times 10^{-6} T = 6.0583 \times 10^{-4} + 1.6906 \times 10^{-6}\left(\frac{T_W - T_I}{2} + T_I\right)$$

$$k_{Air} = 6.0583 \times 10^{-4} + 1.6906 \times 10^{-6}\left(\frac{74.7 - 30}{2} + 30\right) = 6.943 \times 10^{-4}$$

$$\overline{Nu}_b = \left[\left(\frac{24}{Ra_b}\right)^2 + \left(\frac{1}{0.59 Ra_b^{1/4}}\right)^2\right]^{-1/2} = \left[\left(\frac{24}{53.903}\right)^2 + \left(\frac{1}{0.59(53.903)^{1/4}}\right)^2\right]^{-1/2} = 1.304$$

$$h = \left(\frac{k_{Air}}{b_{opt}}\right)\overline{Nu}_b = \left(\frac{6.943 \times 10^{-4}}{0.323}\right)(1.304) = 2.801 \times 10^{-3}\ \text{W}/\left(\text{in.}^2 \cdot {}^\circ\text{C}\right)$$

$$T_W = \frac{(Q/2)}{hWL} + T_I = \frac{(25/2)}{(2.801 \times 10^{-3})(10)(10)} + 30 = 74.627\,{}^\circ\text{C}$$

Thus we see that this computed $T_W = 74.627{}^\circ$C is very close to the $T_W = 74.7{}^\circ$C that we began with in this iteration. We need to continue no further with the isothermal model.

Table 8.10. Application Example 8.7: Summary of iterative results for isothermal model.

T_W [$^\circ$C]	b_{opt} [in.]	Calculated T_W [$^\circ$C]
40	0.451	95.078
75	0.323	74.552
74.7	0.323	74.627

Next we make the same b_{opt} calculation except that we use the symmetric isoflux model. When we combine Eqs. (8.3) and (8.21) for the isoflux model, we get

$$\frac{b_{opt}}{L^{1/5}} = 1.4715\left[\frac{(T_I + 273.16)}{(\gamma/\beta)\Delta T}\right]^{1/5} = 1.4715\left[\frac{(T_I + 273.16)}{5.454 \times 10^5 \exp\left(-9.254 \cdot 10^{-3} T\right)}\right]^{1/5} \frac{1}{\Delta T^{1/5}}$$

$$= 1.4715\left[\frac{(T_I + 273.16)}{5.454 \times 10^5 \exp\left(-9.254 \times 10^{-3}\left(T_I + \Delta T/2\right)\right)}\right]^{1/5} \frac{1}{\Delta T^{1/5}}$$

$$\frac{b_{opt}}{L^{1/5}} = 1.4715\left\{\frac{(T_I + 273.16)}{5.454 \times 10^5 \exp\left(-9.254 \times 10^{-3}\left[T_I + (T_W - T_I)/2\right]\right)}\right\}^{1/5} \frac{1}{(T_W - T_I)^{1/5}}$$

As was the situation with the isothermal model, in order to calculate b_{opt} we see that we also need the temperature rise $T_W - T_I$. We must begin with an estimate of T_W, calculate b_{opt}, and recalculate T_W. In order to calculate T_W, we need to combine Eqs. (8.3) with (8.15) and (8.16).

$$Ra_b^* \equiv \left(\frac{g\beta}{v^2}\right)Pr\left(\frac{b^5}{L}\right)\left(\frac{q}{k}\right) = \gamma\left(\frac{b^5}{L}\right)\left(\frac{q}{k}\right) = (\gamma/\beta)\beta\left(\frac{b_{opt}^5}{L}\right)\left(\frac{q}{k_{Air}}\right)$$

$$Ra_b^* = \frac{5.454\times10^5 \exp\left[-9.254\times10^{-3}\left(\dfrac{T_W - T_I}{2} + T_I\right)\right]}{(T_I + 273.16)}\left(\frac{b_{opt}^5}{L}\right)\left(\frac{q}{k_{Air}}\right)$$

$$k_{Air} = 6.0583\times10^{-4} + 1.6906\times10^{-6}T = 6.0583\times10^{-4} + 1.6906\times10^{-6}\left(\frac{T_W - T_I}{2} + T_I\right)$$

$$Nu_b(L) = \left[\frac{48}{Ra_b^*} + \frac{2.51}{\left(Ra_b^*\right)^{0.4}}\right]^{-1/2}, \quad h = \left(\frac{k_{Air}}{b}\right)Nu_b(L), \quad T_W = \frac{(Q/2)}{\left(\dfrac{k_{Air}}{b}\right)Nu_b(L)WL} + T_I$$

A total of five iterations were used. We will use the model for $Nu_b(L)$ first, showing the arithmetic for the final iteration, which begins with $T_W = 135°C$.

$$k_{Air} = 6.0583\times10^{-4} + 1.6906\times10^{-6}T = 6.0583\times10^{-4} + 1.6906\times10^{-6}\left(\frac{T_W - T_I}{2} + T_I\right)\times$$

$$k_{Air} = 6.0583\times10^{-4} + 1.6906\times10^{-6}\left(\frac{135 - 30}{2} + 30\right) = 7.453\times10^{-4}$$

$$b_{opt} = 1.4715\left\{\frac{(T_I + 273.16)}{5.454\times10^5 \exp\left(-9.254\times10^{-3}\left[T_I + (T_W - T_I)/2\right]\right)}\right\}^{1/5}\left[\frac{L}{(q/k_{Air})}\right]^{1/5}$$

$$= 1.4715\left\{\frac{(30 + 273.16)}{5.454\times10^5 \exp\left(-9.254\times10^{-3}\left[30 + (135 - 30)/2\right]\right)}\right\}^{1/5}\left[\frac{10}{\left(0.125/7.453\times10^{-4}\right)}\right]^{1/5}$$

$$b_{opt} = 0.218\,\text{in.}$$

$$Ra_b^* = \frac{5.454\times10^5 \exp\left[-9.254\times10^{-3}\left(\dfrac{T_W - T_I}{2} + T_I\right)\right]}{(T_I + 273.16)}\left(\frac{b_{opt}^5}{L}\right)\left(\frac{q}{k_{Air}}\right)$$

$$Ra_b^* = \frac{5.454\times10^5 \exp\left[-9.254\times10^{-3}\left(\dfrac{135 - 30}{2} + 30\right)\right]}{(30 + 273.16)}\left[\frac{(0.218)^5}{10}\right]\left(\frac{0.125}{7.453\times10^{-4}}\right) = 6.900$$

$$Nu_b(L) = \left[\frac{48}{6.900} + \frac{2.51}{(6.900)^{0.4}}\right]^{-1/2} = 0.351$$

$$h = \left(\frac{k_{Air}}{b}\right)Nu_b(L) = \left(\frac{7.453\times10^{-4}}{0.218}\right)(0.351) = 0.00120\,\text{W}/\left(\text{in.}^2 \cdot °C\right)$$

$$T_W = \frac{(Q/2)}{hWL} + T_I = \frac{(25/2)}{(0.00120)(10)(10)} + 30 = 134.11°C$$

This computed value of $T_W - T_I = 134.11°C$ is sufficiently close to the starting value of $T_W = 135°C$ that we can consider our problem of calculating b_{opt} completed. We see that this $b_{opt} = 0.218$ in. is close enough to the earlier estimated value of 0.198 in. The initial and final values of T_W for the iterations of this isoflux problem are listed in Table 8.11.

Table 8.11. Application Example 8.7: Summary of iterative results for isoflux model for $Nu(L)$.

T_W [°C]	b_{opt} [in.]	Calculated T_W [°C]
50	0.197	134.342
75	0.203	134.184
100	0.209	134.103
125	0.215	134.095
135	0.218	134.110

Plots in Figure 8.13 comparing isoflux and isothermal models are quite different. We expect that the $Nu_b(L)$ plot would be above the plot for $Nu_b(L/2)$ as the former is a measure of the wall temperature at channel exit and the latter is at the wall half-height. It is also apparent that as the wall spacing decreases, the wall temperature increases at a greater rate for the top half of the card than for the lower half. However, the b_{opt} for the isoflux model is misleading in that a considerably lower wall temperature rise is possible by increasing the spacing. The optimum spacing concept for the isoflux model does not appear very useful from a temperature viewpoint. The b_{opt} for the isotherm model seems useful in that a wall temperature decrease is not very likely gained by increasing b beyond b_{opt}. Note also that the isoflux $Nu_b(L/2)$ temperature profile is not very different than that for the isotherm model at spacing greater than b_{opt} computed for an isotherm wall model. The reader will certainly question whether the isoflux or isothermal model should be used. We will postpone answering this question until the completion of an enclosure with circuit boards example in Section 8.10.

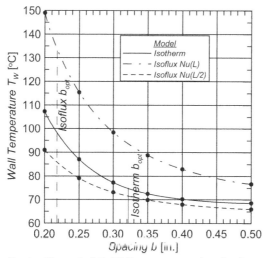

Figure 8.13. Application Example 8.7: Wall temperature plots for three channel models.

8.8 RECOMMENDED USE OF VERTICAL CHANNEL MODELS IN SEALED AND VENTED ENCLOSURES

An end view of a set of circuit boards is shown in Figure 8.14. As you will see after reading the next section, we will be able to calculate only one temperature for each chamber in a sealed or vented enclosure. This presents a problem as to what to use for a *PCB* inlet temperature T_I. There are three temperatures that one might use either alone or in some combination: (1) the calculated internal air T_{Air}, (2) the calculated enclosure surface temperature T_{Wall}, and (3) the ambient air T_A. The author has determined that an average of T_{Air} and T_{Wall} seems to be the most satisfactory compromise.

$$T_I = (T_{Air} + T_{Wall})/2 \qquad\qquad PCB \text{ inlet air } (8.22)$$

The reader should note that from this point on, we shall denote the circuit boards by the subscript *PCB* and any enclosure wall surface temperatures by the subscript *Wall* or *W*.

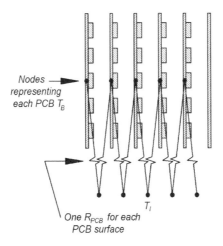

Figure 8.14. Circuit boards with thermal resistances R_{PCB} representing either isothermal or isoflux model with board channel inlet temperature T_I. Room ambient $T_{Amb} \equiv T_A$.

8.9 CONVERSION OF ISOTHERMAL WALL CHANNEL HEAT TRANSFER COEFFICIENTS FROM REFERENCED-TO-INLET AIR TO REFERENCED-TO-LOCAL AIR

I am repeating this section from Chapter 7, but discussing it in the present context. You may not have noticed, but the vertical channel heat transfer coefficients outlined in Section 8.6 all referenced the channel inlet air temperature T_I. There may be occasions where you wish to adapt your model to board temperatures referenced to a local air temperature and use the channel heat transfer coefficients that you have learned about in this chapter. Figure 8.15 illustrates the two situations.

In Chapter 7 we derived a thermal conductance $C_I = 1/R_{Local}$, where the heat transfer coefficient h_{Local} was referenced to local air, and the conductance C_I was specifically referenced to inlet air (thus the subscript on C_I).

$$C_I / h_{Local} A_s = (1 - e^{-\beta})/\beta, \quad \beta = h_{Local} A_s / C_f$$

The conductance is clearly written as $C_I = h_I A_s$ for a convecting channel area A_s,

$$C_I / h_{Local} A_S = \frac{1 - e^{-h_{Local} A_S / C_f}}{(h_{Local} A_S / C_f)} \quad \text{or} \quad h_I A_S / h_{Local} A_S = \frac{1 - e^{-h_{Local} A_S / C_f}}{(h_{Local} A_S / C_f)}$$

which is readily solved to give

$$\boxed{h_{Local} = \frac{C_f}{A_S} \ln\left\{1 \Big/ \left[1 - \frac{h_I}{(C_f / A_S)}\right]\right\} = \frac{\rho G c_P}{A_S} \ln\left\{1 \Big/ \left[1 - \frac{h_I}{(\rho G c_P / A_S)}\right]\right\}} \quad h_I \text{ to } h_{Local} \; (8.23)$$

If we examine Eq. (8.23), we note that there is a problem if $h_l/(\rho Gc_P/A_S) > 1$ because in this situation, h_L becomes negative. This is readily explained if we compare the air and wall temperature rises above the inlet air.

$$\Delta T_{PCB} = T_{PCB} - T_I = Q/(h_l A_S), \quad \Delta T_{Air} = T_E - T_I = Q/(\rho c_P G)$$

Then
$$\frac{\Delta T_{Air}}{\Delta T_{PCB}} = \frac{Q/(\rho c_P G)}{Q/(h_l A_S)} = \frac{h_l}{(\rho c_P G/A_S)}$$

or
$$\frac{h_l}{(\rho c_P G/A_S)} = \frac{\Delta T_{Air}}{\Delta T_{PCB}}$$

Now we also know that the air temperature rise is less than that of the PCB so that $h_l/(\rho Gc_P/A_S)$ is less than one and we should not have a problem with Eq. (8.23).

Finally, you might find it curious that although we are addressing natural convection problems, we must know the airflow rate G to get h_L. We got into this dilemma because we used Eq. (8.23), the derivation of which was for forced air flow. This means that we can make our conversion if we are addressing a vented enclosure wherein we must compute G during the course of our problem solution. However, if we are considering a sealed enclosure, we can't apply Eq. (8.23). The next section is an application example of all the principles that we have learned in this chapter: horizontal and vertical flat plate convective heat transfer.

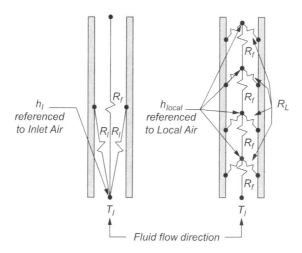

Figure 8.15. PCB channel convection thermal resistances R_I referenced to inlet air and resistances R_L referenced to local air. R_f are one-way, thermal fluid resistances (see Section 5.4).

8.10 APPLICATION EXAMPLE: ENCLOSURE WITH CIRCUIT BOARDS - ENCLOSURE TEMPERATURES ONLY

Our example is illustrated in Figure 8.16 with the various dimensions. The inlet and exit grill patterns, when relevant, are identical. The left- and right-hand sides of the card cage are open channels, which we will largely ignore in our *first-order analysis*. Structural support members for the card cage are not shown and are neglected in the airflow and thermal analysis. The enclosure dimensions are $D = 7$ in., $W = 7$ in., $H = 7$ in., $W_I = 5$ in., $H_I = 1.0$ in., $f_I = 0.35$, $D_{PCB} = 5$ in., $L = 5.0$ in., $b = 0.75$ in., $a = 0.25$ in., $w = l = 1.0$ in. The heat dissipation for each circuit board is $Q = 2.0$ W and the room ambient temperature is $T_A = 20°C$.

There is not much that we can do about analyzing airflow patterns in the sealed enclosure version in our network modeling, but we can perform a modest airflow simulation for the vented box.

Using what we hope is good judgment, the airflow pattern is shown in a side view of the vented enclosure in Figure 8.17. Clearly we are basing the flow resistance model on a direct inlet-to-exit path and not including anything concerning internal recirculation that we know must exist.

The reader should study Figure 8.18, the author's network representation of the airflow pattern suggested in Figure 8.17. The *Expand-to-Cards* and *Contract-to-Exit* elements are used to incorporate the fact that "something" must happen between the cards and the inlet/exit regions. A 90-degree turn would seem more reasonable, but such an element is considerably more resistive than the author has actually encountered in other simulations.

Our first step is to calculate the total system airflow resistance R_{af} consistent with Figure 8.18.

$$R_{af} = R_{Inlet} + R_{Expan-to-Cards} + R_{Cont-to-Cards} + R_{Card\,Cage} + R_{Expan-from-Cards} + R_{Cont-to-Exit} + R_{Exit}$$

$$R_{Inlet} = \text{Inlet grill}$$

$$R_{Expan-to-Cards} = \text{Expansion from inlet grill region to card cage inlet envelope}$$

$$R_{Cont-to-Cards} = \text{Contraction into card cage}$$

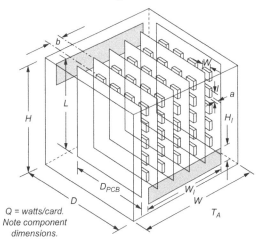

Figure 8.16. Application Example 8.10: Geometry for nonfan cooled enclosure in an ambient air T_A.

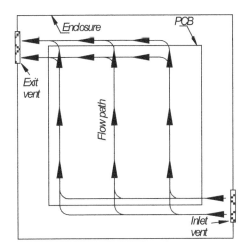

Figure 8.17. Application Example 8.10: Proposed airflow paths for vented enclosure.

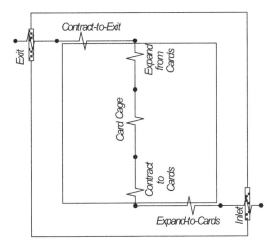

Figure 8.18. Application Example 8.10: Airflow circuit for vented enclosure.

$$R_{Card\,Cage} = \text{Entire card cage}$$

$$R_{Expan-from-Cards} = \text{Expansion from card cage interior to card cage exit envelope}$$

$$R_{Cont-to-Exit} = \text{Contraction from card cage envelope to exit grill region}$$

$$R_{Exit} = \text{Exit grill}$$

There is considerable judgment required concerning both the selection and application of the airflow resistance elements. If your ideas are different than those resulting here, you might try your own calculations.

We refer to Figure 3.13 to obtain the formula that enables us to calculate the inlet and exit grill resistances.

$$R_{Inlet} = R_{Exit} = \frac{1.9\times10^{-3}}{A_{Inlet}^2} = \frac{1.9\times10^{-3}}{\left(f_I W_I H_I\right)^2} = \frac{1.9\times10^{-3}}{\left(0.35\times5.0\,\text{in.}\times1.0\,\text{in.}\right)^2} = 6.20\times10^{-4}\,\frac{\text{in.H}_2\text{O}}{\text{CFM}^2}$$

Then calculate $R_{Expan-to-Cards}$, $R_{Cont-toCards}$ (note that we use A_1, A_2 in context with the element under consideration).

$$R_{Expan-to-Cards} = 1.29\times10^{-3}\left[\frac{1}{A_1}\left(1-\frac{A_1}{A_2}\right)\right]^2 = 1.29\times10^{-3}\left[\frac{1}{W_I H_I}\left(1-\frac{W_I H_{I1}}{5D_{PCB}\left(b+a\right)}\right)\right]^2$$

$$R_{Expan-to-Cards} = 1.29\times10^{-3}\left[\frac{1}{5\times1}\left(1-\frac{5\cdot1}{5\left(5.0\right)\left(0.75+0.25\right)}\right)\right]^2 = 3.30\times10^{-5}\,\text{in.H}_2\text{O}/\text{CFM}^2$$

$$R_{Cont-to-Cards} = \frac{0.5\times10^{-3}}{A_2^2}\left(1-\frac{A_2}{A_1}\right)^{3/4} = \frac{0.5\times10^{-3}}{\left[5D_{PCB}\left(b+a\right)-4\times5\left(wa\right)\right]^2}\left[1-\frac{5D_{PCB}\left(b+a\right)-4\times5\left(wa\right)}{5D_{PCB}\left(b+a\right)}\right]^{3/4}$$

$$= \frac{0.5\times10^{-3}}{\left[5.0\,\text{in.}\times5\times\left(0.75\,\text{in.}+0.25\,\text{in.}\right)-4\times5\left(1.0\,\text{in.}\times0.25\,\text{in.}\right)\right]^2}$$

$$\cdot\left[1-\frac{5\times5.0\,\text{in.}\times\left(0.75\,\text{in.}+0.25\,\text{in.}\right)-4\times5\left(1.0\,\text{in.}\times0.25\,\text{in.}\right)}{5\times5.0\,\text{in.}\times\left(0.75\,\text{in.}+0.25\,\text{in.}\right)}\right]^{3/4}$$

$$R_{Cont-to-Cards} = 3.74\times10^{-7}\,\text{in.H}_2\text{O}/\text{CFM}^2$$

The card cage resistance is calculated next. We first note the free area ratio and then select our best fit formula from Table 3.2.

$$A_{Cards-free} = \frac{5D_{PCB}\left(b+a\right)-4\times5\left(wa\right)}{5D_{PCB}\left(b+a\right)} = \frac{5\times5.0\,\text{in.}\left(0.75\,\text{in.}+0.25\,\text{in.}\right)-4\times5\left(1.0\,\text{in.}\times0.25\,\text{in.}\right)}{5\left(5.0\,\text{in.}\right)\left(0.75\,\text{in.}+0.25\,\text{in.}\right)} = 0.8\,\text{in.}^2$$

$$R_{Card-Cage} = \frac{3.08L\times10^{-4}}{\left[5D_{PCB}\left(b+a\right)\right]^2} = \frac{3.08\left(5.0\,\text{in.}\right)\times10^{-4}}{\left[5\left(5.0\,\text{in.}\right)\left(0.75\,\text{in.}+0.25\,\text{in.}\right)\right]^2} = 2.46\times10^{-6}\,\frac{\text{in.H}_2\text{O}}{\text{CFM}^2}$$

We continue by calculating the expansion resistance from the card cage to the region just following the card cage, then the resistance for contraction from the region near the card cage exit to the envelope of the exit grill.

$$R_{Expan-from-Cards} = 1.29\times10^{-3}\left[\frac{1}{A_1}\left(1-\frac{A_1}{A_2}\right)\right]^2 = 1.29\times10^{-3}\left\{\left[\frac{1}{5D_{PCB}\left(b+a\right)-4\times5\left(wa\right)}\right]\times\left[1-\frac{5D_{PCB}\left(b+a\right)-4\times5\left(wa\right)}{5D_{PCB}\left(b+a\right)}\right]\right\}^2$$

$$R_{Expan-from-Cards} = 1.29 \times 10^{-3} \left\{ \frac{1}{5(5.0\,\text{in.})(0.75\,\text{in.} + 0.25\,\text{in.}) - 4 \times 5(1.0\,\text{in.} \times 0.25\,\text{in.})} \times \left[1 - \frac{5(5.0\,\text{in.})(0.75\,\text{in.} + 0.25\,\text{in.}) - 4 \times 5(1.0\,\text{in.})(0.25\,\text{in.})}{5(5.0\,\text{in.})(0.75\,\text{in.} + 0.25\,\text{in.})} \right] \right\}^2$$

$$R_{Expan-from-Cards} = 1.29 \times 10^{-7} \,\text{in.H}_2\text{O}/\text{CFM}^2$$

$$R_{Cont-to-Exit} = \frac{0.5 \times 10^3}{A_2^2}\left(1 - \frac{A_2}{A_1}\right)^{3/4} = \frac{0.5 \times 10^{-3}}{\left(W_E H_E\right)^2}\left[1 - \frac{W_E H_E}{5 D_{PCB}(b+a)}\right]^{3/4} = \frac{0.5 \times 10^{-3}}{(5.0\,\text{in.} \times 1.0\,\text{in.})^2}$$

$$\times \left[1 - \frac{5.0\,\text{in.} \times 1.0\,\text{in.}}{5(5.0\,\text{in.})(0.75\,\text{in.} + 0.25\,\text{in.})}\right]^{3/4} = 1.69 \times 10^{-5} \,\text{in.H}_2\text{O}/\text{CFM}^2$$

The total system airflow resistance is given by the series total.

$$R_{af} = R_{Inlet} + R_{Expan-to-Cards} + R_{Cont-to-Cards} + R_{Card\,Cage} + R_{Expan-from-Cards} + R_{Cont-to-Exit} + R_{Exit}$$

$$= 6.20 \times 10^{-4} + 3.30 \times 10^{-5} + 3.74 \times 10^{-7} + 2.46 \times 10^{-6} + 1.29 \times 10^{-7} + 1.69 \times 10^{-5} + 6.20 \times 10^{-4}$$

$$R_{af} = 1.29 \times 10^{-3} \,\text{in.H}_2\text{O}/\text{CFM}^2$$

We won't use this result until we analyze the enclosure in the vented condition.

Now we refer to Figure 8.19, the thermal circuit for our enclosure problem. The circuit boards are not represented in this illustration. We know that the total heat Q has convected from the board components to the air, which we represent by the node at temperature T_{Air}. Implicit in this last statement is the fact that we are neglecting internal radiation. We can assume the multiple circuit boards inhibit much of the radiation exchange from the boards to the internal enclosure walls in the same way that concentric cylinders in a Thermos inhibit radiation. Furthermore, if this were a metal-walled enclosure, it is very likely that the internal wall surfaces would have a low emissivity.

Another approximation that is implicit in the thermal circuit is that the wall conduction resistance from interior to exterior is neglected. This would be a very reasonable approximation if, again, the walls were metal. If the walls were made of something like plastic, i.e., poor conducting, we

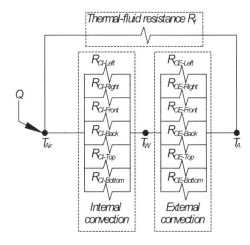

Figure 8.19. Application Example 8.10: Thermal circuit for nonfan cooled enclosure in an ambient air T_A.

could include a resistance from an internal wall node to an external wall node. It is also a simple procedure to calculate an estimated increase in T_{Air} due to a wall conduction resistance without explicitly placing a resistance in the circuit.

The perceptive reader will notice that a single node represents the entire internal air temperature, when we know that the air increases in temperature from the lower portions of the enclosure to the upper portions. This issue is a limitation of the lumped parameter method. We will discuss this later.

All six of the enclosure walls are also represented by a single node. You might therefore interpret the final results as an average wall temperature. The author has explored more detailed versions of this model where all six walls are included. Other model variations such as including multi-node walls with in-plane wall conduction resistances have been explored. Neither of these two variations has a significant effect on the computed value of T_{Air}.

It is now necessary to calculate the various resistances in Figure 8.19, or at least as much as possible of each resistance, and proceed with the problem solution. The solution must be carried out in an iterative process due to the dependence of the convection resistances on surface temperature rise.

We begin with the airdraft portion of the problem using the buoyancy draft Eq. (5.4).

$$G = 1.53 \times 10^{-2} \left(d_H Q_d / R_{af} \right)^{\frac{1}{3}} = 1.53 \times 10^{-2} \left[(5.0 \text{ in.}) Q_d / 1.29 \times 10^{-3} \right]^{\frac{1}{3}} = 0.24 Q_d^{1/3}$$

In this application of the draft equation, we have estimated the dissipation height $d_H = 5.0$ in. This is the vertical distance between the midpoints of the inlet and exit grills. Q_d will be the *heat carried away by the draft*, not the total dissipation Q. The thermal-fluid resistance R_f is calculated using the simple formula for C_f, Eq. (5.5).

$$R_f = 1/C_f = 1.76/G$$

The internal and external convection resistances are calculated using the laminar flow flat plate formulae from Table 8.7. Since the internal convection resistances are all in parallel with one another, it is more sensible to calculate the total internal convection conductance and then invert to get the total resistance. If we do this, factoring out the common temperature rise variable ΔT_{Air-W}, we easily arrive at the following:

$$R_{CI} = \frac{1}{C_{CI}} = \frac{1}{\left(h_{Top} WD + h_{Bottom} WD + 2h_{Left} DH + 2h_{Front} WH \right)}$$

$$= \frac{1}{\left\{ \begin{array}{l} 0.0022 WD \left[2(W+D)/(WD) \right]^{0.25} + 0.0011 WD \left[2(W+D)/(WD) \right]^{0.25} \\ +2(0.0024) HDH^{-0.25} + 2(0.0024) HWH^{-0.25} \end{array} \right\} \Delta T_{Air-W}^{0.25}}$$

$$= \frac{1}{\left\{ \begin{array}{l} 0.0022 (7.0 \text{ in.})^2 \left[2(14.0 \text{ in.})/(7.0 \text{ in.})^2 \right]^{0.25} + 0.0011 (7.0 \text{ in.})^2 \\ \times \left[2(14.0 \text{ in.})/(7.0 \text{ in.})^2 \right]^{0.25} + 2(0.0024)(7.0 \text{ in.})^2 (7.0 \text{ in.})^{-0.25} \\ +2(0.0024)(7.0 \text{ in.})^2 (7.0 \text{ in.})^{-0.25} \end{array} \right\} \Delta T_{Air-Wall}^{0.25}}$$

$$R_{CI} = 2.327 / \Delta T_{Air-Wall}^{0.25}$$

The external convection resistance is exactly the same as the preceding, except for the change in the temperature rise subscript.

$$R_{CE} = 2.327 / \Delta T_{Wall-A}^{0.25}$$

The total system thermal resistance from the internal air to the external ambient is the series sum of the convection resistances R_{CI} and R_{CE}, with the result in parallel to the thermal-fluid resistance R_{Tf}.

$$R_x = R_{CI} + R_{CE}$$
$$R_{Total} = R_x R_f / (R_x + R_f)$$

Table 8.12 is prepared using the results for the various resistances. Where relevant, each column in the iteration table is labeled with the necessary formula for this problem. The iteration is started for the sealed version where we use $Q_d = 1 \times 10^{-20}$, i.e., zero. The arbitrary values of $\Delta T_{Wall-A} = 25°C$ and $\Delta T_{Air-A} = 50°C$ are used to begin the sealed box iterations. The final sealed box result is used to begin the vented box solution. Note that in the latter problem, the initial value of Q_d is a guess. We see that the final results are the internal air and wall temperature rises of 25°C and 17°C for the sealed and vented enclosure variations, respectively. The reader is strongly advised to try the iterations for himself/herself following the table footnotes exactly as indicated, but perhaps beginning with different initial values.

Table 8.12. Application Example 8.10: Iteration table for sealed and vented versions of non-fan cooled enclosure in an ambient air T_A.

It.	ΔT_{Wall-A}	ΔT_{Air-A}	$\Delta T_{Air-Wall}$ $= \Delta T_{Air-A} - \Delta T_{Wall-A}$	Q_d $= \Delta T_{Air-A} G / 1.76$	G $= 0.24 Q_d^{1/3}$	R_f $= 1.76/G$	R_{CE} $= 2.327/\Delta T_{Wall-A}^{0.25}$	ΔT_{Wall-A} $= R_{CE}(Q - Q_d)$	R_{CI} $= 2.327/\Delta T_{Air-Wall}^{0.25}$	R_x $= R_{CI} + R_{CE}$	R_{Total} $= R_f R_x /(R_f + R_x)$	ΔT_{Air-A} $= R_{Total} Q$
1	25*	50*	25	10^{-20}	0	10^7	1.04	10.41	1.04	2.08	2.08	20.81
2	10.41**	20.08**	9.67	10^{-20}	0	10^7	1.30	13.00	1.30	2.62	2.62	26.15
3	13.00**	20.81**	7.08	10^{-20}	0	10^7	1.23	12.25	1.23	2.65	2.65	26.52
4	12.28**	26.52**	14.27	10^{-20}	0	10^7	1.24	12.44	1.24	2.44	2.44	24.41
5	12.44**	24.41**	11.97	10^{-20}	0	10^7	1.24	12.39	1.25	2.49	2.49	24.90
1	12.39**	24.90**	12.51	5*	0.411	4.29	1.24	6.20	1.24	2.48	1.57	15.70
2	6.20**	15.70**	9.50	3.67***	0.370	4.75	1.48	9.34	1.33	2.80	1.76	17.62
3	9.34**	17.02**	7.68	3.74***	0.358	4.80	1.33	8.55	1.40	2.73	1.74	17.39
4	8.55**	17.39**	8.84	3.54***	0.366	4.81	1.36	8.79	1.35	2.71	1.73	17.34
5	8.75**	17.41**	8.66	3.66***	0.370	4.76	1.37	8.68	1.37	2.74	1.74	17.40

Note: * Initial value is a guess; ** Use most recent value from previous iteration; *** Use most recent value of G from previous iteration.

8.11 APPLICATION EXAMPLE: ENCLOSURE WITH CIRCUIT BOARDS - CIRCUIT BOARD TEMPERATURES ONLY

We continue with the preceding enclosure problem, but now focus on the circuit boards. We shall neglect any end effects, by which we mean that we will use a symmetric temperature or flux model. It is certainly reasonable to apply a symmetric model if each circuit board has heat dissipation Q and we assume that the two sides of a *PCB* convect equally. The single circuit board and thermal model is illustrated in Figure 8.20.

Isothermal PCB in Sealed Enclosure

We shall begin by calculating the circuit board temperature for the sealed enclosure using Eqs. (8.3), (8.11), and (8.10). The first step is to estimate the board inlet air temperature using Eq. (8.22).

$$T_I = \left(T_{Air} + T_{Wall}\right)/2 = \left[(24.90 + 20) + (12.39 + 20)\right]/2 = 38.65\,^{\circ}\text{C}$$

The iterative solution can be accomplished in three iterations when we begin with $T_{PCB} = 60^{\circ}\text{C}$, but we shall show only intermediate results for the third and final iteration which we begin with $T_{PCB} = 52.0^{\circ}\text{C}$:

$$\beta = 1/\left(T_I + 273.16\right) = 1/\left(38.65 + 273.16\right) = 3.21 \times 10^{-3}\ \text{K}^{-1}$$

$$\gamma/\beta = \left(g/v^2\right)Pr = 5.454 \times 10^5\ \exp\left[-9.254 \times 10^{-3}\left(\frac{T_{PCB} + T_I}{2}\right)\right]$$

$$\gamma/\beta = 5.454 \times 10^5\ \exp\left[-9.254 \times 10^{-3}\left(\frac{52.0 + 38.77}{2}\right)\right] = 3.59 \times 10^5$$

The Rayleigh number is

$$Ra_b = \frac{g\beta\left(T_{PCB} - T_I\right)b^4}{v^2 L}Pr = \left(\frac{g\beta}{v^2}\right)Pr\left(\frac{b^4}{L}\right)\left(T_{PCB} - T_I\right) = \gamma\left(b^4/L\right)\left(T_{PCB} - T_I\right)$$

$$Ra_b = \left(\gamma/\beta\right)\beta\left(b^4/L\right)\left(T_{PCB} - T_I\right) = \left(3.59 \times 10^5\right)\left(3.21 \times 10^{-3}\right)\left[(0.75)^4/5.0\right]\left(52.0 - 38.65\right) = 972$$

from which we calculate the Nusselt number.

$$Nu_b = \left[\left(\frac{24}{Ra_b}\right)^2 + \left(\frac{1}{0.59 Ra_b^{1/4}}\right)^2\right]^{-1/2} = \left\{\left(\frac{24}{972}\right)^2 + \left[\frac{1}{0.59(972)^{0.25}}\right]^2\right\}^{-1/2} = 3.28$$

We use the thermal conductivity curve fit formula for air to calculate the heat transfer coefficient.

$$k_{Air} = 6.0583 \times 10^{-4} + 1.6906 \times 10^{-6}\left[\left(T_{PCB} + T_I\right)/2\right] = 6.0583 \times 10^{-4} + 1.6906 \times 10^{-6}\left[(52.0 + 38.77)/2\right]$$

$$k_{Air} = 6.83 \times 10^{-4}\ \text{W}/\left(\text{in.}\cdot{}^{\circ}\text{C}\right)$$

$$h_b = \left(k_{Air}/b\right)Nu_b = \left(6.83 \times 10^{-4}/0.75\right)(3.28) = 0.00299\ \text{W}/\left(\text{in.}^2\cdot{}^{\circ}\text{C}\right)$$

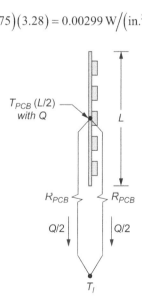

Figure 8.20. Application Example 8.11: Thermal circuit for circuit board in nonfan cooled enclosure in an ambient air T_A.

We calculate the board temperature remembering to use only half of the total board dissipation.

$$T_{PCB} = R_{PCB}(Q/2) + T_I = \frac{1}{hLD_{PCB}}(Q/2) + T_I = \frac{1}{(0.00299)(5.0\,\text{in.})^2}(2.0\,\text{W}/2) + 38.65 = 52.0\,^\circ\text{C}$$

where our result is equal to the value of 52.0°C for which we began the iteration.

Isothermal PCB in Vented Enclosure

The solution for the vented box is similarly calculated except we have different values of $T_{Air} = 37.4$°C and $T_{Wall} = 28.7$°C.

$$T_I = (T_{Air} + T_{Wall})/2 = (37.4 + 28.7)/2 = 33.0\,^\circ\text{C}$$

The solution can be accomplished in two iterations when we begin with $T_{PCB} = 60$°C, but we shall show intermediate results for the third and final iteration which we begin with $T_{PCB} = 46.5$°C.

$$\beta = 1/(T_I + 273.16) = 1/(33.1 + 273.16) = 3.27 \times 10^{-3}\,\text{K}^{-1}$$

$$\gamma/\beta = (g/v^2)Pr = 5.460 \times 10^5 \exp\left[-9.2817 \times 10^{-3}\left(\frac{T_{PCB} + T_I}{2}\right)\right]$$

$$\gamma/\beta = 5.460 \times 10^5 \exp\left[-9.2817 \times 10^{-3}\left(\frac{46.5 + 33.1}{2}\right)\right] = 3.78 \times 10^5$$

$$Ra_b = \frac{g\beta(T_{PCB} - T_I)b^4}{v^2 L}Pr = \left(\frac{g\beta}{v^2}\right)Pr\left(\frac{b^4}{L}\right)(T_{PCB} - T_I) = \gamma\left(b^4/L\right)(T_{PCB} - T_I)$$

$$= (\gamma/\beta)\beta\left(b^4/L\right)(T_{PCB} - T_I) = (3.78 \times 10^5)(3.27 \times 10^{-3})\left[(0.75)^4/5.0\right](46.5 - 33.1)$$

$$Ra_b = 1050$$

$$Nu_b = \left[\left(\frac{24}{Ra_b}\right)^2 + \left(\frac{1}{0.59 Ra_b^{1/4}}\right)^2\right]^{-1/2} = \left\{\left(\frac{24}{1050}\right)^2 + \left[\frac{1}{0.59(1050)^{0.25}}\right]^2\right\}^{-1/2} = 3.35$$

$$k_{Air} = 6.0583 \times 10^{-4} + 1.6906 \times 10^{-6}\left[(T_{PCB} + T_I)/2\right]$$

$$= 6.0583 \times 10^{-4} + 1.6906 \times 10^{-6}\left[(46.5 + 33.1)/2\right]$$

$$k_{Air} = 6.73 \times 10^{-4}\,\text{W}/(\text{in.}\cdot{}^\circ\text{C})$$

$$h_b = (k_{Air}/b)Nu_b = (6.73 \times 10^{-4}/0.75)(3.34) = 0.00301\,\text{W}/(\text{in.}^2\cdot{}^\circ\text{C})$$

$$T_{PCB} = R_{PCB}(Q/2) + T_I = \frac{1}{hLD_{PCB}}(Q/2) + T_I = \frac{1}{(0.00301)(5.0\,\text{in.})^2}(2.0\,\text{W}/2) + 33.0 = 46.4\,^\circ\text{C}$$

Isoflux PCB in Sealed Enclosure

The PCB inlet temperature $T_I = 38.65$°C is applicable. The solution is calculated in two iterations beginning with $T_{PCB} = 60$°C, but the results are shown here for only the second and last iteration where we begin with $T_{PCB} = 51$°C. Equations (8.3), (8.16), and (8.15) are used:

$$\beta = 1/(T_I + 273.16) = 1/(38.77 + 273.16) = 3.21 \times 10^{-3}\,\text{K}^{-1}$$

$$\gamma/\beta = (g/v^2)Pr = 5.454 \times 10^5 \exp\left[-9.254 \times 10^{-3} \times 0.5 \times (T_{PCB} + T_I)\right]$$

$$\gamma/\beta = 5.454 \times 10^5 \exp\left[-9.254 \times 10^{-3} \times 0.5 \times (51.0 + 38.65)\right] = 3.60 \times 10^5$$

The isoflux Rayleigh number requires a heat flux density, and air thermal conductivity.

$$q = (Q/2)/LD_{PCB} = (2.0\,W/2)/(5.0\,in.)(5.0\,in.) = 0.040\,W/in.^2$$

$$k_{Air} = 6.0583 \times 10^{-4} + 1.6906 \times 10^{-6} \left[(T_{PCB} + T_I)/2 \right]$$

$$= 6.0583 \times 10^{-4} + 1.6906 \times 10^{-6} \left[(51.0 + 38.77)/2 \right]$$

$$k_{Air} = 6.82 \times 10^{-4}\,W/(in. \cdot {}^{\circ}C)$$

and the isoflux Rayleigh number is calculated to be

$$Ra_b^* = \left(\frac{g\beta}{v^2} \right) Pr \left(\frac{b^5}{L} \right) \left(\frac{q}{k_{Air}} \right) = \gamma \left(\frac{b^5}{L} \right) \left(\frac{q}{k_{Air}} \right) = \left(\frac{\gamma}{\beta} \right) \beta \left(\frac{b^5}{L} \right) \left(\frac{q}{k_{Air}} \right)$$

$$Ra_b^* = \left(3.60 \times 10^5 \right) \left(3.21 \times 10^{-3} \right) \left[\frac{(0.75)^5}{5.0} \right] \left(\frac{0.040}{6.82 \times 10^{-4}} \right) = 3217$$

The isoflux Nusselt number $Nu_b(L/2)$ and heat transfer coefficient are calculated as

$$Nu_b \left(\frac{L}{2} \right) = \left[\frac{12}{Ra_b^*} + \frac{1.85}{\left(Ra_b^* \right)^{0.4}} \right]^{-1/2} = \left[\frac{12}{3217} + \frac{1.85}{(3217)^{0.4}} \right]^{-1/2} = 3.61$$

$$h_b = (k_{Air}/b) Nu_b (L/2) = (6.82 \times 10^{-4}/0.75)(3.61) = 0.00328\,W/(in.^2 \cdot {}^{\circ}C)$$

with the final result for the circuit board temperature being

$$T_{PCB} = R_{PCB} (Q/2) + T_I = \frac{1}{hLD_{PCB}} (Q/2) + T_I = \frac{1}{(0.00328)(5.0\,in.)^2} (2.0\,W/2) + 38.65 = 50.9\,{}^{\circ}C$$

Isoflux PCB in Vented Enclosure

The PCB inlet temperature $T_I = 33.1{}^{\circ}C$ is applicable here. This solution is also calculated in two iterations beginning with $T_{PCB} = 60{}^{\circ}C$ with the results shown for the second and last iterations where we begin with $T_{PCB} = 45{}^{\circ}C$. Equations (8.3), (8.16), and (8.15) are used:

$$\beta = 1/(T_I + 273.16) = 1/(33.1 + 273.16) = 3.27 \times 10^{-3}\,K^{-1}$$

$$\gamma/\beta = (g/v^2) Pr = 5.454 \times 10^5 \exp \left[-9.254 \times 10^{-3} \left(\frac{T_{PCB} + T_I}{2} \right) \right]$$

$$\gamma/\beta = 5.454 \times 10^5 \exp \left[-9.254 \times 10^{-3} \left(\frac{45.0 + 33.1}{2} \right) \right] = 3.80 \times 10^5$$

The isoflux Rayleigh number requires a heat flux and air thermal conductivity.

$$q = (Q/2)/LD_{PCB} = (2.0\,W/2)/(5.0\,in.)(5.0\,in.) = 0.040\,W/in.^2$$

$$k_{Air} = 6.0583 \times 10^{-4} + 1.6906 \times 10^{-6} \left[(T_{PCB} + T_I)/2 \right]$$

$$= 6.0583 \times 10^{-4} + 1.6906 \times 10^{-6} \left[(45.0 + 33.0)/2 \right]$$

$$k_{Air} = 6.72 \times 10^{-4}\,W/(in. \cdot {}^{\circ}C)$$

The isoflux Rayleigh number is calculated to be

$$Ra_b^* = \left(\frac{g\beta}{\nu^2}\right)Pr\left(\frac{b^5}{L}\right)\left(\frac{q}{k_{Air}}\right) = \gamma\left(\frac{b^5}{L}\right)\left(\frac{q}{k_{Air}}\right) = \left(\frac{\gamma}{\beta}\right)\beta\left(\frac{b^5}{L}\right)\left(\frac{q}{k_{Air}}\right)$$

$$Ra_b^* = \left(3.80\times10^5\right)\left(3.27\times10^{-3}\right)\left[\frac{(0.75)^5}{5.0}\right]\left(\frac{0.040}{6.72\times10^{-4}}\right) = 3508$$

The isoflux Nusselt number $Nu_b(L/2)$ and heat transfer coefficient are calculated as

$$Nu_b\left(\frac{L}{2}\right) = \left[\frac{12}{Ra_b^*} + \frac{1.85}{\left(Ra_b^*\right)^{0.4}}\right]^{-1/2} = \left[\frac{12}{3508} + \frac{1.85}{(3508)^{0.4}}\right]^{-1/2} = 3.67$$

$$h_b = (k_{Air}/b)Nu_b(L/2) = \left(6.72\times10^{-4}/0.75\right)(3.67) = 0.00329 \text{ W}/\left(\text{in.}^2 \cdot {}^\circ\text{C}\right)$$

The final result for the circuit board temperature is

$$T_{PCB} = R_{PCB}(Q/2) + T_I = \frac{1}{hLD_{PCB}}(Q/2) + T_I = \frac{1}{(0.00329)(5.0\,\text{in.})^2}(2.0\,\text{W}/2) + 33.1 = 45.2\,^\circ\text{C}$$

The intermediate and final results for the two circuit board models are listed in Table 8.13.

Table 8.13. Application Example 8.11: Circuit board temperatures only.

Model	Iter.	$T_{PCB}[^\circ\text{C}]$	γ/β	β	Ra_b	Nu_b
Sealed isotherm.	1	60	3.45×10^5	3.21×10^{-3}	1497	3.66
	2	51	3.60×10^5	3.21×10^{-3}	903	3.22
	3	52	3.59×10^5	3.21×10^{-3}	972	3.28
Vented isotherm.	1	60	3.55×10^5	3.27×10^{-3}	1975	3.93
	2	42	3.85×10^5	3.27×10^{-3}	714	3.03
	3	46.5	3.78×10^5	3.27×10^{-3}	1050	3.35
Sealed isoflux	1	60	3.45×10^5	3.21×10^{-3}	3057	3.57
	2	51	3.60×10^5	3.21×10^{-3}	3217	3.61
Vented isoflux	1	60	3.55×10^5	3.27×10^{-3}	3211	3.61
	2	45	3.80×10^5	3.27×10^{-3}	3508	3.67

Table 8.13 continued. Application Example 8.11: Circuit board temperatures only.

Model	Iter.	$T_{PCB}[^\circ\text{C}]$	k_{Air}	h_b	$T_{PCB}[^\circ\text{C}]$
Sealed isotherm.	1	60	6.89×10^{-4}	0.00337	50.5
	2	51	6.81×10^{-4}	0.00293	52.3
	3	52	6.83×10^{-4}	0.00299	52.0
Vented isotherm.	1	60	6.85×10^{-4}	0.00359	44.2
	2	42	6.69×10^{-4}	0.00271	47.8
	3	46.5	6.73×10^{-4}	0.00301	46.4
Sealed isoflux	1	60	6.89×10^{-4}	0.00328	50.9
	2	51	6.82×10^{-4}	0.00328	51.9
Vented isoflux	1	60	6.85×10^{-4}	0.00329	45.2
	2	45	6.72×10^{-4}	0.00329	45.2

8.12 APPLICATION EXAMPLE: ENCLOSURE WITH CIRCUIT BOARDS, COMPARISON OF SECTIONS 8.10 AND 8.11 APPROXIMATE RESULTS WITH *CFD*

An elementary computational fluid dynamics model has been set up and solved for the sealed and vented enclosure problem with circuit boards using the FLOTHERM™ software. The inlet/exit vent resistances were converted to equivalent loss coefficients in the *CFD* program and the external convective heat transfer coefficients used in our preceding calculations were also used for the external surfaces in the *CFD* model. The results are compared in Table 8.14.

The reader should draw his/her own conclusions regarding the adequacy of the text results that we have completed. The discrepancy between the text and *CFD* models is the greatest for the wall temperature rise. The author has used the methods presented here for many years and notes that this discrepancy in computed wall temperatures has usually been of the approximate accuracy (or lack of) indicated in Table 8.14. The computed internal air temperatures have usually been of adequate accuracy compared with test data. What is more confusing is the fact that using identical external heat transfer coefficients in the text and *CFD* models leads to such a large discrepancy.

Table 8.14. Application Example 8.10, 8.11: Enclosure with *PCBs*. Approximate text results compared with *CFD* model.

Model	Variable	Sealed Enclosure		Vented Enclosure	
		Text Result	*CFD* Result	Text Result	*CFD* Result
	T_A	20	20	20	20
	T_{Wall}	32.4	40	29	36
			(45+29+4x41)/6 [1]		(47+24+4x37)/6 [1]
	T_I	38.7	34	33.0	29
	T_{Air}	45	49, 44, 34 [2]	37	45, 40, 29 [2]
Isotherm.	$T_{PCB}-T_I$	13.3	18	13.1	13
Isotherm.	T_{PCB}	52.2	52	46.4	48
Isoflux	$T_{PCB}-T_I$	12.3	18	12.1	13
Isoflux	T_{PCB}	51.0	52	45.2	48

Note: [1] Center top, bottom, four side panels; [2] Air at card channel center, top, mid-height, bottom.

With regard to the internal air temperatures, the two computational methods seem to indicate that the test results correlate best with the air at about the mid-height of the circuit boards. This is actually a bit of an exception to comparison of network models with test data. The author's usual interpretation of the network model is that the computed air temperature represents the average of the air across the top of the circuit board exit regions, i.e., the "average-maximum" air temperature.

8.13 ILLUSTRATIVE EXAMPLE: METAL-WALLED ENCLOSURE, TEN *PCBS*

The geometry for this enclosure problem is illustrated in Figure 8.21. The dimensions are $H = 5.0$ in., $D = 14.5$ in., $W = 7.5$ in., $L = 4.0$ in., $D_{PCB} = 10.0$ in., $f_I = 0.25$; free vent areas $A_I = 9.0$ in.^{2}, $A_2 = 11.0$ in.^{2}, $A_3 = 9.0$ in.^{2}; $\varepsilon = 0.8$, $f_{PCB} = 0.90$ with $T_A = 20°C$. The calculated model results are shown in Table 8.15 with a comparison with test data, but the actual solution is left as an exercise for the student.

The test data compares quite well with the calculations, although the computed values are somewhat less than the "average maximum" circuit board air temperature.

Table 8.15. Illustrative Example 8.13: Enclosure w/10 *PCBs*. Calculated and measured air ΔT above ambient.

Inlet	Exit		Power Dissipation $Q[W]$					
Free Vent Areas [in.^{2}]			25		50		100	
A_I	A_2	A_3	Calc.	Meas.	Calc.	Meas.	Calc.	Meas.
0	0	0	28°C	27 ± 3°C	51°C	62 ± 6°C	90°C	100 ± 2°C
11	11	0	15	21 ± 4	24	36 ± 6	39	55 ± 9
11	11	9	14		22	27 ± 4	36	41 ± 5

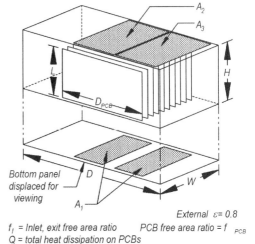

Figure 8.21. Illustrative Example 8.14:
Metal-walled enclosure with ten circuit boards.

Bottom panel displaced for viewing

External $\varepsilon = 0.8$

f_I = Inlet, exit free area ratio   PCB free area ratio = f_{PCB}
Q = total heat dissipation on PCBs

8.14 ILLUSTRATIVE EXAMPLE: METAL-WALLED ENCLOSURE WITH HEAT DISSI-PATION PROVIDED BY SEVERAL WIRE-WOUND RESISTORS

This example is illustrated in Figure 8.22. The dimensions are $d_H = 8.0$ in., $W = 16.0$ in., $H = 10.0$ in., and $D = 19.0$ in. The emissivity is $\varepsilon = 0.1$. The inlet and exit free area ratios are $f_I = f_E = 0.25$. The *unperforated* total inlet area is $A_I = 136$ in.^{2} and the *unperforated* exit area is $A_E = 98$ in.^{2}. There are no circuit boards, but instead, the heat dissipation is provided by several wire-wound resistors distributed throughout the enclosure. As in the preceding example, solution of the problem is left as an exercise for the student.

The calculated results are compared with test data in Table 8.16. The test data was acquired by determining the "average-maximum" air temperature. Certainly the calculated and measured air temperature rises compare very well. One can speculate that this problem has better theoretical-experimental correlation than some others because the absence of circuit boards enhances the internal recirculation of the air and diminishes the occurrence of hot spots.

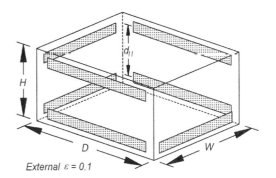

Figure 8.22. Illustrative Example 8.16:
Metal-walled enclosure with heat dissipation
provided by wire-wound resistors.

External $\varepsilon = 0.1$

Table 8.16. Illustrative Example 8.14: Metal-walled enclosure with
heat dissipation provided by wire-wound resistors.

Power	Calculated			Measured
Dissipation Q[W]	Q_d[W]	G[CFM]	ΔT [°C]	ΔT [°C]
25	17.9	4.4	6.9	7.8
50	35.5	5.5	11.4	12.5
100	74.2	7.1	18.4	19.8

EXERCISES

8.1 Simplified vertical flat plate heat transfer coefficients for natural convection are summarized in Table 8.7. However, the ambient temperature used for this table is $T_A = 20°C$. You need to see for yourself what the formulae would be for $T_A = 50°C$. Therefore, calculate the coefficients of $(\Delta T/P)^n$ for a vertical plate, a horizontal plate facing up, and a horizontal plate facing down for $T_A = 50°C$ and ΔT equal to about 50°C. Compare your results with those in Table 8.7. Hint: Use a $\Delta T = 60°C$ so that you can use physical data from the table without interpolation error.

8.2 A small vertical plate-type heat sink is illustrated. The plate dimensions arc $H = 1.0$ in., $W = 2.0$ in. The ambient is $T_A = 20°C$. The emissivity is less than 0.1 so we shall neglect radiation. Calculate the total heat that may be dissipated for average surface temperature rises above ambient of $\Delta T = 50$, 75, and 100°C. Perform a complete set of calculations first using the *small device* heat transfer coefficient and then repeat the calculations using the classical heat transfer coefficient. What is the temperature rise of the sink above ambient for a $Q = 2.0$ W? Use the small device heat transfer coefficient.

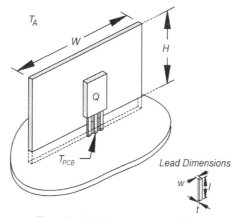

Exercise 8.2. Plate-type heat sink.

8.3 Consider Exercise 8.2, but this time assume the power transistor with $Q = 2.0$ W has leads that are attached to the *PCB* at a surface temperature $T_{PCB} = 70°C$. The ambient is $T_A = 30°C$. Each of the three leads has dimensions of $w = 0.1$ in., $l = 0.4$ in., $t = 0.05$ in. and has a thermal conductivity of $k = 9$ W/(in. · °C). Neglect hot spot effects on the heat sink plate and circuit board (we have not yet studied these issues). You are asked to (1) draw and label a thermal circuit, (2) derive a formula for the heat sink temperature consistent with the thermal circuit, (3) calculate the heat sink temperature, and (4) make a comment concerning a comparison of T_{Sink} with T_{PCB}. Hints: To accomplish (2), write a thermal energy balance on the heat sink "node," then solve for T_{Sink}; you will need to iterate the temperature rise for part (3).

8.4 This problem is very similar to Exercises 8.2 and 8.3, except that we now show the PCB with finite length and width dimensions L_{PCB} and W_{PCB}, respectively. All dimensions and properties are identical to Exercise 8.3, but now we use $L_{PCB} = 2.0$ in. and $W_{PCB} = 3.0$ in. Include convection from both sides of the heat sink *and* the circuit board. Neglect hot spot effects on the heat sink plate and circuit board (we have not yet studied these issues). You are asked to (1) draw and label a thermal circuit, (2) derive formulae for the heat sink and *PCB* temperatures following the thermal circuit, and (3) calculate the heat sink and *PCB* temperatures. Hints: To accomplish (2), write a thermal energy balance on the heat sink "node" and also on the *PCB* "node," then solve for T_{Sink} and T_{PCB}; you will need to iterate the temperature rise of both the heat sink and the circuit board for part (3); begin the iterations with 50°C for both the heat sink and *PCB* temperatures, using each calculated result as the guess for the next iteration (after several iterations, you may wish to plot the results and estimate the temperatures toward which the two nodes are converging). The reader is cautioned that there are two approximations that cause the solution to be optimistic: (1) the hot spot effect beneath the power transistor is neglected, (2) the hot spots on the *PCB* where the transistor package leads are soldered into the *PCB*.

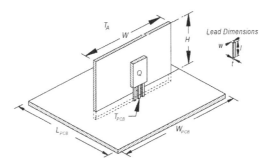

Exercise 8.4. Plate-type heat sink and horizontal circuit board coupon.

8.5 Derive an expression that relates the Nusselt numbers Nu_m and Nu where the former references the local mean air temperature T_m and the latter references the inlet air temperature T_I. Hint: Use the first of the two Nusselt number interpretations in Eq. (6.7).

8.6 Application Examples 8.10 and 8.11 solved enclosure and circuit board problems. However, the component surface-to-opposing board distance was fixed at $b = 0.75$ in. You are asked to use Figures 8.10 and 8.11 to estimate the optimum board spacing b_{opt} for this same problem using the (a) isothermal symmetric and (b) isoflux symmetric theories for both the sealed and vented versions of the enclosure. You should use the board channel inlet temperatures T_I, "Text Result," listed in Table 8.14. Hint: Use the computed $T_{PCB} - T_I$ for the $T_W - T_I$, "Text Result," where needed, listed in Table 8.14.

8.7 Application Examples 8.10 and 8.11 solved an enclosure problem and circuit board problems, respectively. However, the component surface-to-opposing board distance was fixed at $b = 0.75$ in. You are asked to use Eqs. 8.19 and 8.21 to calculate the optimum board spacing b_{opt} for this same problem using the (a) isothermal symmetric and (b) isoflux symmetric theory for both the sealed and vented versions of the enclosure. You should use the board channel inlet temperatures T_I, "Text Result," listed in Table 8.14. If you also solved Exercise 8.6, compare the results with the same in the current problem. Hint: Use the computed $T_{PCB} - T_I$ for the $T_W - T_I$, "Text Result," where needed, listed in Table 8.14.

8.8 Application Example 8.11 described a board temperature calculation using a symmetric isothermal wall model. However, we understand that the two outer boards may not be well modeled using this symmetric board model. Using the "sealed enclosure" model results, you are asked to recalculate the outer board temperature using (a) the isothermal, symmetric wall model, (b) the isothermal, one heated wall, one unheated adiabatic wall model, and (c) the isothermal, unequal wall model from the text. The part (a) will be a repeat of the text, but this should give you a baseline result to help ensure that you are calculating parts (b) and (c) correctly. Compare the three calculated board temperatures. Hint for all parts: You will have to iterate at least a couple of times to get a *PCB* temperature. Hint for Part (c): Use $r = (T_{Wall} - T_I)/(T_{PCB} - T_I)$, otherwise you will get a negative Rayleigh number.

8.9 Application Example 8.11 described a board temperature calculation using a symmetric isothermal wall model. However, we understand that the two outer boards may not be well modeled using this symmetric board model. Using the "vented enclosure" model results, you are asked to recalculate the outer board temperature using (a) the isothermal, symmetric wall model, (b) the isothermal, one heated wall, one unheated adiabatic wall model, and (c) the isothermal, unequal wall model from the text. The part (a) will be a repeat of the text, but this should give you a baseline result to help ensure that you are calculating parts (b) and (c) correctly. Compare the three calculated board temperatures. Hint for all parts: You will have to iterate at least a couple of times to get a PCB temperature. Hint for Part (c): Use $r = (T_{Wall} - T_I)/(T_{PCB} - T_I)$, otherwise you will get a negative Rayleigh number.

8.10 Application Examples 8.10 and 8.11 solved enclosure problems and circuit board problems. However, the component surface-to-opposing board distance was fixed at $b = 0.75$ in. This exercise is a multipart problem where you are asked to analyze the entire sealed enclosure for three different heat dissipation values and card sets. Therefore, do the following: (a) Solve the sealed enclosure problem of Application Example 8.10 for 2 W/card, using five cards, seven cards, ten cards, and twelve cards. The five-card problem is the same as the text problem, but you need to establish a correct baseline result to ensure that your method will give the correct results for the two remaining portions of the problem; (b) Using the results from (a), calculate the correct board channel inlet temperature T_I for the three dissipations, then calculate the board temperature T_{PCB} for the three dissipations. Use the symmetric isothermal board model; (c) Make two simple graphs of T_{Air} and T_{PCB}. What conclusions can you make regarding your results?

8.11 Application Examples 8.10 and 8.11 solved an enclosure problem and circuit board problems, respectively. However, the component surface-to-opposing board distance was fixed at $b = 0.75$ in. This exercise is a multipart problem where you are asked to analyze the entire vented enclosure for three different heat dissipation values and card sets. Therefore, do the following: (a) Solve the sealed enclosure problem of Application Example 8.10 for a 2 W/card, using five cards, seven cards, ten cards, and twelve cards. The five-card problem is the same as the text problem, but you need to establish a correct baseline result to ensure that your method will give the correct results for the two remaining portions of the problem; (b) Using the results from (a), calculate the correct board channel inlet temperature T_I for the three dissipations, then calculate the board temperature T_{PCB} for the three dissipations. Use the symmetric isothermal board model; (c) Make two simple graphs of T_{Air} and T_{PCB}. What conclusions can you make regarding your results?

8.12 Section 8.13 describes an enclosure with ten circuit boards and the computed air temperature results are compared with experimental data in Table 8.15. Your problem is to solve the three different cases (sealed and two different venting conditions) for $Q = 25$ W and compare your results with those in Table 8.15.

8.13 Section 8.13 describes an enclosure with ten circuit boards and the computed air temperature results are compared with experimental data in Table 8.15. Your problem is to solve the three different cases (sealed and two different venting conditions) for $Q = 50$ W and compare your results with those in Table 8.15.

8.14 Section 8.13 describes an enclosure with ten circuit boards and the computed air temperature results are compared with experimental data in Table 8.15. Your problem is to solve the three different cases (sealed and two different venting conditions) for $Q = 100$ W and compare your results with those in Table 8.15.

8.15 Section 8.14 describes an enclosure with heat dissipation provided by several wire-wound resistors. The enclosure surfaces are of low emissivity so that radiation may be neglected. The computed results are compared with test data in Table 8.16. Solve for the internal air temperature rise, the heat Q_d carried away by the airdraft, and the draft airflow G for a $Q_{Box} = 25$ W. Compare your results with the calculated values in Table 8.16.

8.16 Section 8.14 describes an enclosure with heat dissipation provided by several wire-wound resistors. The enclosure surfaces are of low emissivity so that radiation may be neglected. The computed results are compared with test data in Table 8.16. Solve for the internal air temperature rise, the heat Q_d carried away by the airdraft, and the draft airflow G for a $Q_{Box} = 50$ W. Compare your results with the calculated values in Table 8.16.

8.17 Section 8.14 describes an enclosure with heat dissipation provided by several wire-wound resistors. The enclosure surfaces are of low emissivity so that radiation may be neglected. The computed results are compared with test data in Table 8.16. Solve for the internal air temperature rise, the heat Q_d carried away by the airdraft, and the draft airflow G for a $Q_{Box} = 100$ W. Compare your results with the calculated values in Table 8.16.

8.18 Section 8.14 describes an enclosure with heat dissipation provided by several wire-wound resistors. The external surfaces of the enclosure have an emissivity $\varepsilon = 0.8$ so that radiation may not be neglected. Solve for the internal air temperature rise, the heat Q_d carried away by the airdraft, and the draft airflow G for a $Q_{Box} = 25$ W. Compare your results with the calculated values in Table 8.16.

8.19 Section 8.14 describes an enclosure with heat dissipation provided by several wire-wound resistors. The external surfaces of the enclosure have an emissivity $\varepsilon = 0.8$ so that radiation may not be neglected. Solve for the internal air temperature rise, the heat Q_d carried away by the airdraft, and the draft airflow G for a $Q_{Box} = 50$ W. Compare your results with the calculated values in Table 8.16.

8.20 Section 8.14 describes an enclosure with heat dissipation provided by several wire-wound resistors. The external surfaces of the enclosure have an emissivity $\varepsilon = 0.8$ so that radiation may not be neglected. Solve for the internal air temperature rise, the heat Q_d carried away by the airdraft, and the draft airflow G for a $Q_{Box} = 100$ W. Compare your results with the calculated values in Table 8.16.

8.21 A vented, metal-walled enclosure in ambient air at $T_A = 20°C$ containing several circuit boards is illustrated. The total heat dissipation of all circuit boards is $Q_{Box} = 125$ W. The given dimensions are $a = 0.20$ in., $l = 1.0$ in., $w = 1.0$ in., $p = 0.5$ in., $W = 11$ in., $D = 12$ in., $H = 10$ in., $D_{PCB} = 10$ in., $L = 9$ in., $H_I = 1.0$ in. The exit and inlet vents have identical dimensions and each has a free area ratio $f_I = 0.40$ and has an emissivity of 0.1 or less. You are asked to do the following: (1) Calculate the internal air temperature rise above ambient for the vented enclosure, (2) Calculate the estimated circuit board air inlet temperature T_I, (3) Using N_{Cards}, calculate the board temperatures using a symmetric, isothermal wall model, (4) Calculate the optimum b using the calculated T_I, T_{PCB}. Hints: (1) Use the airflow circuit, with appropriate variable values, from Application Example 8.10; (2) Begin the airflow circuit model with circuit boards on a pitch $p = 0.5$ in.; (3) Calculate the starting number of cards from $N_{Cards} = (W/p)-1$.

8.22 An enclosure with negligible external and internal radiation is illustrated in the figure for Exercise 8.22. The dimensions are $W = 9.0$ in., $H = 9.0$ in., $D = 18$ in., $W_I = 7.0$ in., $W_E = 7.0$ in., $H_I = 2.0$ in., $f_I = 0.35$, $f_E = 0.35$, $N_{Cards} = 7$, $Q_{Total} = 35$ W, $D_{PCB} = 16$ in., $L = 5.0$ in., $f_{PCB} = 0.5$ (free area for PCB channel), $b = 0.5$ in. (component surface to opposing PCB). You are asked to solve for the enclosure wall temperature rise, internal air temperature rise, circuit board inlet air temperature, circuit board temperature, and the venting airdraft and heat carried away by the draft (where relevant) for both the sealed and vented versions of the problem. Hint: Suggested airflow circuit is series sum of inlet vent, expansion from inlet region to $D \times W_I$ entrance to PCBs, contraction to PCBs, PCBs, expansion from PCBs, contraction from PCB exit region to exit region (illustration on previous page).

8.23 An enclosure with negligible external and internal radiation is illustrated in the figure for Exercise 8.23. The dimensions are $W = 3.0$ in., $H = 10.0$ in., $D = 5$ in., $W_I = 3.0$ in., $W_E = 3.0$ in.,

H_I = 1.0 in., f_I = 0.50, f_E = 0.50, N_{Cards} = 2, Q_{Total} = 20 W, D_{PCB} = 5 in., L = 8.0 in., f_{PCB} = 0.80 (free area ratio for *PCB* channel), a = 0.25 in., b = 0.75 in. (component surface to opposing *PCB*), s = 1.0 in. You are asked to solve for the enclosure wall temperature rise, internal air temperature rise, circuit board inlet air temperature, circuit board temperature, and the venting airdraft and heat carried away by the draft (where relevant) for both the sealed and vented versions of the problem. Hints: (1) Use a card cage inlet cross-sectional area that does not include the smooth channel of width s because there will be some down draft (due to recirculation) in this channel (the same applies to the other side with components but ignore this aspect); (2) Suggested airflow circuit is series sum of inlet vents, contraction to *PCBs*, *PCBs*, expansion from *PCBs*, exit vents.

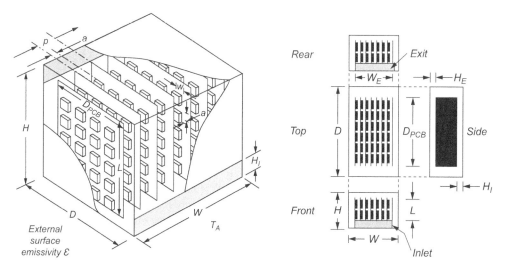

Exercise 8.21. Vented enclosure.   Exercise 8.22. Vented enclosure.

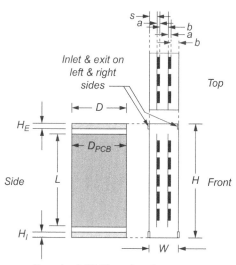

Exercise 8.23. Vented enclosure.

8.26 Another natural convection cooled enclosure is illustrated in the figure for Exercise 8.26. The dimensions are D = 10.0 in., W = 10.0 in., H = 2.5 in., D_{PCB} = 8.0 in., $W_I = W_E$ = 8.0 in., H_I = 0.5 in., H_E = 0.5 in., $f_I = f_E$ = 0.35, a (component height) = 0.2 in., $l = w$ = 1.5 in. (component length and width), b – 1.68 in.. The emissivity is less than 0.1 so radiation should not be included in the model. The total heat dissipation on the circuit board is Q_{PCB} = 50 W. Calculate the temperature rise of the enclosure wall and internal air above ambient for both the sealed and vented enclosure. Also calculate the heat carried away by the venting draft and also the vent draft, where appropriate. Hints: (1) Suggested airflow circuit is series sum of inlet vent, slot resistance in plane of *PCB* and at front, contraction to card, card, expansion from card, exit; (2) Use $d_H = a$.

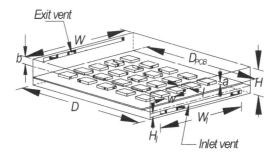

Exercise 8.24. Vented enclosure.

8.25 Solve Exercise 8.24 for the same enclosure but with an external emissivity of 0.8.

8.26 A natural convection cooled enclosure is illustrated in the figure for Exercise 8.26. The dimensions and other data are $H = 23.0$ in., $D = 8.0$ in., $W = 4.0$ in., $D_{PCB} = 8.0$ in., $L = 15.0$ in., $W_I = W_E = 4.0$ in., $H_I = H_E = 4.0$ in., $f_I = f_E = f_{PS} = 0.4$, $a = 0.2$ in., $b = 0.6$ in., $s_x = 0.2$ in., $l = 0.775$ in., $Q_{PS} = 10$ W, $R_{PS} = 2.5 \times 10^{-5}$ in. H_2O/CFM^2, $Q_{cc} = 40$ W (total for the four *PCBs*). You are asked to solve the sealed box problem for the enclosure wall temperature rise and the internal air temperature rise. Next you are asked to solve the vented box problem for the enclosure wall temperature rise, the internal air temperature rise, the venting airdraft and heat carried away by the draft, the circuit board inlet air temperature, and the circuit board temperature. Hint: (1) Use a card cage inlet cross-sectional area that does not include the smooth channel of width s because there will be some down draft (due to recirculation) in this channel (the same applies to the other side with components but ignore this aspect); (2) Suggested airflow circuit is series sum of inlet vent, expansion from inlet region to board inlet region, contraction to *PCBs*, *PCBs*, expansion from *PCBs*, power supply inlet, power supply, and exit vent.

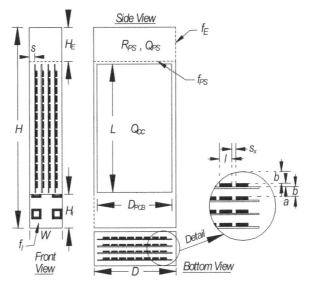

Exercise 8.26. Vented enclosure.

8.27 Repeat the solution of Exercise 8.26, but use an external surface emissivity of 0.8.

CHAPTER 9

Natural Convection Heat Transfer II: Heat Sinks

We continue our study of natural convection with details concerning heat sinks. This is a very short chapter as only vertically oriented heat sinks are covered here. Since radiation heat transfer has not been addressed in any detail up to this point, the reader is advised to study the material concerning U-channel radiation in Chapter 10. Furthermore, thermal conduction spreading effects (source hot spots) are not included in this chapter. Please consult the appropriate section of spreading resistance in Chapter 12 to learn how to include the spreading effects into your analysis.

9.1 HEAT SINK GEOMETRY AND SOME NOMENCLATURE

A basic heat sink is illustrated in Figure 9.1. You should understand that it is possible to fabricate a heat sink consisting of a base plate with both finned and unfinned regions and it is not uncommon for a heat sink to be finned on both sides of the base. Once you understand the principles required to analyze any given fin section, you should be able to analyze a more complex fin system.

The interior (between-fin region) surfaces are described as rectangular U-channels. As a practical matter, real heat sinks have fins that usually decrease in thickness, t_f, from the fin root (base) to the tip. Except for extreme situations, we ignore this variable fin thickness issue.

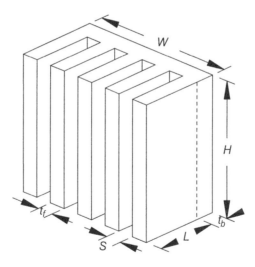

Figure 9.1. Geometry and nomenclature for vertical heat sink.

9.2 A RECTANGULAR U-CHANNEL CORRELATION FROM VAN DE POL AND TIERNEY

Some authors, designers, and thermal analysts are predisposed to use parallel flat plate correlations for vertical fin arrays. Heat sinks with large L/H ratios where effects in the region of the fin root are negligible may well be analyzed as a set of parallel plates. However, as pointed out by Van de Pol and Tierney (1974) in their classic paper, heat sink designs may consist of a series of relatively short fins and cannot be accurately approximated by parallel flat plates. The Van de Pol and Tierney correlation is built on a study of several different heat sink configurations. Those experimental results are combined into a single correlation for laminar flow:

$$Nu_r = \frac{Ra_r^*}{\psi}\left\{1 - \exp\left[-\psi\left(\frac{0.5}{Ra_r^*}\right)^{3/4}\right]\right\}$$

$$\psi = \frac{24\left[1 - 0.483e^{(-0.17/a)}\right]}{\left\{\left[1 + \frac{a}{2}\right]\left[1 + \left(1 - e^{-0.83a}\right)\left(9.14a^{1/2}e^{VS} - 0.61\right)\right]\right\}^3}$$

$$Ra_r^* = \left(\frac{r}{H}\right)Gr_r Pr,\ r = 2LS/(2L + S)$$

$$a = S/L,\ V = -11.8\left(\text{in.}^{-1}\right)$$

Van de Pol (9.1)
and Tierney

We see that the characteristic length is $r = 2LS/(2L + S)$. The Rayleigh number Ra_r^* is a modified channel Rayleigh number. All physical properties except β are evaluated at the surface temperature. β should be evaluated at the fluid temperature. It is left as an exercise for you to prove that the channel heat transfer coefficient h_c obtained from Eq. (9.1) is such that as the fin length is shortened, $L/S \to 0$, $h_c/h_H \to 1.0$. h_H is a vertical flat plate heat transfer coefficient listed in the first row of Table 8.1, i.e., $Nu_H = 0.59(Gr_H Pr)^{1/4}$.

Equations (9.1) are sufficiently complicated to inhibit many design engineers using them. The next section therefore contains numerous design curves as a substitute for using the correlation formulae.

9.3 DESIGN PLOTS REPRESENTING THE VAN DE POL AND TIERNEY CORRELATION

There are so many independent variables in Eqs. (9.1) that a single graph, or even two or three, is not sufficient for reasonable analyis, thus the extensive set of plots for h_c / h_H follows. You will

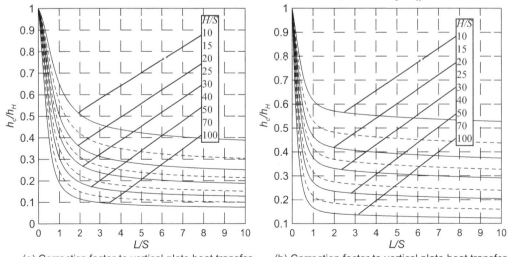

(a) Correction factor to vertical plate heat transfer coefficient for finned heat sink. $\triangle T$=10°C, S=0.20 in.

(b) Correction factor to vertical plate heat transfer coefficient for finned heat sink. $\triangle T$=10°C, S=0.25 in.

Figure 9.2. Van de Pol and Tierney correlation. h_c = U-channel convective heat transfer coefficient. h_H = vertical flat plate convective heat transfer coefficient.

find each figure is for a specific surface temperature rise ΔT and fin spacing S. The temperature-dependent parameters are determined according to the requirements indicated at the end of Section 9.2. However, an ambient temperature of $T_A = 20°C$ is used for all plots. Ambient temperatures greater than this value will not have a significant effect on the results, particularly when one understands that graph reading is also subject to some inaccuracy.

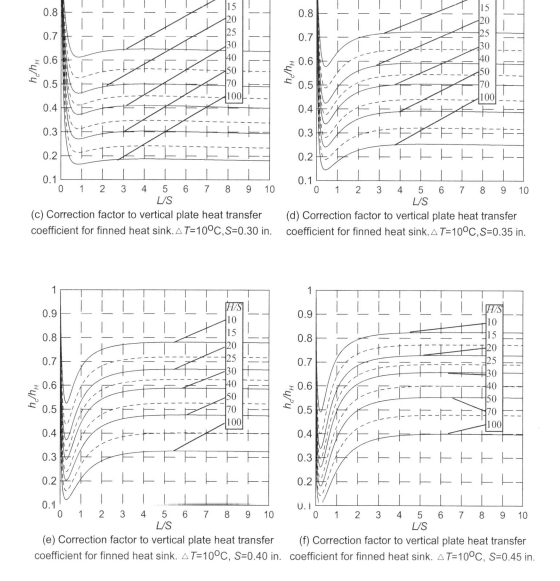

(c) Correction factor to vertical plate heat transfer coefficient for finned heat sink. $\triangle T = 10°C$, $S = 0.30$ in.

(d) Correction factor to vertical plate heat transfer coefficient for finned heat sink. $\triangle T = 10°C$, $S = 0.35$ in.

(e) Correction factor to vertical plate heat transfer coefficient for finned heat sink. $\triangle T = 10°C$, $S = 0.40$ in.

(f) Correction factor to vertical plate heat transfer coefficient for finned heat sink. $\triangle T = 10°C$, $S = 0.45$ in.

Figure 9.2 (continued). Van de Pol and Tierney correlation. h_c = U-channel convective heat transfer coefficient. h_H = vertical flat plate convective heat transfer coefficient.

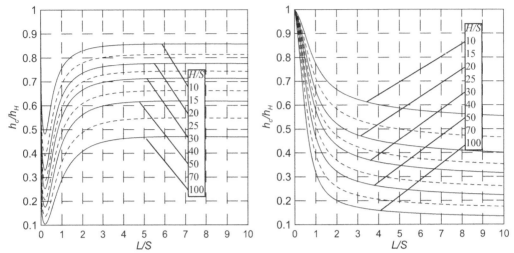

(g) Correction factor to vertical plate heat transfer coefficient for finned heat sink. $\triangle T$=10°C, S=0.50 in.

(h) Correction factor to vertical plate heat transfer coefficient for finned heat sink. $\triangle T$=25°C, S=0.20 in.

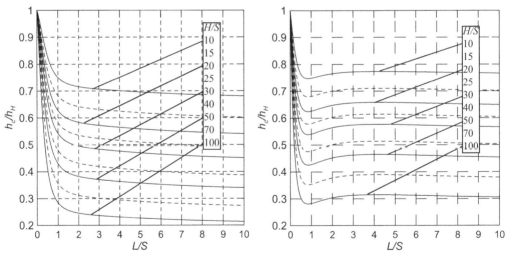

(i) Correction factor to vertical plate heat transfer coefficient for finned heat sink. $\triangle T$=25°C, S=0.25 in.

(j) Correction factor to vertical plate heat transfer coefficient for finned heat sink. $\triangle T$=25°C, S=0.30 in.

Figure 9.2 (continued). Van de Pol and Tierney correlation. h_c = U-channel convective heat transfer coefficient. h_H = vertical flat plate convective heat transfer coefficient.

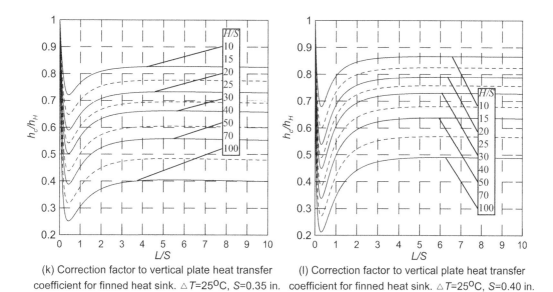

(k) Correction factor to vertical plate heat transfer coefficient for finned heat sink. $\triangle T$=25°C, S=0.35 in.

(l) Correction factor to vertical plate heat transfer coefficient for finned heat sink. $\triangle T$=25°C, S=0.40 in.

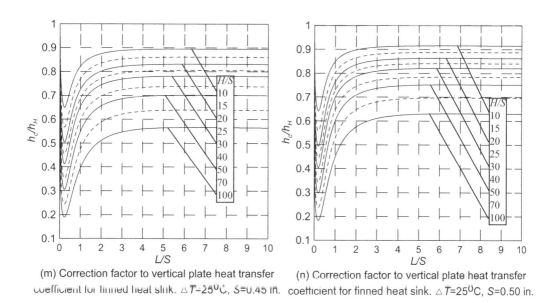

(m) Correction factor to vertical plate heat transfer coefficient for finned heat sink. $\triangle T$=25°C, S=0.45 in.

(n) Correction factor to vertical plate heat transfer coefficient for finned heat sink. $\triangle T$=25°C, S=0.50 in.

Figure 9.2 (continued). Van de Pol and Tierney correlation. h_c = U-channel convective heat transfer coefficient. h_H = vertical flat plate convective heat transfer coefficient.

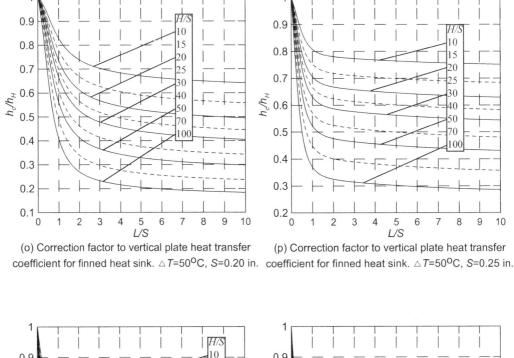

(o) Correction factor to vertical plate heat transfer coefficient for finned heat sink. $\triangle T=50^{\circ}C$, $S=0.20$ in.

(p) Correction factor to vertical plate heat transfer coefficient for finned heat sink. $\triangle T=50^{\circ}C$, $S=0.25$ in.

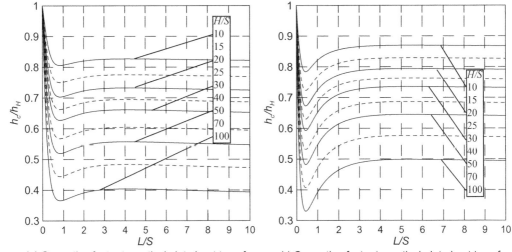

(q) Correction factor to vertical plate heat transfer coefficient for finned heat sink. $\triangle T=50^{\circ}C$, $S=0.30$ in.

(r) Correction factor to vertical plate heat transfer coefficient for finned heat sink. $\triangle T=50^{\circ}C$, $S=0.35$ in.

Figure 9.2 (continued). Van de Pol and Tierney correlation. h_c = U-channel convective heat transfer coefficient. h_H = vertical flat plate convective heat transfer coefficient.

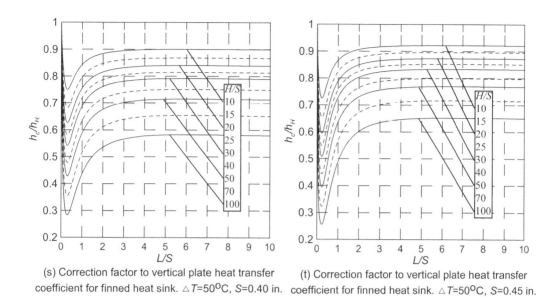

(s) Correction factor to vertical plate heat transfer coefficient for finned heat sink. $\triangle T$=50°C, S=0.40 in.

(t) Correction factor to vertical plate heat transfer coefficient for finned heat sink. $\triangle T$=50°C, S=0.45 in.

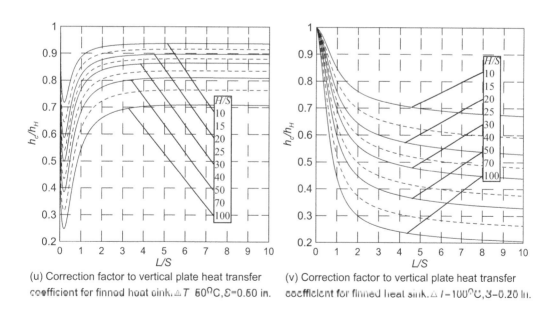

(u) Correction factor to vertical plate heat transfer coefficient for finned heat sink. $\triangle T$ 50°C, S=0.50 in.

(v) Correction factor to vertical plate heat transfer coefficient for finned heat sink. $\triangle T$=100°C, S=0.20 in.

Figure 9.2 (continued). Van de Pol and Tierney correlation. h_c = U-channel convective heat transfer coefficient. h_H = vertical flat plate convective heat transfer coefficient.

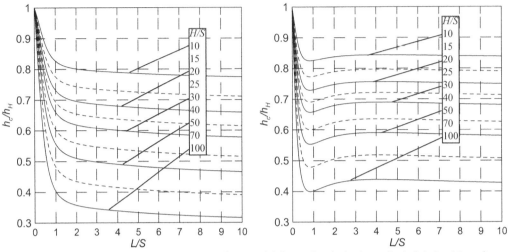

(w) Correction factor to vertical plate heat transfer coefficient for finned heat sink. $\triangle T$=100°C, S=0.25 in.

(x) Correction factor to vertical plate heat transfer coefficient for finned heat sink. $\triangle T$=100°C, S=0.30 in.

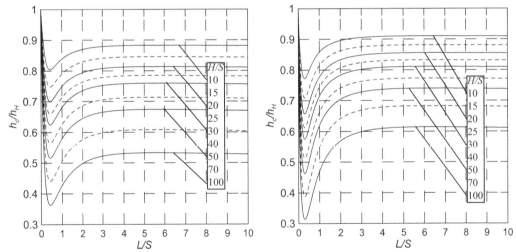

(y) Correction factor to vertical plate heat transfer coefficient for finned heat sink. $\triangle T$=100°C, S=0.35 in.

(z) Correction factor to vertical plate heat transfer coefficient for finned heat sink. $\triangle T$=100°C, S=0.40 in.

Figure 9.2 (continued). Van de Pol and Tierney correlation. h_c = U-channel convective heat transfer coefficient. h_H = vertical flat plate convective heat transfer coefficient.

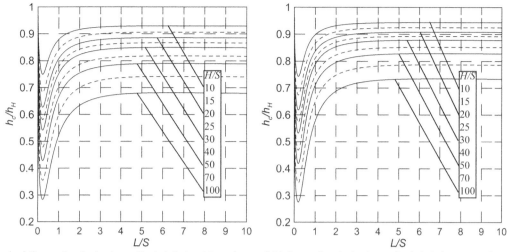

(aa) Correction factor to vertical plate heat transfer coefficient for finned heat sink. $\triangle T$=100°C, S=0.45 in.

(bb) Correction factor to vertical plate heat transfer coefficient for finned heat sink. $\triangle T$=100°C, S=0.50 in.

Figure 9.2 (continued). Van de Pol and Tierney correlation. h_c= U-channel convective heat transfer coefficient. h_H = vertical flat plate convective heat transfer coefficient.

9.4 A FEW USEFUL FORMULAE

A common error is often made when calculating one of the various heat sink dimensions such as fin spacing, number of fins, etc. The following formula based on Figure 9.1 is readily written.

$$\boxed{W = N_f t_f + \left(N_f - 1\right)S}$$

Heat sink dimensions (9.2) in spanwise direction

Most heat sink analyses are based on a fixed width W. A typical situation is that you are evaluating a heat sink described in a vendor's catalog where W and t_f are given and the number of fins N_f is obvious. You would use Eq. (9.2) to calculate the fin spacing S.

You will also need to calculate the interior (used with h_c) and exterior (used with h_H) surface areas A_I, A_E, respectively. When calculating A_I, it is usually reasonable to include the fin tip area as a matter of convenience. When the fins are tapered, it is not actually possible to obtain a meaningful tip area.

$$A_I = 2\left(N_f - 1\right)LH + WH$$
$$\boxed{A_I = \left[2\left(N_f - 1\right)L + W\right]H}$$
$$\boxed{A_E = 2HL}$$

Front-side areas (9.3)

You will note that A_E does not include the flat, backside surface area.

The internal and external conductances are therefore written as

$$\boxed{C_I = \eta_I\left(h_c/h_H\right)h_H A_I; \quad C_E = \eta_E h_H A_E}$$

Convection (9.4) conductances

We have included the fin efficiencies η_I and η_E where the first is used for the inner fins and the heat transfer coefficient h_c is the same on both sides of the fin. η_E is used for the end fins where one side uses h_c and the other side uses h_H. If the heat sink is such that the two heat transfer coefficients are either very different or the heat sink has so few fins that the two end fins are significant, then we should use $\eta_I = \eta_E$. The factor $(h_c/h_H)h_H$ is used because you should first calculate h_H, obtain (h_c/h_H) from Figure 9.2, then multiply the two together to obtain h_c. If you are not familiar with calculating fin efficiency you should peruse the example in Section 7.7 or read the more detailed discussion later in this book.

9.5 APPLICATION EXAMPLE: NATURAL CONVECTION COOLED, VERTICALLY ORIENTED HEAT SINK

Figure 9.3 illustrates a top view of a heat sink with dimensions W = 1.86 in., H = 5.0 in., L = 1.0 in., t_f = 0.06 in., and ΔT = 50°C. Our problem is to calculate the amount of heat that we can convect from the finned side only.

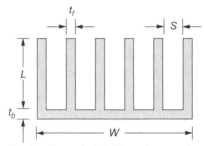

Figure 9.3. Application Example 9.4: Natural convection cooled heat sink.

The first steps are to calculate the exact fin spacing and other required geometric ratios.

$$W = N_f t_f + (N_f - 1)S$$

$$S = (W - N_f t_f)/(N_f - 1) = [1.86\,\text{in.} - (6)(0.06\,\text{in.})]/(6-1) = 0.30\,\text{in.}$$

$$L/S = 1.0\,\text{in.}/0.3\,\text{in.} = 3.33, \qquad H/S = 5.0\,\text{in.}/0.3\,\text{in.} = 16.67$$

Next we calculate h_H using the formula from the first line of Table 8.7.

$$h_H = 0.0024(\Delta T_{Sink}/H)^{0.25} = 0.0024(50°\text{C}/5.0\,\text{in.})^{0.25} = 0.00427\,\text{W}/(\text{in.}^2 \cdot °\text{C})$$

Using the geometric ratios L/S, H/S, and the heat sink temperature rise of 50°C, we refer to Figure 9.2 (q) to get $h_c/h_H = 0.76$, followed by the calculation of h_c.

$$h_c = (h_c/h_H)h_H = (0.76)(0.00427) = 0.0032\,\text{W}/(\text{in.}^2 \cdot °\text{C})$$

The interior (channel) and exterior (two outer fins) surface areas are

$$A_I = \left[2(N_f - 1)L + W\right]H = \left[2(6-1)(1.0\,\text{in.}) + 1.86\,\text{in.}\right](5.0\,\text{in.}) = 59.3\,\text{in.}^2$$

$$A_E = 2HL = 2(5.0\,\text{in.})(1.0\,\text{in.}) = 10.0\,\text{in.}^2$$

Now we calculate the fin efficiency "primitives" and then the fin efficiency.

$$R_k = L/(k_{fin}Ht_f) = 1.0\,\text{in.}/\left\{\left[5.0\,\text{W}/(\text{in.} \cdot °\text{C})\right](5.0\,\text{in.})(0.06\,\text{in.})\right\} = 0.67\,°\text{C/W}$$

$$R_c = 1/(2h_c LH) = 1/\left\{2\left[0.0032\,\text{W}/(\text{in.}^2 \cdot °\text{C})\right](1.0\,\text{in.})(5.0\,\text{in.})\right\} = 31.25\,°\text{C/W}$$

$$\eta = \sqrt{\frac{R_c}{R_k}}\,\tanh\sqrt{\frac{R_k}{R_c}} = \sqrt{\frac{31.25}{0.67}}\,\tanh\sqrt{\frac{0.67}{31.25}} = 0.99$$

The interior, exterior, and total conductances are

$$C_I = \eta h_c A_I = (0.993)(0.0032)(59.30) = 0.191\,\text{W}/^\circ\text{C}$$

$$C_E = \eta h_H A_E = (0.993)(0.00427)(10.0) = 0.0424\,\text{W}/^\circ\text{C}$$

$$C = C_I + C_E = 0.191 + 0.0424 = 0.233\,\text{W}/^\circ\text{C}$$

The amount of heat that can be convected from this heat sink at a uniform temperature rise of 50°C is

$$Q = C\Delta T = (0.233\,\text{W}/^\circ\text{C})(50\,^\circ\text{C}) = 11.67\,\text{W}$$

9.6 APPLICATION EXAMPLE: NATURAL CONVECTION COOLED, NINE-FIN HEAT SINK WITH CALCULATIONS COMPARED TO TEST DATA

The heat sink model for this problem is based on the geometry of Figure 9.3, except that there are nine fins. Also, the actual heat sink has tapered fins, which is not shown in the illustration. An average spacing and fin thickness are therefore used. The dimensions are $W = 4.15$ in., $H = 4.0$ in., $L = 2.62$ in., $S = 0.35$ in., $t_f = 0.15$ in., and $t_b = 0.63$ in. The heat sink temperature rise above ambient is $\Delta T = 50^\circ\text{C}$. We shall use an aluminum thermal conductivity of $k = 5.0$ W/(in. · °C). The calculations proceed in a fashion nearly identical to the Application Example 9.5,

$$L/S = 2.62\,\text{in.}/0.35\,\text{in.} = 7.5, \qquad H/S = 4.0\,\text{in.}/0.35\,\text{in.} = 11.4,$$

with the exception that we shall use the small device heat transfer coefficient formulae from Table 8.8.

$$h_H = 0.0022\left(\Delta T_{Sink}/H\right)^{0.35} = 0.0022\left(50\,^\circ\text{C}/4.0\,\text{in.}\right)^{0.35} = 0.0053\,\text{W}/\left(\text{in.}^2 \cdot {}^\circ\text{C}\right)$$

Using the geometric ratios L/S, H/S, and the heat sink temperature rise of 50°C, we refer to Figure 9.2 (r) to get $h_c/h_H = 0.86$, followed by the calculation of h_c.

$$h_c = \left(h_c/h_H\right)h_H = (0.86)(0.0053) = 0.0046\,\text{W}/\left(\text{in.}^2 \cdot {}^\circ\text{C}\right)$$

Calculations for other values of the heat sink height H and temperature rise are not shown here. Since we will compare the calculations with test data, we must be careful to include all of the convecting surface area such as the flat backside and the various edges.

$$A_I = \left[2\left(N_f - 1\right)L + W\right]H = \left[2(9-1)(2.62\,\text{in.}) + 4.15\,\text{in.}\right](4.0\,\text{in.}) = 184.28\,\text{in.}^2$$

$$A_E = 2H\left(L + t_b\right) + WH + 2t_b W + 2N_f t_f L$$

$$A_E = 2(4.0\,\text{in.})(2.62\,\text{in.} + 0.63\,\text{in.}) + (4.15\,\text{in.})(4.0\,\text{in.}) + 2(0.63\,\text{in.})(4.15\,\text{in.})$$
$$+2(9)(0.15\,\text{in.})(2.62\,\text{in.}) = 54.90\,\text{in.}^2$$

Next we calculate the fin efficiency "primitives" R_k, R_c and then the fin efficiency.

$$R_k = L/\left(k_{fin} H t_f\right) = 2.62\,\text{in.}/\left\{\left[5.0\,\text{W}/\left(\text{in.} \cdot {}^\circ\text{C}\right)\right](4.0\,\text{in.})(0.15\,\text{in.})\right\} = 0.87\,^\circ\text{C}/\text{W}$$

$$R_c = 1/\left(2h_c LH\right) = 1/\left\{2\left[0.0046\,\text{W}/\left(\text{in.}^2 \cdot {}^\circ\text{C}\right)\right](2.62\,\text{in.})(4.0\,\text{in.})\right\} = 10.46\,^\circ\text{C}/\text{W}$$

$$\eta = \sqrt{\frac{R_c}{R_k}}\,\tanh\sqrt{\frac{R_k}{R_c}} = \sqrt{\frac{10.46}{0.87}}\,\tanh\sqrt{\frac{0.87}{10.46}} = 0.97$$

The interior, exterior, and total conductances are

$$C_I = \eta h_c A_I = (0.97)(0.0046)(184.28) = 0.82 \, \text{W}/^\circ\text{C}$$

$$C_E = \eta h_H A_E = (0.97)(0.0053)(54.9) = 0.28 \, \text{W}/^\circ\text{C}$$

$$C = C_I + C_E = 0.82 + 0.28 = 1.10 \, \text{W}/^\circ\text{C}$$

The amount of heat that can be convected from this heat sink at a uniform temperature rise of 50°C is

$$Q = C\Delta T = (1.10 \, \text{W}/^\circ\text{C})(50^\circ\text{C}) = 55.1 \, \text{W}$$

which gives us one data point for the plots in Figure 9.4. The other calculated data points are left as exercises for the student (computed data points connected with curves). A few test-data points are selected from Spoor (1974) and shown with symbols. The greatest experimental-calculated discrepancy is about 15%.

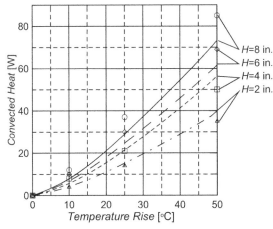

Figure 9.4. Application Example 9.6: Natural convection cooled, nine-fin heat sink. Calculations for: $H = 6.0$ in., 8.0 in. used the simplified classical h, Table 8.7; $H = 2.0$ in., 4.0 in. used the simplified "small device" h, Table 8.8. Test data represented by symbols.

EXERCISES

9.1 Prove that the channel heat transfer coefficient h_c obtained from Eq. (9.1) is such that as the fin length is shortened or $L/S \rightarrow 0$, $h_c/h_H \rightarrow 1.0$, where h_H is a vertical flat plate heat transfer coefficient listed in the first row of Table 8.1, i.e., $h_H = 0.59 (Gr_H Pr)^{1/4}$.

9.2 Perform the calculations of Application Example 9.6 for $H = 4.0$ in., $\Delta T = 25^\circ\text{C}$.

9.3 Perform the calculations of Application Example 9.6 for $H = 4.0$ in., $\Delta T = 10^\circ\text{C}$.

9.4 Perform the calculations of Application Example 9.6 for $H = 2.0$ in., $\Delta T = 50^\circ\text{C}$.

9.5 Perform the calculations of Application Example 9.6 for $H = 2.0$ in., $\Delta T = 25^\circ\text{C}$.

9.6 Perform the calculations of Application Example 9.6 for $H = 2.0$ in., $\Delta T = 10^\circ\text{C}$.

9.7 Perform the calculations of Application Example 9.6 for $H = 6.0$ in., $\Delta T = 50^\circ\text{C}$.

9.8 Perform the calculations of Application Example 9.6 for $H = 6.0$ in., $\Delta T = 25^\circ\text{C}$.

9.9 Perform the calculations of Application Example 9.6 for $H = 6.0$ in., $\Delta T = 10°C$.

9.10 Perform the calculations of Application Example 9.6 for $H = 8.0$ in., $\Delta T = 50°C$.

9.11 Perform the calculations of Application Example 9.6 for $H = 8.0$ in., $\Delta T = 25°C$.

9.12 Perform the calculations of Application Example 9.6 for $H = 8.0$ in., $\Delta T = 10°C$.

9.13 Use the *calculated* data points from Figure 9.4 to obtain a dependence of thermal resistance R vs. height sink H. Hints: (1) Construct individual log-log plots of resistance R vs. height H for temperature rise $\Delta T - 10$, 25, and $50°C$; (2) From the plots, obtain the exponent n and factor a for the formula $R = a / H^n$; (3) The constant a will be different for each plot, but the exponent n should be nearly the same for each; (4) The author obtained an average $n = 0.448$ from the three plots.

9.14 Use the geometry shown in Figure 9.3, with dimensions and data of $W = 4.0$ in., $H = 5.0$ in., $L = 1.0$ in., $t_f = 0.10$ in., a negligible radiation emissivity, a heat sink temperature rise of $\Delta T = 50°C$, an ambient $T_A = 25°C$, and an aluminum thermal conductivity of $k = 5.0$ W/(in. · °C). Calculate the following: (1) the optimum number of fins, (2) the amount of heat that can be convected from the heat sink with the optimum number of fins, neglecting heat conduction spreading. You should include the effects of fin efficiency. Neglect convection effects from the top and bottom fin edges, the base edges, and do not include the flat backside. Hint: (1) Use the "optimum board spacing" formula or graph from Chapter 8 to determine S_{Opt}; (2) Use the h_c / h_H graphs from Figure 9.2.

9.15 The illustrated heat sink has a vertically oriented base (normal to plane of page) and is symmetric about a centerline through and parallel to the base thickness. The dimensions and data are $S_o = 0.35$ in., $S_i = 1.5$ in., $W = 2.6$ in., $H = 3.0$ in., $L = 0.75$ in., $t_f = 0.10$ in., there is a negligible radiation emissivity, and a heat sink temperature rise above ambient of $\Delta T = 50°C$, an ambient $T_A = 25$ °C, and an aluminum thermal conductivity of $k = 5.0$ W/(in. · °C). Use a single fin efficiency based on the average of h_c for S_o and h_H. The purpose of $S_i > S_o$ is to provide room for placement of the power transistor. Calculate the total thermal resistance and amount of heat that can be convected from the heat sink, neglecting heat conduction spreading. You should include the effects of fin efficiency. Neglect convection effects from the top and bottom fin and base edges. Hints: (1) Use the h_c / h_H graphs from Figure 9.2; (2) Calculate the result for one half of the heat sink (four fins) and then halve the resistance.

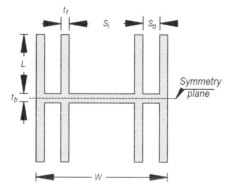

Exercise 9.15. Symmetric heat sink.

CHAPTER 10

Thermal Radiation Heat Transfer

Thermal energy transport by radiation is unique compared to conduction and convection in that a transport medium is not required. In fact, heat transfer by radiation between two surfaces is greater when there is no intervening material. Fortunately the absorption of radiation between surfaces within many electronic enclosures is such that this absorption can be neglected throughout this text.

We had a very brief discussion of radiation heat transfer in Chapter 1 and a couple of application examples in Chapter 8. In this chapter, we will go into greater detail in order to enhance your understanding of the subject. We will begin with the basics of radiation emission and absorption for both ideal and actual surfaces, followed by a discussion of the geometric aspects of the phenomenon, and two methodologies useful when multiple, i.e., more than two, surfaces are involved. The chapter concludes with a detailed evaluation of radiation from rectangular U-channels simulating plate-fin heat sink radiation.

10.1 BLACKBODY RADIATION

Radiation from solid objects such as components, cabinet walls, and heat sinks is a surface-related phenomenon. The radiation is electromagnetic in character with a theoretical basis found in the realm of the quantum physics of solids. Briefly, however, it may be thought of as originating at microscopic electronic oscillators emitting radiation over an extremely broad bandwidth. The maximum monochromatic radiation flux per unit wavelength interval into a half space, or monochromatic emissive power, is described by Planck's radiation law, Eq. (10.1), and is referred to as *blackbody radiation*.

$$E_{\lambda b} = \frac{c_1}{\lambda^5 \left(e^{c_2/\lambda T'} - 1 \right)}$$

Planck's radiation law (10.1)

where

$$E_{\lambda b} \left[W/\left(m^2 \cdot \mu m \right) \right] \equiv \text{monochromatic emissive power}$$

$$\lambda[\mu] = \text{wavelength in microns} \left(1\,\mu \text{ or } 1\,\mu m = 10^{-6}\,m \right)$$

$$T'\left[K \right] \equiv \text{surface temperature} = T\left[{}^\circ C \right] + 273.16$$

$$c_1 \equiv \text{first radiation constant} = 3.7415 \times 10^8$$

$$c_2 \equiv \text{second radiation constant} = 1.4388 \times 10^4$$

Equation (10.1) is plotted in Figure 10.1 for several different surface temperatures in a range of values that one encounters in electronic equipment. Most, if not all, standard heat transfer textbooks plot Eq. (10.1) for much greater temperatures (usually in the vicinity of a thousand K) than shown in our graph. Note the shift toward shorter wavelengths as the temperature is increased. The very small, relatively speaking, visible region of the spectrum is indicated. Clearly, the maximum emissive power does not occur in the visible portion of the spectrum for temperatures of interest to us. The wavelength for which the maximum radiation occurs is found by setting the derivative of Eq. (10.1) equal to zero.

$$\frac{\partial E_{\lambda b}}{\partial \lambda} = \frac{c_1 c_2 e^{c_2/\lambda_M T'}}{\lambda_M^7 T' \left(e^{c_2/\lambda_M T'} - 1 \right)^2} - \frac{5 c_1}{\lambda_M^6 \left(e^{c_2/\lambda_M T'} - 1 \right)} = 0$$

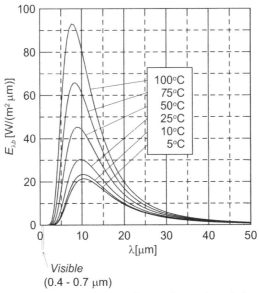

Figure 10.1. Spectral distribution of monochromatic emissive power.

Performing some algebraic manipulation,

$$\frac{5c_1}{\lambda_M^6\left(e^{c_2/\lambda_M T'}-1\right)}=\frac{c_1c_2 e^{c_2/\lambda_M T'}}{\lambda_M^7 T'\left(e^{c_2/\lambda_M T'}-1\right)^2}$$

$$5=\frac{c_2 e^{c_2/\lambda_M T'}}{\lambda_M T'\left(e^{c_2/\lambda_M T'}-1\right)}$$

Setting $z=c_2/\lambda_M T'$

$$5=\frac{ze^z}{\left(e^z-1\right)},1=\frac{z/5}{\left(1-e^{-z}\right)},\qquad\left(1-e^{-z}\right)=z/5$$

$$f_1(z)=\left(1-z/5\right),\ f_2(z)=e^{-z},\ f_1(z)=f_2(z)$$

The equation $f_1(z)=f_2(z)$ is a transcendental equation that we cannot solve explicitly. We must resort to a numerical or graphical method. The latter is used to obtain the value of z for which f_1 and f_2 are identically equal. The graph is shown in Figure 10.2 from which we obtain

$$z=4.9653 \text{ at intersection}$$
$$1.4388\times10^4/\lambda_M T'=4.9653$$
$$\lambda_M T'=2997.7 \qquad \text{Wien's displacement law (10.2)}$$

Using the temperatures in Figure 10.1 in Eq. (10.2), we can calculate the exact wavelength at which the maximum blackbody radiation is obtained.

$$T=100\,^\circ\text{C},\ \lambda_M=7.765\,\mu\text{m};\ T=25\,^\circ\text{C};\ \lambda_M=9.719\,\mu\text{m}$$
$$T=75\,^\circ\text{C},\ \lambda_M=8.323\,\mu\text{m};\ T=10\,^\circ\text{C},\ \lambda_M=10.233\,\mu\text{m}$$
$$T=50\,^\circ\text{C},\ \lambda_M=8.967\,\mu\text{m};\ T=5\,^\circ\text{C},\ \lambda_M=10.417\,\mu\text{m}$$

which are consistent with Figure 10.1.

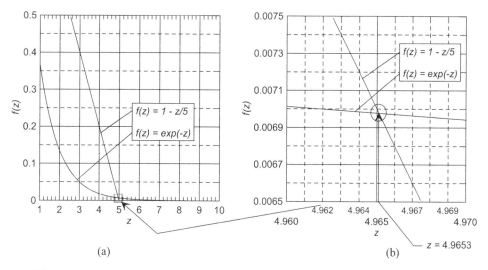

Figure 10.2. (a) Plot of $(1 - z / 5)$ and (e^{-z}). (b) Exploded view of intersecting functions.

The total blackbody energy radiated at any temperature is obtained by integrating Eq. (10.1), i.e., calculation of the area under one of the curves in Figure 10.1

$$E_b = \int_0^\infty E_{\lambda b} d\lambda = \int_0^\infty \frac{c_1}{\lambda^5 \left(e^{c_2/\lambda T'} - 1\right)} d\lambda \text{ Total emissive power } (10.3)$$

With some algebraic manipulation of Eq. (10.3), we can accomplish the desired integration.

$$E_b = \int_0^\infty \frac{c_1}{\lambda^5 \left(e^{c_2/\lambda T'} - 1\right)} d\lambda = -c_1 \int_0^\infty \frac{\lambda^2}{\lambda^5 \left(e^{c_2/\lambda T'} - 1\right)} d\left(\frac{1}{\lambda}\right)$$

where we have used $d\lambda = -\lambda^2 d\left(1/\lambda\right)$ so that

$$E_b = -c_1 \int_{\lambda=0}^{\lambda=\infty} \frac{1}{\lambda^3 \left(e^{c_2/\lambda T'} - 1\right)} \left(\frac{T'}{c_2}\right) d\left(\frac{c_2}{\lambda T'}\right) = -c_1 \int_{\lambda=0}^{\lambda=\infty} \frac{T'^4 c_2^3}{\lambda^3 T'^3 c_2^4 \left(e^{c_2/\lambda T'} - 1\right)} d\left(\frac{c_2}{\lambda T'}\right)$$

$$E_b = -c_1 \int_{\lambda=0}^{\lambda=\infty} \left(\frac{c_2}{\lambda T'}\right)^3 \frac{T'^4}{c_2^4 \left(e^{c_2/\lambda T'} - 1\right)} d\left(\frac{c_2}{\lambda T'}\right) = -c_1 \left(\frac{T'}{c_2}\right)^4 \int_{\lambda=0}^{\lambda=\infty} \left(\frac{c_2}{\lambda T'}\right)^3 \frac{1}{\left(e^{c_2/\lambda T'} - 1\right)} d\left(\frac{c_2}{\lambda T'}\right)$$

$$E_b = -c_1 \left(\frac{T'}{c_2}\right)^4 \int_{\lambda-\infty}^{\lambda=0} x^3 \frac{1}{\left(e^x - 1\right)} dx = c_1 \left(\frac{T'}{c_2}\right)^4 \int_{x-0}^{x=\infty} x^3 \frac{1}{\left(e^x - 1\right)} dx$$

The reader may find the integral (which looks like a candidate for contour integration in the complex plane) in a good handbook. The author used a computer math program to obtain the total blackbody emissive power.

$$E_b = c_1 \left(\frac{T'}{c_2}\right)^4 \int_{x=0}^{x=\infty} \frac{x^3}{\left(e^x - 1\right)} dx = c_1 \left(\frac{T'}{c_2}\right)^4 \left(\frac{\pi^4}{15}\right)$$

$$\boxed{E_b = \sigma T'^4} \qquad \text{Stefan-Boltzmann equation } (10.4)$$

σ is known as the Stefan-Boltzmann constant and has the value

$$\sigma = \frac{c_1}{15}\left(\frac{\pi}{c_2}\right)^4 = 5.6696 \times 10^{-8} \left(\frac{W}{m^2 \cdot K^4}\right); \qquad \sigma = 3.6578 \times 10^{-11}\left(\frac{W}{in.^2 \cdot K^4}\right)$$

so that E_b has the units of W / m^2 or W / in.^{2}

It is interesting to adapt Eqs. (10.3) and (10.4) for estimation of the flux radiated at some temperature T' and between wavelengths λ_1 and λ_2. The fraction of the total emissive power is given by Eq. (10.5) and plotted (as a percent) in Figure 10.3. We see that at surface temperatures of about 80°C and greater and $\lambda_1 \cong 0$ to $\lambda_2 = 20\,\mu m$ and greater, a minimum of 80% of the total emissive power is radiated. The significance of this will be more relevant when we discuss real surfaces, i.e., not blackbody radiators, later in this chapter.

$$r = \int_{\lambda_1}^{\lambda_2} E_{\lambda b} d\lambda \left/ \int_0^\infty E_{\lambda b} d\lambda = \frac{1}{\sigma T'^4} \int_{\lambda_1}^{\lambda_2} \frac{c_1}{\lambda^5 \left(e^{c_2/\lambda T'} - 1\right)} d\lambda \right. \qquad \begin{array}{l}\text{Fraction of total}\quad(10.5)\\ \text{emissive power}\end{array}$$

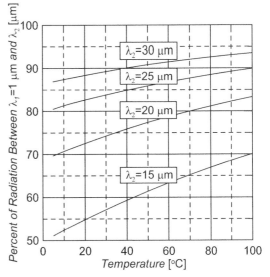

Figure 10.3. Percent of total emissive power radiated between wavelengths $\lambda_1 \cong 0$ and λ_2.

10.2 SPACIAL EFFECTS AND THE VIEW FACTOR

Radiation emitted by a surface propagates in all directions. Similarly, radiation incident upon a surface may arrive from many or all directions. These directional issues are addressed by the concept of *radiation intensity*. We can use a spherical coordinate system illustrated in Figure 10.4 (a) to describe directional effects. Figure 10.4 (b) is the same coordinate system with a solid angle $d\Omega$ subtended by the infinitesimal area element dA_n where the latter lies on a spherical (hemispherical in this instance) surface with radius r. The usefulness of the solid angle will be demonstrated in the following mathematical development. Radiation rays are emitted from the area element dA_1 and pass through the area element dA_n, which subtends the solid angle element $d\Omega$. The math definition of the solid angle element is

$$d\Omega \equiv dA_n/r^2$$

where r is the distance from dA_1 to dA_n. Note that since the area is proportional to r^2, the ratio dA_n/r^2, i.e., $d\Omega$ is independent of r. The infinitesimal dA_n can be written in the spherical coordinates as

$$dA_n = r^2 \sin\theta d\theta d\phi$$

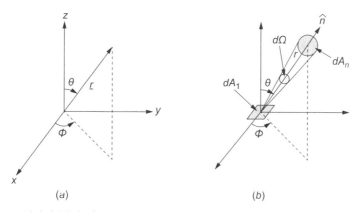

Figure 10.4. (a) Spherical coordinate system. (b) Solid angle $d\Omega$ subtended by dA_n.

Then the solid angle element is

$$d\Omega = dA_n / r^2 = \sin\theta d\theta d\phi$$

and has dimensions of *steradians* (sr).

The next issue is in regard to the rate of radiation emitted by dA_1 and passing through dA_n. We need to introduce the concept of *spectral intensity* I_λ.

$I_\lambda \equiv$ rate of energy per unit area emitted by dA_1 at wavelength λ in the (θ, ϕ) direction, passing through dA_n, per unit solid angle, per unit wavelength interval $d\lambda$,

or
$$I_\lambda = dQ / dA_n d\Omega d\lambda \quad \left[\text{W} / \left(\text{m}^2 \cdot \text{sr} \cdot \mu \right) \right]$$

and from Figure 10.5 we see that $dA_n = dA_1 \cos\theta$ so that

$$I_\lambda = dQ / \left(dA_1 \cos\theta d\Omega d\lambda \right)$$
$$dQ / d\lambda = I_\lambda dA_1 \cos\theta d\Omega \tag{10.6}$$
$$dQ / dA_1 d\lambda = I_\lambda \cos\theta \sin\theta d\theta d\phi \tag{10.7}$$

If we integrate Eq. (10.7) over an entire hemisphere, i.e., $\theta = 0 \to \pi/2, \phi = 0 \to 2\pi$, we get the spectral (or monochromatic) emissive power E_λ,

$$E_{\lambda b} = \int_{\theta=0}^{\theta=\pi/2} \int_{\phi=0}^{\phi=2\pi} I_\lambda \cos\theta \sin\theta d\theta d\phi \quad \text{Spectral emissive power} \tag{10.8}$$

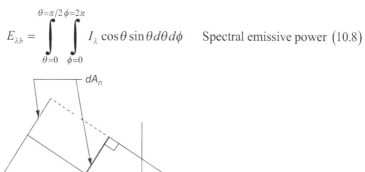

Figure 10.5. Relationship between dA_n and dA_1.

The next step is very important with regard to the physics of the model that we are developing. We define that

> *A diffuse surface radiates with a uniform intensity in all directions*

so that we remove I_λ from beneath the integral sign in Eq. (10.8) to get

$$E_{\lambda b} = I_\lambda \int_{\theta=0}^{\theta=\pi/2} \int_{\phi=0}^{\phi=2\pi} \cos\theta \sin\theta d\theta d\phi$$

which is easily integrated by using the substitution $u = \sin\theta$. The result is

$$I_\lambda = E_{\lambda b}/\pi \qquad \text{Radiation intensity} \quad (10.9)$$

so that after substituting Eq. (10.9) into Eq. (10.6), we arrive at

$$\frac{dQ}{d\lambda} = \frac{E_{\lambda b}}{\pi} dA_1 \cos\theta d\Omega \qquad (10.10)$$

Our next consideration is when we have two surfaces exchanging radiation. We rewrite Eq. (10.10) to clarify the nomenclature defining the two interacting surfaces.

$$\frac{dQ_{1,2}}{d\lambda} = \frac{E_{\lambda b1}}{\pi} dA_1 \cos\theta_1 d\Omega = \frac{E_{\lambda b1}}{\pi} dA_1 \cos\theta_1 \frac{dA_n}{r^2}$$

where the variables are subscripted according to Figure 10.6. But dA_n, when seen from surface 2 is

$$dA_n = \cos\theta_2 dA_2$$

and

$$\frac{dQ_{1,2}}{d\lambda} = \frac{E_{\lambda b1}}{\pi} dA_1 \cos\theta_1 \left(\frac{dA_n}{r^2}\right) = \frac{E_{\lambda b1}}{\pi} A_1 \cos\theta_1 \left(\frac{\cos\theta_2 dA_2}{r^2}\right)$$

$$\frac{dQ_{1,2}}{d\lambda} = \frac{E_{\lambda b1}}{\pi r^2} \cos\theta_1 \cos\theta_2 dA_1 dA_2 \qquad (10.11)$$

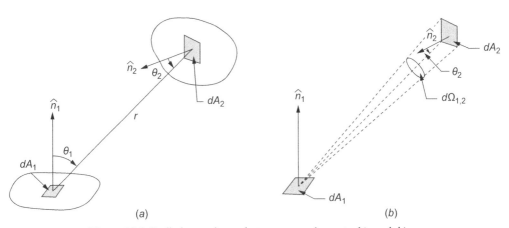

(a)   (b)

Figure 10.6. Radiation exchange between area elements dA_1 and dA_2.

Integrating Eq. (10.11) for $\lambda = 0 \to \infty$:

$$\Delta Q_{1,2} = \frac{1}{\pi} \int_{\lambda=0}^{\lambda=\infty} E_{\lambda b1} d\lambda \frac{\cos\theta_1 \cos\theta_2 dA_1 dA_2}{r^2}$$

But

$$E_{b1} = \int_{\lambda=0}^{\lambda=\infty} E_{\lambda b1} d\lambda$$

When we have diffusely radiating surfaces (radiate uniformly in all directions)

$$\Delta Q_{1,2} = \frac{E_{b1} \cos\theta_1 \cos\theta_2 dA_1 dA_2}{\pi r^2} \qquad \begin{array}{l} \text{Blackbody radiation} \\ \text{from } dA_1 \text{ to } dA_2 \end{array} \quad (10.12)$$

By precisely the same kind of arguments we can also write

$$\Delta Q_{2,1} = \frac{E_{b2} \cos\theta_2 \cos\theta_1 dA_2 dA_1}{\pi r^2} \qquad \begin{array}{l} \text{Blackbody radiation} \\ \text{from } dA_2 \text{ to } dA_1 \end{array} \quad (10.13)$$

The *net* radiation exchange is just the difference between Eqs. (10.12) and (10.13).

$$\Delta Q_{Net} = \frac{\left(E_{b1} - E_{b2}\right)\cos\theta_1 \cos\theta_2 dA_1 dA_2}{\pi r^2} \qquad \begin{array}{l} \text{Blackbody radiation} \\ \text{between } dA_1 \text{ and } dA_2 \end{array} \quad (10.14)$$

We integrate over the total areas A_1 and A_2 to obtain

$$Q_{Net} = \left(E_{b1} - E_{b2}\right) \int_{A_1} \int_{A_2} \frac{\cos\theta_1 \cos\theta_2 dA_1 dA_2}{\pi r^2} \qquad \begin{array}{l} \text{Blackbody radiation} \\ \text{between } A_1 \text{ and } A_2 \end{array} \quad (10.15)$$

or more generally,

$$Q_{ij} = \left(E_{bi} - E_{bj}\right) \int_{A_i} \int_{A_j} \frac{\cos\theta_i \cos\theta_j dA_i dA_j}{\pi r_{ij}^2} \qquad \begin{array}{l} \text{Blackbody radiation} \\ \text{between } A_i \text{ and } A_j \end{array} \quad (10.16)$$

We can rewrite Eq. (10.16)

$$\boxed{Q_{ij} = \left(E_{bi} - E_{bj}\right) A_i F_{ij}}$$

Net radiation exchange (10.17)
between surfaces i, j

where

$$\boxed{F_{ij} = \frac{1}{A_i} \int_{A_i} \int_{A_j} \frac{\cos\theta_i \cos\theta_j dA_i dA_j}{\pi r_{ij}^2}}$$

View factor (10.18)

Just as Eq. (10.18) is

$$A_i F_{ij} = \int_{A_i} \int_{A_j} \frac{\cos\theta_i \cos\theta_j dA_i dA_j}{\pi r_{ij}^2}$$

we can exchange the indices i, j on the right-hand side to obtain

$$A_j F_{ji} = \int_{A_j} \int_{A_i} \frac{\cos\theta_j \cos\theta_i dA_j dA_i}{\pi r_{ji}^2}$$

We see by inspection that

$$\boxed{A_i F_{ij} = A_j F_{ji}}$$

Reciprocity of FA product (10.19)

Closed enclosure problems utilize the fact that the radiation leaving any given surface must be intercepted by all of the other surfaces, so we write

$$\boxed{\sum_{j} F_{ij} = 1.0}$$ Interior of closed enclosure (10.20)

An additional tool that is useful in *view factor algebra* is that

The total view factor is the sum of its parts

so that for example, $A_1 F_{1,(2+3)} = A_1 F_{1,2} + A_1 F_{1,3}$

The net heat transfer between *black* surfaces, i.e., those surfaces that emit according to Planck's blackbody radiation formula, is obtained by rewriting Eq. (10.17) as

$$\boxed{Q_{ij} = \sigma F_{ij} A_i \left(T_i'^4 - T_j'^4 \right)}$$ Blackbody radiation (10.21)
                                                           exchange

Most heat transfer textbooks contain a variety of view factor formulae and graphs. The book by John R. Howell is perhaps the most complete. The formulae, Eqs. (10.22) and (10.23), for parallel and perpendicular plates, respectively, are provided here according to the geometry of Figure 10.7. Plots for these equations are shown as Figures 10.8 and 10.9.

$$x = L/S, \qquad y = W/S$$

$$\left(\frac{\pi xy}{2} \right) F_{1,2} = \ln \left[\frac{\left(1+x^2\right)\left(1+y^2\right)}{1+x^2+y^2} \right]^{1/2} + y\sqrt{1+x^2}\ \tan^{-1}\left(\frac{y}{\sqrt{1+x^2}} \right)$$ Parallel plates (10.22)

$$+ x\sqrt{1+y^2}\ \tan^{-1}\left(\frac{x}{\sqrt{1+y^2}} \right) - y\tan^{-1} y - x\tan^{-1} x$$

$$x = L_2/W, \qquad y = L_1/W, \qquad z = x^2 + y^2$$

$$\left(\pi y\right) F_{1,2} = \frac{1}{4} \ln \left\{ \left[\frac{\left(1+x^2\right)\left(1+y^2\right)}{1+z} \right] \left[\frac{y^2\left(1+z\right)}{\left(1+y^2\right)z} \right]^{y^2} \left[\frac{x^2\left(1+z\right)}{\left(1+x^2\right)z} \right]^{x^2} \right\}$$ Perpendicular plates (10.23)

$$+ y\tan^{-1}\left(1/y\right) + x\tan^{-1}\left(1/x\right) - \sqrt{z}\ \tan^{-1}\left(1/\sqrt{z}\right)$$

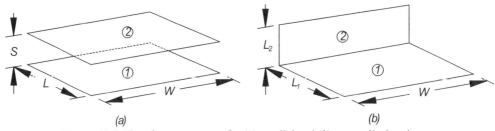

(a)   (b)

Figure 10.7. View factor geometry for (*a*) parallel and (*b*) perpendicular plates.

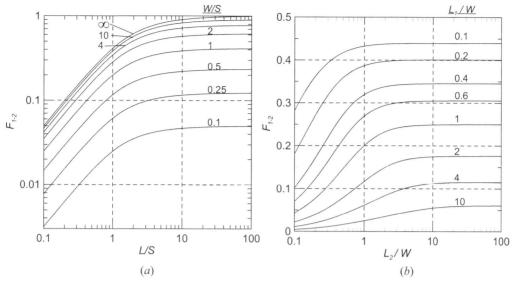

Figure 10.8. View factors for (*a*) parallel and (*b*) perpendicular plates.

The Hottel crossed-string method is useful for calculating view factors for two surfaces that extend to infinity in one direction. Figure 10.9 indicates the parameters for such a system. The view factor is given by

$$F_{1,2} = (\text{sum of crossed strings}) - (\text{sum of uncrossed strings}) / 2 \times \text{length of shape 1}$$

Thus following the preceding statement and Figure 10.9, where neither L_1 nor L_2 are necessarily straight lines, we have the formula

$$F_{1,2} = \frac{(L_3 + L_4) - (L_5 + L_6)}{2L_1}$$ Crossed strings   (10.24)

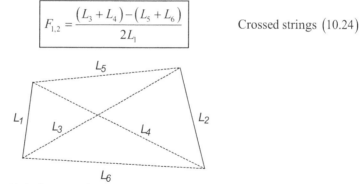

Figure 10.9. Geometry for Hottel crossed-string method.

10.3 APPLICATION EXAMPLE: VIEW FACTORS FOR FINITE PARALLEL PLATES

Using the nomenclature of Figure 10.7 (a), we calculate a view factor for two parallel plates. Then a plot is shown for increasingly large values of W. The result for a very large value of W will be compared to a calculation using the method of crossed strings.

The dimensions that we use are those that one might expect for circuit boards cooled by natural convection, but perhaps with a little greater spacing. The dimensions are therefore $W = 10.0$ in., $L = 10.0$ in., and $S = 1.0$ in., remembering that S should represent the component surface-to-opposing PCB as the actual spacing. The board-to-board view factor is calculated from Eq. (10.22).

$$x = L/S = 10.0/1.0 = 10.0, \quad y = W/S = 10.0/1.0 = 10.0$$

$$F_{1,2} = \frac{2}{\pi xy} \left\{ \ln \left[\frac{\left(1+x^2\right)\left(1+y^2\right)}{1+x^2+y^2} \right]^{1/2} + y\sqrt{1+x^2}\, \tan^{-1}\left(\frac{y}{\sqrt{1+x^2}} \right) + x\sqrt{1+y^2}\, \tan^{-1}\left(\frac{x}{\sqrt{1+y^2}} \right) \right. \\ \left. -y\tan^{-1}y - x\tan^{-1}x \right\}$$

$$F_{1,2} = \frac{2}{\pi(10.0)(10.0)} \left\{ \ln \left[\frac{\left(1+(10.0)^2\right)\left(1+(10.0)^2\right)}{1+(10.0)^2+(10.0)^2} \right]^{1/2} \right. \\ \left. +2(10.0)\sqrt{1+(10.0)^2}\, \tan^{-1}\left(\frac{10.0}{\sqrt{1+(10.0)^2}} \right) - 2(10.0)\tan^{-1}(10.0) \right\}$$

$$F_{1,2} = \frac{2}{10^2\,\pi} \left\{ 1.96 - 157.36 - 29.43 \right\} = 0.83$$

(more exactly, 0.82699)

If you use Figure 10.8 (*a*) you will get the same answer, but not with the same precision as above.

Since both plates are square, we need only calculate one view factor, $F_{1,E}$, from plate to edge. The remaining view factors are the same.

$$x = L_2/W = 1.0/10.0 = 0.10, \, y = L_1/W = 10.0/10.0 = 1.0, \, z = x^2 + y^2 = (0.10)^2 + (1.0)^2 = 1.01$$

$$F_{1,E} = \frac{1}{4\pi y} \ln \left\{ \left[\frac{\left(1+x^2\right)\left(1+y^2\right)}{1+z} \right] \left[\frac{y^2\left(1+z\right)}{\left(1+y^2\right)z} \right]^{y^2} \left[\frac{x^2\left(1+z\right)}{\left(1+x^2\right)z} \right]^{x^2} \right\} + \quad \text{Perpendicular plates (10.23)}$$

$$\frac{1}{\pi}\tan^{-1}\left(1/y\right) + \frac{x}{\pi y}\tan^{-1}\left(1/x\right) - \frac{\sqrt{z}}{\pi y}\tan^{-1}\left(1/\sqrt{z}\right)$$

$$F_{1,E} = \frac{1}{4\pi(1.0)} \ln \left\{ \left[\frac{\left(1+(0.10)^2\right)\left(1+(1.0)^2\right)}{1+1.01} \right] \left[\frac{(1.0)^2\left(1+1.01\right)}{\left(1+(1.0)^2\right)(1.01)} \right]^{1.0} \left[\frac{(0.10)^2\left(1+1.01\right)}{\left(1+(0.10)^2\right)(1.01)} \right]^{0.01} \right\} + $$

$$\frac{1}{\pi} \left[\tan^{-1}\left(1/1.0\right) + \left(0.10/1.0\right)\tan^{-1}\left(1/0.10\right) - \left(\sqrt{1.01}/1.0\right)\tan^{-1}\left(1/\sqrt{1.01}\right) \right]$$

$$F_{1,E} = \frac{1}{4\pi} \ln \left[(1.005)(.995)(0.197)^{0.01} \right] + $$

$$\frac{1}{\pi} \left[\tan^{-1}(1.0) + (0.10)\tan^{-1}(10.0) - \sqrt{1.01}\,\tan^{-1}\left(1/\sqrt{1.01}\right) \right] = 0.0433 \, \text{(more exactly, 0.04325)}$$

which you could also obtain from Figure 10.8 (*b*), though not to the same accuracy.

We can check our results by determining if $F_{1,2}$ and $F_{1,E}$ satisfy Eq. (10.20) according to

$$\sum_{j=2}^{5} F_{1,j} = F_{1,2} + 4F_{1,E} = 0.83 + 4(0.043) = 1.00$$

which it does. The view factor problem tells us that 83% of the radiation is intercepted by the opposing circuit board. Of course, our problem probably doesn't really have surfaces that behave like blackbody radiators, an issue that we will address later in this chapter.

The last item to look at in this example is the application of the crossed-strings method using Eq. (10.24). We only need to substitute the appropriate variables into the equation,

$$F_{1,2} = \left[(L_3 + L_4) - (L_5 + L_6)\right]/2L_1 = \left[\left(2\sqrt{L_5^2 + L_1^2}\right) - (L_5 + L_6)\right]/2L_1$$

which becomes, using the variable names in our problem,

$$F_{1,2} = \frac{(L_3 + L_4) - (L_5 + L_6)}{2L_1} = \frac{\left(2\sqrt{S^2 + L^2}\right) - (2S)}{2L} = \frac{2\sqrt{(1.0)^2 + (10.0)^2} - 2(1.0)}{2(10.0)} = 0.91$$

Note that we first obtained $F_{1,2} = 0.83$ as the exact value and now we get $F_{1,2} = 0.91$ with the crossed-strings method. Remember, however that the crossed-strings method presumes that in our problem $W \to \infty$. In Figure 10.10, a comparison is made of the result for the exact formula, Eq. (10.22), and that for the crossed-strings method, Eq. (10.24). We see that for our problem, Eq. (10.22) gives a view factor that is 91% of the infinite width plate and a view factor that is 99% of the infinite width plate for a plate width that is ten times our $W = 10$ in. Perhaps more importantly, the simple crossed-strings method does indeed give correct results for problems where the width is very large.

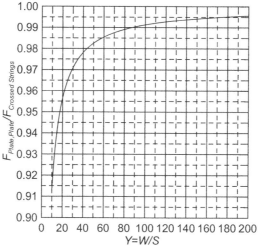

Figure 10.10. Application Example 10.3: $F_{Plate-Plate} / F_{Crossed\ Strings}$ for $S = 1.0$ in., $L = 10.0$ in. and various values of W.

10.4 NONBLACK SURFACES

Surfaces encountered in actual practice do not radiate precisely in the manner described by the blackbody equations. The actual monochromatic emissive power of a real surface is always less than $E_{\lambda b}$, although sometimes not by much. This deviation from the ideal is due to surface conditions such as roughness and oxidation. The ratio of the actual to blackbody monochromatic emissive power defines the monochromatic emissivity.

$$\varepsilon_\lambda = E_\lambda / E_{\lambda b} \qquad\qquad \text{Monochromatic emissivity} \quad (10.25)$$

The total emissivity is

$$\varepsilon = E / E_b = E / \sigma T'^4 \qquad\qquad \text{Total emissivity} \quad (10.26)$$

Table 10.1 lists emissivity for several common materials. Where values for several wavelengths were available, the region of about 10 μm was selected. A more detailed compilation may be found in Appendix *ii*.

A particularly interesting example of the importance of the interrelationship between monochromatic emissivity and (total) emissivity is illustrated by the plot of ε_λ for anodized aluminum in Figure 10.11 as obtained by Dunkle (1954). Although this material has an ε_λ that varies considerably with wavelength, the graph indicates that at temperatures commonly encountered in electronic equipment (25 - 100°C), the dominant spectral region is 5 - 15 μm. This would suggest that according to Figure 10.11, anodized aluminum heat sinks may be characterized by an emissivity of 0.4 to 0.9. Dunkle used

to calculate $\varepsilon = 0.7$.

$$\varepsilon = \int_0^\infty \varepsilon_\lambda E_{\lambda b} d\lambda \bigg/ \int_0^\infty E_{\lambda b} d\lambda \qquad \text{Average emissivity} \quad (10.27)$$

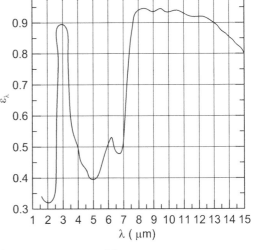

Figure 10.11. Monochromatic emittance for anodized aluminum at 282°C. Plot data extracted from Dunkle (1954).

Table 10.1. Emissivity of some common materials.

Surface	Emissivity
Metals	
Aluminum, polished	0.05
Aluminum, lightly oxidized	0.1
Aluminum, anodized	0.7-0.9
Copper, polished	0.06
Copper, oxidized	0.25-0.7
Gold	0.02
Iron, polished	0.06
Iron, bright cast	0.2
Iron, oxidized cast	0.6
Paint	
Varnish	0.89
Lacquer, clear on bright copper	0.07
Lacquer, white	0.9
Lacquer, flat black	0.95
Enamel, most colors	0.9
Oil, most colors	0.9

10.5 THE RADIATION HEAT TRANSFER COEFFICIENT

We used the notion of a radiation heat transfer coefficient in both Chapters 1 and 9, but without much explanation or derivation. Our background knowledge is now sufficient to clarify this subject. In Eq. (10.21) we have an expression for radiation exchange between blackbodies. You will see later in the present chapter that if we have the situation of a *surface A_S with emissivity ε surrounded by a much larger surface*, one may write

$$Q_r = \varepsilon \sigma A_S \left(T_S'^4 - T_A'^4 \right) = \varepsilon \sigma A_S \left[\left(T_S + 273.15 \right)^4 - \left(T_A + 273.15 \right)^4 \right]$$

$$Q_r = \varepsilon \sigma A_S \frac{\left(T_S'^4 - T_A'^4 \right)}{\left(T_S - T_A \right)} \left(T_S - T_A \right)$$

$$\boxed{Q_r = \varepsilon A_S h_r \left(T_S - T_A \right)}$$

Small surface S sur- (10.28)
rounded by a much
larger surface

Using a binomial expansion formula, the radiation heat transfer coefficient is

$$h_r = \sigma \frac{\left(T_S'^4 - T_A'^4 \right)}{T_S - T_A} = \sigma \frac{\left(T_S'^3 + T_S'^2 T_A' + T_S' T_A'^2 + T_A'^3 \right)\left(T_S' - T_A' \right)}{T_S - T_A}$$

$$h_r = \sigma \frac{\left(T_S'^3 + T_S'^2 T_A' + T_S' T_A'^2 + T_A'^3 \right)\left(T_S' - T_A' \right)}{T_S' - T_A'}$$

$$\boxed{h_r = 3.657 \times 10^{-11} \left(T_S'^3 + T_S'^2 T_A' + T_S' T_A'^2 + T_A'^3 \right) \ \text{W}/\left(\text{in.}^2 \cdot {}^\circ\text{C} \right)} \qquad (10.29)$$

Equation (10.29) is plotted in Figure 10.12 for several different values of T_A. This plot indicates that for a T_A of about 20 to 30°C and modest temperature rises, $T_S - T_A$, of less than about 10°C, and $h_r = 0.004$ W/(in.$^2 \cdot$ °C). Using Eqs. (10.28) and (10.29), the radiation resistance for the small surface surrounded by a much larger one is $R = 1/(\varepsilon h_r A_S)$.

It is suggested that the reader take time to review the problem in Section 8.4 where we analyzed a radiating and convecting plate. We solved the problem by adding the convection and radiation *conductances* to obtain $C = C_c + C_r = h_c A_S + \varepsilon h_r A_S = \left(h_c + \varepsilon h_r \right) A_S$. Many applications of the simple formula (10.28) are such that the second surface is the ambient at temperature T_A which represents the room walls.

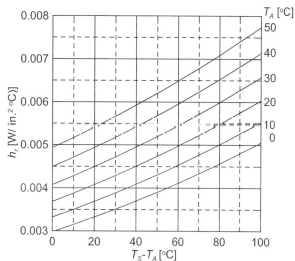

Figure 10.12. Radiation heat transfer coefficient.

10.6 APPLICATION EXAMPLE: RADIATION AND NATURAL CONVECTION COOLED ENCLOSURE WITH CIRCUIT BOARDS - ENCLOSURE TEMPERATURES ONLY

This problem is identical to the natural convection cooled problem solved in Section 8.10 except that we now add radiation heat loss from the enclosure exterior. The problem geometry is identical to that shown in Figure 8.16 so it will not be repeated here. The change from the earlier problem is shown in Figure 10.13 as a *radiation resistance* R_r for an emissivity $\varepsilon = 0.8$. Since radiation is not dependent on the panel orientation, we need add only the one resistance that uses the enclosure total surface area.

Formulae for the total internal convection, external convection, and thermal-fluid resistances, plus the air draft formula were developed with the result

$$R_{CI} = 2.327/\Delta T_{Air-Wall}^{0.25}, \ R_{CE} = 2.327/\Delta T_{Wall-A}^{0.25}, \ R_f = 1.76/G, \ G = 0.24 Q_d^{1/3}$$

The radiation resistance is more complicated than the convection resistance formula.

$$R_r = 1/(\varepsilon h_r A_S) = 1/(\varepsilon h_r 6WH) = 1 \left/ \left\{ 0.8\sigma (6WH) \left[\begin{array}{l} (T_W+273)^3 + (T_W+273)^2(T_A+273) \\ + (T_W+273)(T_A+273)^2 + (T_A+273)^3 \end{array} \right] \right\} \right.$$

$$R_r = 1 \left/ \left\{ 0.8(3.657 \times 10^{-11})(6 \times 7 \times 7) \left[\begin{array}{l} (T_W+273)^3 + (T_W+273)^2(T_A+273) \\ + (T_W+273)(T_A+273)^2 + (T_A+273)^3 \end{array} \right] \right\} \right.$$

$$R_r = 1 \left/ \left\{ 8.60 \times 10^{-9} \left[\begin{array}{l} (T_W+273)^3 + (T_W+273)^2(T_A+273) \\ + (T_W+273)(T_A+273)^2 + (T_A+273)^3 \end{array} \right] \right\} \right.$$

The total external resistance and wall temperature rise, following Figure 10.13, is

$$R_E = R_{CE}R_r/(R_{CE}+R_r); \ \Delta T_{W-A} = R_E(Q-Q_d)$$

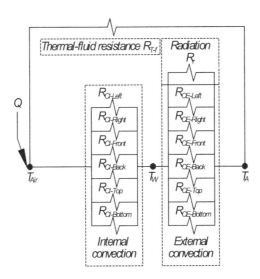

Figure 10.13. Natural convection and external radiation cooled enclosure (modified version of Figure 8.19).

and the sum of the internal convection and total external resistances is

$$R_x = R_{CI} + R_E$$

which results in a total enclosure thermal resistance from the internal air at T_{Air} to the external ambient at T_A. The iterated solution is shown in Table 10.2. Compare these results with those in Table 8.12 to see the effects of the external radiation.

Table 10.2. Application Example 10.6: Iteration table for sealed and vented versions of nonfan cooled enclosure in an ambient air $T_{A'}$.

It.	ΔT_{Wall-A}	ΔT_{Air-A}	Q_d $= 0.24 Q_d^{1/3}$	G $= 1.76/G$	R_{T-f} $= 2.327/\Delta T_{Wall-A}^{0.25}$	R_{CE} $= \dfrac{R_{CE} R_r}{R_{CE} + R_r}$	R_r $= R_{CE}(Q-Q_d)$	R_E	ΔT_{Wall-A} $= 2.327/\Delta T_{Air-Wall}^{0.25}$	R_{CI} $= R_{CI} + R_E$	R_X $= R_f R_x/(R_f + R_x)$	R_{Total} $= R_{Total} Q$	ΔT_{Air-A}
1	25*	50*	10^{-20}	0	10^7	1.041	1.016	0.514	5.141	1.041	1.555	1.555	15.555
2	5.141**	15.56**	10^{-20}	0	10^7	1.545	1.124	0.651	6.506	1.295	1.946	1.946	19.457
3	6.506**	19.46**	10^{-20}	0	10^7	1.457	1.116	0.632	6.319	1.227	1.858	1.858	18.584
4	6.319**	18.58**	10^{-20}	0	10^7	1.467	1.117	0.634	6.342	1.243	1.878	1.878	18.777
5	6.342**	18.78**	10^{-20}	0	10^7	1.466	1.117	0.634	6.340	1.239	1.873	1.873	18.729
1	6.34**	18.73**	5*	0.411	4.287	1.466	1.117	0.634	3.170	1.240	1.874	1.304	13.040
2	3.17**	13.04**	3.05***	0.348	5.057	1.744	1.135	0.688	4.782	1.313	2.000	1.433	14.333
3	4.78**	14.33**	2.83***	0.340	5.180	1.574	1.126	0.656	4.703	1.324	1.980	1.432	14.324
4	4.70**	14.32**	2.77***	0.337	5.222	1.580	1.126	0.658	4.757	1.321	1.979	1.435	14.350

Note: * Initial value is a guess; ** Use most recent value from previous iteration; *** Use most recent value of G from previous iteration.

10.7 RADIATION FOR MULTIPLE GRAY-BODY SURFACES

Figure 10.14 illustrates radiation of unity magnitude incident upon some material. A fraction ρ is reflected, α absorbed, and τ transmitted. Thus for a Q total energy rate incident, conservation of energy is applicable:

$$Q = Q_R + Q_A + Q_T, \quad 1 = \frac{Q_R}{Q} + \frac{Q_A}{Q} + \frac{Q_T}{Q}, \quad 1 = \rho + \alpha + \tau$$

Opaque surfaces transmit nothing, thus

$$\tau = 0, \quad \alpha = 1 - \rho$$

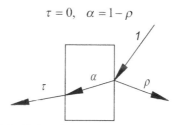

Figure 10.14. Radiation absorption, reflection, and transmission.

The reflectivity and absorptivity will also be assumed to be wavelength independent, or are *gray-bodies*. We will continue to assume only diffuse surfaces (incident radiation is reflected equally in all directions) and not specular (smooth surfaces that obey Snell's law of angle of incidence = angle of reflection).

Our next step is to prove Kirchhoff's identity. Figure 10.15 illustrates an internal gray-body in equilibrium (steady-state) with a blackbody enclosure. Conservation of energy allows us to write

Energy radiated by internal gray − body = Energy absorbed by internal gray − body

$$E\Lambda = q_i \Lambda \alpha$$

Next we can replace the internal gray-body with an internal blackbody of the identical geometry *and temperature* and let it come to equilibrium with the enclosure. Since this latter internal blackbody is therefore a perfect absorber, we write

Energy radiated by internal blackbody = Energy absorbed by internal blackbody

$$E_b A = q_i A$$

Dividing the first of the two preceding equations by the second we obtain

$$E/E_b = \alpha$$

However, we also have by Eq. (10.26) that $E/E_b = \varepsilon$. Thus we arrive at

$$\boxed{\varepsilon = \alpha}$$ Kirchhoff's identity (10.30)

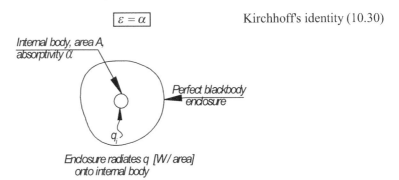

Figure 10.15. Geometry used to derive Kirchhoff's identity.

10.8 HOTTEL SCRIPT F ($\mathcal{F}$) METHOD FOR GRAY-BODY RADIATION EXCHANGE

Now we begin our study of radiation when more than one surface is involved. If we always considered only blackbody surfaces, we could stop here, but we seldom have that luxury. And of course we always have two or more surfaces. We have solved a few problems for just two surfaces, but we didn't really prove Eq. (10.28).

We begin our analysis by considering radiation, as shown in Figure 10.16, arriving at surface 1 and coming from some other surface 2, where the existence of surface 2 is not shown. The quantities in Figure 10.16 are

H_1 : total incident irradiance [W]
$\rho_1 H_1$: total reflected irradiance [W]
E_1 : total emissive power due to
        surface 1 at temperature T_1'
J_1 : radiosity of surface 1 of area A_1

From an examination of Figure 10.16, we see that we may write

$$J_1 A_1 = E_1 A_1 + \rho_1 H_1 \qquad (10.31)$$

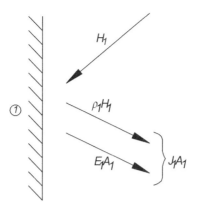

Figure 10.16. Radiation incident upon gray-body
surface 1, and diffusely reflected and emitted.

The irradiation of surface 1 by surface 2 is

$$H_1 = F_{2-1}A_2J_2 \tag{10.32}$$

Substituting Eq. (10.32) into Eq. (10.31) we have

$$J_1A_1 = \varepsilon_1 E_{b1}A_1 + \rho_1 F_{2-1}A_2J_2 \tag{10.33}$$

Equation (10.33) is easily generalized to a system of N surfaces, including surface i:

$$J_iA_i = \varepsilon_i E_{bi}A_i + \sum_{j=1}^{N}\rho_i F_{ji}A_jJ_j \tag{10.34}$$

The reciprocity of $F_{ji}A_j$ as shown in Eq. (10.19) allows us to write Eq. (10.34) as

$$J_iA_i = \varepsilon_i E_{bi}A_i + \rho_i A_i\sum_{j=1}^{N}F_{ij}J_j$$

$$\boxed{J_i = \varepsilon_i E_{bi} + \rho_i\sum_{j=1}^{N}F_{ij}J_j} \tag{10.35}$$

We continue with

$$J_i - \rho_i\sum_{j=1}^{N}F_{ij}J_j = \varepsilon_i E_{bi}$$

which is multiplied by (A_i/ρ_i) to obtain

$$(A_i/\rho_i)\left(J_i - \rho_i\sum_{j=1}^{N}F_{ij}J_j\right) = (\varepsilon_i A_i/\rho_i)E_{bi}$$

$$\sum_{j=1}^{N}\left[(A_i/\rho_i)(\delta_{ij} - \rho_i F_{ij})\right]J_j = (\varepsilon_i A_i E_{bi}/\rho_i) \tag{10.36}$$

where we have taken advantage of the *Kronecker delta* δ_{ij}, which allows a more compact form of the equation.

$$\delta_{ij} \equiv \text{Kronecker delta} = \begin{cases} 0 \text{ for } i \neq j \\ 1 \text{ for } i = j \end{cases}$$

Now we write the [...] term in Eq. (10.36) as the element

$$e_{ij} = [\mathbf{E}]_{ij} = (A_i/\rho_i)(\delta_{ij} - \rho_i F_{ij}) \tag{10.37}$$

of the matrix $\mathbf{E}$ and $[\mathbf{J}]_i$, $[\mathbf{T}]_i$ are elements of the column vectors $\mathbf{J}, \mathbf{T}$ written as

$$[\mathbf{J}]_i = J_i$$
$$[\mathbf{T}]_i = \varepsilon_i A_i E_{bi}/\rho_i \qquad (10.38)$$

so that following Eq. (10.36) we have the matrix equation (actually a set of N simultaneous equations)

$$\mathbf{EJ} = \mathbf{T} \qquad (10.39)$$

which can be solved for the radiosity by an inversion of $\mathbf{E}$ and a matrix multiplication:

$$\mathbf{J} = \mathbf{E}^{-1}\mathbf{T} \qquad (10.40)$$

In simultaneous equation form, the radiosity in Eq. (10.40) is written as

$$J_i = \sum_{j=1}^{N} e_{ij}^{-1} \varepsilon_j A_j E_{bj}/\rho_j \qquad (10.41)$$

i.e., e_{ij}^{-1} is the ij element of the inverse of the matrix $\mathbf{E}$.
Returning to Eq. (10.31) for surface i,

$$J_i A_i = A_i E_i + \rho_i H_i = A_i \varepsilon_i E_{bi} + \rho_i H_i = A_i \varepsilon_i \sigma T_i'^4 + \rho_i H_i$$

and solving for H_i,

$$H_i = (A_i/\rho_i)(J_i - \varepsilon_i \sigma T_i'^4)$$

If surface i has an absorptivity α_i, the net radiative heat transfer $Q_{i\,Net}$ from surface i is

$$Q_{i\,Net} = \text{Radiation leaving surface } i - \text{Radiation into surface } i$$
$$Q_{i\,Net} = (\varepsilon_i A_i \sigma T_i'^4 + \rho_i H_i) - H_i = \varepsilon_i A_i \sigma T_i'^4 + (\rho_i - 1) H_i \qquad (10.42)$$

We can write
$$\alpha_i + \rho_i = 1, \text{ or } (\rho_i - 1) = -\alpha_i$$
$$(\rho_i - 1) = -\varepsilon_i$$

where Kirchhoff's identity was used in the last step. Equation (10.42) becomes

$$Q_{i\,Net} = \varepsilon_i A_i \sigma T_i'^4 - \varepsilon_i H_i = -\varepsilon_i (H_i - A_i \sigma T_i'^4) = -\varepsilon_i \left[(A_i/\rho_i)(J_i - \varepsilon_i \sigma T_i'^4) - A_i \sigma T_i'^4 \right]$$
$$Q_{i\,Net} = -(A_i \varepsilon_i/\rho_i)\left[(J_i - \varepsilon_i \sigma T_i'^4) - \rho_i \sigma T_i'^4 \right]$$

$$Q_{i\,Net} = -(A_i \varepsilon_i/\rho_i)\left[J_i - (\rho_i + \varepsilon_i) \sigma T_i'^4 \right] \qquad (10.43)$$

Substituting Eq. (10.41) for J_i in Eq. (10.43),

$$Q_{i\,Net} = -(A_i \varepsilon_i/\rho_i)\left[\sum_{j=1}^{N} e_{ij}^{-1}(\varepsilon_j A_j/\rho_j)\sigma T_j'^4 - (\rho_i + \varepsilon_i)\sigma T_i'^4 \right]$$

$$Q_{i\,Net} = -(A_i \varepsilon_i/\rho_i)\sum_{j \neq i}^{N} \left\{ \begin{array}{l} e_{ij}^{-1}(\varepsilon_j A_j/\rho_j)\sigma T_j'^4 \\ -\left[\rho_i + \varepsilon_i - e_{ii}^{-1}(\varepsilon_i A_i/\rho_i) \right]\sigma T_i'^4 \end{array} \right\} \qquad (10.44)$$

Note that the summation over j in Eq. (10.44) does not include the ith term, as it is explicitly written. Now we note that for the equilibrium condition where $T'_j = T'_i$ for all j, $Q_{i\,Net} = 0$ and we conclude from Eq. (10.44) that

$$\rho_i + \varepsilon_i - e_{ii}^{-1}\left(\varepsilon_i A_i/\rho_i\right) = \sum_{j\neq i}^{N} e_{ij}^{-1}\left(\varepsilon_j A_j/\rho_j\right) \qquad (10.45)$$

which can be substituted back into Eq. (10.44) to result in

$$Q_{i\,Net} = -\left(\varepsilon_i A_i/\rho_i\right)\sum_{j\neq i}^{N} e_{ij}^{-1}\left(\varepsilon_j A_j/\rho_j\right)\sigma\left(T'^4_j - T'^4_i\right)$$

$$Q_{i\,Net} = \sum_{j\neq i}^{N} e_{ij}^{-1}\left(\varepsilon_i\varepsilon_j A_i A_j/\rho_i\rho_j\right)\sigma\left(T'^4_i - T'^4_j\right) \qquad (10.46)$$

We define

$$\mathcal{F}_{ij} = \left(\frac{\varepsilon_i\varepsilon_j}{\rho_i\rho_j}\right)A_j e_{ij}^{-1}, \quad i\neq j$$

Then assuming that $\mathcal{F}_{ij}$ exists for $i = j$, but with no present concern as to what it actually is, we write Eq. (10.46) as

$$Q_{i\,Net} = \sum_{j=1}^{N} \mathcal{F}_{ij} A_i\sigma\left(T'^4_i - T'^4_j\right) \qquad (10.47)$$

Note that since

$$Q_{ij} = \mathcal{F}_{ij} A_i\sigma\left(T'^4_i - T'^4_j\right)$$

it is also true that

$$Q_{ji} = \mathcal{F}_{ji} A_j\sigma\left(T'^4_j - T'^4_i\right)$$

and since

$$Q_{ji} = -Q_{ij}$$

$$\mathcal{F}_{ji} A_j\sigma\left(T'^4_j - T'^4_i\right) = -\mathcal{F}_{ij} A_i\sigma\left(T'^4_i - T'^4_j\right)$$

$$\mathcal{F}_{ji} A_j\sigma\left(T'^4_j - T'^4_i\right) = \mathcal{F}_{ij} A_i\sigma\left(T'^4_j - T'^4_i\right)$$

we obtain the important result $\boxed{\mathcal{F}_{ji} A_j = \mathcal{F}_{ij} A_i}$ Reciprocity (10.48)

which is markedly similar to the reciprocity law, Eq. (10.19), that we obtained for view factors.

There is another "theorem" that is necessary for us to develop. Suppose that all our $T'_j = 0$. Then

$$Q_{i\,Net} = \sum_{j=1}^{N} \mathcal{F}_{ij} A_i\sigma T'^4_i = A_i\sigma T'^4_i\sum_{j=1}^{N} \mathcal{F}_{ij}$$

But if all, surface i is the only radiator and radiates an amount $T'_j = 0$

$$Q_{i\,Net} = \sigma\varepsilon_i A_i T'^4_i$$

Setting the right side of the two preceding equations equal,

$$A_i\sigma T'^4_i\sum_{j=1}^{N} \mathcal{F}_{ij} = \sigma\varepsilon_i A_i T'^4_i$$

$$\boxed{\sum_{j=1}^{N} \mathcal{F}_{ij} = \varepsilon_i} \qquad (10.49)$$

which also resembles an earlier result, Eq. (10.20).

The next few paragraphs are devoted to finding a formula for $\mathcal{F}_{ij}$, $i = j$. We begin with Eq. (10.49).

$$\sum_{j=1}^{N} \mathcal{F}_{ij} = \varepsilon_i$$

$$\sum_{j=i}^{N} \mathcal{F}_{ij} = \sum_{j\neq i}^{N} \mathcal{F}_{ij} + \mathcal{F}_{ii}$$

$$\sum_{j\neq 1}^{N} \mathcal{F}_{ij} + \mathcal{F}_{ii} = \varepsilon_i$$

$$\mathcal{F}_{ii} = \varepsilon_i - \sum_{j\neq i}^{N} \mathcal{F}_{ij} = \varepsilon_i - \sum_{j\neq i}^{N} \left(\frac{\varepsilon_i \varepsilon_j}{\rho_i \rho_j} \right) A_j e_{ij}^{-1} = \varepsilon_i - \frac{\varepsilon_i}{\rho_i} \sum_{j\neq i}^{N} \left(\frac{\varepsilon_j}{\rho_j} \right) A_j e_{ij}^{-1}$$

We can use Eq. (10.45) for the last term on the right.

$$\sum_{j\neq i}^{N} e_{ij}^{-1} \left(\frac{\varepsilon_j A_j}{\rho_j} \right) = \rho_i + \varepsilon_i - e_{ii}^{-1} \left(\frac{\varepsilon_i A_i}{\rho_i} \right)$$

$$\mathcal{F}_{ii} = \varepsilon_i - \frac{\varepsilon_i}{\rho_i} \left[\sum_{j\neq i}^{N} \left(\frac{\varepsilon_j}{\rho_j} \right) A_j e_{ij}^{-1} \right]$$

$$= \varepsilon_i - \frac{\varepsilon_i}{\rho_i} \left[\rho_i + \varepsilon_i - e_{ii}^{-1} \left(\frac{\varepsilon_i A_i}{\rho_i} \right) \right]$$

$$= \varepsilon_i - \left(\frac{\varepsilon_i}{\rho_i} \right) \rho_i - \left(\frac{\varepsilon_i}{\rho_i} \right) \left[\varepsilon_i - e_{ii}^{-1} \left(\frac{\varepsilon_i A_i}{\rho_i} \right) \right] = \left(\frac{\varepsilon_i}{\rho_i} \right) \left[e_{ii}^{-1} \left(\frac{\varepsilon_i A_i}{\rho_i} \right) - \varepsilon_i \right]$$

$$\mathcal{F}_{ii} = \left(\frac{\varepsilon_i}{\rho_i} \right) \left(\frac{\varepsilon_i A_i}{\rho_i} \right) \left[e_{ii}^{-1} - \varepsilon_i \left(\frac{\rho_i}{\varepsilon_i A_i} \right) \right] = \left(\frac{\varepsilon_i}{\rho_i} \right) \left(\frac{\varepsilon_i A_i}{\rho_i} \right) \left[e_{ii}^{-1} - \left(\frac{\rho_i}{A_i} \right) \right]$$

Then $\mathcal{F}_{ii}$ may be written as

$$\mathcal{F}_{ij} = \left(\frac{\varepsilon_i}{\rho_i} \right) \left(\frac{\varepsilon_j A_j}{\rho_j} \right) \left[e_{ij}^{-1} - \left(\frac{\rho_i}{A_i} \right) \right], \quad i = j$$

$$\mathcal{F}_{ij} = \left(\frac{\varepsilon_i \varepsilon_j}{\rho_i \rho_j} \right) A_j \left[e_{ij}^{-1} - \left(\frac{\rho_i}{A_i} \right) \right], \quad i = j$$

In summary,

$$\boxed{\begin{aligned} \mathcal{F}_{ij} &= \left(\frac{\varepsilon_i \varepsilon_j}{\rho_i \rho_j} \right) A_j e_{ij}^{-1}, & i \neq j \\[2ex] \mathcal{F}_{ij} &= \left(\frac{\varepsilon_i \varepsilon_j}{\rho_i \rho_j} \right) A_j \left[e_{ij}^{-1} - \left(\frac{\rho_i}{A_i} \right) \right], & i = j \end{aligned}}$$

Hottel script F (10.50)

plus the corresponding "theorems," Eqs. (10.48) and (10.49). Equation (10.50) is a remarkable result, considering the complexities of the multi-surface problem with reflections at each surface. If we combine the definition of the radiation heat transfer coefficient, Eq. (10.29) and a single, two-surface interaction from Eq. (10.47), we see that we may write

$$Q_{ij} = \mathcal{F}_{ij} h_r A_i \left(T_i - T_j \right)$$

$$\boxed{Q_{ij} = C_{ij} \left(T_i - T_j \right), C_{ij} = \mathcal{F}_{ij} h_r A_i}$$

Multi-surface (10.51)
radiation conductance

This author regards the Hottel script F method as a beautiful example of the power of mathematics, algebra in this instance, to manage a very complex physical problem. Although there are other methods (which are often built into high-level thermal analysis software) for computing radiation exchange, an understanding of the Hottel method is well within the capability of all readers of this book. The Hottel method is ideally applicable to thermal network modeling and though the arithmetic may be somewhat tedious when calculating view factors and the various required matrix elements, it is nonetheless quite amazing.

This writer cannot take credit for the method as most of the derivations in this section were found in a computer program manual. When the indicated references for the method were examined, the Hottel method was not existent. So, unfortunately, proper and full credit is not presently possible, although we shall assume that credit should go to Hottel.

10.9 APPLICATION EXAMPLE: GRAY-BODY CIRCUIT BOARDS ANALYZED AS INFINITE PARALLEL PLATES

This problem is actually a continuation of the parallel plate problem in Section 10.3, where we calculated view factors. Those results could be used when the plates, or circuit boards, behaved as nearly blackbodies. Now we consider the same problem, but model the boards as gray-bodies. This gives us the opportunity to use the Hottel script F method that we learned about in Section 10.8. As with most examples, this should enhance your understanding of the theory and its richness.

The two parallel circuit boards are illustrated in Figure 10.17. Surfaces 1 and 2 are the inner surfaces of the two boards, 3 and 4 are the end "surfaces," and 5 and 6 are side "surfaces." Since all of the radiation that passes to ambient from the board's inner-channel region must pass through surfaces 3-6, we can use these as the ambient surfaces. However, surfaces 3-6 are not really relevant in this problem because of the infinite board extent (we will use them in the next example).

The dimensions are $W = 10.0$ in., $L = 10.0$ in., and $S = 1.0$ in. The inner board surfaces are given emissivities of $\varepsilon_1 = 1.0$ and $\varepsilon_2 = 0.5$, respectively. The emissivities are arbitrarily selected as unequal to make the problem a bit more interesting. Note that the board dimensions are finite. Our radiation model is for infinite parallel boards because we will assume a view factor of $F_{1-2} = 1.0$. This is obviously an approximation of reality. We begin our model development with some calculations.

$$A_1 = A_2 = WL; \quad F_{1,2}A_1 = F_{2,1}A_2 = F_{2,1}WL = (1.0)(10.0\,\text{in.})(10.0\,\text{in.}) = 100.0\,\text{in.}^2$$

Since both surfaces 1 and 2 are flat, $F_{1,1} = 0.0$, $F_{2,2} = 0.0$.

The **E** matrix definition is given in Eq. (10.37).

$$e_{ij} = [\mathbf{E}]_{ij} = (A_i/\rho_i)(\delta_{ij} - \rho_i F_{ij})$$

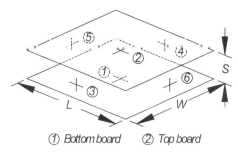

① Bottom board   ② Top board

Figure 10.17. Application Examples 10.9 and 10.10: Radiation between gray-body circuit boards.

$$e_{1,1} = \left(A_1/\rho_1 \right)\left(\delta_{1,1} - \rho_1 F_{1,1} \right) = \left[A_1/\left(1-\varepsilon_1 \right) \right]\left[1-\left(1-\varepsilon_1 \right)F_{1,1} \right]$$

$$e_{1,2} = \left(A_1/\rho_1 \right)\left(\delta_{1,2} - \rho_1 F_{1,2} \right) = \left[A_1/\left(1-\varepsilon_1 \right) \right]\left[\delta_{1,2}-\left(1-\varepsilon_1 \right)F_{1,2} \right]$$

$$e_{2,1} = \left(A_2/\rho_2 \right)\left(\delta_{2,1} - \rho_2 F_{2,1} \right) = \left[A_2/\left(1-\varepsilon_2 \right) \right]\left[\delta_{2,1}-\left(1-\varepsilon_2 \right)F_{2,1} \right]$$

$$e_{2,2} = \left(A_2/\rho_2 \right)\left(\delta_{2,2} - \rho_2 F_{2,2} \right) = \left[A_2/\left(1-\varepsilon_2 \right) \right]\left[1-\left(1-\varepsilon_2 \right)F_{2,2} \right]$$

We see that when we have an emissivity of value unity, we have a problem in the denominator of the $e_{1,1}$ and $e_{1,2}$. The solution is to use an emissivity of nearly 1.0, e.g., 0.999.

$$e_{1,1} = \left[A_1/\left(1-\varepsilon_1 \right) \right]\left[1-\left(1-\varepsilon_1 \right)F_{1,1} \right] = \left[100.0/\left(1-0.999 \right) \right]\left[1-\left(1-0.999 \right)\left(0.0 \right) \right] = 10^5$$

$$e_{1,2} = \left[A_1/\left(1-\varepsilon_1 \right) \right]\left[\delta_{1,2}-\left(1-\varepsilon_1 \right)F_{1,2} \right] = \left[A_1/\left(1-\varepsilon_1 \right) \right]\left[0-\left(1-\varepsilon_1 \right)F_{1,2} \right] = -F_{1,2}A_1 = -100.0$$

$$e_{2,1} = \left[A_2/\left(1-\varepsilon_2 \right) \right]\left[\delta_{2,1}-\left(1-\varepsilon_2 \right)F_{2,1} \right] = \left[A_2/\left(1-\varepsilon_2 \right) \right]\left[0-\left(1-\varepsilon_2 \right)F_{2,1} \right] = -F_{2,1}A_2 = -100.0$$

$$e_{2,2} = \left[A_2/\left(1-\varepsilon_2 \right) \right]\left[1-\left(1-\varepsilon_2 \right)F_{2,2} \right] = \left[100.0/\left(1-0.5 \right) \right]\left[1-\left(1-0.5 \right)\left(0.0 \right) \right] = 200.0$$

The preceding written in matrix form is

$$\mathbf{E} = \begin{pmatrix} e_{1,1} & e_{1,2} \\ e_{2,1} & e_{2,2} \end{pmatrix} = \begin{pmatrix} 10^5 & -100 \\ -100 & 200 \end{pmatrix}$$

We need $\mathbf{E}^{-1}$, the inverse of $\mathbf{E}$. We know from linear algebra that the inverse of $\mathbf{E}$ is the transpose $\mathbf{C}^T$ the matrix $\mathbf{C}$ of cofactors of the elements of $\mathbf{E}$ divided by the determinant $\left| \mathbf{E} \right|$.

$$\left| \mathbf{E} \right| = \left(e_{1,1}e_{2,2} - e_{1,2}e_{2,1} \right) = 2\times10^7 - 1\times10^4 = 1.999\times10^7$$

$$\mathbf{C} = \begin{pmatrix} e_{2,2} & -e_{2,1} \\ -e_{1,2} & e_{1,1} \end{pmatrix}, \ \mathbf{C}^T = \begin{pmatrix} e_{2,2} & -e_{1,2} \\ -e_{2,1} & e_{1,1} \end{pmatrix} = \begin{pmatrix} 200 & 100 \\ 100 & 10^5 \end{pmatrix}$$

The elements e_{ij}^{-1} are

$$e_{1,1}^{-1} = e_{2,2}/\left| \mathbf{E} \right| = 200/1.999\times10^7 = 1.0005\times10^{-5}$$

$$e_{1,2}^{-1} = -e_{1,2}/\left| \mathbf{E} \right| = 100/1.999\times10^7 = 5.0025\times10^{-6}$$

$$e_{2,1}^{-1} = -e_{2,1}/\left| \mathbf{E} \right| = 100/1.999\times10^7 - 5.0025\times10^{-6}$$

$$e_{2,2}^{-1} = e_{1,1}/\left| \mathbf{E} \right| = 10^5/1.999\times10^7 = 5.0025\times10^{-3}$$

The $\mathcal{F}_{ij}$ from Eq. (10.50) are

$$\mathcal{F}_{1,1} = \left(\frac{\varepsilon_1}{1-\varepsilon_1} \right)^2 A_1\left[e_{1,1}^{-1}-\left(\frac{1-\varepsilon_1}{A_1} \right) \right] = \left(\frac{0.999}{1-0.999} \right)^2 \left(100.0 \right)\left[1.0005\times10^{-5}-\left(\frac{1-0.999}{100.0} \right) \right] = 0.499$$

$$\mathcal{F}_{1,2} = \frac{\varepsilon_1\varepsilon_2}{\left(1-\varepsilon_1 \right)\left(1-\varepsilon_2 \right)} A_2 e_{1,2}^{-1} = \frac{\left(0.999 \right)\left(0.5 \right)}{\left(1-0.999 \right)\left(1-0.5 \right)}\left(100.0 \right)\left(5.0025\times10^{-6} \right) = 0.500$$

$$\mathcal{F}_{2,1} = \frac{\varepsilon_2\varepsilon_1}{\left(1-\varepsilon_2 \right)\left(1-\varepsilon_1 \right)} A_1 e_{2,1}^{-1} = \frac{\left(0.5 \right)\left(0.999 \right)}{\left(1-0.5 \right)\left(1-0.999 \right)}\left(100.0 \right)\left(5.0025\times10^{-6} \right) = 0.500$$

$$\mathcal{F}_{2,2} = \left(\frac{\varepsilon_2}{1-\varepsilon_2} \right)^2 A_2\left[e_{2,2}^{-1}-\left(\frac{1-\varepsilon_2}{A_2} \right) \right] = \left(\frac{0.5}{1-0.5} \right)^2 \left(100.0 \right)\left[5.0025\times10^{-3}-\left(\frac{1-0.5}{100.0} \right) \right] = 2.50\times10^{-4}$$

A check using Eq. (10.49):

$$\sum_{j=1}^{2} \mathcal{F}_{1j} = \varepsilon_1 : \qquad \sum_{j=1}^{2} \mathcal{F}_{1j} = \mathcal{F}_{11} + \mathcal{F}_{12} = 0.499 + 0.499 = 0.998$$

$$\sum_{j=1}^{2} \mathcal{F}_{2j} = \varepsilon_2 : \qquad \sum_{j=1}^{2} \mathcal{F}_{2j} = \mathcal{F}_{21} + \mathcal{F}_{22} = 0.499 + 2.50 \times 10^{-4} = 0.499$$

which are correct.

The $\mathcal{F}A$ matrix is

$$\mathcal{F}A = \begin{pmatrix} 49.925 & 49.975 \\ 49.975 & 0.025 \end{pmatrix}$$

which we see is symmetric as required by Eq. (10.48).

10.10 APPLICATION EXAMPLE: GRAY-BODY CIRCUIT BOARDS ANALYZED AS FINITE PARALLEL PLATES

This example also uses the geometry shown in Figure 10.17. The difference between this and the problem in Section 10.9 is that we will not assume a unity view factor between the two plates, but will instead use an exact calculated value. We will also include the view factor problems for radiation exchange between the plates and the open ends. The only additional data required is an emissivity for the end and side regions. Since it will be interesting to compare the results from this section with those in Section 10.9, we should try to use an emissivity that is consistent with the earlier problem. The previous model used $F_{1,2} = 1.0$, which implies that all radiation is confined to the region between the two circuit boards. This suggests that a very high reflectivity is appropriate for the open ends (surfaces 3, 4) and sides (surfaces 5, 6):

$$\varepsilon_3 = \varepsilon_4 = \varepsilon_5 = \varepsilon_6 = 1 - \rho = 1 - 0.99999 = 1 \times 10^{-5}$$

It seems rather pointless to detail all of the calculations since you have seen an adequate example of that in the previous problem. The various results are presented in their native matrix format to an accuracy of five decimal places. Many elements that are indicated as zero are actually nonzero, but require more decimal places to display.

$$\mathbf{FA} = \begin{pmatrix} 0 & 82.69945 & 4.32514 & 4.32514 & 4.32514 & 4.32514 \\ 82.69945 & 0 & 4.32514 & 4.32514 & 4.32514 & 4.32514 \\ 4.32514 & 4.32514 & 0 & 0.24925 & 0.55024 & 0.55024 \\ 4.32514 & 4.32514 & 0.24925 & 0 & 0.55024 & 0.55024 \\ 4.32514 & 4.32514 & 0.55024 & 0.55024 & 0 & 0.24925 \\ 4.32514 & 4.32514 & 0.55024 & 0.55024 & 0.24925 & 0 \end{pmatrix}$$

$$\mathbf{E} = \begin{pmatrix} 10000000.00005 & -82.69945 & -4.32514 & -4.32514 & -4.32514 & -4.32514 \\ -82.69945 & 200 & -4.32514 & -4.32514 & -4.32514 & -4.32514 \\ -4.32514 & -4.32514 & 10.0001 & -0.24925 & -0.55024 & -0.55024 \\ -4.32514 & -4.32514 & -0.24925 & 10.0001 & -0.55024 & -0.55024 \\ -4.32514 & -4.32514 & -0.55024 & -0.55024 & 10.0001 & -0.24925 \\ -4.32514 & -4.32514 & -0.55024 & -0.55024 & -0.24925 & 10.0001 \end{pmatrix}$$

$$
\mathbf{E}^{-1} =
\begin{pmatrix}
0 & 0 & 0 & 0 & 0 & 0 \\
0 & 0.00523 & 0.00261 & 0.00261 & 0.00261 & 0.00261 \\
0 & 0.00261 & 0.10203 & 0.00446 & 0.00717 & 0.00717 \\
0 & 0.00261 & 0.00446 & 0.10203 & 0.00717 & 0.00717 \\
0 & 0.00261 & 0.00717 & 0.00717 & 0.10203 & 0.00446 \\
0 & 0.00261 & 0.00717 & 0.00717 & 0.00446 & 0.10203
\end{pmatrix}
$$

$$
\mathcal{F} =
\begin{pmatrix}
0.52259 & 0.47739 & 0 & 0 & 0 & 0 \\
0.47739 & 0.02261 & 0 & 0 & 0 & 0 \\
0.00001 & 0 & 0 & 0 & 0 & 0 \\
0.00001 & 0 & 0 & 0 & 0 & 0 \\
0.00001 & 0 & 0 & 0 & 0 & 0 \\
0.00001 & 0 & 0 & 0 & 0 & 0
\end{pmatrix}
$$

$$
\mathcal{F}\mathbf{A} =
\begin{pmatrix}
52.25934 & 47.73936 & 0.00007 & 0.00007 & 0.00007 & 0.00007 \\
47.73936 & 2.26053 & 0.00003 & 0.00003 & 0.00003 & 0.00003 \\
0.00007 & 0.00003 & 0 & 0 & 0 & 0 \\
0.00007 & 0.00003 & 0 & 0 & 0 & 0 \\
0.00007 & 0.00003 & 0 & 0 & 0 & 0 \\
0.00007 & 0.00003 & 0 & 0 & 0 & 0
\end{pmatrix}
$$

The results of the two-surface and six-surface models are listed in Table 10.3. The matrix elements e_{ij} and $e^{-1}{}_{ij}$ have quite different values, particularly for the 1,1 element. However, the 1,1 and 1,2 $\mathcal{F}$ and $\mathcal{F}\mathbf{A}$ element values are quite similar, as we would expect since the plate spacing S is small compared to W and L.

Table 10.3. Application Examples 10.9 and 10.10: Parallel plate circuit boards.

i,j	Model	F_{ij}	$[FA]_{ij}$	e_{ij}	$e^{-1}{}_{ij}$	$\mathcal{F}_{ij}$	$[\mathcal{F}A]_{ij}$
1,1	2-surface	0	0	10^5	1.0005×10^{-5}	0.49925	49.92501
	6-surface	0	0	10^7	1.0000×10^{-7}	0.52259	52.25934
1,2	2-surface	1.0	100	-100	5.0025×10^{-6}	0.49975	49.97499
	6-surface	0.82699	82.69945	-82.69945	4.77398×10^{-8}	0.47739	47.73936

10.11 THERMAL RADIATION NETWORKS

In this section we will work out the mathematics of a thermal network method that is discussed in most heat transfer texts. This writer has not used it directly in systems thermal analysis, but has found it useful for a couple of stand-alone solutions. The reader may find it useful in other applications not discussed in this book.

We begin the discussion with Eq. (10.35) and proceed from there.

$$J_i = \varepsilon_i E_{bi} + \rho_i \sum_{j=1}^{N} F_{ij} J_j$$

$$\rho_i \sum_{j=1}^{N} F_{ij} J_j + \varepsilon_i E_{bi} - J_i = 0$$

Add and substract $\varepsilon_i J_i$ to get

$$\rho_i \sum_{j=1}^{N} F_{ij} J_j + \varepsilon_i E_{bi} - J_i + \varepsilon_i J_i - \varepsilon_i J_i = 0$$

$$\rho_i \sum_{j=1}^{N} F_{ij} J_j + \varepsilon_i E_{bi} - \left(1 - \varepsilon_i\right) J_i - \varepsilon_i J_i = 0$$

Now using Kirchhoff's identity, $\alpha = \varepsilon$ so that $\rho = 1 - \alpha = 1 - \varepsilon$ and multiplying each term by A_i we get

$$\rho_i A_i \sum_{j=1}^{N} F_{ij} J_j + \varepsilon_i A_i E_{bi} - \rho_i A_i J_i - \varepsilon_i A_i J_i = 0$$

Since $\sum_{j=1}^{N} F_{ij} = 1$, the third term of the preceding may be multiplied by this summation.

$$\rho_i A_i \sum_{j=1}^{N} F_{ij} J_j + \varepsilon_i A_i E_{bi} - \rho_i A_i J_i \sum_{j=1}^{N} F_{ij} - \varepsilon_i A_i J_i = 0$$

$$\rho_i A_i \sum_{j=1}^{N} F_{ij} \left(J_j - J_i\right) - \varepsilon_i A_i \left(J_i - E_{bi}\right) = 0$$

$$\frac{\left(J_i - E_{bi}\right)}{\left(\rho_i / \varepsilon_i A_i\right)} = \sum_{j=1}^{N} \frac{\left(J_j - J_i\right)}{\left(A_i F_{ij}\right)^{-1}} \qquad \text{Kirchhoff's law for (10.52)}$$
$$\text{Nodal current conservation}$$

Equation (10.52) is given the name of a conservation law because we can interpret it in that manner if we define the two numerators as "potential differences" and the denominators as "resistances." Thus the left-hand term is a "current" and the right-hand side is a sum of "currents." It seems reasonable to define the following:

$$\boxed{R_i = \left(1 - \varepsilon_i\right) / \left(\varepsilon_i A_i\right)} \qquad \text{Surface resistance (10.53)}$$

$$\boxed{R_{ij} = 1 / \left(A_i F_{ij}\right)} \qquad \text{Spatial resistance (10.54)}$$

Combining Eqs. (10.52), (10.53), and (10.54),

$$\boxed{\frac{\left(E_{bi} - J_i\right)}{R_i} = \sum_{j=1}^{N} \frac{\left(J_i - J_j\right)}{R_{ij}}} \qquad \text{Kirchhoff's law for (10.55)}$$
$$\text{Nodal current conservation}$$

It is helpful in our understanding of this method if we make an effort to better identify the exact nature of the left- and right-hand sides of Eq. (10.55). Consider a surface i for which we have a net radiative heat rate loss:

Net radiative heat rate loss from surface i =
Total radiative heat rate out of i - Total heat rate into i

which is written in mathematical form as

$$Q_{i\,Net} = J_i A_i - H_i \qquad (10.56)$$

Earlier in this chapter, we noted Eq. (10.31), which we can generalize to surface i.

$$J_i A_i = E_i A_i + \rho_i H_i = \varepsilon_i E_{bi} A_i + \rho_i H_i$$

Solving for H_i,

$$H_i = \left(1/\rho_i\right) J_i A_i - \left(\varepsilon_i/\rho_i\right) E_{bi} A_i \qquad (10.57)$$

Substituting H_i in Eq. (10.57) into Eq. (10.56),

$$Q_{iNet} = J_i A_i - \left(1/\rho_i\right) J_i A_i + \left(\varepsilon_i/\rho_i\right) E_{bi} A_i = \left(1 - \frac{1}{\rho_i}\right) J_i A_i + \left(\frac{\varepsilon_i}{\rho_i}\right) E_{bi} A_i$$

$$Q_{iNet} = \left(\frac{\rho_i - 1}{\rho_i}\right) J_i A_i + \left(\frac{\varepsilon_i}{\rho_i}\right) E_{bi} A_i$$

and since $\rho_i - 1 = -\varepsilon_i, \rho_i = 1 - \varepsilon_i$,

$$Q_{iNet} = -\left(\frac{\varepsilon_i}{1-\varepsilon_i}\right) J_i A_i + \left(\frac{\varepsilon_i}{1-\varepsilon_i}\right) E_{bi} A_i = \left(\frac{\varepsilon_i A_i}{1-\varepsilon_i}\right) \left(E_{bi} - J_i\right)$$

$$\boxed{Q_{iNet} = \left(E_{bi} - J_i\right)/R_i ; \quad R_i = \left(1-\varepsilon_i\right)/\varepsilon_i A_i} \qquad (10.58)$$

which is identical to the left-hand side of Eq. (10.55), which we now know is the *net radiative heat loss from surface i*.

Now we begin again with Figure 10.18 where we consider

The net radiative heat exchange between surfaces i and j =
Radiative heat rate from surface i that is intercepted by surface j -
Radiative heat rate from surface j that is intercepted by surface i

We write this statement mathematically as

$$Q_{ij} = F_{ij} J_i A_i - F_{ji} J_j A_j = F_{ij} A_i J_i - F_{ji} A_j J_j = F_{ij} A_i \left(J_i - J_j\right)$$

where we have invoked the reciprocity of the FA, Eq. (10.19). Then we have

$$\boxed{Q_{ij} = \left(J_i - J_j\right)/R_{ij}; \quad R_{ij} = 1/\left(F_{ij} A_i\right)} \qquad (10.59)$$

thus each term in the right-hand side summation of Eq. (10.55) is a *net radiative heat rate exchange between surfaces i and j*. Putting the results of Eqs. (10.58) and (10.59) together, we are led to the additional interpretation of Eq. (10.55) as *the net heat rate loss "current" into node i = the sum of the heat rate loss "currents" from node i*.

Equations (10.58) and (10.59) are sufficient to enable us to solve some radiation exchange problems. A three-surface problem is illustrated in Figure 10.19. Then according to Eqs. (10.58) and (10.59), the various resistances are

$$R_1 = \left(1-\varepsilon_1\right)/\left(\varepsilon_1 A_1\right); \quad R_2 = \left(1-\varepsilon_2\right)/\left(\varepsilon_2 A_2\right); \quad R_3 = \left(1-\varepsilon_3\right)/\left(\varepsilon_3 A_3\right)$$

$$R_{1,2} = 1/\left(A_1 F_{1,2}\right), \quad R_{2,3} = 1/\left(A_2 F_{2,3}\right), \quad R_{1,3} = 1/\left(A_1 F_{1,3}\right)$$

Figure 10.18. Radiative heat exchange between two surfaces.

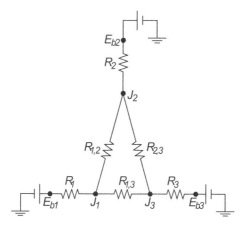

Figure 10.19. Three-surface radiation network according to Section 10.11.

The "source currents" $(E_{bi} - J_i)/R_i$ flow from each of the blackbody "potentials" E_{bi} and the internode currents $(J_i - J_j)/R_{ij}$ flow between each of the i, j nodes.

The next example, a two-surface radiation exchange problem, is important because the mathematical treatment results in an important and useful formula. As in the preceding example, the "source currents" $(E_{bi} - J_i)/R_i$ flow from the two blackbody "potentials" E_{b1} and E_{b2} and the internode currents $(J_i - J_j)/R_{ij}$ flow between nodes 1 and 2. The resistances in Figure 10.20 are

$$R_1 = (1-\varepsilon_1)/(\varepsilon_1 A_1); \quad R_3 = (1-\varepsilon_2)/(\varepsilon_2 A_2); \quad R_{1,2} = 1/(A_1 F_{1,2})$$

The derivation that we are interested in begins by writing down the series total resistance from node 1 to node 2.

$$R = R_1 + R_{1,2} + R_3 = \frac{1-\varepsilon_1}{\varepsilon_1 A_1} + \frac{1}{A_1 F_{1,2}} + \frac{1-\varepsilon_2}{\varepsilon_2 A_2}$$

The net radiative heat exchange is given by

$$Q_{Net} = (E_{b1} - E_{b2})/R \text{ and } Q_{Net} = \sigma \mathcal{F}_{1,2} A_1 \left(T_1'^4 - T_2'^4\right)$$

so that the identification is made for the script F.

$$Q_{Net} = (E_{b1} - E_{b2})/R = \sigma \left(T_1'^4 - T_2'^4\right)/R$$

$$Q_{Net} = \sigma \mathcal{F}_{1,2} A_1 \left(T_1'^4 - T_2'^4\right)$$

$$\sigma \left(T_1'^4 - T_2'^4\right)/R = \sigma \mathcal{F}_{1,2} A_1 \left(T_1'^4 - T_2'^4\right)$$

$$\mathcal{F}_{1,2} = 1/(A_1 R)$$

$$\boxed{\mathcal{F}_{1,2} = \frac{1}{\dfrac{1-\varepsilon_1}{\varepsilon_1} + \left(\dfrac{1-\varepsilon_2}{\varepsilon_2}\right)\left(\dfrac{A_1}{A_2}\right) + \dfrac{1}{F_{1,2}}}} \quad \text{Two-surface radiation} \quad (10.60)$$

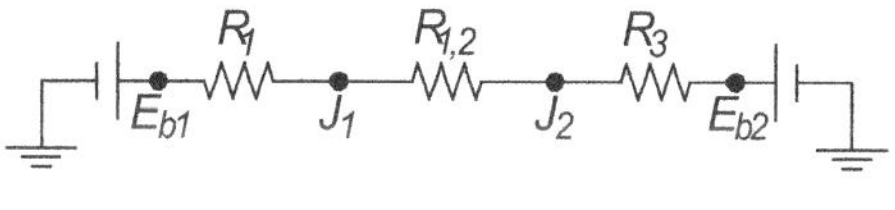

Figure 10.20. Two-surface radiation network.

Equation (10.60) is useful in those problems where we are willing to make the approximation of two surfaces, e.g., the electronics enclosure in a room. In this instance we have the situation that if nodes 1 and 2 represent the external surfaces of the enclosure and the room walls, respectively, Eq. (10.61) certainly applies.

$$\mathcal{F}_{1,2} = \cfrac{1}{\cfrac{1-\varepsilon_1}{\varepsilon_1} + \left(\cfrac{1-\varepsilon_2}{\varepsilon_2}\right)\left(\cfrac{A_1}{A_2}\right) + \cfrac{1}{F_{1,2}}} = \cfrac{1}{\left(\cfrac{1-\varepsilon_1}{\varepsilon_1} + 1\right)} = \cfrac{1}{\left(\cfrac{1-\varepsilon_1 + \varepsilon_1}{\varepsilon_1}\right)}$$

$$\boxed{\mathcal{F}_{1,2} = \varepsilon_1} \quad A_1 \lll A_2, \Gamma_{1,2} - 1.0 \text{ or } \varepsilon_2 - 1, \Gamma_{1,2} - 1.0 \quad (10.61)$$

Note that Eq. (10.61) is applicable if $\varepsilon_2 = 1.0$. Both the approximations $A_1 \ll A_2$ or $\varepsilon_2 = 1.0$ and $F_{1,2} = 1.0$ conform to the same physical interpretation where radiation from surface 2 does not find its way back to surface 1.

10.12 THERMAL RADIATION SHIELDING FOR RECTANGULAR U-CHANNELS (FINS)

This section results in a formula and several design curves for the interior of a finned channel radiating to a perfectly absorbing ambient. The following paragraphs are also a useful introduction to a practical application of the radiation network theory in Section 10.11, which you may find useful in applying to other problems.

The succeeding treatment of the overall $\mathcal{F}$ follows Ellison (1979). All surfaces of the heat sink are assumed to be gray and diffusely emitting/reflecting, with the fins and base at a uniform temperature. In practice, none of these conditions apply perfectly, but for most engineering calculations they are adequate. A single U-channel with appropriately identified interior surfaces is shown in Figure 10.21. Numerals 1, 3, and 4 identify the heat sink surfaces, whereas numerals 2, 5, and 6 refer to a nonreflecting ambient. An equivalent thermal radiation circuit for the U-channel is illustrated in Figure 10.22.

The equivalent surface resistance formula, Eq. (10.58), for the heat sink interior surfaces of emissivity ε is applicable between the radiosity and blackbody potentials J and E_b.

$$R_1 = R_{19} = (1-\varepsilon)/\varepsilon A_3, \quad R_2 = (1-\varepsilon)/\varepsilon A_1$$

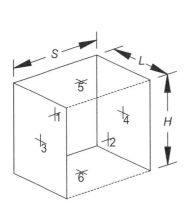

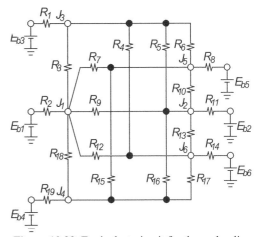

Figure 10.21. Surface identification of U-channel interior (From: Ellison, G.N. *IEEE Trans. CHMT*, vol. 2, no. 4, 517-522, © 1979 *IEEE*. With permission).

Figure 10.22. Equivalent circuit for thermal radiation from U-channel interior (From: Ellison, G.N. *IEEE Trans. CHMT*, vol. 2, no. 4, 517-522, © 1979 *IEEE*. With permission).

The ambient (surfaces 2, 5, 6) is taken as a non-reflecting surface, thus

$$R_8 = R_{11} = R_{14} = 0$$

The spatial resistances interconnecting the radiosity nodes utilize appropriate view factors and Eq. (10.59):

$$R_3 = R_{18} = 1/F_{1,3}A_1; \quad R_4 = R_6 = R_{15} = R_{17} = 1/F_{3,5}A_3; \quad R_7 = R_{12} = 1/F_{1,5}A_1;$$
$$R_9 = 1/F_{1,2}A_1; \quad R_5 = R_{16} = 1/F_{3,2}A_3 = 1/F_{3,1}A_3 = 1/F_{1,3}A_1$$

Since the two fins and base, surfaces 3, 4, and 1, respectively, are at identical temperatures,

$$E_{b1} = E_{b4} = E_{b3}$$

and we have a single ambient temperature, then

$$E_{b2} = E_{b5} = E_{b6}$$

which results in

$$J_2 = J_5 = J_6$$

such that there is no radiation exchange between surfaces 2, 5, and 6 so that R_{10} and R_{13} are open circuit. Finally a brief examination of Figure 10.22 shows us that there is symmetry in the upper and lower halves of the circuit where the symmetry line is through the J_1 - J_2 nodes. The result is the circuit displayed in Figure 10.23.

The circuit in Figure 10.23 is redrawn as Figure 10.24 where the resistances are now

$$R_a = R_1, \quad R_b = 2R_2, \quad 1/R_c = 1/R_4 + 1/R_5 + 1/R_6$$
$$R_e = R_3, \quad 1/R_d = 1/R_7 + 1/2R_9$$

The net radiation resistance, R_{Net} is found from

$$R_{Net} = (E_{b1} - E_{b2})/q_3$$

A set of three simultaneous equations is written using the "loop currents" q_1, q_2, and q_3 and resistances in Figure 10.24:

$$(R_a + R_b + R_e)q_1 - R_e q_2 - R_b q_3 = 0$$
$$-R_e q_1 + (R_c + R_d + R_e)q_2 - R_d q_3 = 0$$
$$-R_b q_1 - R_d q_2 + (R_b + R_d)q_3 = E_{b1} - E_{b2}$$

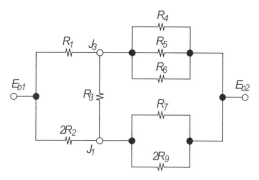

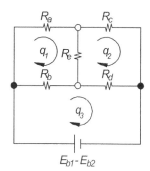

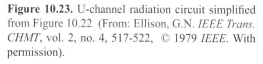

Figure 10.23. U-channel radiation circuit simplified from Figure 10.22 (From: Ellison, G.N. *IEEE Trans. CHMT*, vol. 2, no. 4, 517-522, © 1979 *IEEE*. With permission).

Figure 10.24. Final circuit for U-channel radiation (From: Ellison, G.N. *IEEE Trans. CHMT*, vol. 2, no. 4, 517-522, © 1979 *IEEE*. With permission).

The preceding loop equations are solved for q_3 (solution detail is given in Appendix iv) which is used to get the radiation resistance

$$R_{Net} = \left(E_{b1} - E_{b2}\right)/q_3 \text{ or } C_{Net} = q_3/\left(E_{b1} - E_{b2}\right)$$

Next it is important to remember that we used symmetry to solve for only half of the circuit, thus half of the total heat flow. Then in our problem,

$$Q_{Net} = \mathcal{F}A_{U-chan}\left(E_{b1} - E_{b2}\right)$$

and

$$\mathcal{F} = 2/R_{Net}A_{U-chan} = 2C_{Net}/\left[H\left(S+2L\right)\right]$$

The final results are
$$\boxed{\mathcal{F} = 2C_{Net}/\left[H\left(S+2L\right)\right]}$$
U-channel $\mathcal{F}$ (10.62)

where
$$C_{Net} = \frac{\left[\left(R_a + R_b + R_e\right)\left(R_c + R_d + R_e\right) - R_e^2\right]}{\left\{\begin{array}{l}\left(R_b + R_d\right)\left[\left(R_a + R_b + R_e\right)\left(R_c + R_d + R_e\right) - R_e^2\right]\\ -R_b\left[R_b\left(R_c + R_d + R_e\right) + R_e R_d\right] - R_d\left[R_d\left(R_a + R_b + R_e\right) + R_b R_e\right]\end{array}\right\}}$$

$$R_a = \left(1-\varepsilon\right)/\left(\varepsilon A_3\right); \quad R_b = 2\left(1-\varepsilon\right)/\left(\varepsilon A_1\right); \quad R_c = 1/\left(A_1 F_{1,3} + 2A_3 F_{3,5}\right)$$
$$R_d = 2/\left(A_1 F_{1,2} + 2A_1 F_{1,5}\right); \quad R_e = 1/\left(A_1 F_{1,3}\right)$$

and $\mathcal{F}$ is plotted in Figures 10.25 through 10.35 for several values of ε. These design curves are not plotted directly from Eq. (10.62), but instead from a form that uses dimensionless L/S and H/L. This author analyzed 18 variations of a rectangular U-channel with a constant fin length of $L = 1.0$ in. The $\mathcal{F}$ was calculated using Eq. (10.62) and thermal analyzer software written by Hultberg and O'Brian (1971). The software uses linear algebra techniques to calculate the surface radiation exchange, and input consists of surface dimensions and view factors. The latter were the same for both Eq. (10.62) and the software model. The results are tabulated in Table 10.4. The $\mathcal{F}$ calculations from Eq. (10.62) as reported by Ellison (1979) were accomplished using an electronic calculator, but those listed in Table 10.4 were performed with Mathcad™. Clearly the thermal analyzer results are a sufficient verification of the circuit analysis algebra.

Table 10.4. Equation (10.62) $\mathcal{F}$ compared with values from TAS (Hultberg and O'Brien, 1971).

H (in.)	S (in.)	ε	$\mathcal{F}$-Eq. (10.62)	$\mathcal{F}$-TAS	H (in.)	S (in.)	ε	$\mathcal{F}$-Eq. (10.62)	$\mathcal{F}$-TAS
1.0	1.0	1.0	0.60006	0.60007	10.0	1.0	1.0	0.36656	0.36656
		0.6	0.42860	0.42854			0.6	0.29400	0.29399
		0.1	0.09375	0.09375			0.1	0.08517	0.08518
	0.5	1.0	0.43390	0.43392		0.5	1.0	0.22588	0.22590
		0.6	0.33655	0.33647			0.6	0.19625	0.19625
		0.1	0.08846	0.08846			0.1	0.07446	0.07445
	0.2	1.0	0.23171	0.23170		0.25	1.0	0.12821	0.12819
		0.6	0.20070	0.20070			0.6	0.11811	0.11810
		0.1	0.07510	0.07510			0.1	0.05952	0.05952

Source: Ellison, G.N. *IEEE Trans. CHMT*, vol. 2, no. 4, 517-522, © 1979 *IEEE*. With permission.

Another comparison is possible using results for an *effective emittance* $\hat{\varepsilon}$ published in the book by Kraus and Bar-Cohen (1995) who provide an overview of independent work by Bilitsky (1986), which is nearly identical to that of Ellison (1979). Bilitsky, Kraus and Bar-Cohen were apparently unaware of this work. The effective emissivity is used with a projected area so that

$$Q_r = \hat{\varepsilon} H S \sigma \left(T_1'^4 - T_2'^4 \right)$$

The derivation of the conversion from $\mathcal{F}$ to $\hat{\varepsilon}$, Eq. (10.63), is left as a student exercise. A comparison is listed in Table 10.5, with results for $\hat{\varepsilon}$ given to only two decimals due to the limiting accuracy of reading graphical results from Kraus and Bar-Cohen. The two independent calculations are in excellent agreement.

$$\hat{\varepsilon} = \mathcal{F}(1 + 2L/S) \qquad\qquad \text{Effective emittance} \quad (10.63)$$

Table 10.5. Comparison of $\hat{\varepsilon}$ conversion from Eq. (10.63)
with values from Kraus and Bar-Cohen and Bilitsky.

			Eq. (10.62)	Eq. (10.63)	Kraus/Bar-Cohen, Bilitsky
ε	H/L	L/S	$\mathcal{F}$	$\hat{\varepsilon}$	$\hat{\varepsilon}$
0.1	1	1	0.0938	0.28	0.28
		10	0.0599	1.26	1.26
	100	1	0.0834	0.25	0.25
		10	0.0337	0.72	0.72
1.0	1	1	0.6001	1.80	1.83
		10	0.1300	2.76	2.76
	100	1	0.3367	1.01	1.02
		10	0.04846	1.02	1.03

The design curves, Figures 10.25 through 10.35, are self-explanatory, although the reader should note that each plot has the expected result that as $L \to 0$, the gray-body $\mathcal{F} \to \varepsilon$. An important issue is the accuracy of Eq. (10.62) and thus the plots for the six-surface model. The various parameters that have impact on the accuracy are the emissivity, the assumptions of diffusely emitting and absorbing surfaces, and fin temperature uniformity. The first two assumptions are adequate for engineering approximations and the third item is not a problem for many design situations we encounter where the fin efficiency is often found to be not very different from unity.

Rea and West (1976) computed the apparent emittance for a single U-channel where the surfaces were sectioned into a large number of differential-sized strips and assumed diffuse, gray, uniform-temperature surfaces. Radiation enclosure theory was used here also. The Rea and West calculated results are compared with calculations from Eq. (10.62) and plotted in Figures 10.36 and 10.37 for high and low values of emissivity. The apparent emittance from Rea and West is converted to a script F using Eq. (10.63). Rea and West show excellent agreement with their heat sink temperature measurements in vacuum. The plots in Figure 10.36 (ε = 0.8) show excellent agreement between Eq. (10.62) and the Rea and West results. However, the agreement between the two methods is not as good for ε = 0.08 as shown in Figure 10.37. These results suggest that reflection and absorption from lower emissivity surfaces has a bearing on the adequacy or lack of large, single surface view factors, an issue that we shall not examine here. High emissivity, air cooled heat sinks will not usually radiate more than about 25% of the total heat transfer and for these situations the design formulae/plots are quite adequate. The low emissivity heat sinks will not radiate sufficient heat for us to be concerned about the view factor/script F accuracy. Of course, when a heat sink is in vacuum, we must be more concerned about the accuracy of radiation calculations, and in these situations, our assumptions of gray-diffuse surfaces may also be inadequate.

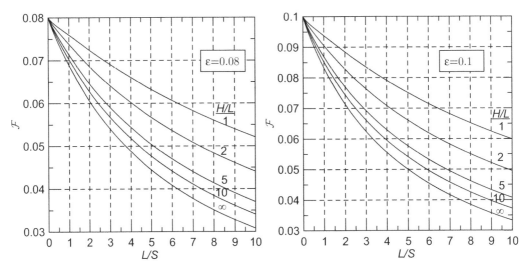

Figure 10.25. Equation (10.62) plotted.

Figure 10.26. Equation (10.62) plotted (From: Ellison, G.N. *IEEE Trans. CHMT*, vol. 2, no. 4, 517-522, © 1979 *IEEE*. With permission).

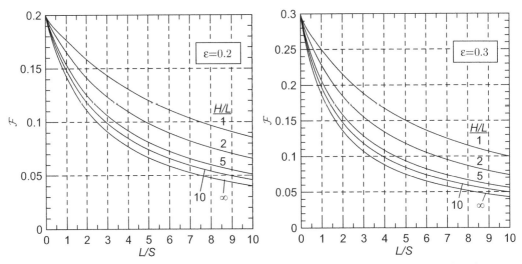

Figure 10.27. Equation (10.62) plotted (From: Ellison, G.N. *IEEE Trans. CHMT*, vol. 2, no. 4, 517-522, © 1979 *IEEE*. With permission).

Figure 10.28. Equation (10.62) plotted.

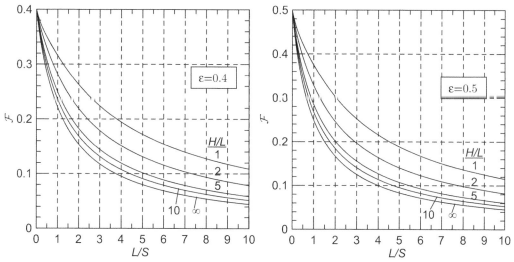

Figure 10.29. Equation (10.62) plotted (From: Ellison, G.N. *IEEE Trans. CHMT*, vol. 2, no. 4, 517-522, © 1979 *IEEE*. With permission).

Figure 10.30. Equation (10.62) plotted.

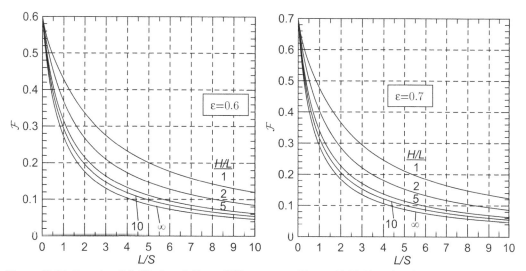

Figure 10.31. Equation (10.62) plotted (From: Ellison, G.N. *IEEE Trans. CHMT*, vol. 2, no. 4, 517-522, © 1979 *IEEE*. With permission).

Figure 10.32. Equation (10.62) plotted.

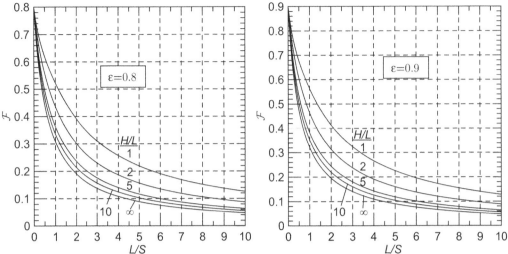

Figure 10.33. Equation (10.62) plotted (From: Ellison, G.N. *IEEE Trans. CHMT*, vol. 2, no. 4, 517-522, © 1979 *IEEE*. With permission).

Figure 10.34. Equation (10.62) plotted.

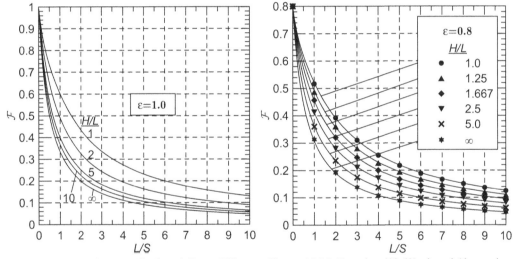

Figure 10.35. Equation (10.62) plotted (From: Ellison, G.N. *IEEE Trans. CHMT*, vol. 2, no. 4, 517-522, © 1979 *IEEE*. With permission).

Figure 10.36. Equation (10.62) plotted (data points adapted from Rea and West, 1976).

10.13 APPLICATION EXAMPLE: NATURAL CONVECTION AND RADIATION COOLED, VERTICALLY ORIENTED HEAT SINK (SEE SECTION 9.4)

We originally solved this problem for a heat sink cooled only by natural convection. For the sake of convenience, the heat sink is shown again as Figure 10.38. The dimensions are $W = 1.86$ in., $H = 5.0$ in., $L = 1.0$ in., $t_f = 0.06$ in., and $\Delta T = 50°C$. The results from Section 9.4 were that the heat sink convected $Q_c = 11.5$ W. The areas for the interior U-channels and two exterior fins are $A_I = 59.3$ in.^{2} and $A_E = 10.0$ in.^{2}, respectively. The calculated convective heat transfer coefficients for the interior and two outer fins were $h_c = 0.00324$ and $h_H = 0.00427$ W/(in.$^2 \cdot °C$), respectively. Our current problem is to calculate the amount of heat that is convected and radiated from the finned side only. We shall analyze this heat sink in detail for an emissivity of 0.1 and give only a result for an emissivity of 0.8.

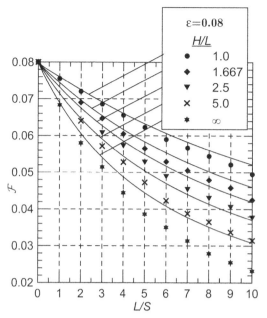

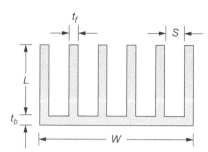

Figure 10.38. Application Example 10.13: Natural convection and radiation cooled heat sink.

Figure 10.37. Equation (10.62) plotted (data points adapted from Rea and West, 1976).

The radiation calculations are separated into (1) the surfaces of the channel interior and (2) the surfaces of the two end fins. The first requires that we determine $\mathcal{F}$ and the second only requires the emissivity ε. In Section 9.4 we found that $S = 0.30$ in. Then referring to Figure 10.26 using

$$L/S = 1.0/0.30 = 3.33; H/L = 5.0/1.0 = 5.0$$

we obtain $\mathcal{F} = 0.067$. Figure 10.12 is used to obtain an adequately accurate radiation heat transfer coefficient for the $T_A = 20°C$ and $T_S - T_A = 50°C$ so that $h_r = 0.0047$ W/(in.$^2 \cdot$ °C). The total heat transfer coefficients for the interior and exterior surfaces are

$$h_I = h_c + \mathcal{F}h_r = 0.00324 + (0.067)(0.0047) = 0.00356 \text{ W}/\left(\text{in.}^2 \cdot °C\right)$$

$$h_E = h_H + \varepsilon h_r = 0.00427 + (0.1)(0.0047) = 0.00474 \text{ W}/\left(\text{in.}^2 \cdot °C\right)$$

The fin efficiencies will be different for the inner and outer fins, but it is not worth the trouble to distinguish between the two values. We shall use the smaller of the two total heat transfer coefficients for this calculation.

$$R_k = L/\left(k_{Al}Ht_f\right) = 1.0/\left[(5.0)(5.0)(0.06)\right] = 0.667 \text{ °C/W}$$

$$R_c = 1/\left(2HLh_I\right) = 1/\left[2(5.0)(1.0)(0.00356)\right] = 24.99 \text{ °C/W}$$

$$\eta = \sqrt{R_c/R_k} \tanh \sqrt{R_k/R_c} = \sqrt{24.99/0.667} \tanh \sqrt{0.667/24.99} = 0.992$$

The internal and external total conductances are calculated separately and then added as two conductances in parallel.

$$C_I = \eta h_I A_I = (0.992)(0.00356)(59.3) = 0.210 \text{ W}/°C$$

$$C_E = \eta h_E A_E = (0.992)(0.00474)(10.0) = 0.047 \text{ W}/°C$$

$$C = C_I + C_E = 0.210 + 0.047 = 0.257 \text{ W}/°C$$

The total heat transfer is

$$Q = C\Delta T = (0.257)(50) = 12.83\,\mathrm{W}$$

This is a modest increase over the convection-only case (Section 9.4) of only

$$\text{Increase} = \left[(12.83 - 11.67)/11.67\right] \times 100 = 9.9\%$$

It is left as an exercise for the student to analyze this same problem for $\varepsilon = 0.8$, but the result is $\mathcal{F} = 0.16$ and $Q = 15.76\,\mathrm{W}$ for an increase of

$$\text{Increase} = \left[(15.67 - 11.67)/11.67\right] \times 100 = 34.3\%$$

10.14 APPLICATION EXAMPLE: NATURAL CONVECTION AND RADIATION COOLED NINE-FIN HEAT SINK - CALCULATIONS COMPARED TO TEST DATA

In this section we shall reanalyze the problem in Section 9.5 but include radiation, and finally compare with test data from Spoor (1974). The dimensions are $W = 4.15$ in., $H = 4.0$ in., $L = 2.62$ in., $S = 0.35$ in., $t_f = 0.15$ in., and $t_b = 0.63$ in. The heat sink temperature rise above ambient is $\Delta T = 50\,^\circ\mathrm{C}$. We shall use an aluminum thermal conductivity of $k = 5.0\,\mathrm{W}/(\mathrm{in.}\cdot{}^\circ\mathrm{C})$. The heat sink is black anodized so we shall use $\varepsilon = 0.8$. The calculations proceed in a fashion nearly identical to the preceding example.

The heat sink analyzed in detail in Section 9.5 for $H = 4.0$ in. resulted in the following results which we will use here:

$$L/S = 2.62/0.35 = 7.5;\, H/L = 4.0/2.62 = 1.53,\, A_E = 54.90\,\mathrm{in.}^2,\, A_I = 184.28\,\mathrm{in.}^2$$

$$h_c = 0.0046\,\mathrm{W}/(\mathrm{in.}^2\cdot{}^\circ\mathrm{C}),\, h_H = 0.0053\,\mathrm{W}/(\mathrm{in.}^2\cdot{}^\circ\mathrm{C})$$

where although the Van de Pol and Tierney correlation was used to determine h_c / h_H, the *small device* formula was used to obtain h_H. The external area A_E includes the flat backside, base edges, and the top and bottom fin edges. Using the above values of L/S and H/L and Figure 10.33, we obtain $\mathcal{F} = 0.127$. Then

$$h_I = h_c + \mathcal{F}h_r = 0.0046 + (0.127)(0.0047) = 0.0052$$
$$h_E = h_H + \varepsilon h_r = 0.0053 + (0.8)(0.0047) = 0.0091$$

The fin efficiency must be recalculated to include radiation (using h_r from Section 10.13).

$$R_k = L/(k_{Al}Ht_f) = 2.62/\left[(5.0)(4.0)(0.15)\right] = 0.873\,^\circ\mathrm{C}/\mathrm{W}$$
$$R_c = 1/\left[2HL(h_c + \mathcal{F}h_r)\right] = 1/\left[2(4.0)(2.62)(0.0046 + (0.127)(0.0047))\right] = 9.24\,^\circ\mathrm{C}/\mathrm{W}$$
$$\eta = \sqrt{R_c/R_k}\,\tanh\sqrt{R_k/R_c} = \sqrt{9.24/0.873}\,\tanh\sqrt{0.873/9.24} = 0.97$$

The required conductances are

$$C_I = \eta h_I A_I = (0.97)(0.0052)(184.28) = 0.922\,\mathrm{W}/{}^\circ\mathrm{C}$$
$$C_E = \eta h_E A_E = (0.97)(0.0091)(59.9) = 0.31\,\mathrm{W}/{}^\circ\mathrm{C}$$
$$C = C_I + C_E = 0.922 + 0.31 = 1.23\,\mathrm{W}/{}^\circ\mathrm{C}$$

and $\quad Q = C\Delta T = (0.1.23)(50) = 61.5\,\mathrm{W}$

which provides one data point for Figure 10.39. Calculations of the remaining data points are left as exercises for the student. Several test-data points are selected from Spoor (1974) and shown with symbols. The theoretical-experimental discrepancy is less than about 10% for most of the data points.

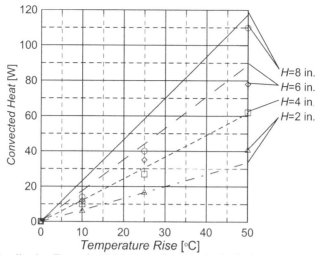

Figure 10.39. Application Example 10.14: Natural convection/radiation cooled, nine-fin heat sink. Calculations for: $\varepsilon = 0.8$, $H = 6.0$ in., 8.0 in. used the simplified classical h, Table 8.7; $H = 2.0$ in., 4.0 in. used the simplified "small device" h, Table 8.8. Calculated and test data both represented by symbols, solid lines drawn through calculated results, test data typically displaced from calculated.

10.15 APPLICATION EXAMPLE: NATURAL CONVECTION AND RADIATION COOLED NINE-FIN HEAT SINK ANALYZED FOR A TEMPERATURE RISE NOT INCLUDED IN FIGURE 9.2.

Figure 9.2 is a set of several plots for $\Delta T = 10, 25, 50$, and $100°C$. These are fine when we wish to analyze a heat sink with one of these temperature rises. It is just as common for the heat sink dissipation to be a specified value requiring *calculation* of ΔT, in which case we have two choices: iterate or construct a table. Figure 9.2 is not convenient for iterating. We shall reexamine Application Example 10.14 using a total heat dissipation of $Q = 100$ W where the heat sink height is $H = 8$ in. Convection and radiation areas include only the interior U-channel area, fin tips, and the two outer surfaces of the end fins.

We shall calculate the heat dissipation for the values $\Delta T = 10, 25, 50$, and $100°C$ and then interpolate to get our answer. Some of the details are shown for the first value of the temperature rise and then several of the intermediate results are tabulated in Table 10.4. The internal and external areas are

$$W = N_f t_f + \left(N_f - 1\right)S = (9)(0.15) + (9-1)(0.35) = 4.15 \text{ in.}$$

$$A_I = \left[2\left(N_f - 1\right)L + W\right]H = \left[2(9-1)(2.62) + 4.15\right](8.0) = 368.56 \text{ in.}^2$$

$$A_E = 2HL = 2(8.0)(2.62) = 41.92 \text{ in.}^2$$

$$L/S = 2.62/0.35 = 7.49, H/S = 8.0/0.35 = 22.86$$

The $h_c/h_H = 0.57$ from Figure 9.2(d) for $\Delta T = 10°C$ and $S = 0.35$ in. Then

$$h_H = 0.0024\left(\Delta T/H\right)^{0.25} = 0.0024(10.0/8.0)^{0.25} = 0.0025 \text{ W}/\left(\text{in.}^2 \cdot °C\right)$$

$$h_c = \left(h_c/h_H\right)h_H = (0.57)(0.0025) = 0.0015 \text{ W}/\left(\text{in.}^2 \cdot °C\right)$$

The radiation properties are calculated:

$$\varepsilon = 0.8 \text{ and using } L/S = 2.62/0.35 = 7.49, H/L = 8.0/2.62 = 3.05$$

We get $\mathcal{F} = 0.1$ from Figure 10.33. The radiation heat transfer coefficient follows as

$$h_r = 3.657 \times 10^{-11} \left[\begin{array}{l} \left(T_A + \Delta T + 273.16\right)^3 + \left(T_A + \Delta T + 273.16\right)^2 \left(T_A + 273.16\right) \\ + \left(T_A + \Delta T + 273.16\right)\left(T_A + 273.16\right)^2 + \left(T_A + 273.16\right)^3 \end{array} \right]$$

$$= 3.657 \times 10^{-11} \left[\begin{array}{l} \left(20.0 + 10.0 + 273.16\right)^3 + \left(20.0 + 10.0 + 273.16\right)^2 \left(20.0 + 273.16\right) \\ + \left(20.0 + 10.0 + 273.16\right)\left(20.0 + 273.16\right)^2 + \left(20.0 + 273.16\right)^3 \end{array} \right]$$

$$h_r = 0.0039 \ \mathrm{W} / \left(\mathrm{in.}^2 \cdot {}^\circ\mathrm{C}\right)$$

The fin efficiency is calculated using the heat transfer coefficients and fin properties:

$$R_k = L / \left(k_{Al} H t_f\right) = 2.62 / \left[(5.0)(8.0)(0.15)\right] = 0.44 \ {}^\circ\mathrm{C/W}$$

$$R_c = 1 / \left[2\left(h_c + \mathcal{F}h_r\right)HL\right] = 1 / \left\{2\left[0.0015 + (0.1)(0.0039)\right](8.0)(2.62)\right\} = 13.01 \ {}^\circ\mathrm{C/W}$$

$$\eta = \sqrt{R_c/R_k} \ \tanh \sqrt{R_k/R_c} = \sqrt{13.01/0.44} \ \tanh \sqrt{0.44/13.01} = 0.99$$

The conductances are

$$C_I = \eta\left(h_c + \mathcal{F}h_r\right)A_I = (0.99)\left[0.0015 + (0.1)(0.0039)\right](368.56) = 0.67 \ \mathrm{W}/{}^\circ\mathrm{C}$$

$$C_E = \eta\left(h_H + \varepsilon h_r\right)A_E = (0.99)\left[0.0025 + (0.8)(0.0039)\right](41.92) = 0.11 \ \mathrm{W}/{}^\circ\mathrm{C}$$

$$C = C_I + C_E = 0.67 + 0.11 = 0.78 \ \mathrm{W}/{}^\circ\mathrm{C}$$

Finally, the total heat convected and radiated is

$$Q = C\Delta T = (0.78)(10.0) = 7.8 \ \mathrm{W}$$

We see from the plotted (Figure 10.40) table (Table 10.4) data, that the temperature rise is $\Delta T = 66\,{}^\circ\mathrm{C}$ for $Q = 100$ W.

Table 10.4. Application Example 10.15: Intermediate results for 8.0-in. heat sink cooled by natural convection and radiation.

$\Delta T \left[{}^\circ\mathrm{C}\right]$	$\dfrac{h_c}{h_H}$	$h_c \left[\dfrac{\mathrm{W}}{\mathrm{in.}^2 \cdot {}^\circ\mathrm{C}}\right]$	$h_r \left[\dfrac{\mathrm{W}}{\mathrm{in.}^2 \cdot {}^\circ\mathrm{C}}\right]$	η	$C\left[\mathrm{W}/{}^\circ\mathrm{C}\right]$	$Q[\mathrm{W}]$
10	0.57	0.0015	0.0039	0.99	0.78	7.8
25	0.71	0.0023	0.0042	0.98	1.11	27.6
50	0.77	0.0029	0.0047	0.98	1.38	69.1
100	0.79	0.0036	0.0060	0.98	1.68	168.3

10.16 ILLUSTRATIVE EXAMPLE: NATURAL CONVECTION AND RADIATION COOLED NINE-FIN HEAT SINK ANALYZED FOR OPTIMUM NUMBER OF FINS

The geometry in this problem is identical with that in the examples studied in Sections 10.14 and 10.15 with dimensions of $W = 4.15$ in., $H = 8.0$ in., $L = 2.62$ in., $t_f = 0.15$ in. The heat sink temperature rise above ambient is $\Delta T = 50\,{}^\circ\mathrm{C}$. The thermal resistance, including both natural convection and radiation ($\varepsilon = 0.8$), has been calculated for the finned side for several different numbers of fins N_f. A plot of the results is shown in Figure 10.41 and Table 10.5. We note that the optimum number of fins is about nine. If we apply Figure 8.10 which, strictly speaking, applies only to natural convection, exclusive of radiation, we can estimate the optimum fin spacing.

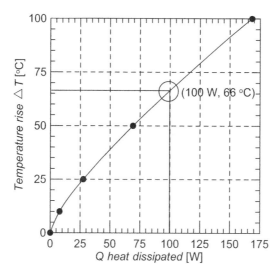

Figure 10.40. Application Example 10.15: Table 10.4 results plotted.

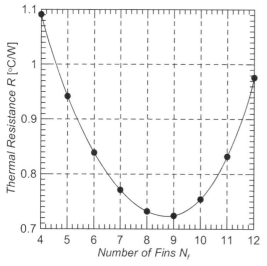

Figure 10.41. Illustrative Example 10.16: Thermal resistance vs. number of fins.

Table 10.5. Illustrative Example 10.16: Fin spacing and calculated thermal resistance for various numbers of fins N_f.

N_f	S [in.]	R [°C/W]
4	1.183	1.091
5	0.850	0.942
6	0.650	0.839
7	0.517	0.771
8	0.421	0.732
9	0.350	0.724
10	0.294	0.754
11	0.250	0.832
12	0.214	0.976

The formula

$$S_{opt} = 0.17H^{1/4} = 0.17(8.0)^{1/4} = 0.29\,\text{in.}$$

also gives us a value of N_{opt} equal to about nine fins.

EXERCISES

10.1 Figure 9.3 illustrates a top view of a heat sink with dimensions $W = 1.86$ in., $H = 5.0$ in., $L = 1.0$ in., $t_f = 0.06$ in., and $\Delta T = 50\,°C$ in an ambient of $T_A = 20°C$. Use the number of fins $N_f = 5$ as suggested in the figure. Suppose that the heat sink interior and exterior surfaces radiate as perfect blackbodies. Calculate (1) the total view factor of an interior channel, (2) the total heat radiated from all of the interior channels, (3) the heat radiated from the outer surfaces of the two end fins, and (4) the total heat radiated. Hints: (1) It is very easy to get confused when translating the heat sink geometry variables into names used for the parallel and perpendicular plate formulae; therefore, it is suggested that you use the drawings below and fill in the exact dimensions for each figure; (2) The *Shape Factor x Area* for the total interior $= F_{Base\text{-}Front}A_{Base} + F_{Base\text{-}Top}(2A_{Base})$.

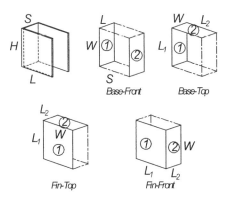

Exercises 10.1 - 10.5. Correctly labeled dimensions according to Figures 10.7 and 10.8.

10.2 Consider the heat sink analyzed in Section 9.5. The dimensions are $H = 2.0$ in., $W = 4.15$ in., $L = 2.62$ in., $S = 0.35$ in., $t_f = 0.15$ in., and $t_b = 0.63$ in. The heat sink temperature rise above ambient is $\Delta T = 50°C$. Suppose that the heat sink interior and exterior surfaces radiate as perfect blackbodies. Calculate (1) the total view factor of an interior channel, (2) the total heat radiated from all of the interior channels, (3) the heat radiated from the outer surfaces of the two end fins, and (4) the total heat radiated. Follow the hints for Exercise 10.1.

10.3 Consider the heat sink analyzed in Section 9.5. The dimensions are $H = 4.0$ in., $W = 4.15$ in., $L = 2.62$ in., $S = 0.35$ in., $t_f = 0.15$ in., and $t_b = 0.63$ in. The heat sink temperature rise above ambient is $\Delta T = 50°C$. Suppose that the heat sink interior and exterior surfaces radiate as perfect blackbodies. Calculate (1) the total view factor of an interior channel, (2) the total heat radiated from all of the interior channels, (3) the heat radiated from the outer surfaces of the two end fins, and (4) the total heat radiated. Follow the hints for Exercise 10.1.

10.4 Consider the heat sink analyzed in Section 9.5. The dimensions are $H = 6.0$ in., $W = 4.15$ in., $L = 2.62$ in., $S = 0.35$ in., $t_f = 0.15$ in., and $t_b = 0.63$ in. The heat sink temperature rise above ambient is $\Delta T = 50°C$. Suppose that the heat sink interior and exterior surfaces radiate as perfect blackbodies. Calculate (1) the total view factor of an interior channel, (2) the total heat radiated from all of the interior channels, (3) the heat radiated from the outer surfaces of the two end fins, and (4) the total heat radiated. Follow the hints for Exercise 10.1.

10.5 Consider the heat sink analyzed in Section 9.5. The dimensions are $H = 8.0$ in., $W = 4.15$ in., $L = 2.62$ in., $S = 0.35$ in., $t_f = 0.15$ in., and $t_b = 0.63$ in. The heat sink temperature rise above ambient is $\Delta T = 50°C$. Suppose that the heat sink interior and exterior surfaces radiate as perfect blackbodies. Calculate (1) the total view factor of an interior channel, (2) the total heat radiated from all of the interior channels, (3) the heat radiated from the outer surfaces of the two end fins, and (4) the total heat radiated. Follow the hints for Exercise 10.1.

10.6 A two-fin heat sink is used to radiate heat from two power transistors. The dimensions according to the illustration for this exercise are $S = 0.25$ in., $H = 2.0$ in., and $L = 1.0$ in. The heat sink is at a uniform temperature, including the fins, with a temperature rise above ambient of $\Delta T = 50\,^\circ C$ in an ambient $T_A = 20^\circ C$. Calculate the total heat, neglecting the flat backside, that is radiated to ambient if the heat sink acts as a perfect blackbody.

Exercise 10.6. Two transistors on a heat sink.

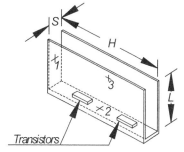

10.7 Perform an approximate integration of Figure 10.11 using Eq. (10.27) to determine an average value of the emissivity of the anodized aluminum. Using a temperature of 100°C, perform the calculation using (1) E_b for wavelength from 1 to 15 μm and (2) E_b for wavelength from zero to infinity. Hints: (1) Use Figure 10.3 for calculation (1); (2) Use wavelength intervals of (1.5-2.6), (2.6-3.5), (3.5-4.8), (4.8-6.6), (6.6-8.0), (8.0-12.5), and (12.5-15.0) μm with the mean value of $\overline{E}_{\lambda b}$ in each interval.

10.8 (Problem 8.22, modified.) A vented, metal-walled enclosure in ambient air at $T_A = 20^\circ C$ containing several circuit boards is illustrated (as Exercise 8.22). The total heat dissipation of all circuit boards is $Q_{Box} = 125$ W. The given dimensions are $a = 0.20$ in., $l = 1.0$ in., $w = 1.0$ in., $p = 0.5$ in., $W = 11$ in., $D = 12$ in., $H = 10$ in., $D_{PCB} = 10$ in., $L = 9$ in., $H_I = 1.0$ in. The exit and inlet vents have identical dimensions and each has a free area ratio $f_I = 0.40$ and has an emissivity of 0.8. Calculate the internal air temperature rise above ambient for the vented enclosure. Hints: (1) Use the airflow circuit, with appropriate variable values, from Application Example 8.10; (2) Begin the airflow circuit model with circuit boards on a pitch $p = 0.5$ in.; (3) Calculate the starting number of cards from $N_{Cards} = (W/p)-1$.

10.9 (Problem 8.23, modified.) An enclosure (Figure Exercise 8.23) has dimensions of $W = 9.0$ in., $H = 9.0$ in., $D = 18$ in., $W_I = 7.0$ in., $W_E = 7.0$ in., $H_I = 2.0$ in., $f_I = 0.35, f_E = 0.35, N_{Cards} = 7$, $Q_{Total} = 35$ W, $D_{PCB} = 16$ in., $L = 5.0$ in., $f_{PCB} = 0.5$ (free area for PCB channel), $b = 0.5$ in. (component surface to opposing PCB) and an external emissivity is 0.8. You are asked to solve for the enclosure wall temperature rise, internal air temperature rise, the venting airdraft and the heat carried away by the draft (where relevant). Hint: Suggested airflow circuit is series sum of inlet vent, expansion from inlet region to $D \cdot W_I$ entrance to $PCBs$, contraction to $PCBs$, $PCBs$, expansion from $PCBs$, contraction from PCB exit region to exit region.

10.10 Solve Application Example 10.9: $W = 10.0$ in., $L = 10.0$ in., and $\varepsilon_1 = 1.0, \varepsilon_2 = 1.0$. Use a view factor $F_{1,2}$ for infinite parallel plates.

10.11 Solve Application Example 10.9: $W = 10.0$ in., $L = 10.0$ in., and $\varepsilon_1 = 0.7, \varepsilon_2 = 0.5$. Use a view factor $F_{1,2}$ for infinite parallel plates.

10.12 Solve Application Example 10.10: $W = 10.0$ in., $L = 10.0$ in., $S = 1.0$ in., and $\varepsilon_1 = 1.0, \varepsilon_2 = 1.0$. Consider radiation exchange between all six surfaces. Calculate the view factors using Figure 10.8 or the exact formulae, Eqs. (10.22) and (10.23).

10.13 Solve Application Example 10.10: $W = 10.0$ in., $L = 10.0$ in., $S = 1.0$ in., and $\varepsilon_1 = 0.7, \varepsilon_2 = 0.5$. Consider radiation exchange between all six surfaces. Calculate the view factors using Figure 10.8 or the exact formulae, Eqs. (10.22) and (10.23).

CHAPTER 11

Conduction I: Basics

Our detailed study of conduction is organized into two chapters. This first chapter includes many basic topics one would expect, such as Fourier's law of heat conduction, the differential equations for heat conduction, and properties of materials. It then concludes with thermal interface resistance. The next chapter is also concerned with conduction, but concentrates on thermal spreading resistance.

11.1 FOURIER'S LAW OF HEAT CONDUCTION

We begin with a review of the conduction discussion in Chapter 1. The pertinent aspects of heat conduction are demonstrated by a one-dimensional solid element, as in Figure 11.1, for which it is assumed that convection and radiation are not present. The convention of positive heat flow in the positive x-axis direction is used. The heat flows through a path length $x_2 - x_1$ and a cross-sectional area A_k that is perpendicular to the axis. In this case, therefore, the temperature $T(x_1)$ at x_1 is greater than the temperature $T(x_2)$ at x_2.

Fourier's law is used to quantify conductive heat flow. In one dimension the law is written as

$$\boxed{Q_k = -kA_k \frac{dT}{dx}\bigg|_x}$$
Fourier's law (11.1)

$Q_k = -kA_k \dfrac{dT}{dx}\bigg|_x$ $Q_k \equiv$ heat transferred, watts

$A_k \equiv$ cross-sectional area of heat flow path, $\left[\text{cm}^2\right], \left[\text{m}^2\right],$ or $\left[\text{in.}^2\right]$

$\dfrac{dT}{dx} \equiv$ temperature gradient, $\left[\text{°C/cm}\right], \left[\text{°C/m}\right],$ or $\left[\text{°C/in.}\right]$

$k \equiv$ thermal conductivity, $\left[\text{W}/\left(\text{°C·cm}\right)\right], \left[\text{W}/\left(\text{°C·m}\right)\right],$

    or $\left[\text{W}/\left(\text{°C·in.}\right)\right].$

where we will continue to use in. and in.² for length and area units, respectively. The gradient and the cross-sectional area are defined at the same point x. The positive heat flow through a decreasing temperature, i.e., a negative temperature gradient, necessitates the minus sign immediately preceding the thermal conductivity factor in Eq. (11.1).

The thermal conductivity may be considered a constant of proportionality that is dependent only on the specific material involved. While it is not uncommon for k to be temperature-dependent, this usually causes only a minor source of error in engineering design problems. When the temperature

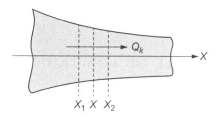

Figure 11.1. Heat conduction in a one-dimensional solid element.

dependence of k becomes important, digital computer program techniques may be required. Otherwise, we settle for using k based on the average temperature of our device. The range of values of k that we can encounter in our study of electronic devices is illustrated in Figure 1.2, where we note that k can differ by about five orders of magnitude, between the smallest and largest k values. A table of conductivities is provided in Appendix *iv*.

Fourier's law for heat conduction may be immediately integrated resulting in a common application formula. Rearranging the variables in Eq. (11.1) and remembering that $T_1 > T_2$,

$$dT = -\left(Q_k / kA_k\right) dx$$

$$\int_{T_1}^{T_2} dT = -Q_k \int_{x_1}^{x_2} dx / \left(kA_k\right)$$

$$\int_{T_2}^{T_1} dT = Q_k \int_{x_1}^{x_2} dx / \left(kA_k\right)$$

which is written as

$$\Delta T = R_k Q_k \qquad \Delta T = T_1 - T_2 \qquad (11.2)$$

and

$$R_k = \int_{x=0}^{x=L} dx / \left(kA_k\right) \qquad \text{Conduction resistance} \quad (11.3)$$

The formula for the one-dimensional conduction resistance, Eq. (11.3), is very general. If neither k nor A_k vary over the length L of a one-dimensional bar, we arrive at

$$R_k = L / \left(kA_k\right) \qquad \text{Conduction resistance} \quad (11.4)$$

Multiple resistors in series or parallel are easily added following Figure 11.2 and Eq. (11.5).

$$\text{Parallel: } \frac{1}{R} = \frac{1}{R_1} + \frac{1}{R_2}; \qquad \text{Series: } R = R_1 + R_2 \qquad (11.5)$$

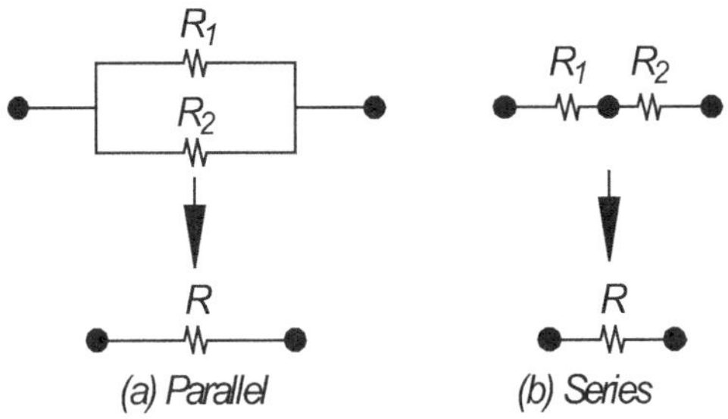

Figure 11.2. Addition of resistors.

11.2 APPLICATION EXAMPLE: MICA INSULATOR WITH THERMAL PASTE

This is a trivial example, but nevertheless is an adequate illustration of adding conduction resistances. Suppose a mica washer is used to provide electrical insulation between the base of a power transistor and a heat sink. Each side of the washer has a thin layer of filled thermal compound. The mica has a cross-sectional area of about 1.0 in.² and a thickness of 0.002 in. We assume each layer of thermal compound has a thickness of 0.0005 in. We wish to know the total thermal resistance through the two thin thermal compound layers and the mica washer. Clearly we have a situation of three thermal resistances in series:

$$R = R_{TC1} + R_{Mica} + R_{TC2} = \frac{t_{TC1}}{k_{TC1}A} + \frac{t_M}{k_M A} + \frac{t_{TC2}}{k_{TC2}A} = \frac{0.0005\,\text{in.}}{\left[0.02\,\text{W}\big/\left(\text{in.}\cdot{}^\circ\text{C}\right)\right]\left(1.0\,\text{in.}^2\right)}$$

$$+ \frac{0.002\,\text{in.}}{\left[0.02\,\text{W}\big/\left(\text{in.}\cdot{}^\circ\text{C}\right)\right]\left(1.0\,\text{in.}^2\right)} + \frac{0.0005\,\text{in.}}{\left[0.02\,\text{W}\big/\left(\text{in.}\cdot{}^\circ\text{C}\right)\right]\left(1.0\,\text{in.}^2\right)}$$

$$R = 0.15\,{}^\circ\text{C/W}$$

A resistance of 0.15 °C/W doesn't seem to be a very large number, but the important effect of the temperature rise depends on the quantity of heat conducted through the structure.

11.3 THERMAL CONDUCTION RESISTANCE OF SOME SIMPLE STRUCTURES

It is sometimes useful to have formulae for solid conduction structures that are simple, but not as simple as the case of a uniform cross-section structure with the resistance given by Eq. (11.4). The elements in Figure 11.3 illustrate three such structures.

The structure in Figure 11.3 (a) with a uniform width has a cross-sectional area A_k given by

$$A_k = w\left[t_1 + \left(t_2 - t_1\right)\left(x/L\right)\right]$$

where x is an axis drawn parallel to the length L, beginning at the face with thickness t_1. The thermal resistance from face 1 to face 2 is derived by beginning with Eq. (11.3).

$$R_k = \int_{x=0}^{x=L} dx/\left(kA_k\right) = \left(1/kw\right)\int_{x=0}^{x=L} dx/\left[t_1 + \left(t_2 - t_1\right)\left(x/L\right)\right]$$

Making a substitution we obtain

$$u = t_1 + \left(t_2 - t_1\right)\left(x/L\right), \quad du = \left(t_2 - t_1\right)\left(dx/L\right)$$

$$R_k = \left[L/kw\left(t_2 - t_1\right)\right]\int_{u=t_1}^{u=t_2} \frac{du}{u}$$

$$\boxed{R_k = \frac{L\ln\left(t_2/t_1\right)}{kw\left(t_2 - t_1\right)}} \qquad\qquad \text{Uniform (11.6)}\\ \text{width wedge}$$

The cross-sectional area A_k for the variable width wedge shown as Figure 11.3 (b) is

$$A_k = \left[t_1 + \left(t_2 - t_1\right)\left(x/L\right)\right]\left[w_1 + \left(w_2 - w_1\right)\left(x/L\right)\right]$$

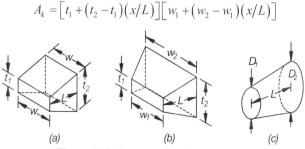

Figure 11.3. Some conduction structures.

$$A_k = t_1 w_1 + t_1 \left(w_2 - w_1 \right) \left(x/L \right) + w_1 \left(t_2 - t_1 \right) \left(x/L \right) + \left(t_2 - t_1 \right) \left(w_2 - w_1 \right) \left(x^2/L \right)$$

$$= t_1 w_1 + \left(t_1 \Delta w + w_1 \Delta t \right) \left(x/L \right) + \Delta t \Delta w \left(x^2/L^2 \right)$$

$$A_k = a + bx + cx^2$$

$$a = t_1 w_1, \quad b = \left(t_1 \Delta w + w_1 \Delta t \right) \left(x/L \right), \quad c = \Delta t \Delta w / L^2, \quad \Delta w = w_2 - w_1, \quad \Delta t = t_2 - t_1$$

$$R_k = \left(1/k \right) \int_{x=0}^{x=L} dx / \left(a + bx + cx^2 \right) = \left(1/k\sqrt{-q} \right) \ln \left(\frac{2cx + b - \sqrt{-q}}{2cx + b + \sqrt{-q}} \right) \Bigg|_{x=0}^{x=L}, \quad q = 4ac - b^2$$

The final result is

$$\boxed{R_k = \left[\frac{L}{k \left(t_1 \Delta w - w_1 \Delta t \right)} \right] \ln \left[\frac{\left(\Delta w / w_1 \right) + 1}{\left(\Delta t / t_1 \right) + 1} \right]}$$

Variable (11.7)
width wedge

Figure 11.3 (*c*) illustrates a structure with a circular, variable cross-sectional area. This area A_k is given by

$$A_k = \pi \left[r_1 + \left(r_2 - r_1 \right) \left(x/L \right) \right]^2$$

$$R_k = \left(1/\pi k \right) \int_{x=0}^{x=L} dx / \left[r_1 + \left(r_2 - r_1 \right) \left(x/L \right) \right]^2$$

$$u = r_1 + \left(r_2 - r_1 \right) \left(x/L \right), \quad du = \left(r_2 - r_1 \right) \left(dx/L \right)$$

$$R_k = \left[L/\pi k \left(r_2 - r_1 \right) \right] \int_{u=r_1}^{u=r_2} \frac{du}{u^2} = \left[L/\pi k \left(r_2 - r_1 \right) \right] \left[-1/u \right]_{u=r_1}^{u=r_2}$$

$$\boxed{R_k = 4L / \left(\pi k D_1 D_2 \right)}$$

Variable area (11.8)

11.4 THE ONE-DIMENSIONAL DIFFERENTIAL EQUATION FOR HEAT CONDUCTION

There are circumstances where one-dimensional solutions are appropriate. We begin by referring to Figure 11.4 for a thin bar of thickness t and an element of length $\Delta x = x_2 - x_1$ on which we perform an energy balance.

Heat into Δx - heat out of $\Delta x = 0$

$$\left[-kA_k \frac{dT}{dx} \bigg|_{x_1} + Q_V \Delta x A_k \right] - \left[-kA_k \frac{dT}{dx} \bigg|_{x_2} + 2h \left(w + t \right) \Delta x T \right] = 0$$

The average element temperature T is referenced to a zero ambient. This means that T is the temperature rise above ambient. Q_V is the internal source density in units of watts per unit volume. Dividing the preceding equation by $k \Delta x A_k$ and rearranging,

$$\frac{1}{\Delta x} \left[\frac{dT}{dx} \bigg|_{x_2} - \frac{dT}{dx} \bigg|_{x_1} \right] - \frac{2h \left(w + t \right)}{kA_k} T = -\frac{Q_V}{k}$$

Taking the limit as $\Delta x \to 0$ and recognizing that

$$\lim_{\Delta x \to 0} \frac{1}{\Delta x}\left[\left.\frac{dT}{dx}\right|_{x_2} - \left.\frac{dT}{dx}\right|_{x_1}\right] \to \frac{d^2T}{dx^2}$$

we find the result is
$$\frac{d^2T}{dx^2} - \theta^2 T = -\frac{Q_V}{k}$$
One-dimensional (11.9)
DEQ

$$\theta^2 = R_k / L^2 R_S , R_k = L/kA_k , R_S = 1/hA_S$$

where A_S is the convective/radiative surface area for the entire bar and A_k is the cross-sectional area for conduction from left to right. Although we originally specified a thin bar, the only restriction is that temperature gradients perpendicular to the x-axis are negligible. The general solution to Eq. (11.9) for a nonzero θ and Q_V uniform over L is

$$T = c_1 \cosh\theta x + c_2 \sinh\theta x + \left(\alpha/\theta^2\right), \ \alpha = Q_V/k \tag{11.10}$$

Our first solution is the conduction only (negligible convection/radiation), one-dimensional bar illustrated in Figure 11.5 where the heat source is distributed uniformly over the length. Remembering that we used an ambient T_A, the sink temperature T_0 is actually a temperature rise above ambient. First we note that $h = 0$ so that R_S is infinite, thus $\theta = 0$ and Eq. (11.9) becomes

$$\frac{d^2T}{dx^2} = -\frac{Q_V}{k}$$

which is immediately integrated to

$$T = -Q_V x^2 / 2k + c_1 x + c_2$$

The constants c_1 and c_2 are obtained from two boundary conditions at $x = 0$.

$$T = T_0, \ -kA_k \, dT/dx = -Q_V A_k L, \ Q = Q_V A_k L$$

The second boundary condition stems from the condition that all of the source heat must pass into $x = 0$ with the result that
$$T(x) = x\left(Q/kA_k\right)\left(1 - x/2L\right) + T_0$$

The thermal resistance is defined as the hottest temperature above T_0, divided by the total heat dissipation.

$$R = \left[T\left(x = L\right) - T_0\right]/Q = \left(LQ/2kA_k\right)/Q$$

$$\boxed{R = \left(1/2\right)R_k}$$
Uniform source,
conduction only, (11.11)
one end sinked

Our second possibility is illustrated in Figure 11.6. This is similar to the preceding problem except that we have added convection and/or radiation from the bar to T_A. Our general solution, Eq. (11.10) is applicable. The boundary conditions are applied at $x=0$ and $x = L$.

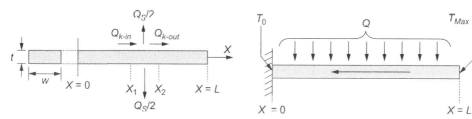

Figure 11.4. Geometry for performing an energy balance on $\Delta x = x_2 - x_1$.

Figure 11.5. Equation (11.11): Uniform source, conduction to heat sinked end.

$$x = 0: T(x=0) = T_0, \quad c_1 = T_0 - \alpha/\theta^2$$

$$x = L: -kA_k \, dT/dx = 0, \quad c_2 = -\left(T_0 - \alpha/\theta^2\right)\tanh\theta L$$

and since

$$\alpha/\theta^2 = \left(Q_V/k\right)/\left(R_k/L^2 R_s\right) = QR_s$$

then

$$T(x) = \left(T_0 - QR_s\right)\left(\cosh\theta x - \tanh\theta L \sinh\theta x\right) + QR_s$$

Defining

$$R = T(x=L)/Q$$

$$\boxed{\frac{R}{R_s} = \left[\left(\frac{T_0/Q}{R_s} - 1\right)\middle/\cosh\sqrt{R_k/R_s}\right] + 1}$$

Uniform source,
cond./conv·, (11.12)
one end sinked

which is plotted in Figure 11.9 (a). It is left as an exercise to show that the heat Q_0 into the sink end is given by

$$\boxed{\frac{Q_0}{Q} = \left(\frac{T_0/Q}{R_s} - 1\right)\frac{\tanh\sqrt{R_k/R_s}}{\sqrt{R_k/R_s}}}$$

Uniform source,
cond./conv., (11.13)
one end sinked

which is plotted in Figure 11.9 (b).

The third case, a bar with an end source and conduction/convection, is illustrated in Figure 11.7. It should be emphasized that neither end is heat sinked so that all losses occur via convection/radiation to the ambient T_A. The solution is again given by Eq. (11.10) to which we apply boundary conditions:

$$x = 0: -kA_k \, dT/dx = Q, \, Q = |Q|, \quad c_2 = -Q/\left(kA_k\theta\right)$$

$$x = L: -kA_k \, dT/dx = 0, \quad\quad c_1 = c_2 \coth\theta L$$

so that the temperature as a function of x is

$$T(x) = Q\sqrt{R_k R_s}\left\{\coth\sqrt{R_k/R_s}\cosh\left[\sqrt{R_k/R_s}\,(x/L)\right] - \sinh\left[\sqrt{R_k/R_s}\,(x/L)\right]\right\}$$

The resistance is defined with respect to a zero ambient temperature so that

$$R = T(x=0)/Q$$

$$\boxed{R/R_s = \sqrt{R_k/R_s}\,\coth\sqrt{R_k/R_s}}$$

End source, (11.14)
cond./conv.

which is plotted in Figure 11.10.

The last case, a one-dimensional bar with an end source, end sinking, and convection/radiation, is illustrated in Figure 11.8. Using our general solution, Eq. (11.10), we have the boundary conditions that permit solution for the constants c_1 and c_2.

$$x = 0: T = T_0; \quad\quad x = L: -kA_k \, dT/dx = -Q, \, Q = |Q|$$

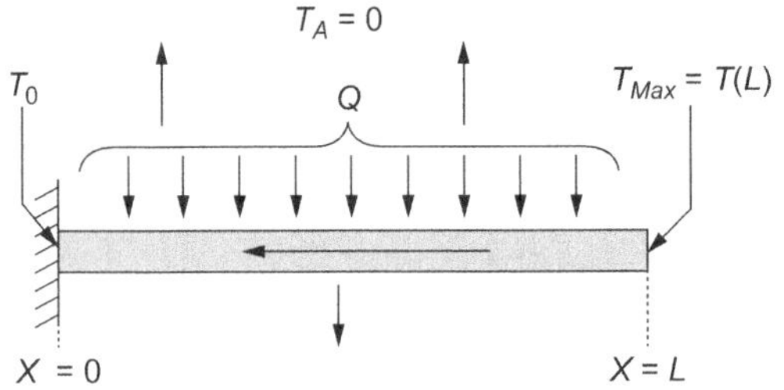

Figure 11.6. Equations (11.12) and (11.13): Uniform source, conduction, convection, one end sinked.

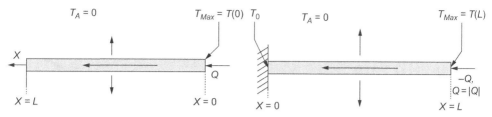

Figure 11.7. Equation (11.14): End source, conduction and convection/radiation.

Figure 11.8. Equations (11.15) and (11.16): End source, opposite end sinked, conduction and convection/radiation.

The result is
$$T(x) = T_0 \left(\cosh \theta x - \tanh \theta L \sinh \theta x \right) + \left[Q \sinh \theta x / \left(\theta k A_k \cosh \theta L \right) \right]$$

$$\boxed{\frac{R}{R_s} = \left[\frac{(T_0/Q)}{R_s} + \sqrt{\frac{R_k}{R_s}} \sinh \sqrt{\frac{R_k}{R_s}} \right] / \cosh \sqrt{\frac{R_k}{R_s}}}$$

End source, opp. end sinked, (11.15) cond./conv.

which is plotted in Figure 11.11. It is left as an exercise to show that the heat Q_0 conducted into the sink end is given by

$$\boxed{\frac{Q_0}{Q} = \frac{\tanh \sqrt{R_k/R_s}}{\sqrt{R_k/R_s}} \left(\frac{T_0/Q}{R_s} - \frac{\sqrt{R_k/R_s}}{\sinh \sqrt{R_k/R_s}} \right)}$$

End source, opp. end sinked, (11.16) cond./conv.

which is plotted in Figure 11.12.

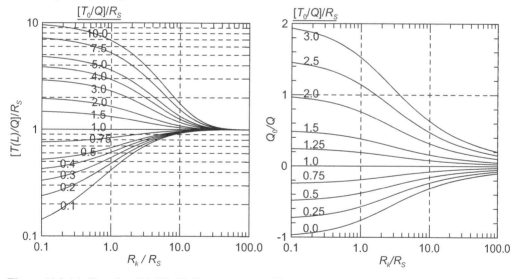

Figure 11.9 (a). Equation (11.12): Uniform source, conduction, convection, one end sinked.

Figure 11.9 (b). Equation (11.13): Uniform source, conduction, convection, one end sinked.

Thus far, we have used end-boundary conditions that denote either sinked to a given temperature T_0 or adiabatic. We need to consider one variation of this: the case where we have one end source and an unsinked end that is allowed to convect. The heat transfer coefficient along the length is h and the heat transfer coefficient at the convecting end ($x = L$) is h_L. The applicable illustration is Figure 11.7 (remembering the convecting end). The applicable differential equation and general solution are Eqs. (11.9) and (11.10), respectively, with Q_V, α both zero. The boundary conditions applied to Eq. (11.10) are

$$x = 0: \quad T = T_0, \quad c_1 = T_0$$

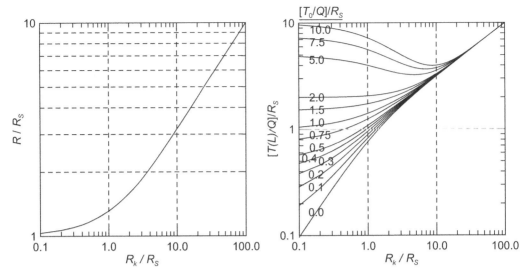

Figure 11.10. Equation (11.14): End source, conduction, convection.

Figure 11.11. Equation (11.15): End source, opposite end sinked, conduction, convection.

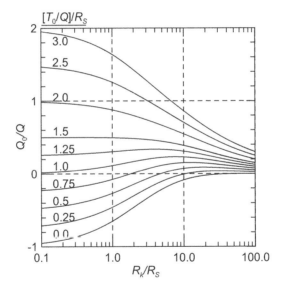

Figure 11.12. Equation (11.16): End source, end sink, conduction, convection.

$$x = L: \quad -k \, dT/dx = h_L T$$

$$-k\theta \left(T_0 \sinh \theta x + c_2 \cosh \theta x\right)\Big|_{x=L} = h_L \left(T_0 \cosh \theta x + c_2 \sinh \theta x\right)\Big|_{x=L}$$

$$-k\theta \left(T_0 \sinh \theta L + c_2 \cosh \theta L\right) = h_L \left(T_0 \cosh \theta L + c_2 \sinh \theta L\right)$$

$$-k\theta T_0 \sinh \theta L - k\theta c_2 \cosh \theta L = h_L T_0 \cosh \theta L + h_L c_2 \sinh \theta L$$

$$c_2 \left(h_L \sinh \theta L + k\theta \cosh \theta L\right) = -T_0 \left(h_L \cosh \theta L + k\theta \sinh \theta L\right)$$

$$c_2 = -T_0 \left(\frac{h_L \cosh \theta L + k\theta \sinh \theta L}{h_L \sinh \theta L + k\theta \cosh \theta L}\right)$$

The solution for $T(x)$ is

$$T(x) = T_0 \left[\cosh\theta x - \left(\frac{h_L \cosh\theta L + k\theta \sinh\theta L}{h_L \sinh\theta L + k\theta \cosh\theta L} \right) \sinh\theta x \right]$$

Now we derive equations for the heat Q_0 conducted into the bar at $x = 0$.

$$Q_0 = -kA_k \, dT/dx\big|_{x=0} = -kA_k\theta T_0 \left[\sinh\theta x - \left(\frac{h_L \cosh\theta L + k\theta \sinh\theta L}{h_L \sinh\theta L + k\theta \cosh\theta L} \right) \cosh\theta x \right]_{x=0}$$

$$Q_0 = kA_k\theta T_0 \left(\frac{h_L \cosh\theta L + k\theta \sinh\theta L}{h_L \sinh\theta L + k\theta \cosh\theta L} \right)$$

$$Q_0 = \sqrt{R_k/R_s}\,(kA_k/L)T_0 \left[\frac{h_L A_k \cosh\theta L + \sqrt{R_k/R_s}\,(kA_k/L)\sinh\theta L}{h_L A_k \sinh\theta L + \sqrt{R_k/R_s}\,(kA_k/L)\cosh\theta L} \right]$$

$$\theta = \sqrt{R_k/R_s}\big/L, \quad k\theta = k\sqrt{R_k/R_s}\big/L, \quad R_L = 1/h_L A_k$$

$$Q_0 = \frac{T_0}{\sqrt{R_k R_s}} \left\{ \frac{\left(\dfrac{\cosh\sqrt{R_k/R_s}}{R_L} \right) + \left(\dfrac{\sinh\sqrt{R_k/R_s}}{\sqrt{R_k R_s}} \right)}{\left(\dfrac{\sinh\sqrt{R_k/R_s}}{R_L} \right) + \left(\dfrac{\cosh\sqrt{R_k/R_s}}{\sqrt{R_k R_s}} \right)} \right\} \qquad \begin{array}{l} x = 0 \text{ cond., } (11.17) \\ \text{end conv.,} \\ \text{cond./conv.} \end{array}$$

The heat Q_L convected from the bar tip at $x = L$ is

$$Q_L = -kA_k \, dT/dx\big|_{x=L} = -kA_k\theta T_0 \left[\sinh\theta x - \left(\frac{h_L \cosh\theta L + k\theta \sinh\theta L}{h_L \sinh\theta L + k\theta \cosh\theta L} \right) \cosh\theta x \right]_{x=L}$$

$$Q_L = -kA_k\theta T_0 \left[\sinh\theta L - \left(\frac{h_L \cosh\theta L + k\theta \sinh\theta L}{h_L \sinh\theta L + k\theta \cosh\theta L} \right) \cosh\theta L \right]$$

$$Q_L = -\frac{T_0}{\sqrt{R_k R_s}} \left\{ \sinh\sqrt{R_k/R_s} - \left[\frac{\left(\dfrac{\cosh\sqrt{R_k/R_s}}{R_L} \right) + \left(\dfrac{\sinh\sqrt{R_k/R_s}}{\sqrt{R_k R_s}} \right)}{\left(\dfrac{\sinh\sqrt{R_k/R_s}}{R_L} \right) + \left(\dfrac{\cosh\sqrt{R_k/R_s}}{\sqrt{R_k R_s}} \right)} \right] \cosh\sqrt{R_k/R_s} \right\}$$

$$x = 0 \text{ cond., end cond./conv. } (11.18)$$

The heat Q_s convected/radiated along the bar surface from $x = 0$ to $x = L$, excluding the tip, is

$$Q_s = Q_0 - Q_L$$

$$Q_s = \frac{T_0}{\sqrt{R_k R_s}} \left[\sinh\sqrt{R_k/R_s} + \left(1 - \cosh\sqrt{R_k/R_s}\right) \frac{\dfrac{\cosh\sqrt{R_k/R_s}}{R_L} + \dfrac{\sinh\sqrt{R_k/R_s}}{\sqrt{R_k R_s}}}{\dfrac{\sinh\sqrt{R_k/R_s}}{R_L} + \dfrac{\cosh\sqrt{R_k/R_s}}{\sqrt{R_k R_s}}} \right]$$

$$x = 0 \text{ cond., end cond./conv. } (11.19)$$

Equations (11.17), (11.18), and (11.19) will be used when we look at fin efficiency later in this chapter.

11.5 APPLICATION EXAMPLE: ALUMINUM CORE
BOARD WITH NEGLIGIBLE AIR COOLING

Consider an electronic package where several circuit boards (Figure 11.13) are in a totally sealed box. The circuit boards are sufficiently close together that convection cannot be considered and the boards are expected to be so identical in temperature that radiation will not be effective. Thus the designer is forced to use a metal core board, for which we will use a thickness $t = 0.0625$ in., length $L = 6.0$ in., width $W = 4.0$ in., and conductivity $k = 5.0$ W/(in. $^\circ$C). One end of each board is heat sinked at a temperature T_0. The problem is to determine how many watts may be dissipated on each circuit board such that the temperature rise over the length L is limited to 20°C.

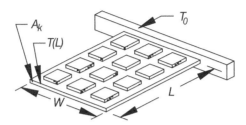

Figure 11.13. Application Example 11.5: Aluminum core board with negligible air cooling.

$$T(L) - T_0 = RQ$$

$$Q = \left[T(L) - T_0\right]/R = \left[T(L) - T_0\right]/(R_k/2) = \left[T(L) - T_0\right]2kA_k/L = \left[T(L) - T_0\right]2kWt/L$$

$$Q = (20\,^\circ\text{C})2\left[5.0\,\text{W}/(\text{in.}\cdot\,^\circ\text{C})\right](4.0\,\text{in.})(0.0625\,\text{in.})/(6.0\,\text{in.}) = 8.33\,\text{W}$$

11.6 APPLICATION EXAMPLE: ALUMINUM CORE BOARD WITH FORCED AIR
COOLING

We re-examine the preceding problem, but now suppose that we have forced air flow in the W-direction at a speed of $V = 300$ ft/min. We require $T(L) - T_0 = 20$°C and a boundary condition of $T_0 - T_A = 40$°C.

Equation (6.15) is used to obtain the heat transfer coefficient.

$$h_c = 0.374\left(\frac{k_{Air}}{L}\right)\left(\frac{VL}{5v}\right)^{0.607} = 0.374\left(\frac{6.9\times10^{-4}}{4.0}\right)\left[\frac{(300)(4.0)}{5(0.029)}\right]^{0.607} = 0.013\,\text{W}/\left(\text{in.}^2\cdot\,^\circ\text{C}\right)$$

We need to use either Eq. (11.12) or Figure 11.9 (a) and begin with the $Q = 8.33$ W result from Section 11.5.

$$R_k = L/(kWt) = (6.0)/\left[(5.0)(4.0)(0.0625)\right] = 4.80\,^\circ\text{C}/\text{W}$$

$$R_S = 1/(2hWL) = 1/\left[2(0.013)(4.0)(6.0)\right] = 1.59\,^\circ\text{C}/\text{W}$$

$$R_k/R_S = 3.03;\quad (T_0/Q)/R_S = (40.0/8.33)/1.59 = 3.03$$

$$R = \left[\left(\frac{T_0/Q}{R_S} - 1\right)\bigg/\cosh\sqrt{R_k/R_S}\right]R_S + R_S = \left[(3.03-1)/\cosh\sqrt{3.03}\right](1.59) + 1.59 = 2.68\,^\circ\text{C}/\text{W}$$

$$Q = \left[T(L) - T_A\right]/R = \left\{\left[T(L) - T_0\right] + (T_0 - T_A)\right\}/R = (20+40)/2.68 = 22.38\,\text{W}$$

The calculation began with $Q = 8.33$ W, but we just recalculated $Q = 22.38$ W. We continue by revising the Q used in Eq. (11.12) until the calculated Q changes by a very small amount. Table 11.1 summarizes the necessary iterations.

Table 11.1. Application Example 11.6: Aluminum core board with forced air cooling.

It.	Q [W]	R [°C/W]	Q[W]
1	8.33	2.68	22.38
2	22.38	1.65	36.30
3	36.30	1.42	42.22
4	42.22	1.37	43.85
5	43.85	1.36	44.24
6	44.24	1.35	44.33

The heat conducted into the heat sink temperature T_0 should be calculated using Eq. (11.13).

$$Q_0 = Q\left(\frac{T_0/Q}{R_S} - 1\right)\frac{\tanh\sqrt{R_k/R_S}}{\sqrt{R_k/R_S}} = (44.33)\left(\frac{20/44.33}{1.59} - 1\right)\frac{\tanh\sqrt{3.03}}{\sqrt{3.03}} = -10.32\,\text{W}$$

Figure 11.6 must be consulted to understand that the negative Q_0 means the 10.32 W is directed *from* the heat sink wall.

11.7 APPLICATION EXAMPLE: SIMPLE HEAT SINK

Equation (11.14) is sometimes useful for heat sink analysis, as this example will show. One-half of a heat sink is shown in Figure 11.14, where symmetry is used to calculate a resistance for the entire heat sink for which only half need be analyzed. A power source is shown in a region void of fins. The total heat dissipation for the complete source is Q. We will use a source value of $Q/2$ in our half-model. The heat sink dimensions are $W = 4.0$ in., $L = 2.5$ in., $t_f = 0.1$ in., $S = 0.3$ in., $S_s = 1.5$ in., $H = 0.75$ in., $t = 0.2$ in., $Q = 100$ W, and a forced air approach speed $V_f = 400$ ft/min. We will calculate the source-to-ambient thermal resistance and temperature rise for the heat sink surface beneath the power source. The flat backside does not convect.

We should probably use duct flow to determine the convective heat transfer coefficient, but instead, as an optimistic solution, we will use Eq. (6.15) to obtain the heat transfer coefficient for a flat plate.

$$h_c = 0.374\left(\frac{k_{Air}}{L}\right)\left(\frac{VL}{5\nu}\right)^{0.607} = 0.374\left(\frac{6.9\times10^{-4}}{2.5}\right)\left[\frac{(400)(2.5)}{5(0.029)}\right]^{0.607} = 0.022\,\text{W}/\left(\text{in.}^2\cdot{}^\circ\text{C}\right)$$

The fin efficiency is calculated for a single fin, but first we calculate the *primitives*, R_k and R_c as shown next.

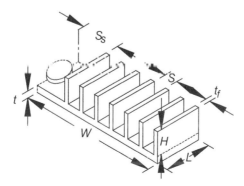

Figure 11.14. Application Example 11.7: Simple heat sink; one-half is analyzed using symmetry.

$$R_k = H/kLt_f = 0.75/\big[(5.0)(2.5)(0.1)\big] = 0.60 \ ^\circ\text{C}/\text{W}$$

$$R_c = 1/2hHL = 1/\big[2(.022)(0.75)(2.5)\big] = 12.08 \ ^\circ\text{C}/\text{W}$$

$$\eta = \sqrt{R_c/R_k}\,\tanh\sqrt{R_k/R_c} = \sqrt{12.08/0.60}\,\tanh\sqrt{0.60/12.08} = 0.98$$

A surface conductance C_S for this problem must be calculated for four different pieces of the convection problem: (1) fin convection with the fin efficiency included, (2) six interfin spaces, (3) the fin-less source region, and (4) seven fin tips.

$$C_S = \eta\big(2N_f HL\big)h + \big[\big(N_f - 1\big)SL + S_S L + N_f t_f L\big]h = (0.98)\big[2(7)(0.75)(2.5)\big](0.022)$$

$$+ \big[(6)(0.3)(2.5) + (1.50)(2.5) + (7)(0.2)(2.5)\big](0.022) = 0.79 \ \text{W}/^\circ\text{C}$$

Then using Eq. (11.14) we calculate the resistance from source to ambient.

$$R_S = 1/C_S = 1/0.79 = 1.27 \ ^\circ\text{C}/\text{W}$$

$$R_k = W/(kLt) = 4.0/\big[(5.0)(2.5)(0.2)\big] = 1.60 \ ^\circ\text{C}/\text{W}$$

$$R = R_S\sqrt{R_k/R_S}\,\coth\sqrt{R_S/R_k} = (1.27)\sqrt{1.60/1.27}\,\coth\sqrt{1.27/1.60}$$

$$R = 2.00 \ ^\circ\text{C}/\text{W}$$

$$T = R(Q/2) = (2.00)(100/2) = 100\,^\circ\text{C} \text{ above ambient}$$

Had we considered only the convection resistance and also not included the conduction over the length W, the heat sink resistance would have been $R_S = 1.60$ °C/W, instead of the more appropriate value of $R = 2.00$ °C/W. We *have not* included the *thermal spreading resistance*, a contribution due to temperature gradients spreading outward from the base of the heat source. We will consider this contribution in Chapter 12.

11.8 FIN EFFICIENCY

Fin efficiency has been used several times in earlier chapters, but the derivation of any formulae has been avoided. It is now appropriate to address this matter in more detail using some of the results from Section 11.4. We shall continue to address only fins with a uniform cross-sectional area A_k.

The fin efficiency η is defined as

$$\eta = \frac{\text{Actual heat dissipation}}{\text{Maximum possible heat dissipation}} = \frac{Q}{Q_{Max}}$$

$$\eta = Q/\big(hA_{fin}T_0\big) \qquad\qquad \text{Fin, insulated tip} \quad (11.20)$$

$$\eta_{fin} = Q_S/\big(hA_{fin}T_0\big) \qquad\qquad \text{Fin, convecting tip}$$

$$\eta_L = Q_L/\big(h_L A_k T_0\big) \qquad\qquad \text{Tip, convecting tip} \qquad (11.21)$$

where A_{fin} excludes the fin tip, h_L is the heat transfer coefficient for the fin tip, and A_k is the fin tip area, which is also the fin cross-sectional area for conduction. The problem of deriving a fin efficiency becomes one of finding an expression for Q or Q_S and Q_L.

The most common formula is for a fin with an insulated tip. We will use the resistance given by Eq. (11.14) and the fin efficiency from Eq. (11.20).

$$R = R_S\sqrt{R_k/R_S}\,\coth\sqrt{R_k/R_S} = \sqrt{R_k R_S}\,\coth\sqrt{R_k/R_S}$$

$$\eta = Q/\big(hA_{fin}T_0\big) = (T_0/R)/\big(hA_{fin}T_0\big) = R_S/R = R_S/\sqrt{R_k R_S}\,\coth\sqrt{R_k/R_S}$$

Then our insulated fin tip efficiency is

$$\boxed{\eta = \sqrt{R_S/R_k}\ \tanh\sqrt{R_k/R_S}}\qquad\text{Insulated tip}\ (11.22)$$

The other case that you should be aware of is for those instances where the fin tip area is not negligible, and the heat transfer coefficient h_L is different from the channel h. When the problem is natural convection driven, the best choice for h_L is probably the vertical flat plate heat transfer coefficient which we have denoted as h_H at other places in this book. We begin by determining η_s, the fin efficiency excluding the tip. Using Eqs. (11.19) for Q_s and (11.21) for η_{fin}, where A_{fin} is the surface area from $x = 0$ to $x = L$ excluding the tip, we have

$$\eta_{fin} = Q_S\Big/\big(hA_{fin}T_0\big)$$

$$\eta_{fin} = \frac{1}{hA_{fin}T_0}\left(\frac{T_0}{\sqrt{R_kR_S}}\right)\left[\sinh\sqrt{\frac{R_k}{R_S}} + \left(1-\cosh\sqrt{\frac{R_k}{R_S}}\right)\left(\frac{\dfrac{\cosh\sqrt{\dfrac{R_k}{R_S}}}{R_L} + \dfrac{\sinh\sqrt{\dfrac{R_k}{R_S}}}{\sqrt{R_kR_S}}}{\dfrac{\sinh\sqrt{\dfrac{R_k}{R_S}}}{R_L} + \dfrac{\cosh\sqrt{\dfrac{R_k}{R_S}}}{\sqrt{R_kR_S}}}\right)\right]$$

The algebra is not shown, but the result is

$$\boxed{\eta_{fin} = \sqrt{\frac{R_S}{R_k}}\left\{\frac{\sqrt{\dfrac{R_S}{R_k}}\sinh\sqrt{\dfrac{R_k}{R_S}} - \dfrac{R_S}{R_L}\left(1-\cosh\sqrt{\dfrac{R_k}{R_S}}\right)}{\dfrac{R_S}{R_L}\sinh\sqrt{\dfrac{R_k}{R_S}} + \sqrt{\dfrac{R_S}{R_k}}\cosh\sqrt{\dfrac{R_k}{R_S}}}\right\}}\qquad\begin{array}{l}\text{Fin, convect-}(11.23)\\ \text{ing tip}\end{array}$$

Using Eqs. (11.18) and (11.22) and $R_L = 1/h_L A_k$, we have

$$\eta_L = Q_L\Big/\big(h_L A_k T_0\big) = -\left(\frac{1}{h_L A_k T_0}\right)\left(\frac{T_0}{\sqrt{R_kR_S}}\right)\left\{\sinh\sqrt{\frac{R_k}{R_S}} - \left[\frac{\left(\dfrac{\cosh\sqrt{\dfrac{R_k}{R_S}}}{R_L}\right) + \left(\dfrac{\sinh\sqrt{\dfrac{R_k}{R_S}}}{\sqrt{R_kR_S}}\right)}{\left(\dfrac{\sinh\sqrt{\dfrac{R_k}{R_S}}}{R_L}\right) + \left(\dfrac{\cosh\sqrt{\dfrac{R_k}{R_S}}}{\sqrt{R_kR_S}}\right)}\right]\cosh\sqrt{\frac{R_k}{R_S}}\right\}$$

$$\boxed{\eta_L = \frac{1}{\left(\dfrac{R_S}{R_L}\right)\sqrt{\dfrac{R_k}{R_S}}\sinh\sqrt{\dfrac{R_k}{R_S}} + \cosh\sqrt{\dfrac{R_k}{R_S}}}}\qquad\begin{array}{l}\text{Tip, convect-}\ (11.24)\\ \text{ing tip}\end{array}$$

So that for one fin in a $T_A = 0$ ambient,

$$\boxed{Q = \left(\eta_{fin} hA_{fins} + \eta_L h_L A_k\right)T_0}$$ Convecting tip (11.25)

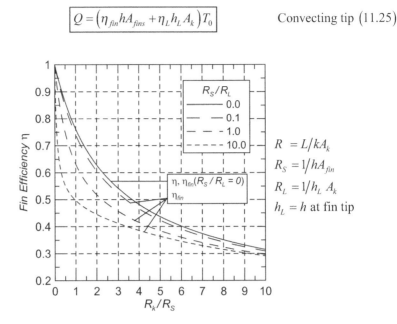

Figure 11.15. Fin efficiency η, Eq. (11.22), and η_{fin}, Eq. (11.23).

The fin efficiency formula, Eq. (11.23), for a fin with a convecting tip becomes Eq. (11.22), the efficiency for a fin with a non-convecting tip when $R_L \to \infty$, just as we would expect. However, Eq. (11.24), for the tip of a convecting tip fin does not converge to zero. The result for the total heat transfer in this latter condition, Eq. (11.25), is correct because $h_L = 0$. It is very unlikely that the reader will find reason to use Eqs. (11.23)-(11.25) because the tip area is most likely to be negligible. However, should one write software for a user base that might consider short, stubby fins where the tip is not negligible, the efficiency equations are available. We won't delve into an application of Eq. (11.22) as several examples are shown in earlier chapters.

We consider one final aspect of the fin efficiency problem. Suppose that we set a criteria for a fin efficiency $\eta \geq 0.95$. Then using appropriate variable names according to Figure 9.1, and following Figure 11.15, we can state

$$R_k / R_S \leq 0.2, \left(\frac{L}{kHt_f}\right) \Big/ \left(\frac{1}{2hHL}\right) \leq 0.2, \frac{2hL^2}{kt_f} \leq 0.2$$

$$t_f \geq 10hL^2/k$$ Min. fin thickness (11.26)

Suppose we consider a fin with a total heat transfer coefficient of about $h = 0.005$ W/(in.$^2 \cdot$ °C), roughly appropriate for natural convection and radiation, $L = 1.0$ in., and $k = 5.0$ W/(in. $\cdot$ °C) for aluminum.

$$t_f \geq 10hL^2/k = 10(0.005)(1.0)^2/5.0 = 0.01 \text{in.}$$

Extrusions will have fin thicknesses that are at least 10 times this computed value, so the fin thickness would not be a constraint. However, in the case of metal stampings or other manufacturing methods, you should be aware of the issue.

11.9 DIFFERENTIAL EQUATIONS FOR MORE THAN ONE DIMENSION

A small volume element $\Delta V = \Delta x \Delta y \Delta z$ is illustrated in Figure 11.16 which has a heat generation per unit volume with units of W/in.[3]. We apply Fourier's law of conduction, Eq. (11.1), to each of the six faces a, b, c, d, e, and f.

$$Q_x = -k_x \Delta y \Delta z (\partial T / \partial x) \text{ at } c$$
$$Q_{x+\Delta x} = -k_x \Delta y \Delta z (\partial T / \partial x) \text{ at } d$$
$$Q_y = -k_y \Delta x \Delta z (\partial T / \partial y) \text{ at } a$$
$$Q_{y+\Delta y} = -k_y \Delta x \Delta z (\partial T / \partial y) \text{ at } b$$
$$Q_z = -k_z \Delta x \Delta y (\partial T / \partial z) \text{ at } e$$
$$Q_{z+\Delta z} = -k_z \Delta x \Delta y (\partial T / \partial z) \text{ at } f$$

Application of conservation of energy requires that, in a steady-state condition, i.e., no energy storage, *heat in = heat out*. Adopting the convention that a source density of *generation*, Q_V, is positive, conservation of energy requires

$$\left[-k_x \Delta y \Delta z (\partial T / \partial x) \right]_c - \left[-k_x \Delta y \Delta z (\partial T / \partial x) \right]_d +$$
$$\left[-k_y \Delta x \Delta z (\partial T / \partial y) \right]_a - \left[-k_y \Delta x \Delta z (\partial T / \partial y) \right]_b +$$
$$\left[-k_z \Delta x \Delta y (\partial T / \partial z) \right]_e - \left[-k_z \Delta x \Delta y (\partial T / \partial z) \right]_f + Q = 0$$

Dividing both sides by $\Delta x \Delta y \Delta z$, taking the limit of $\Delta V \to 0$, and recalling from basic calculus that

$$\lim_{\Delta x \to 0} (1/\Delta x)(k_x \, \partial T / \partial x) \Big|_c^d = \lim_{\Delta x \to 0} (1/\Delta x)(k_x \, \partial T / \partial x) \Big|_{x,y,z}^{x+\Delta x, y, z} = \frac{\partial}{\partial x} \left(k_x \frac{\partial T}{\partial x} \right)$$
$$\text{and } Q_V = \lim_{\Delta V \to 0} (Q / \Delta V)$$

The energy balance with the limit $\Delta V \to 0$ leads us to the partial differential equation for steady-state heat conduction with orthotropic conductivity.

$$\boxed{\frac{\partial}{\partial x} \left(k_x \frac{\partial T}{\partial x} \right) + \frac{\partial}{\partial y} \left(k_y \frac{\partial T}{\partial y} \right) + \frac{\partial}{\partial z} \left(k_z \frac{\partial T}{\partial z} \right) = -Q_V}$$

Steady-state *PDE* (11.27)

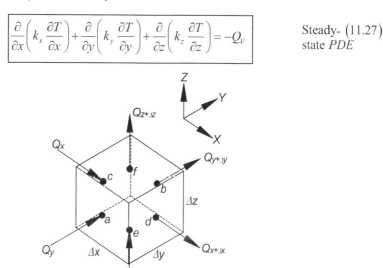

Figure 11.16. Volume element used for heat balance.

Not all practical problems are steady-state. The temperature/time relationship of a component or instrument may be of interest in some time-dependent fashion. The theoretical basis of quantifying this is accomplished by recognizing that the steady-state energy balance must be modified to account for energy storage or discharge for a volume density ρ and specific heat capacity c, i.e.,

$$\frac{\partial}{\partial x}\left(k_x \frac{\partial T}{\partial x}\right) + \frac{\partial}{\partial y}\left(k_y \frac{\partial T}{\partial y}\right) + \frac{\partial}{\partial z}\left(k_z \frac{\partial T}{\partial z}\right) = -Q_V + \rho c \frac{\partial T}{\partial t}$$

Time (11.28)
dep. *PDE*

Solutions of Eqs. (11.27) and (11.28) are addressed in later chapters.

For the sake of completeness, the steady-state heat conduction (for isotropic k) *PDEs* in spherical and cylindrical coordinates are written as Eqs. (11.29) and (11.30), respectively.

$$\frac{1}{r^2}\frac{\partial}{\partial r}\left(r^2 \frac{\partial T}{\partial r}\right) + \frac{1}{r^2 \sin\theta}\frac{\partial}{\partial \theta}\left(\sin\theta \frac{\partial T}{\partial \theta}\right) + \frac{1}{r^2 \sin^2\theta}\frac{\partial^2 T}{\partial \phi^2} = -\frac{Q_V}{k}$$

Spher. (11.29)
coord.

$$\frac{1}{\rho}\frac{\partial}{\partial \rho}\left(\rho \frac{\partial T}{\partial \rho}\right) + \frac{1}{\rho^2}\frac{\partial^2 T}{\partial \phi^2} + \frac{\partial^2 T}{\partial z^2} = -\frac{Q_V}{k}$$

Cyl. coord., (11.30)
iso. k

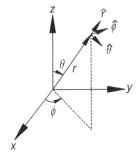

Figure 11.17. Spherical coordinates.

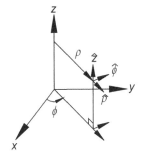

Figure 11.18. Cylindrical coordinates.

11.10 PHYSICS OF THERMAL CONDUCTIVITY OF SOLIDS

Thermal design often requires little more knowledge of thermal conductivity than either a single value or a few values covering a small temperature range. However, the reader may find himself/herself considering low temperature, i.e., cryogenic applications where at the very least, some knowledge of the cause and effects of temperature is useful. In the following few paragraphs you will read about this topic.

We were reminded in Chapter 1 that we should expect to encounter materials with thermal conductivities that range from a low value of about 1×10^{-4} to an upper value of about 10 W/(in. · °C), but we didn't explore the variation with temperature. For example, Figure 11.19 displays the conductivity for temperatures from at least as low as 20K to about 500K. We see that copper and aluminum have maximum conductivities between 20 and 30K but are also rather temperature-insensitive at room temperature and above. A more detailed plot of conductivity for copper with impurities shows a very pronounced peak at about 20K and again, temperature insensitivity near room temperature. A little explanation is in order. Readers interested in more detail should consult Rosenberg (1963).

When considering metallic solids, it is known that free electrons are the principal heat carriers. In this case, it is more convenient to work with the thermal resistivity defined as

$$\rho = 1/k$$

Resistance to heat flow in a metal is found to be described by "Matthiessen's rule":

$$\rho = \rho_l + \rho_i$$

$$\rho_l \equiv \text{defect resistivity}$$

$$\rho_i \equiv \text{intrinsic resistivity}$$

where it seems reasonable that when two or more effects are in play, the resistances would combine in a *series fashion*. The defect resistivity is due to scattering of the electronic heat carriers (electrons) from lattice defects and impurities and is typically described by

$$\rho_l = A/T'$$

where A is a constant and the temperature T is primed to denote an absolute scale. The intrinsic resistivity is due to scattering of the electronic heat carriers from the lattice and is typically described by (B is a constant)

$$\rho_i = BT'^2$$

The net thermal conductivity is therefore

$$k = \frac{1}{\rho} = \frac{1}{\left(A/T'\right) + BT'^2} \qquad \text{Metal conductivity} \quad (11.31)$$

which is used to describe the portion of the thermal conductivity curve over the range shown in Figure 11.20 for copper. The defect portion dominates conduction from the lowest temperature region up to the vicinity of the conductivity maximum. The purer the metal, the greater the peak, i.e., the maximum conductivity decreases as defects and impurities are added to the metal. The effects of defects and lattice scattering are equal at the temperature where the conductivity maximum occurs. The lattice scattering term dominates at temperatures from the conductivity maximum and greater and as we have seen, the conductivity changes little from about 100 to 300K. It is usually very difficult to fit a single formula such as Eq. (11.31) over the full range of 4 to 300K, but it is usually possible to fit $k(T')$ up to something less than room temperature, e.g., 100K for copper.

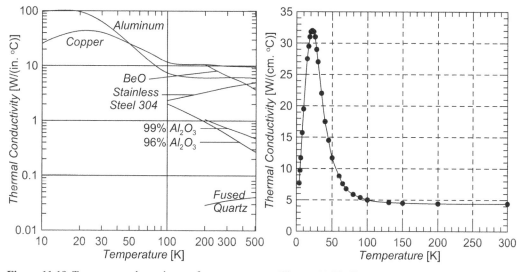

Figure 11.19. Temperature dependence of some common materials. Data from a variety of sources.

Figure 11.20. Temperature dependence of copper with imperfections at low temperature. The filled circles are data from Childs et al. (1973).

Heat conduction in a nonmetallic, crystalline material is due to transfer of the lattice vibrational energy from a higher to a lower temperature. Solid state physicists describe the vibrational nature in terms of a quantized quantity called a *phonon*. Just as electromagnetic propagation can be described by photons, vibrational energy in a crystal can be described by phonons.

The resistance to thermal energy transfer by phonons in the lattice is largely due to two mechanisms: (1) geometrical scattering of the phonons from the crystal boundary and also from lattice imperfections; and (2) scattering of phonons by other phonons. It can be shown that if the forces between atomic entities in the crystal are those of a harmonic oscillator, i.e., the restorative force is $F = -a\Delta x$, then there is no mechanism for collisions between different phonons and the thermal resistivity is determined only by collisions of a phonon with a crystal boundary or imperfection. If the forces between atomic entities in the crystal are those of an anharmonic oscillator, e.g.,

$$F = -a\Delta x - b\Delta x^2 - c\Delta x^3 - \dots$$

then there is a phonon-phonon interaction and a contribution to the thermal resistivity of the crystal. Mathematical formulation of this problem is difficult for all but the simplest of structures. Nevertheless, the temperature dependence of nonmetallic insulators follows a formula something like Eq. (11.32).

$$k = 1 / \left[\left(\alpha / T'^3 \right) + \beta T' \right] \qquad \text{Nonmetallic crystal } (11.32)$$

The α / T'^3 term is due to phonon scattering with the crystal boundary and lattice defects and is important in a rather low-temperature region. The $\beta T'$ term is due to phonon-phonon scattering and is dominant in a higher temperature region, i.e., above the temperature where the thermal conductivity maximum occurs. Figure 11.21 shows the low-temperature conductivity temperature dependence for some well-known materials and Figure 11.22 is a plot of the conductivity for *undoped* silicon and germanium. Semiconductors are reported to have the same heat conduction mechanism as dielectrics, i.e., heat transfer via the lattice or phonons.

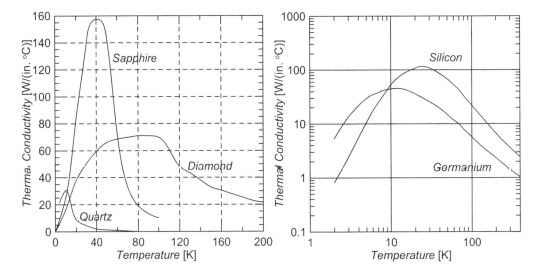

Figure 11.21. Low-temperature thermal conductivity for three nonmetallic crystals.

Figure 11.22. Low-temperature thermal conductivity for *undoped* silicon and germanium. Data from *Thermophysical Properties of Matter, TPRC Data Series*, ed. Touloukian, Y.S. et al. (1970).

11.11 THERMAL CONDUCTIVITY OF CIRCUIT BOARDS (EPOXY-GLASS LAMINATES)

The role of circuit board thermal conductivity is important and sometimes overlooked. The effect of conductivity becomes very apparent when calculating a component temperature where heat is conducted to the board, but we will not study that aspect here as it would be a distraction from our main topic.

There have been numerous papers published on epoxy-metal-laminate conductivities. Two different methods are examined here; both assume that whatever the laminate structure may be, we wish to incorporate the effects of both the low conductivity epoxy-glass and the high conductivity metal, which we assume is copper. Because of the typically layered structure of the materials, the conductivity is different for the in-plane and through-plane directions.

The first method we list is due to Graebner and Azar (1995), where they provide empirically based correlations.

Graebner and Azar conclusions are as follows:

1. The thermal resistance at the copper and epoxy-glass interfaces is negligible.
2. The effect of continuous layers of copper, either on the surface or embedded, was found to be far more important than that of topical circuitry.
3. The conductivity within each copper layer is close to bulk copper, i.e., $k_{Cu-layer} = 9.78 \pm 0.38 \, \text{W}/(\text{in.} \cdot {}^\circ\text{C})$.
4. The conductivity of epoxy-glass is $k_{E-G} = 0.015 \, \text{W}/(\text{in.} \cdot {}^\circ\text{C})$.
5. Graebner and Azar recommend a k_P (planar or in-plane) and k_N (through-plane or normal) so that

$$k_P \left[\text{W}/(\text{in.} \cdot {}^\circ\text{C}) \right] = 0.0203 + 8.89 \left(t_{Cu}/t \right) \tag{11.33}$$

$$k_N \left[\text{W}/(\text{in.} \cdot {}^\circ\text{C}) \right] = \frac{1}{66.53 \left(1 - t_{Cu}/t \right) + 0.10 \left(t_{Cu}/t \right)} \tag{11.34}$$

t_{Cu} = total thickness of all continous Cu layers, t = total PCB thickness

A second method is one where the metallic and dielectric materials are combined in a parallel or series resistive/conductive fashion based on the volume fraction that each material occupies. The arithmetic can be tedious, but the method is very flexible. We have three cases to consider:

1. In-plane and through-plane, one effective layer k_e combining two layered materials k_1, k_2 of equal thickness in a single layer t:

$$\left(k_e A/t \right) = \left(k_1 A_1/t \right) + \left(k_2 A_2/t \right), k_e = \left(k_1 A_1/A \right) + \left(k_2 A_2/A \right)$$

$$k_e = k_1 f_1 + k_2 f_2 \qquad \text{One layer, two} \quad (11.35)$$
$$\text{materials}$$

where $f_1 = A_1/A$, $f_2 = A_2/A$ may be considered volume fractions.

2. In-plane, one effective layer k_P combining multiple layers of differing thickness over identical distance l, width w:

$$\left(k_P wt/l \right) = \left(k_1 wt_1/l \right) + \left(k_2 wt_2/l \right)$$

$$k_P = k_1 f_1 + k_2 f_2 \qquad \text{Multiple layers,} \quad (11.36)$$
$$\text{differing thickness}$$

where $t = t_1 + t_2$, $f_1 + f_2 = 1$, and $f_1 = t_1/t$, $f_2 = t_2/t$, may be considered volume fractions.

3. Through-plane, one effective layer k_N combining multiple layers of differing conductivity and thickness:

$$t/\left(k_N A \right) = \left[t_1/\left(k_1 A \right) \right] + \left[t_2/\left(k_2 A \right) \right]$$

$$1/k_N = \left(t_1/t \right)/k_1 + \left(t_2/t \right)/k_2 = \left(f_1/k_1 \right) + \left(f_2/k_2 \right)$$

$$k_N = 1/\left[\left(f_1/k_1 \right) + \left(f_2/k_2 \right) \right] \qquad \text{Multiple layers,} \quad (11.37)$$
$$\text{differing thickness}$$
$$\text{and} \quad t = t_1 + t_2, f_1 + f_2 = 1.$$

A third method, sometimes used by Yovanovich, uses a harmonic mean conductivity for a single value that does not distinguish between the in-plane and through-plane directions.

$$k_M = 2k_P k_N / (k_P + k_N)$$ Harmonic mean (11.38)

Application of Eq. (11.38) still requires that the in-plane and through-plane conductivities be calculated using one of the previously described methods.

Equation (11.35) is best used for combining two or more materials considered to be in the same single layer. However, Eq. (11.33) with (11.34) or Eq. (11.36) with (11.37) are best used for combining several layers into the single pair k_P, k_N for the entire circuit board. Note that using k_P, k_N requires that your model and software admit use of an orthotropic thermal conductivity, which is easily done with a thermal network methodology and most, if not all, commercial finite element software. You would only use Eq. (11.38) if an orthotropic conductivity model is not available to you. These issues are examined in a single example in the next section.

11.12 APPLICATION EXAMPLE: EPOXY-GLASS CIRCUIT BOARD WITH COPPER

In Figure 11.23, several different conduction models are illustrated. Figures 11.23 *(a)* and *(e)* indicate the exact layer arrangements: *(a)* layer 1 with 50% copper "*Cu*" and 50% epoxy-glass "*E-G*", layer 2 with 100% *E-G*, layer 3 with 100% *Cu*, layer 4 with 100% *E-G*; *(e)* layer 1 with 100% *E-G*, layer 2 with 100% *Cu*, layer 3 with 100% *E-G*, layer 4 with 50% *Cu* and 50% *E-G*. Figures *(a)* and *(e)* were solved nearly exactly with the *TAMS* program (Ellison, 1984a) for a 1.0 in. × 1.0 in. square *E-G* coupon with a center square surface source of different source/*PCB* edge length ratios $\Delta x/a =$ 0.01 to 1.0. The exact solutions exactly modeled the layered structure and used both top/bottom heat transfer coefficients either $h = 0.005$ or 1.0 W/(in.² · °C).

Figure 11.23 *(b)* is applicable to a thermal network model with top/bottom surface nodes using Eq. (11.35) for layer 1 and Eqs. (11.36) and (11.37) for k_P, k_N, respectively, to combine all layers. Figure 11.23 *(c)* is applicable to an *FEA* model with layers of orthotropic elements, and Figure 11.23 *(d)* is applicable to an *FEA* model with one or more layers of isotropic elements or a thermal network model with one layer of nodes.

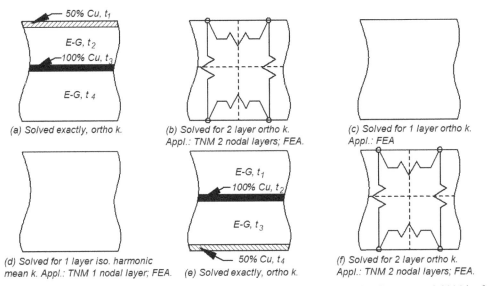

(a) Solved exactly, ortho k.

(b) Solved for 2 layer ortho k.
Appl.: TNM 2 nodal layers; FEA.

(c) Solved for 1 layer ortho k.
Appl.: FEA

(d) Solved for 1 layer iso. harmonic
mean k. Appl.: TNM 1 nodal layer; FEA.

(e) Solved exactly, ortho k.

(f) Solved for 2 layer ortho k.
Appl.: TNM 2 nodal layers; FEA.

Figure 11.23. Edge view of *PCB* thermal conductivity model. *(a)* Exact solution for *Cu* top: $t_1 = 0.0014$ in. for 50% *Cu*, $t_2 = 0.03$ in., $t_3 = 0.0014$ in. for 100% *Cu*, $t_4 = 0.03$ in. *(b)* - *(d)* models applicable to *Cu* top. *(e)* Exact solution for *Cu* bottom: $t_1 = 0.03$ in., $t_2 = 0.0014$ in. for 100% *Cu*, $t_3 = 0.03$ in., $t_4 = 0.0014$ in. for 50% *Cu*. *(c)*, *(d)* and, *(f)* models applicable to *Cu* bottom.

We shall calculate the model conductivities out of order. The calculations for k_P, k_N, and k_M are first calculated for models (c) and (d). Using the Azar method, Eqs. (11.33) and (11.34) we have

$$k_P = 0.0203 + 8.89\left(t_{Cu}/t\right) = 0.0203 + 8.89\left[\frac{0.0014 + (0.5)(0.0014)}{0.0628}\right] = 0.318\,\mathrm{W/(in.\cdot°C)}$$

$$k_N = \frac{1}{66.53\left(1 - t_{Cu}/t\right) + 0.10\left(t_{Cu}/t\right)}$$

$$k_N = \frac{1}{66.53\left[1 - \dfrac{0.0014 + (0.5)(0.0014)}{0.0628}\right] + 0.10\left[1 - \dfrac{0.0014 + (0.5)(0.0014)}{0.0628}\right]} = 0.016\,\mathrm{W/(in.\cdot°C)}$$

Using the "parallel-series" method, Eqs. (11.35), (11.36), and (11.37) we calculate

Layer 1 - $k_1 = k_{E-G1}f_{E-G1} + k_{Cu1}f_{Cu1} = (0.015)(0.5) + (10.0)(0.5) = 5.01\,\mathrm{W/(in.\cdot°C)}$

Layer 2 - $k_2 = 0.015\,\mathrm{W/(in.\cdot°C)}$; Layer 3 - $k_3 = 10.0\,\mathrm{W/(in.\cdot°C)}$
Layer 4 - $k_4 = 0.015\,\mathrm{W/(in.\cdot°C)}$

Now putting it together for the entire PCB,

$$t = t_1 + t_2 + t_3 + t_4 = 0.0014 + 0.030 + 0.0014 + 0.03 = 0.063$$
$$f_1 = t_1/t = 0.0014/0.063 = 0.022;\quad f_2 = t_2/t = 0.03/0.063 = 0.478$$
$$f_3 = t_3/t = 0.0014/0.063 = 0.022;\quad f_4 = t_4/t = 0.03/0.063 = 0.478$$
$$k_P = k_1f_1 + k_2f_2 + k_3f_3 + k_4f_4 = (5.01)(0.022) + (0.015)(0.478)$$
$$+ (10.0)(0.022) + (0.015)(0.478) = 0.348\,\mathrm{W/(in.\cdot°C)}$$
$$k_N = \frac{1}{\left(f_1/k_1\right) + \left(f_2/k_2\right) + \left(f_3/k_3\right) + \left(f_4/k_4\right)} = 0.015\,\mathrm{W/(in.\cdot°C)}$$

We see that the k_P for Azar and parallel-series give us 0.318 and 0.348, respectively. The k_N for Azar and parallel-series give us 0.016 and 0.015, respectively. The two methods agree quite favorably. This writer is using an arbitrary preference for the series-parallel results for the remainder of the problem, thus we have all we need for model (c) in Figure 11.23.

The harmonic mean conductivity for model (d) in Figure 11.23 is calculated as

$$k_M = 2k_Pk_N/(k_P + k_N) = 2(0.348)(0.015)/(0.348 + 0.015) = 0.0288\,\mathrm{W/(in.\cdot°C)}$$

Now we return to model (b) for Figure 11.23 where we calculate a separate k_P for each of the two layers where we put the 50% Cu and 50% E-G into a "top" layer and the 100% Cu and 100% E-G into a "bottom" layer.

Top layer, where we use $k_{1\,Top} = 5.01$ from the preceding top 50% Cu layer calculation:

$$t_{1Top} = 0.0014, t_{2Top} = 0.030, t_{Top} = t_{1Top} + t_{2Top} = 0.0314$$
$$f_{1Top} = t_{1Top}/t_{Top} = 0.0014/0.0314 = 0.045$$
$$f_{2Top} = t_{2Top}/t_{Top} = 0.030/0.0314 = 0.995$$
$$k_{PTop} = k_{1Top}f_{1Top} + k_{2Top}f_{2Top} = (5.01)(0.045) + (0.015)(0.995) = 0.237\,\mathrm{W/(in.\cdot°C)}$$

Bottom layer:

$$t_{1Bot} = 0.0014, t_{2Bot} = 0.030, t_{Bot} = t_{1Bot} + t_{2Bot} = 0.0314$$

$$f_{1Bot} = t_{1Bot}/t_{Bot} = 0.0014/0.0314 = 0.045$$

$$f_{2Bot} = t_{2Bot}/t_{Bot} = 0.030/0.0314 = 0.995$$

$$k_{PBot} = k_{1Bot}f_{1Bot} + k_{2Bot}f_{2Bot} = (10.0)(0.045) + (0.015)(0.995) = 0.459 \, \text{W}/(\text{in.} \cdot {}^{\circ}\text{C})$$

We interconnect the top and bottom *PCB* surfaces with the $k_N = 0.015$ from the preceding series-parallel calculation.

In model (*e*), the 50% *Cu*, 50% *E-G* layer has been moved from the top to the bottom surface. Then model (*f*) is appropriate for thermal network models with top/bottom surface nodes. It is not necessary to recalculate k_P and k_N for the top and bottom layers as we can just take the conductivities from model (*b*) and apply them to the reverse layers.

The various model results are compared with the *exact* solution in Figures 11.24 through 11.27. It is difficult to arrive at a single conductivity model that is optimum for the cases considered. The author's opinion is that none of the models agree well with the "exact" calculations for sources that are smaller than the substrate. The results in Figures 11.24 and 11.25 show that when there are highly conducting planes in or very near the source plane, none of the conductivity models work well. However, when any highly conducting planes are remote from the source plane, as we see in Figures 11.26 and 11.27, the one-layer harmonic mean model is best. Remember that with circuit boards, we are dealing with a largely single material with highly conducting planes with a discontinuous conductivity in the normal direction. Perhaps the best option in a network or *FEA* model is to build the substrate spreading resistance into each lead resistance, probably via a diminished lead conductivity.

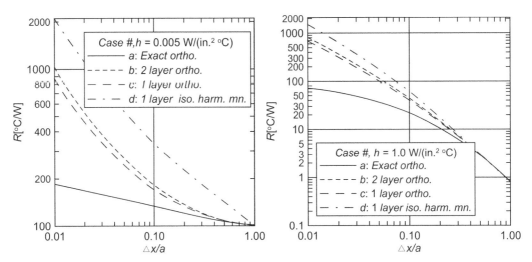

Figure 11.24. Comparison of various thermal conductivity models for a square source on a square substrate (*PCB*) with four layers, 50% *Cu* top layer, $h_{Top} = h_{Bot} = 0.005$ W/(in.$^2 \cdot {}^{\circ}$C).

Figure 11.25. Comparison of various thermal conductivity models for a square source on a square substrate (*PCB*) with four layers, 50% *Cu* top layer, $h_{Top} = 0.005$, $h_{Bot} = 1.0$ W/(in.$^2 \cdot {}^{\circ}$C).

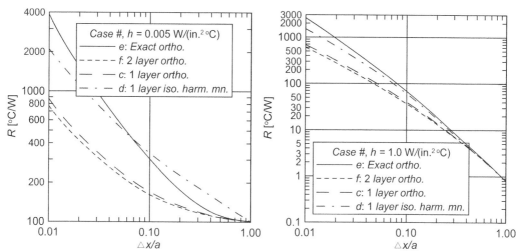

Figure 11.26. Comparison of various thermal conductivity models for a square source on a square substrate (*PCB*) with four layers, 50% *Cu* bottom layer, $h_{Top} = h_{Bot} = 0.005$ W/(in.² · °C).

Figure 11.27. Comparison of various thermal conductivity models for a square source on a square substrate (*PCB*) with four layers, 50% *Cu* bottom layer, $h_{Top} = 0.005$, $h_{Bot} = 1.0$ W/(in.² · °C).

11.13 THERMAL INTERFACE RESISTANCE

When two materials - either of the same kind or different - are placed together, a temperature difference is found to exist across the interface. This interface, illustrated in Figure 11.28, is a complex physical entity with some points of contact between the two materials and areas of noncontact gap regions which may be vacuum, gas, or filled with some sort of interstitial material. The latter is usually some intentionally introduced material intended to enhance the poorly conducting gas gaps. The effect of concave or convex surfaces certainly contributes to the temperature difference across the interface, but is not usually considered within the framework of any mathematical treatment.

Whether we are interested in a theoretical treatment or empirical data, there are a few definitions that are common to both.

$$h_c \equiv \text{contact conductance, } \left[\text{W}/\left(\text{in.}^2 \cdot {}^\circ\text{C} \right) \right]$$

$$h_g \equiv \text{gap conductance, } \left[\text{W}/\left(\text{in.}^2 \cdot {}^\circ\text{C} \right) \right]$$

$$h_i \equiv \text{interface conductance} = h_c h_g / \left(h_c + h_g \right), \left[\text{W}/\left(\text{in.}^2 \cdot {}^\circ\text{C} \right) \right]$$

$$A_a \equiv \text{apparent interface area, } \left[\text{in.}^2 \right]$$

$$C \equiv \text{conductance} = \left(h_c + h_g \right) A_a, \left[\text{W}/{}^\circ\text{C} \right]$$

$$R \equiv \text{resistance} = 1/C = 1/\left(h_c + h_g \right) A_a, \left[{}^\circ\text{C}/\text{W} \right]$$

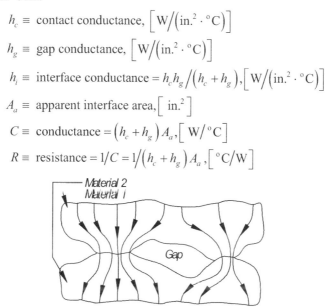

Figure 11.28. Mechanical joint with contact regions and interstitial gaps.

We note that h_c is subscripted the same way that we identified convective heat transfer coefficients in earlier chapters. In fact, even the units of a contact conductance are identical to those used for convective heat transfer coefficients. This should cause little confusion to the reader as the context of the discussion should make the issue clear.

The literature is replete with articles on the interface topic. One of the very best is the review article by Yovanovich (2005) in which he gives a detailed description of one of his lifelong and successful research activities. In the next few paragraphs, the flavor of Yovanovich's work is offered as an introduction to further reading on your part. However, extensive details of Yovanovich's theory is beyond the scope of this book. We shall instead, closely follow a shorter version (Yovanovich et al., 1997) which is limited to rough surfaces.

The contact conductance between two solid surfaces is given by

$$h_c = 9.22 k_M R_a^{-0.598} \left(\frac{P}{H} \right)^{0.95} \quad \frac{\text{W}}{\text{in.}^2 \cdot {}^\circ\text{C}} \quad \text{Contact cond., (11.39)}$$
$$\text{metals}$$

The independent variables are

k_M = harmonic mean thermal conductivity for surfaces 1 and 2

$k_M = 2 k_1 k_2 / (k_1 + k_2), \left[\text{W} / (\text{in.} \cdot {}^\circ\text{C}) \right]$

R_a = combined average roughness

$R_a = \sqrt{R_{a_1}^2 + R_{a_2}^2}, [\text{in.}]$

P = contact pressure

H = surface microhardness

Yovanovich has derived a gap conductance

$$h_g = \frac{k_g}{1.53 R_a \left(\dfrac{P}{H} \right)^{-0.097} + M_0 \left(\dfrac{T'_g}{T'_{g,0}} \right) \left(\dfrac{P_{g,0}}{P_g} \right)} \quad \frac{\text{W}}{\text{in.}^2 \cdot {}^\circ\text{C}} \quad \text{Gap cond., (11.40)}$$
$$\text{metals}$$

where

k_g – conductivity of the gap gas, $\left[\text{W} / (\text{in.} \cdot {}^\circ\text{C}) \right]$

$T'_{g,0}$ = absolute gas reference temperature, $[\text{K}]$

T'_g = absolute gas temperature, $[\text{K}]$

H = surface microhardness

$P_{g,0}$ = gas pressure referenced to $T'_{g,0}$, [atm]

P_g = gas pressure at T_g, [atm]

M_0 = gas parameter at reference values of $P_{g,0}, T'_{g,0}, [\text{in.}]$

Yovanovich et al. (1997) provide some useful estimates of M_0:

Air: $k_g = 6.7 \times 10^{-4} \text{ W} / (\text{in.} \cdot {}^\circ\text{C}), M_0 = 1.469 \times 10^{-5} \text{ in.}$

Helium: $k_g = 3.8 \times 10^{-3} \text{ W} / (\text{in.} \cdot {}^\circ\text{C}), M_0 = 8.07 \times 10^{-5} \text{ in.}$

Thermal grease: $k_g =$ (see Table 11.2), $M_0 = 0$

The full theory for hard materials (Yovanovich et al., 2005) is complex, but different material types, e.g., elastomers, require yet another theory. Utilization of test data, often provided by heat sink and heat sink material vendors may well be the most practical approach for design engineers. There are several excellent articles, easily obtained via the Internet (www.electronics-cooling.com), that the interested reader should consult: Marotta et al. (2002); Blazej (2003); Mahajan et al. (2004); Rantala (2004); and Saums (2007).

The most important resource regarding thermal interface materials is the extensive amount of vendor-specific test data that is readily accessed on the Internet. However, as a start-point, Table 11.2 is provided as a short list of thermal interface conductances (take care to note the units for the interface conductance) for a few materials and metal-to-metal joints. These values are very approximate, thus you must take care if you choose to use any of this data. The joint thickness t and contact pressure P are included where relevant.

11.14 APPLICATION EXAMPLE: INTERFACE RESISTANCE FOR AN ALUMINUM JOINT

Now we will apply the Yovanovich theory from Section 11.13 to an aluminum joint. We will calculate one data point for a contact pressure $p = 100$ lb$_f$/in.^{2} and a combined average roughness $R_a = 120\,\mu$ in.

$$R_a = \sqrt{R_1^2 + R_2^2} = \sqrt{2}R_1 = \sqrt{2}\left(120\times10^{-6}\right) = 1.70\times10^{-4}\,\text{in.}$$

$$k_M = 2k_1 k_2/\left(k_1 + k_2\right) = 2\left[4.0\,\text{W}/\left(\text{in.}\cdot{}^\circ\text{C}\right)\right]^2\Big/\left[4.0\,\text{W}/\left(\text{in.}\cdot{}^\circ\text{C}\right) + 4.0\,\text{W}/\left(\text{in.}\cdot{}^\circ\text{C}\right)\right]$$

$$k_M = 4.0\,W/\left(\text{in.}\cdot{}^\circ\text{C}\right)$$

$$H = 1180\times10^6\,[\text{Pa}]\left(1.45\times10^{-4}\right)\left[\left(\text{lb}_f/\text{in.}^2\right)/\text{Pa}\right] = 1.71\times10^5\,\text{lb}_f/\text{in.}^2$$

$$h_c = 9.22 k_M R_a^{-0.598}\left(\frac{P}{H}\right)^{0.95} = \left(9.22\right)\left(4.0\right)^{-0.598}\left(1.70\times10^{-4}\right)^{-0.257}\left(\frac{100}{1.71\times10^5}\right)^{0.95}$$

$$h_c = 5.723\,\text{W}/\left(\text{in.}^2\cdot{}^\circ\text{C}\right)$$

$$1/h_c = 0.175\left({}^\circ\text{C}\cdot\text{in.}^2\right)/\text{W}$$

$$k_g = 6.7\times10^{-4}\,\text{W}/\left(\text{in.}\cdot{}^\circ\text{C}\right)$$

$$h_g = \cfrac{k_g}{1.53 R_a\left(\dfrac{P}{H}\right)^{-0.097} + M_0\left(\dfrac{T'_g}{T'_{g,0}}\right)\left(\dfrac{P_{g,0}}{P_g}\right)}$$

$$h_g = \cfrac{6.7\times10^{-4}}{1.53\left(1.7\times10^{-4}\right)\left(\dfrac{100}{1.71\times10^5}\right)^{-0.097} + 1.469\times10^{-5}\left(\dfrac{303}{323}\right)\left(\dfrac{P_{g,0}}{P_g}\right)} = 1.222\,\frac{\text{W}}{\text{in.}^2\cdot{}^\circ\text{C}}$$

$$1/h_g = 0.818\,{}^\circ\text{C}\cdot\text{in.}^2/\text{W}$$

$$1/\left(h_g + h_c\right) = 1/\left(1.222 + 5.723\right) = 0.144\,{}^\circ\text{C}\cdot\text{in.}^2/\text{W}$$

Table 11.2 Thermal interface conductance estimates.

Material	t [in.]	Roughness [μ in.]	P [lb$_f$/in.^{2}]	h_c [W/(in.$^2 \cdot$ °C)]
Thermal grease w/ceramic filler	0.002			13
Thermal grease w/metal filler	0.002			9-13
Epoxy w/filler	0.002			5-20
Tapes and pads	0.002			7-38
75S-T6 Aluminum to aluminum		10	0	4.8
		10	100	10.3
		10	200	13.2
		10	300	14.7
		10	400	16.1
		65	0	2.2
		65	100	5.5
		65	200	7.0
		65	300	8.8
		65	400	10.3
		120	0	0.9
		120	100	3.3
		120	200	5.1
		120	300	6.6
		120	400	8.4
Stainless steel to stainless steel		30	0	4.0
		30	100	5.5
		30	200	5.9
		30	300	6.2
		30	400	6.2
		100	0	1.4
		100	100	1.4
		100	200	2.0
		100	300	2.1
		100	400	2.1

Thermal grease estimates are obtained from miscellaneous sources and are not intended to be exact; use with care in design situations. The metal-to-metal interface data is taken from observation of plotted data from Barzelay (1955), for joints at a mean temperature of 93°C.

The values for the thermal resistivities that we just calculated are listed in the first line of Table 11.3. The resistivity $1/(h_g + h_c)$ is plotted in Figure 11.29 for three values of R_a and contact pressures of P from 0 to 500 lb$_f$/in.[2] Some test data points from Table 11.2 are also included in Figure 11.29. The test data and calculations do not agree to the extent that we usually like to see, but for a problem of this complexity, the calculated results are not too bad. They do give us the correct order of magnitude. In addition, the comparison does not reflect the true validity of the work by Yovanovich and his colleagues, as you will learn if you choose to study the theory in more detail. There is a large amount of test data that validates the Yovanovich theory.

The data in Table 11.3 shows that, at least in this case, the total resistivity is largely due to the contact contribution. The gap resistivity also shows a marked dependence on the roughness.

Table 11.3. Yovanovich theory for calculated
$1/(h_c + h_g)$, $1/h_c$, $1/h_g$ at $p = 100$ lb$_f$/in.[2]

R_a RMS [μ in.]	$1/(h_c + h_g)$ [°C·in.[2] / W]	$1/(h_c)$ [°C·in.[2] / W]	$1/(h_g)$ [°C·in.[2] / W]
120	0.144	0.175	0.818
65	0.095	0.121	0.453
10	0.027	0.039	0.087

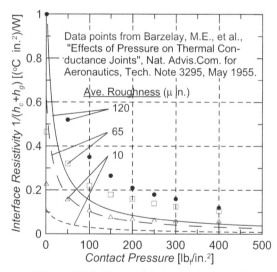

Figure 11.29. Yovanovich theory plotted with test data for thermal interface resistivity.

EXERCISES

11.1 Derive Eq. (11.5) for both parallel and series resistances.

11.2 A thermal network model is illustrated with numbered nodes (1-3) and node dimensions. The illustration shows a way to transition from the small center node to larger nodes. The dimensions are $w_1 = 0.25$ in., $w_2 = 0.75$ in., $l_1 = 0.25$ in., $l_2 = 0.25$ in., $l_3 = 0.5$ in., $t = 0.05$ in. The thermal conductivity is $k = 0.8$ W/(in.·°C). Calculate the thermal resistance from node 1 to node 2 and from node 2 to node 3. Hints: (1) Calculate the resistance from the center of node 1 to the boundary separating nodes 1 and 2, then calculate the resistance from that boundary to the center of node 2, finally adding these two resistances in series to obtain one of the desired results; (2) Use a similar procedure to calculate the node 2 to node 3 resistance.

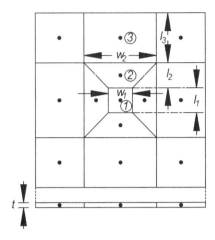

Exercise 11.2. Thermal conduction network model.

11.3 Derive Eq. (11.13).

11.4 Derive Eq. (11.16).

11.5 Thermal network problems, for which cylindrical coordinates are appropriate, use a familiar formula for the thermal conductance from the inner radius r_i to the outer radius r_o where the cylinder has a length or thickness Δz: $C = (2\pi k \Delta z)/\ln(r_0/r_i)$, which is a problem when the inner radius is $r_i = 0$. If the center region is a uniform source Q distributed throughout a volume of radius a and thickness Δz we can use $C = 4\pi k \Delta z$. Your task is to derive the preceding formula. Hint: Integrate the differential equation

$$\frac{1}{r}\frac{d}{dr}\left(r\frac{dT}{dr}\right) = -\frac{Q_V}{k}$$

and apply boundary conditions $\left[dT(r)/dr\right]\big|_{r=0} = 0;\ T(r)\big|_{r=a} = T_0$.

11.6 Derive a formula for the length-averaged temperature of a one-dimensional bar with a uniform source, conduction and convection, one end sinked.

11.7 Derive a formula for the length-averaged temperature of a one-dimensional bar with an end source, opposite end sinked, and conduction and convection.

11.8 In Application Example 11.7, one-half of a heat sink was analyzed, where symmetry was used to calculate a resistance for the entire heat sink, but analyzing only one half. In Figure Exercise 11.8, a column of power transistors is shown in a region void of fins. The total heat dissipation for the complete source is Q. We will use a source value of $Q/2$ in our half-model. The heat sink dimensions are $W = 4.0$ in., $L = 3.5$ in., $t_f = 0.1$ in., $S = 0.3$ in., $S_s = 1.5$ in., $H = 1.0$ in., $t = 0.2$ in., $Q = 250$ W, and a forced air approach speed $V_f = 200$ ft/min. You are asked to calculate the source-to-ambient thermal resistance and temperature rise for the heat sink surface beneath the power source column. Ignore the flat backside and use a flat plate heat transfer coefficient.

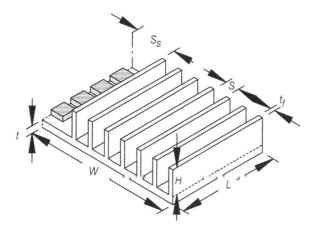

Exercise 11.8. Simple heat sink with column of
power transistors; one-half is analyzed using symmetry.

11.9 Following the derivations for an end source, end convection, length convection and conduction
that led to Eqs. (11.17), (11.18), and (11.19), show that the average temperature over the length L
is given by

$$\bar{T} = \frac{T_0}{\sqrt{R_k/R_s}} \left[\sinh\sqrt{R_k/R_s} - \left(\frac{\dfrac{\cosh\sqrt{R_k/R_s}}{R_L} + \dfrac{\sinh\sqrt{R_k/R_s}}{\sqrt{R_k R_s}}}{\dfrac{\sinh\sqrt{R_k/R_s}}{R_L} + \dfrac{\cosh\sqrt{R_k/R_s}}{\sqrt{R_k R_s}}} \right) \left(\cosh\sqrt{R_k/R_s} - 1 \right) \right]$$

11.10 Consider an array of fins with nomenclature according to Exercise 11.8. Suppose that the dimensions of a fin are $H = 1.0$ in., $L = 4.0$ in., $t_f = 0.1$ in., and $k = 5.0$ W/(in. ·°C). Using Eq. (11.22),
plot the fin efficiency for $h = 0.002, 0.05, 0.1, 0.5,$ and 1.0 W/(in.² ·°C).

11.11 A circuit board has a four-layer structure following the definitions in Figure 11.23 (a) and dimensions and conductivity of Application Example 11.12. One exception to this problem is that the
top surface layer has 40% Cu and 60% E-G. Your problem is: (1) Following the model for Figure
11.23 (b), calculate the top and bottom layer planar conductivities k_p and the normal conductivity
k_N; (2) Also following the model for Figure 11.23 (b), calculate the planar and normal direction *conductances* for nodes that have a width $w = 0.1$ in., length $l = 0.1$ in., putting the top layer Cu/E-G and
the under-layer of E-G into the top planar conductances, and putting the middle Cu layer and the
under-layer of E-G into the lower planar conductances; (3) Following the model in Figure 11.23 (c),
calculate the planar and normal conductivities for the one-layer orthogonal model. In all calculations,
use the series/parallel method for combining layers.

11.12 A circuit board has a four-layer structure following the definitions in Figure 11.23 (e) and
dimensions and conductivity of Application Example 11.12. One exception to this problem is that
the bottom surface layer has 40% Cu and 60% E-G. Your problem is: (1) Following the model for
Figure 11.23 (b), calculate the top and bottom layer planar conductivities k_p and the normal conductivity k_N; (2) Also following the model for Figure 11.23 (b), calculate the planar and normal direction
conductances for nodes that have a width $w = 0.1$ in., length $l = 0.1$ in., putting the top layer E-G and
the under-layer of 100% Cu into the top planar conductances, and putting the bottom 40% Cu layer

and the over-layer (above the 40% Cu) of E-G into the lower planar conductances; (3) Following the model in Figure 11.23 (c), calculate the planar and normal conductivities for the one-layer orthogonal model. In all calculations, use the series/parallel method for combining layers.

11.13 A circuit board has a four-layer structure following the definitions in Figure 11.23 (e) and dimensions and conductivity of Application Example 11.12. One exception to this problem is that the bottom surface layer has 20% Cu and 80% E-G. Your problem is: (1) Following the model for Figure 11.23 (b), calculate the top and bottom layer planar conductivities k_p and the normal conductivity k_N ; (2) Also following the model for Figure 11.23 (b), calculate the planar and normal direction *conductances* for nodes that have a width $w = 0.1$ in., length $l = 0.1$ in., putting the top layer E-G and the under-layer of 100% Cu into the top planar conductances, and putting the bottom 20% Cu layer and the over-layer (above the 20% Cu) of E-G into the lower planar conductances; (3) Following the model in Figure 11.23 (c), calculate the planar and normal conductivities for the one-layer orthogonal model in Figure 11.23 (c). In all calculations use the series/parallel method for combining layers.

11.14 A circuit board has a 20-layer structure. Ten of the layers are 100% Cu, each with a thickness of $t_{Cu} = 0.0014$ in. and a conductivity $k_{Cu} \cong 10.0$ W/(in. $\cdot$ °C). The other ten layers are 100% epoxy-glass, each $t_{E\text{-}G} = 0.0048$ in. Calculate an in-plane conductivity k_p and a through-plane conductivity k_N (1) using the series-parallel method and (2) using the Graebner and Azar method.

11.15 Repeat the thermal interface problem of Application Example 11.14, but use helium as the gap gas. You are asked to (1) validate your understanding by calculating the aluminum with an air gap for the first row of Table 11.3, and (2) construct the equivalent of Table 11.3 for the helium gap gas. Use $Pr = 0.71$, $k_g = 4 \times 10^{-3}$ W/(in. $\cdot$ °C) for helium.

CHAPTER 12

Conduction II: Spreading Resistance

In this chapter you will learn a lot about thermal spreading resistance, the heat conduction from a single heat source centered on a heat conducting substrate. Most results are quoted for a substrate or plate that has an adiabatic boundary condition at the source plane, Newtonian cooling at the opposing plane, and presumes steady-state conditions. This chapter will conform to this description, except that a time-dependent solution that ultimately converges to a steady solution will also be studied.

Our study begins with a couple of mathematically simple models that some design engineers continue to use. Then we look at a heat source on semi-infinite media where the only finite geometric object is a heat source on an otherwise adiabatic plane. The mathematics needed to solve even the semi-infinite media problem are quite complicated, but the resulting formula is useful in understanding the much more complicated finite media problem. The formulae that we end up with in the latter more general problem are too complex to permit solution on an electronic calculator, but are quite easily programmed into a math scratchpad program. You may even wish to write a program using your favorite language. Once you have accomplished this, you can obtain useful, accurate results for the remainder of your engineering or scientific career.

12.1 THE SPREADING PROBLEM

Our problem is illustrated in Figure 12.1 where we see the lines of heat flow that *spread out* from the heat source footprint and ultimately convect and/or radiate from the surface opposite that of the source plane. We all learned early in our educational experience that flow of nearly any type follows the path of least resistance. The heat conduction problem is no exception. Further into our study of this subject, we will see that the surface boundary condition where the Newtonian cooling is specified actually has an effect on the spreading because the boundary effect is part of the total resistance to heat flow. The greater the boundary condition resistance, the greater the conductive spreading. The heat flow lines at the left- and right-hand edges will be consistent with our choice of edge boundary conditions.

12.2 FIXED SPREADING ANGLE THEORIES

The geometry for the most common fixed angle theories is shown in Figure 12.2. The fixed angle methods all assume that the heat flux from the source spreads to fill a cone where ϕ is the angle.

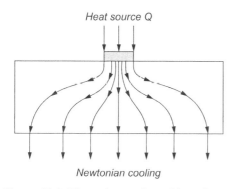

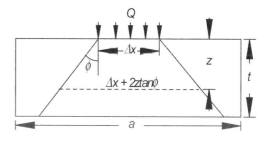

Figure 12.1. Thermal spreading of heat from a surface source.

Figure 12.2. Geometry for fixed angle spreading theories.

A restriction on the method is that the substrate must be sufficiently large in length and width so that the conduction cone is not truncated before the outermost flux lines reach the bottom plane. If the source is square, then the cross-sectional area at a perpendicular distance z from the source surface is

$$A(z) = (\Delta x + 2z \tan \phi)^2 \qquad \begin{array}{l}\text{Cond. cross-sectional area,}\\ \text{square source}\end{array} \qquad (12.1)$$

The first of the two fixed-angle methods that we will look at uses a conduction cross-sectional area that is averaged over the thickness t.

$$\overline{A} = \frac{1}{t} \int_0^t (\Delta x + 2z \tan \phi)^2 \, dz = \Delta x^2 + 2\Delta x t \tan \phi + \frac{4}{3} t^2 \tan^2 \phi \qquad (12.2)$$

Substituting Eq. (12.2) into Eq. (11.3), our general, one-dimensional conduction resistance formula we have

$$R = \int_0^t \frac{dz}{k\overline{A}} = \frac{1}{k\overline{A}} \int_0^t dz = \frac{t}{k\left(\Delta x^2 + 2\Delta x t \tan \phi + \frac{4}{3} t^2 \tan^2 \phi\right)} \qquad (12.3)$$

Equation (12.3) is non-dimensionalized by multiplying R by $k\Delta x$ to arrive at

$$\Theta_{Ave.A} = \frac{1}{\dfrac{1}{\tau_{FA}} + 2 \tan \phi + \dfrac{4}{3} \tau_{FA} \tan^2 \phi} ; \quad \tau_{FA} = \frac{t}{\Delta x} \qquad (12.4)$$

We shall name the other fixed-angle method the *integrated resistance* method because we perform the integration over the entire thickness t.

$$R = \int_0^t \frac{dz}{kA} = \frac{1}{k} \int_0^t \frac{dz}{A} = \frac{1}{k} \int_0^t \frac{dz}{(\Delta x + 2z \tan \phi)^2} = -\frac{1}{2k \tan \phi} \left[\frac{1}{\Delta x + 2z \tan \phi} \right]_{z=0}^{z=t}$$

$$R = -\frac{1}{2k \tan \phi} \left[\frac{1}{(\Delta x + 2t \tan \phi)} - \frac{1}{\Delta x} \right] = \frac{t}{k\Delta x (\Delta x + 2t \tan \phi)} \qquad (12.5)$$

We non-dimensionalize Eq. (12.5) to arrive at the *integrated* result.

$$\Theta_{IR} = \frac{1}{\dfrac{1}{\tau_{FA}} + 2 \tan \psi} ; \quad \tau_{FA} = \frac{t}{\Delta x} \qquad (12.6)$$

Note that both spreading results use a dimensionless thickness $\tau_{FA} = t/\Delta x$ referenced to the source dimension. You will read later in this chapter that more exact spreading theories result in a dimensionless thickness $\tau = t/a$ (thickness/substrate width), thus if we wish to compare results we need to modify Eqs. (12.4) and (12.6).

$$\Theta_{Ave.A} = \frac{1}{\left(\dfrac{\alpha}{\tau}\right) + 2 \tan \phi + \dfrac{4}{3}\left(\dfrac{\tau}{\alpha}\right) \tan^2 \phi} ; \tau = \alpha \tau_{FA}, \alpha = \Delta x/a \qquad (12.7)$$

$$\Theta_{IR} = \frac{1}{\dfrac{\alpha}{\tau} + 2 \tan \phi} ; \tau = \alpha \tau_{FA}, \; \alpha = \Delta x/a \qquad (12.8)$$

Equations (12.7) and (12.8) are plotted in Figures 12.3 and 12.4, respectively, for $\alpha = 0.1$ and three different spreading angles centered about the most popular, $\phi = 45°$. Also plotted in each of these graphs is the thermal spreading resistance $\psi_{Ave.}$ for $h =$ infinite using a very exact spreading theory (covered later in this chapter) for the source-averaged temperature. In order to ensure that the spreading cone is not truncated, we have the condition

$$\tan \phi \le (a - \Delta x)/2t$$

Suppose then that we select $\phi = 45°$, then $\tan \phi = 1$ and

$$(a - \Delta x)/2t \ge 1, \, 1 - (\Delta x/a) \ge 2t/a$$

$$1 - \alpha \ge 2\tau, \, \tau \le (1 - \alpha)/2$$

Using $\alpha = 0.1$ we calculate that $\tau \le 0.45$. The example demonstrated in Figure 12.4, where the exact $\psi_{Ave.}$ deviation from Θ_{IR} for $\tau > 0.45$, is consistent with what we just calculated. Although the reader may find that the fixed spreading angle formulae are still used, particularly for multilayered materials where the formulae can be successively applied to each layer, he/she will ultimately discover that experienced, competent thermal analysts do not use these formulae when other analysis means are available.

Up to this point, we have concerned ourselves with square heat sources. Now we shall look at rectangular sources that are not square and assume a 45° spreading angle, largely because of the popularity of this method. The source has edge dimensions of $\Delta x, \Delta y$ and we begin with the integrated resistance prior to integration for $\phi = 45°$ angle.

$$R = \int_{z=0}^{z=t} \frac{dz}{kA} = \frac{1}{k} \int_{z=0}^{z=t} \frac{dz}{[\Delta x + 2z \tan(\pi/4)][\Delta y + 2z \tan(\pi/4)]}$$

$$= \frac{1}{k} \int_{z=0}^{z=t} \frac{dz}{(\Delta x + 2z)(\Delta y + 2z)} = \frac{1}{2k} \left[\frac{\ln(\Delta y + 2z) - \ln(\Delta y + 2z)}{(\Delta x - \Delta y)} \right]_{z=0}^{z=t}$$

$$R = \frac{1}{2k(\Delta x - \Delta y)} \ln \left[\left(\frac{\Delta x}{\Delta y} \right) \left(\frac{\Delta y + 2t}{\Delta x + 2t} \right) \right] \qquad \text{45° spreading} \quad (12.9) \atop \text{integrated resist.}$$

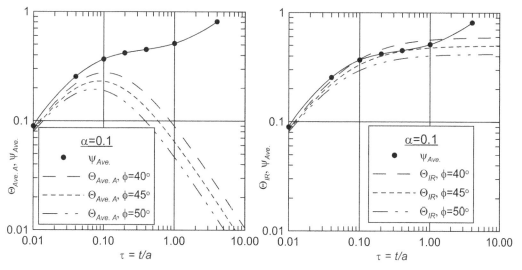

Figure 12.3. Average area spreading resistance $\Theta_{Ave.A}$ and average $\psi_{Ave.} = \overline{\psi}_{Sp} + \psi_{U-Cond.}$ (with data points) versus dimensionless thickness τ. Each $\Theta_{Ave.A}$, Eq. (12.7), for $\alpha = 0.1$ is plotted for $\phi = 40, 45,$ and 50° angles.

Figure 12.4. Integrated spreading resistance Θ_{IR} and average $\psi_{Ave.} = \overline{\psi}_{Sp} + \psi_{U-Cond.}$ (with data points) versus dimensionless thickness τ. Each Θ_{IR}, Eq. (12.8), for $\alpha = 0.1$ is plotted for $\phi = 40, 45,$ and 50° angles.

12.3 CIRCULAR-SOURCE, SEMI-INFINITE MEDIA SOLUTION, UNIFORM FLUX, AVERAGE SOURCE TEMPERATURE, BY CARSLAW AND JAEGER (1986)

The simplest spreading problem that is based on a sound analytical foundation is that of a circular source of radius a and uniform dissipation Q on the semi-infinite media illustrated in Figure 12.5. The applicable differential equation is Eq. (11.3), the heat conduction equation in cylindrical coordinates. The volumetric heat source Q_V is set at zero because we will introduce the surface source via a boundary condition.

$$\frac{1}{\rho}\frac{\partial}{\partial\rho}\left(\rho\frac{\partial T}{\partial\rho}\right)+\frac{1}{\rho^2}\frac{\partial^2 T}{\partial\phi^2}+\frac{\partial^2 T}{\partial z^2}=0 \tag{12.10}$$

Equation (12.10) is simplified when we note that the temperature must be independent of the angle of rotation ϕ about the z - axis so that we then have

$$\frac{1}{\rho}\frac{\partial}{\partial\rho}\left(\rho\frac{\partial T}{\partial\rho}\right)+\frac{\partial^2 T}{\partial z^2}=0$$

$$\frac{\partial^2 T}{\partial\rho^2}+\frac{1}{\rho}\frac{\partial T}{\partial\rho}+\frac{\partial^2 T}{\partial z^2}=0 \tag{12.11}$$

The solution of Eq. (12.11), ignoring multiplicative constants for the time being, is

$$T=R(\rho)Z(z)$$

When this solution for T is substituted into Eq. (12.11) we obtain

$$\frac{1}{R}\frac{d^2 R}{d\rho^2}+\frac{1}{\rho R}\frac{dR}{d\rho}+\frac{1}{Z}\frac{d^2 Z}{dz^2}=0$$

$$\frac{1}{R}\frac{d^2 R}{d\rho^2}+\frac{1}{\rho R}\frac{dR}{d\rho}=-\frac{1}{Z}\frac{d^2 Z}{dz^2}=-\lambda^2$$

$$\frac{d^2 R}{d\rho^2}+\frac{1}{\rho}\frac{dR}{d\rho}+\lambda^2 R=0,\quad \frac{1}{Z}\frac{d^2 Z}{dz^2}-\lambda^2 Z=0$$

where λ is a constant. The differential equation in R is Bessel's equation for $p=0$ (Appendix vi). The solutions R and Z are

$$R(\rho)=J_0(\lambda\rho),\ 0\le\rho \tag{12.12}$$

$$Z(z)=e^{-\lambda z},\qquad 0\le z \tag{12.13}$$

thus

$$T=f(\lambda)e^{-\lambda z}J_0(\lambda\rho) \tag{12.14}$$

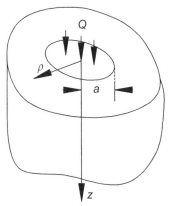

Figure 12.5. Uniform circular-source on semi-infinite media.

The quantity λ is a constant greater than or equal to zero. Since the solution, Eq. (12.14), is valid for all positive values of λ, we must include all of these values to obtain the complete solution.

$$T = \int_{\lambda=0}^{\lambda=\infty} e^{-\lambda z} J_0(\lambda\rho) f(\lambda) d\lambda \tag{12.15}$$

It certainly is not obvious as to what form the function $f(\lambda)$ should take, but Carslaw and Jaeger (1986) mysteriously suggest

$$f(\lambda) = C \frac{J_1(\lambda a)}{\lambda}, \qquad C = \text{an arbitrary constant}$$

so that

$$T = C \int_{\lambda=0}^{\lambda=\infty} e^{-\lambda z} J_0(\lambda\rho) J_1(\lambda a) \frac{d\lambda}{\lambda} \tag{12.16}$$

The *heat flux* is implemented via a flux Newtonian boundary condition.

$$q = -k \frac{\partial T}{\partial z}\bigg|_{z=0}, \qquad 0 \le \rho < a; \quad q = 0, \ \rho > a \tag{12.17}$$

$$q = -k \frac{\partial T}{\partial z}\bigg|_{z=0} = -k \left\{ \frac{\partial}{\partial z}\left[C \int_{\lambda=0}^{\lambda=\infty} e^{-\lambda z} J_0(\lambda\rho) J_1(\lambda a) \frac{d\lambda}{\lambda} \right]_{z=0} \right\}$$

$$q = kC \int_{\lambda=0}^{\lambda=\infty} J_0(\lambda\rho) J_1(\lambda a) d\lambda \tag{12.18}$$

Comparing identities (*vi.*9) and (*vi.*7)

$$\int_{\lambda=0}^{\lambda=\infty} J_0(\lambda\rho) J_1(\lambda a) d\lambda = \frac{1}{a}, \quad \rho < a$$

$$\int_{\lambda=0}^{\lambda=\infty} J_0(\lambda\rho) J_1(\lambda a) d\lambda = 0, \quad \rho > a$$

to Eq. (12.18) it is clear that the boundary conditions, Eq. (12.17), are satisfied by the chosen $f(\lambda)$ so that $q = kC/a$, $C = qa/k$ for $\rho < a$ and $q = 0$ for $\rho > a$.

$$T(\rho,z) = \frac{qa}{k} \int_{\lambda=0}^{\lambda=\infty} e^{-\lambda z} J_0(\lambda\rho) J_1(\lambda a) \frac{d\lambda}{\lambda} \tag{12.19}$$

The result for T given by Eq. (12.19) is useful when we resort to additional steps. For example, suppose we wish to obtain a formula for the source temperature averaged over the source area A_S. Using Eq. (12.19) we write the following:

$$\bar{\bar{T}} = \frac{1}{A_S} \int_{A_S} T dA = \frac{1}{\pi a^2} \int_{\rho=0}^{\rho=a} 2\pi\rho T d\rho = \frac{2q}{ka} \int_{\lambda=0}^{\lambda=\infty} J_1(\lambda a) \frac{d\lambda}{\lambda} \int_{\rho=0}^{\rho=a} J_0(\lambda\rho)\rho d\rho$$

$$\overline{T} = \frac{2q}{ka} \int\limits_{\lambda=0}^{\lambda=\infty} J_1(\lambda a) \left[\int\limits_{\rho=0}^{\rho=a} J_0(\lambda\rho)\rho d\rho \right] \frac{d\lambda}{\lambda} = \frac{2q}{ka} \int\limits_{\lambda=0}^{\lambda=\infty} J_1(\lambda a) \left[\int\limits_{y=0}^{y=\lambda a} J_0(y)\frac{ydy}{\lambda^2} \right] \frac{d\lambda}{\lambda}$$

$$\overline{T} = \frac{2q}{ka} \int\limits_{\lambda=0}^{\lambda=\infty} J_1(\lambda a) \left[\int\limits_{y=0}^{y=\lambda a} J_0(y)ydy \right] \frac{d\lambda}{\lambda^3} \tag{12.20}$$

The integral on the far right-hand side of Eq. (12.20) is evaluated using the identity (*vi.*10) so that

$$\overline{T} = \frac{2q}{ka} \int\limits_{\lambda=0}^{\lambda=\infty} J_1(\lambda a)\left[(\lambda a)J_1(\lambda a)\right]\frac{d\lambda}{\lambda^3} = \frac{2q}{k} \int\limits_{\lambda=0}^{\lambda=\infty} J_1^2(\lambda a)\frac{d\lambda}{\lambda^2} \tag{12.21}$$

and the integral in Eq. (12.21) is evaluated using identity (*vi.*11).

$$\overline{T} = \frac{2q}{k} \int\limits_{\lambda=0}^{\lambda=\infty} J_1^2(\lambda a)\frac{d\lambda}{\lambda^2} = \left(\frac{2q}{k}\right)\left(\frac{4a}{3\pi}\right) = \left(\frac{2Q}{\pi ka^2}\right)\left(\frac{4a}{3\pi}\right)$$

$$\overline{T} = \frac{8Q}{3\pi^2 ka} \tag{12.22}$$

The source to ambient thermal resistance is obtained by dividing both sides of Eq. (12.22) by Q. Nearly all thermal spreading resistance studies found in the general literature use a source area A_S rather than a source radius so that the source-averaged spreading resistance becomes

$$\overline{R}_\infty = \overline{T}/Q = 8/\left(3\pi^2 ka\right) = 8/\left(3\pi^2 k\sqrt{A_S/\pi}\right) = 8/\left(3k\pi^{3/2}\sqrt{A_S}\right)$$

Just as we non-dimensionalized the fixed-angle spreading resistances, we also wish to non-dimensionalize $\overline{R}_\infty$.

$$\overline{\psi}_\infty = k\sqrt{A_S}\,\overline{R}_\infty = k\sqrt{A_S}\,8/\left(3k\pi^{3/2}\sqrt{A_S}\right)$$

$$\boxed{\overline{\psi}_\infty = \frac{8}{3\pi^{3/2}} = 0.4789}$$

Dimensionless, circ. source ave., on semi- (12.23) infinite media

Square sources are readily approximated to a high accuracy by using a circular source with a radius that results in the same area, A_S, as the square source, i.e., $a = \sqrt{A_S/\pi}$. It is quite amazing that after such a considerable mathematical effort, the non-dimensional result is so simple! It is left as an exercise for the student to derive an equivalent result for a spreading resistance based on the peak temperature of the source.

12.4 RECTANGULAR-SOURCE, TIME-DEPENDENT, SEMI-INFINITE MEDIA SOLUTION, UNIFORM FLUX, PEAK SOURCE TEMPERATURE BY JOY AND SCHLIG (1970)

Joy and Schlig (1970) derived a solution to the thermal spreading resistance problem of a cuboid-shaped source that is instantaneously initiated. By letting the solution progress to a steady value, it is possible to get a result that resembles Eq. (12.23). Just as in the preceding section, Joy and Schlig also use work by Carslaw and Jaeger. A formal description of the Joy and Schlig method comes under the topic of time-dependent Green's functions, a topic that is somewhat beyond the scope of what we wish to accomplish here. Nevertheless, we will follow a somewhat formalized method as detailed in the excellent book by Beck et al. (1992, 2011), which readers should consult for further detail on the subject.

We begin with the time-dependent heat conduction equation that also permits a volumetric source Q_V.

$$\nabla^2 T(\underline{r},t) + \frac{1}{k} Q_V(\underline{r},t) = \frac{1}{\alpha} \frac{\partial T(\underline{r},t)}{\partial t} \qquad \begin{array}{l} \text{Time-dependent} \\ \text{heat conduction} \end{array} (12.24)$$

where $\alpha = k/\rho c$ is the thermal diffusivity and Q_V is the volumetric heat source. The temperature $T(\underline{r},t)$ is written using the standard position "vector" r to denote the three-dimensional spatial dependence and t for the time-dependence. ∇^2 is the Laplacian operator.

A general set of initial and boundary conditions are

$$T(\underline{r},0) = F(\underline{r}) \qquad\qquad\qquad \text{initial condition}$$

$$k \frac{\partial T(\underline{r},t)}{\partial n_i} + h_i T(\underline{r},t) = 0, \ \ t > 0 \qquad \text{boundary condition}$$

where n_i denotes the outward directed normal for surface i with a heat transfer coefficient h_i.

The time-dependent Green's function method utilizes the auxiliary equation

$$\nabla^2 G + \frac{1}{\alpha} \delta(\underline{r} - \underline{r}') \delta(t - t') = \frac{1}{\alpha} \frac{\partial G}{\partial t} \qquad \begin{array}{l} \text{Time-dependent} \\ \text{auxiliary equation} \end{array} (12.25)$$

where $\delta(\ )$ is the Dirac delta-function. The Green's function G can be shown to have the important property of reciprocity, proven in Appendix ix.

$$G(\underline{r},t|\underline{r}',t') = G(\underline{r}',-t'|\underline{r},-t) \qquad\qquad \text{Reciprocity of } G \ \ (12.26)$$

The Laplacians ∇^2 and ∇'^2 differentiate with respect to $\underline{r}$ and $\underline{r}'$, respectively.

The Reciprocity Law, Eq. (12.26), is used to convert Eq. (12.25) with the *exchanged variables* $\underline{r}$ by $\underline{r}'$ and t by $-t'$:

$$\nabla'^2 G + \frac{1}{\alpha} \delta(\underline{r}' - \underline{r}) \delta(t' - t) = -\frac{1}{\alpha} \frac{\partial G}{\partial t'}, \qquad t > t'$$

$$G(\underline{r},t|\underline{r}',t') = 0, \ t < t'; \qquad k \frac{\partial G}{\partial n_i'} + h_i G = 0, \quad t > t'$$

The first and second spatial derivatives are unchanged in sign, but the first derivative in time has a sign change. The latter is perhaps clearer by noting that all time dependencies of G have a $t - t'$ dependence.

Now rewrite Eq. (12.24) with the new variables as

$$\nabla'^2 T(\underline{r}',t') + \frac{1}{k} Q_V(\underline{r}',t') = \frac{1}{\alpha} \frac{\partial T(\underline{r}',t')}{\partial t'} \qquad \text{Time-dependent} (12.27)$$

Now multiply Eq. (12.27) by $G(\underline{r},t|\underline{r}',t')$, subtracting the modified form of Eq. (12.26) (using ∇'^2 and the exchanged variables) multiplied by $T(\underline{r}',t')$ to obtain (remembering the T, G arguments)

$$G\nabla'^2 T + \frac{1}{k} G Q_V - T\nabla'^2 G - \frac{1}{\alpha} T \delta(\underline{r}' - \underline{r}) \delta(t' - t) = \frac{1}{\alpha} G \frac{\partial T}{\partial t'} + \frac{1}{\alpha} T \frac{\partial G}{\partial t'}$$

$$\left(G\nabla'^2 T - T\nabla'^2 G \right) + \frac{1}{k} G Q_V - \frac{1}{\alpha} T \delta(\underline{r}' - \underline{r}) \delta(t' - t) = \frac{1}{\alpha} \frac{\partial(TG)}{\partial t'}$$

We integrate the last result over time t' and space V.

$$
\int_{t'=0}^{t'=t+\varepsilon} \int_V \alpha \left(G\nabla'^2 T - T\nabla'^2 G \right) d\underline{r}'dt' + \int_{t'=0}^{t'=t+\varepsilon} \int_V \frac{\alpha}{k} GQ_V d\underline{r}'dt' - \int_{t'=0}^{t'=t+\varepsilon} \int_V T\delta \left(\underline{r}' - \underline{r} \right) \delta \left(t' - t \right) d\underline{r}'dt'
$$

$$
= \int_{t'=0}^{t'=t+\varepsilon} \int_V \frac{\partial (TG)}{\partial t'} d\underline{r}'dt'
$$

The third integral is immediately simplified using Eq. (*vii*.2) for the Dirac delta function. The fourth integral is readily integrated over time. The result of these two operations is

$$
\int_{t'=0}^{t'=t+\varepsilon} \int_V \alpha \left(G\nabla'^2 T - T\nabla'^2 G \right) d\underline{r}'dt' + \int_{t'=0}^{t'=t+\varepsilon} \int_V \frac{\alpha}{k} GQ_V d\underline{r}'dt' - T\left(\underline{r},t \right) = \int_V \left[TG \right]_{t'=0}^{t'=t+\varepsilon} d\underline{r}'
$$

Rearranging, we arrive at a form of the *Green's function solution equation* with the various contributions identified. $T\left(\underline{r},t \right)$ is the temperature at the location r at the time t.

$$
T\left(\underline{r},t \right) = \boxed{-\int_V \left[TG \right]_{t'=0}^{t'=t+\varepsilon} d\underline{r}'} + \boxed{\alpha \int_{t'=0}^{t'=t+\varepsilon} \int_V \left(G\nabla'^2 T - T\nabla'^2 G \right) d\underline{r}'dt'} + \boxed{\frac{\alpha}{k} \int_{t'=0}^{t'=t+\varepsilon} \int_V GQ_V d\underline{r}'dt'} \quad (12.28)
$$

Green's func. sol. eq.

Initial conditions
All boundary conditions
Volumetric source effects

Equation (12.28) can be further simplified. We begin by evaluating the integrand of the first integral. Causality is used for this term to set the integrand at zero for the upper limit because the effect cannot begin before the instantaneous source. Evaluation at the lower limit are sensible initial conditions. Thus we have

$$
\left[GT \right]^{t'=t+\varepsilon} = 0
$$

$$
\left[GT \right]_{t'=0} = -G\left(\underline{r},t \mid \underline{r}',0 \right) T\left(\underline{r}',0 \right) = -G\left(\underline{r},t \mid \underline{r}',0 \right) F\left(\underline{r}' \right)
$$

where the initial conditions on T are explicitly identified as the function $F\left(\underline{r}' \right)$.

The second term in Eq. (12.28) is simplified using Green's theorem from [sort of] elementary calculus to transform the problem from a volume integral to a surface integral. The summation is over all surfaces enclosing the volume in an orthogonal coordinate system. The next step,

$$
\alpha \int_{t'=0}^{t'=t+\varepsilon} \int_V \left(G\nabla'^2 T - T\nabla'^2 G \right) d\underline{r}'dt' = \alpha \int_{t'=0}^{t'=t+\varepsilon} \sum_i \left(G \frac{\partial T}{\partial n_i'} \bigg|_{\underline{r}'=\underline{r}_i'} - T \frac{\partial G}{\partial n_i'} \bigg|_{\underline{r}'=\underline{r}_i'} \right) dS_i'dt'
$$

is commonly employed in the Green's function method, including steady-state problems:

The Green's function and temperature are specified to have the same
homogeneous boundary conditions $k\,\partial T/\partial n_i + h_i T = 0;\ k\,\partial G/\partial n_i + h_i G = 0$.

With the prescribed identically equal boundary conditions for the temperature and the Green's functions, the surface integral term is precisely zero and we take the limit of $t' = t + \varepsilon$ as $\varepsilon \to 0$.

Equation (11.28) becomes

$$T\left(\underline{r},t\right)=\int_{V}G\left(\underline{r},t\big|\underline{r}',0\right)F\left(\underline{r}'\right)d\underline{r}'+\frac{\alpha}{k}\int_{t'=0}^{t'=t}\int_{V}G\left(\underline{r},t\big|\underline{r}',t'\right)Q_{V}\left(\underline{r}',t'\right)d\underline{r}'dt' \qquad (12.29)$$

Green's func. sol. eq.

We have now reached the point where the book by Beck et al. (1992, 2011) becomes an invaluable reference tool in that the authors of that book have created an encyclopedia of a very large number of boundary conditions and associated Green's functions. Beck et al. show that for two- and three-dimensional geometries with infinite, semi-infinite, Dirichlet, Neumann, and Robin boundary conditions, the desired Green's functions may be constructed as products of one-dimensional solutions (this is not possible for spherical coordinates, and generally not possible for steady-state solutions), i.e., $G_{xyz}=G_{x}G_{y}G_{z}$.

When the boundaries of the media are at infinity, a one-dimensional solution may be found in many references such as Beck et al. (1992, 2011) and Morse and Feshbach (1953). The one-dimensional

$$G_{x}=G\left(x,t\big|x',t'\right)=\frac{1}{\left[4\pi\alpha\left(t-t'\right)\right]^{1/2}}\exp\left[\frac{-\left(x-x'\right)^{2}}{4\alpha\left(t-t'\right)}\right] \qquad (12.30)$$

solutions for y and z are obtained by replacing x in Eq. (12.30) by y and z, respectively.

Joy and Schlig (1970) study the problem where the initial condition on the temperature is zero or, for our problem, $F\left(\underline{r}'\right)=0$. The complete three-dimensional solution for the temperature is then

$$T\left(\underline{r},t\right)=\frac{Q}{8\left(\pi\alpha\right)^{3/2}\rho cV}\int_{t'=0}^{t'=t}\frac{dt'}{\left(t-t'\right)^{3/2}}\int_{V}e^{-r^{2}/\left[4\alpha\left(t-t'\right)\right]}dx'dy'dz' \qquad (12.31)$$

where we have assumed a uniform source density $Q_{V}=Q/\Delta x\Delta y\Delta z$ for a source with linear dimensions of $\Delta x,\Delta y,$ and Δz.

Now we are ready to look at the geometry to which Eq. (12.31) is applied. Figure 12.6 shows the actual source at a distance $z=-D$ below the $z=0$ plane. An image source is shown at a distance $z=D$ above the $z=0$ plane. The effect of the image is to create an adiabatic boundary condition at $z=0$. Otherwise, the media extends to infinity in all directions.

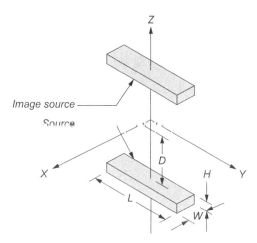

Figure 12.6. Volumetric heat source in semi-infinite media; image source above adiabatic plane is displayed.

Now we have the problem of integrating Eq. (12.31) and we shall largely follow Joy and Schlig (1970). We need concern ourselves only with the mathematics of the lower source, then add the upper source results. Rewriting Eq. (12.31),

$$
T\left[\frac{8\rho c V\left(\pi\alpha\right)^{3/2}}{Q}\right]=\int_{0}^{t}\frac{dt'}{\left(t-t'\right)^{3/2}}\cdot I_{x}\cdot I_{y}\cdot I_{z}
$$

where I_{x}, I_{y}, and I_{z} are written with some modifications made to the basic Joy and Schlig formulation to permit sources not at the center of the coordinate system, i.e., a source centered at x_{s}, y_{s},

$$
I_{x}=\int_{x_{S}-W/2}^{x_{S}+W/2}e^{-\left[\frac{(x-x')^{2}}{4\alpha(t-t')}\right]}dx'
$$

The following substitutions are made:

$$
u=x'-x_{s},\ x'=u+x_{s},\ dx'=du
$$
$$
u_{1}=x_{1}'-x_{s}=\left(x_{s}-W/2\right)-x_{s}=-W/2
$$
$$
u_{2}=x_{2}'-x_{s}=\left(x_{s}+W/2\right)-x_{s}=W/2
$$

with the result,

$$
I_{x}=\int_{-W/2}^{W/2}e^{-\left[\frac{(x-u-x_{S})^{2}}{4\alpha(t-t')}\right]}du
$$

Another substitution is

$$
v=x-u-x_{s},\ dv=-du
$$
$$
v_{1}=x-u_{1}-x_{s}=x-\left(-W/2\right)-x_{s}
$$
$$
v_{1}=x-x_{s}+W/2
$$
$$
v_{2}=x-u_{2}-x_{s}=x-\left(W/2\right)-x_{s}
$$
$$
v_{2}=x-x_{s}-W/2
$$

Then we write

$$
I_{x}=-\int_{x-x_{S}+W/2}^{x-x_{S}-W/2}e^{-\left[\frac{v^{2}}{4\alpha(t-t')}\right]}dv=\int_{x-x_{S}-W/2}^{x-x_{S}+W/2}e^{-\left[\frac{v^{2}}{4\alpha(t-t')}\right]}dv
$$

and with another variable substitution,

$$
\xi=v/\sqrt{4\alpha\left(t-t'\right)},\xi^{2}=v^{2}/\left[4\alpha\left(t-t'\right)\right]
$$
$$
v=\xi\sqrt{4\alpha\left(t-t'\right)},\ dv=\sqrt{4\alpha\left(t-t'\right)}d\xi
$$
$$
\xi_{1}=v_{1}/\sqrt{4\alpha\left(t-t'\right)}=\left(x-x_{s}-W/2\right)/\sqrt{4\alpha\left(t-t'\right)}
$$
$$
\xi_{2}=\left(x-x_{s}+W/2\right)/\sqrt{4\alpha\left(t-t'\right)}
$$

The integral becomes

$$
I_{x}=\sqrt{4\alpha\left(t-t'\right)}\int_{\xi_{1}}^{\xi_{2}}e^{-\xi^{2}}d\xi=\sqrt{4\alpha\left(t-t'\right)}\left[\int_{0}^{\xi_{2}}e^{-\xi^{2}}d\xi-\int_{0}^{\xi_{1}}e^{-\xi^{2}}d\xi\right]
$$

Further progress is achieved by using the *error function erf(x)*:

$$
erf\left(x\right)=\frac{2}{\sqrt{\pi}}\int_{0}^{x}e^{-\xi^{2}}d\xi,\ erf\left(-x\right)=-erf\left(x\right)
$$

With the error function definition our integral becomes

$$I_x = \sqrt{4\alpha(t-t')}\left(\frac{\sqrt{\pi}}{2}\right)\left[erf(\xi_2) - erf(\xi_1)\right]$$

$$I_x = \sqrt{\pi\alpha(t-t')}\left\{erf\left[\frac{x-x_s+W/2}{\sqrt{4\alpha(t-t')}}\right] - erf\left[\frac{x-x_s-W/2}{\sqrt{4\alpha(t-t')}}\right]\right\}$$

and similarly,

$$I_y = \sqrt{\pi\alpha(t-t')}\left\{erf\left[\frac{y-y_s+L/2}{\sqrt{4\alpha(t-t')}}\right] - erf\left[\frac{y-y_s-L/2}{\sqrt{4\alpha(t-t')}}\right]\right\}$$

The I_z integral is broken into two parts, I_{z1}, and I_{z2}, i.e., $I_z=I_{z1}+I_{z2}$. The first integral is over the source for $z < 0$ and the second integral is over the source for $z > 0$. By using these identical two sources, one over and the other under the $z = 0$ plane, we ensure the effect of an adiabatic $z = 0$ plane. This is an application of the *method of images*. First noting the integration limits for I_{z1},

$$z_1' = -(D+H),\ z_2' = -D$$

$$I_{z1} = \int_{z_1'}^{z_2'} e^{-\left[\frac{(z-z')^2}{4\alpha(t-t')}\right]}dz'$$

and then making a variable substitution,

$$v = z-z',\ dv = -dz',\ v_1 = z-z_1',\ v_2 = z-z_2':\qquad I_{z1} = -\int_{v_1}^{v_2} e^{-\left[\frac{v^2}{4\alpha(t-t')}\right]}dv$$

Another variable substitution is required:

$$\xi = v/\sqrt{4\alpha(t-t')},\ v = \sqrt{4\alpha(t-t')}\xi,\ dv = \sqrt{4\alpha(t-t')}d\xi$$
$$\xi_1 = v_1/\sqrt{4\alpha(t-t')} = (z-z_1')/\sqrt{4\alpha(t-t')}$$
$$\xi_2 = (z-z_2')/\sqrt{4\alpha(t-t')}$$

And another:

$$\xi_1 = (z+D+H)/\sqrt{4\alpha(t-t')},\ \xi_2 = (z+D)/\sqrt{4\alpha(t-t')}$$

Then

$$\frac{I_{z1}}{-\sqrt{4\alpha(t-t')}} = \int_{\xi_1}^{\xi_2} e^{-\xi^2}d\xi = \left[\int_0^{\xi_2} e^{-\xi^2}d\xi - \int_0^{\xi_1} e^{-\xi^2}d\xi\right] = \left(\frac{\sqrt{\pi}}{2}\right)\left[erf(\xi_2) - erf(\xi_1)\right]$$

$$I_{z1} = \sqrt{\pi\alpha(t-t')}\left[erf\left(\frac{z+D+H}{\sqrt{4\alpha(t-t')}}\right) - erf\left(\frac{z+D}{\sqrt{4\alpha(t-t')}}\right)\right]$$

Similarly,

$$I_{z2} = \sqrt{\pi\alpha(t-t')}\left[erf\left(\frac{z-D}{\sqrt{4\alpha(t-t')}}\right) - erf\left(\frac{z-D-H}{\sqrt{4\alpha(t-t')}}\right)\right]$$

Then

$$
I_z = \sqrt{\pi \alpha (t-t')} \left\{ \begin{array}{l} \left[erf\left(\dfrac{z+D+H}{\sqrt{4\alpha(t-t')}} \right) - erf\left(\dfrac{z+D}{\sqrt{4\alpha(t-t')}} \right) \right] + \\[3ex] \left[erf\left(\dfrac{z-D}{\sqrt{4\alpha(t-t')}} \right) - erf\left(\dfrac{z-D-H}{\sqrt{4\alpha(t-t')}} \right) \right] \end{array} \right\}
$$

or

$$
I_z = \sqrt{\pi \alpha (t-t')} \left\{ \begin{array}{l} \left[erf\left(\dfrac{z+D+H}{\sqrt{4\alpha(t-t')}} \right) + erf\left(\dfrac{-z-D}{\sqrt{4\alpha(t-t')}} \right) \right] + \\[3ex] \left[erf\left(\dfrac{z-D}{\sqrt{4\alpha(t-t')}} \right) + erf\left(\dfrac{D+H-z}{\sqrt{4\alpha(t-t')}} \right) \right] \end{array} \right\}
$$

and putting it all together:

$$
T = \frac{Q(\pi\alpha)^{3/2}}{8\rho c V (\pi\alpha)^{3/2}} \int_0^t \frac{dt'(t-t')^{3/2}}{(t-t')^{3/2}} \left\{ erf\left[\frac{(W/2)+x-x_s}{\sqrt{4\alpha(t-t')}} \right] + erf\left[\frac{(W/2)-x+x_s}{\sqrt{4\alpha(t-t')}} \right] \right\} \times
$$

$$
\left\{ erf\left[\frac{(L/2)+y-y_s}{\sqrt{4\alpha(t-t')}} \right] + erf\left[\frac{(L/2)-y+y_s}{\sqrt{4\alpha(t-t')}} \right] \right\} \left\{ \begin{array}{l} erf\left[\dfrac{z+D+H}{\sqrt{4\alpha(t-t')}} \right] + erf\left[\dfrac{-z-D}{\sqrt{4\alpha(t-t')}} \right] \\[3ex] +erf\left[\dfrac{z-D}{\sqrt{4\alpha(t-t')}} \right] + erf\left[\dfrac{D+H-z}{\sqrt{4\alpha(t-t')}} \right] \end{array} \right\}
$$

$$
T(x,y,z,t) = \frac{Q}{8\rho c V} \int_0^t dt' \left\{ erf\left[\frac{(W/2)+x-x_s}{\sqrt{4\alpha(t-t')}} \right] + erf\left[\frac{(W/2)-x+x_s}{\sqrt{4\alpha(t-t')}} \right] \right\} \times
$$

$$
\left\{ erf\left[\frac{(L/2)+y-y_s}{\sqrt{4\alpha(t-t')}} \right] + erf\left[\frac{(L/2)-y+y_s}{\sqrt{4\alpha(t-t')}} \right] \right\} \left\{ \begin{array}{l} erf\left[\dfrac{z+D+H}{\sqrt{4\alpha(t-t')}} \right] + erf\left[\dfrac{-z-D}{\sqrt{4\alpha(t-t')}} \right] \\[3ex] +erf\left[\dfrac{z-D}{\sqrt{4\alpha(t-t')}} \right] + erf\left[\dfrac{D+H-z}{\sqrt{4\alpha(t-t')}} \right] \end{array} \right\}
$$

A final variable substitution is made for the time variable.

$$
u = t-t', \ du = -dt', \ u_1 = t-t_1' = t-0 = t, \ u_2 = t-t_2' = t-t = 0
$$

The final analytical result is the following integral:

$$
T(x,y,z,t) = \frac{Q}{8\rho c V} \int_0^t du \tag{12.32}
$$

$$
\times \left\{ erf\left[\frac{(W/2)+x-x_s}{\sqrt{4\alpha u}} \right] + erf\left[\frac{(W/2)-x+x_s}{\sqrt{4\alpha u}} \right] \right\} \left\{ erf\left[\frac{(L/2)+y-y_s}{\sqrt{4\alpha u}} \right] + erf\left[\frac{(L/2)-y+y_s}{\sqrt{4\alpha u}} \right] \right\}
$$

$$
\times \left\{ erf\left[\frac{z+D+H}{\sqrt{4\alpha u}} \right] + erf\left[\frac{-z-D}{\sqrt{4\alpha u}} \right] + erf\left[\frac{z-D}{\sqrt{4\alpha u}} \right] + erf\left[\frac{D+H-z}{\sqrt{4\alpha u}} \right] \right\}
$$

Equation (12.32) cannot be integrated analytically and thus requires a math package. The Maple® software was used to calculate the temperature rise to a steady-state value for a source centered at $x_s = y_s = 0$ with dimensions $D = H = 0$ and the temperature calculated at $x = y = z = 0$. Two results are plotted for varying source aspect ratio L/W in Figure 12.7: One plot is the dimensionless thermal spreading resistance ψ_∞ and the other is a "constant" C_∞, both defined using

$$R_\infty = (T - T_\infty)/Q, \, \psi_\infty = k\sqrt{LW} \, R_\infty, \, C_\infty = 2k\sqrt{LW} \, R_\infty$$

The source aspect ratio L/W is varied from 1 to 100. It is interesting to note that the aspect ratio is not a significant effect until L/W is about 3.0. The square source, maximum, dimensionless, semi-infinite media thermal resistance is

$$\boxed{\psi_\infty = 0.5581} \qquad \begin{array}{l} \text{Dimensionless, sq. source} \quad (12.33) \\ \text{on semi-infinite media} \end{array}$$

We will return to the subject of solutions for semi-infinite media toward the end of this chapter when we can extract more results from finite media solutions.

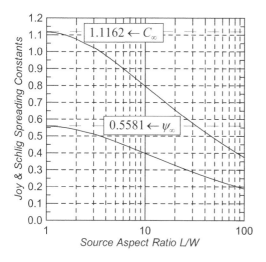

Figure 12.7. Plot of dimensionless thermal spreading resistance properties for rectangular surface source of zero thickness on semi-infinite media according to Joy and Schlig theory (1970).

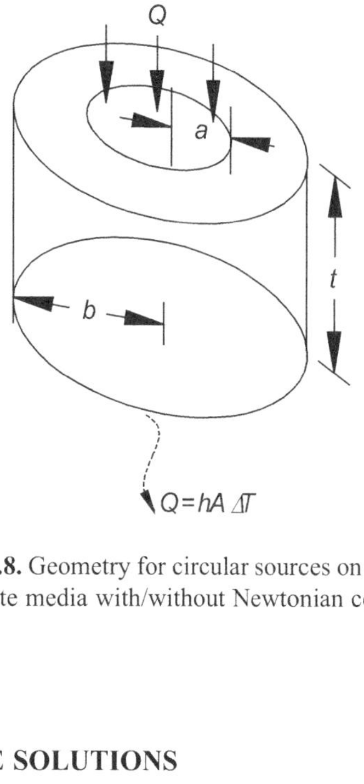

Figure 12.8. Geometry for circular sources on cylindrical, finite media with/without Newtonian cooling at $z = t$.

12.5 OTHER CIRCULAR SOURCE SOLUTIONS

Cylindrical geometry for circular sources on cylindrical, finite media is shown in Figure 12.8. Kennedy (1960) reported what was to become a widely used and referenced solution for the average source temperature, finite media, with an *isothermal base at $z = t$*, which we know is equivalent to $h \to \infty$.

$$\Delta \bar{T}_{Sp} = \left(\frac{4Q}{\pi ak}\right)\left(\frac{b}{a}\right)\sum_{m=1}^{\infty}\tanh\left(\lambda_m\frac{t}{b}\right)\frac{J_1^2\left[\lambda_m(a/b)\right]}{\lambda_m^3 J_o^2(\lambda_m)} \qquad \text{Kennedy} \quad (12.34)$$

$$\Delta T_U = Qt/(k\pi b^2)$$

where the total temperature rise from the substrate base to the source is

$$\Delta \bar{T} = \Delta \bar{T}_{Sp} + \Delta T_U$$

The adiabatic edges are satisfied by the boundary condition, $\partial T / \partial r \big|_{r=b} = 0$. The λ_m are such that the Bessel function $J_1(\lambda_m) = 0$. The reader will also note from Appendix vi, that

$$J_1(u) = 0 \text{ for } n\pi < u < (n+1/2)\pi, \ n = 1, 2, 3 \ldots \infty.$$

Kennedy included many design curves for easy application of Eq. (12.34).

Another highly referenced work was reported by Lee et al. (1995) wherein Newtonian cooling was added to the base, $z = t$. Lee and co-workers' results included both an average and a maximum source temperature rise.

$$\Delta \bar{T}_{Sp} = \left(\frac{4Q}{\pi ak} \right) \left(\frac{b}{a} \right) \sum_{m=1}^{\infty} \left\{ \frac{J_1^2 \left[\lambda_m \left(\dfrac{a}{b} \right) \right]}{\lambda_m^3 J_o^2 (\lambda_m)} \right\} \left\{ \frac{\tanh \left(\lambda_m \dfrac{t}{b} \right) + \dfrac{\lambda_m}{Bi}}{1 + \dfrac{\lambda_m}{Bi} \tanh \left(\lambda_m \dfrac{t}{b} \right)} \right\} \qquad \text{Lee et al. } (12.35)$$

$$Bi \equiv \text{Biot number} = \frac{hb}{k}, \quad J_1(\lambda_m) = 0, \Delta T_U = Q \left[\frac{t}{\pi k b^2} + \frac{1}{\pi h b^2} \right], \Delta \bar{T} = \Delta \bar{T}_{Sp} + \Delta T_U$$

$$\Delta T_{Sp} = \left(\frac{2Q}{\pi ak} \right) \sum_{m=1}^{\infty} \left\{ \frac{J_1 \left[\lambda_m \left(\dfrac{a}{b} \right) \right]}{\lambda_m^2 J_o^2 (\lambda_m)} \right\} \left\{ \frac{\tanh \left(\lambda_m \dfrac{t}{b} \right) + \dfrac{\lambda_m}{Bi}}{1 + \dfrac{\lambda_m}{Bi} \tanh \left(\lambda_m \dfrac{t}{b} \right)} \right\} \qquad \text{Lee et al. } (12.36)$$

$$Bi \equiv \text{Biot number} \equiv \frac{hb}{k}, \quad J_1(\lambda_m) = 0, \Delta T_U = \left(\frac{t}{\pi k b^2} + \frac{1}{h \pi b^2} \right) Q, \Delta T = \Delta T_{Sp} + \Delta T_U$$

Lee (1995) also offers approximations to Eqs. (12.35) and (12.36).

$$\Delta \bar{T}_{Sp} = \left(\frac{Q}{2ka\sqrt{\pi}} \right) \left(1 - \frac{a}{b} \right)^{3/2} \Phi_c, \ \Delta T_{Sp} = \left(\frac{Q}{ka\pi} \right) \left(1 - \frac{a}{b} \right) \Phi_c \qquad \text{Lee et al. } (12.37)$$

$$\text{where } \Phi_c = \frac{\tanh \left(\lambda_c \dfrac{t}{b} \right) + \dfrac{\lambda_c}{Bi}}{1 + \dfrac{\lambda_c}{Bi} \tanh \left(\lambda_c \dfrac{t}{b} \right)} \text{ with } \lambda_c = \pi + \frac{1}{\sqrt{\pi} \left(\dfrac{a}{b} \right)}$$

Design plots for the Kennedy and Lee et al. solutions are not included in this book. An equivalent solution for rectangular sources is included in Section 12.6.

12.6 RECTANGULAR SOURCE ON RECTANGULAR, FINITE-MEDIA WITH ONE CONVECTING SURFACE: THEORY

Ellison (2003) reported a solution to the spreading resistance problem that admits a source and substrate that are rectangular and not necessarily of either square or unity aspect ratios. In that work, the heat conduction equation in Cartesian coordinates was solved for a centered, single, surface-source with Newtonian cooling from the plane at $z = t$ that is opposite the source plane as shown in Figure 12.9.

$$\frac{\partial^2 T}{\partial x^2} + \frac{\partial^2 T}{\partial y^2} + \frac{\partial^2 T}{\partial z^2} = -\frac{Q_V}{k} \qquad \text{Steady-state heat cond. } (12.38)$$

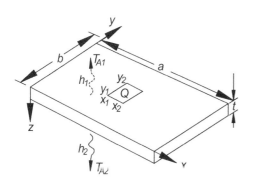

Figure 12.9. Rectangular surface source and substrate for thermal spreading analysis.

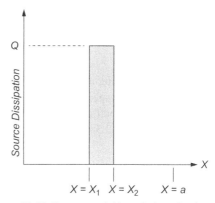

Figure 12.10. Source model in x-Q plane for thermal spreading analysis.

Boundary conditions are specified that permit problem solution as well as being physically sensible.

$$k\frac{\partial T}{\partial x} = 0;\ x = 0,a;\ k\frac{\partial T}{\partial y} = 0;\ y = 0,b \qquad \text{Edge boundary} \atop \text{conditions} \qquad (12.39)$$

$$k\frac{\partial T}{\partial z} = 0 \text{ at } z = 0: \qquad \text{Adiabatic}$$

$$k\frac{\partial T}{\partial z} + hT = 0 \text{ at } z = t: \qquad \text{Newtonian} \qquad (12.40)$$

The heat source is modeled as uniform over the source dimensions, as illustrated in Figure 12.10. A double Fourier cosine series is used to represent the volumetric source Q_V.

$$Q_V(r) = \sum_{l=0}^{\infty}\sum_{m=0}^{\infty}\varepsilon_l\varepsilon_m\phi_{lm}(z)\cos\left(\frac{l\pi x}{a}\right)\cos\left(\frac{m\pi y}{b}\right) \qquad (12.41)$$

$$\varepsilon_l = \begin{bmatrix} 1/2, & l = 0 \\ 1, & l \neq 0 \end{bmatrix} \qquad l = 0,1,2\ldots$$

$$\varepsilon_m = \begin{bmatrix} 1/2, & m = 0 \\ 1, & m \neq 0 \end{bmatrix} \qquad m = 0,1,2\ldots$$

The source location at $z = 0$ is specified mathematically by the Dirac delta function $\delta(z)$, for which a definition and some properties are given in Appendix *vii*. The Fourier coefficients ϕ_{lm} are readily calculated once the source magnitude Q_V and coordinates x_1, x_2, y_1, and y_2 are specified. The derivation of the ϕ_{lm} is given in detail in an appendix, but are listed here as Eq. (12.42). These equations are valid for any rectangular geometry with insulated boundaries and a source centered at $\bar{x}_s = (x_1 + x_2)/2$, $\bar{y}_s = (y_1 + y_2)/2$ with dimensions of $\Delta x_s = x_2 - x_1$, $\Delta y_s = y_2 - y_1$.

$$\phi_{00} = \frac{4q(x,y)\delta(z)\Delta x_s\Delta y_s}{ab}$$

$$\phi_{10} = \frac{8q(x,y)\delta(z)\Delta y_s}{l\pi b}\sin\left(\frac{l\pi\Delta x_s}{2a}\right)\cos\left(\frac{l\pi\bar{x}_s}{a}\right) \qquad (12.42)$$

$$\phi_{0m} = \frac{8q(x,y)\delta(z)\Delta x_s}{m\pi a}\sin\left(\frac{m\pi\Delta y_s}{2b}\right)\cos\left(\frac{m\pi\bar{y}_s}{b}\right)$$

$$\phi_{lm} = \frac{16q(x,y)\delta(z)}{lm\pi^2}\sin\left(\frac{l\pi\Delta x_s}{2a}\right)\sin\left(\frac{m\pi\Delta y_s}{2b}\right)\cos\left(\frac{l\pi\bar{x}_s}{a}\right)\cos\left(\frac{m\pi\bar{y}_s}{b}\right)$$

The temperature T may be immediately written as a Fourier series that satisfies the edge boundary conditions, Eq. (12.39).

$$T(r) = \sum_{l=0}^{\infty} \sum_{m=0}^{\infty} \varepsilon_l \varepsilon_m \Psi_{lm}(z) \cos\left(\frac{l\pi x}{a}\right) \cos\left(\frac{m\pi y}{b}\right) \tag{12.43}$$

where the Ψ_{lm} are Fourier coefficients for the temperature and $\varepsilon_l, \varepsilon_m$ are defined with Eq. (12.41).

Although we have specified a solution that should satisfy the *PDE*, we have yet to determine the temperature Fourier coefficients Ψ_{lm}. We work toward this objective by substituting Eqs. (12.41) and (12.43) into Eq. (12.38).

$$\sum_{l=0}^{\infty} \sum_{m=0}^{\infty} \left\{-\left[\left(\frac{l\pi}{a}\right)^2 + \left(\frac{m\pi}{b}\right)^2\right]\Psi_{lm} + \frac{d^2\Psi_{lm}}{dz^2}\right\} \varepsilon_l \varepsilon_m \cos\left(\frac{l\pi x}{a}\right) \cos\left(\frac{m\pi y}{b}\right)$$

$$= -(1/k)\sum_{l=0}^{\infty} \sum_{m=0}^{\infty} \phi_{lm} \varepsilon_l \varepsilon_m \cos\left(\frac{l\pi x}{a}\right) \cos\left(\frac{m\pi y}{b}\right)$$

Examining this result, we see that we can set equal, the coefficients of like terms of the product, $\varepsilon_l \varepsilon_m \cos(l\pi x/a)\cos(m\pi y/b)$, on the left- and right-hand side so that

$$\frac{d^2\Psi_{lm}}{dz^2} - \gamma_{lm}^2 \Psi_{lm} = -\frac{1}{k}\phi_{lm} \tag{12.44}$$

$$\gamma_{lm}^2 = \alpha_l^2 + \beta_m^2, \ \alpha_l^2 = (l\pi/a)^2, \ \beta_m^2 = (m\pi/b)^2$$

We have reduced our problem of solving a three-dimensional, partial differential equation, to the problem of solving a one-dimensional *PDE*, Eq. (12.44). A Green's function method is used to find Ψ_{lm}. The interested reader is advised to read Appendix *ix* on this subject. In that appendix, it is shown that the Ψ_{lm} we desire is found from the formula

$$\Psi_{lm} = \frac{1}{k} \int_{z'=0}^{z'=t} G_{lm}(z|z')\phi_{lm}(z')dz' \tag{12.45}$$

If you substitute Eq. (12.43) into Eq. (12.40), you will find that the boundary conditions become

$$k\frac{d\Psi_{lm}}{dz} = 0 \text{ at } z = 0: \qquad \text{Adiabatic}$$

$$k\frac{d\Psi_{lm}}{dz} + h\Psi_{lm} = 0 \text{ at } z = t: \qquad \text{Newtonian} \tag{12.46}$$

The Green's function G_{lm} has the properties that it satisfies Eqs. (12.44) and (12.46), and a few other conditions that are derived in Appendix *ix*. Two of those results are

$$\frac{d^2 G_{lm}(z|z')}{dz'^2} - \gamma_{lm}^2 G_{lm}(z|z') = 0, \ z' \neq z \tag{12.47}$$

$$k\frac{dG_{lm}(z|z')}{dz'} = 0 \text{ at } z' = 0: \qquad \text{Adiabatic}$$

$$k\frac{dG_{lm}(z|z')}{dz'} + hG_{lm}(z|z') = 0 \text{ at } z' = t: \qquad \text{Newtonian} \tag{12.48}$$

The first issue to note in looking for a solution to Eq. (12.47) is that we need separate solutions for the two regions separated by $z' = z$.

We also note that there are separate and distinct solutions for $\gamma_{lm} = 0$ and $\gamma_{lm} \neq 0$. Thus we write

$$
\begin{aligned}
G_{lm}\left(z'|z\right) &= Az' + B, \quad z' < z \\
G_{lm}\left(z|z'\right) &= Cz' + D, \quad z' > z
\end{aligned} \quad \gamma_{lm} = 0 \tag{12.49}
$$

$$
\begin{aligned}
G_{lm}\left(z'|z\right) &= E\sinh\left(\gamma_m z'\right) + F\cosh\left(\gamma_m z'\right), \quad z' < z \\
G_{lm}\left(z|z'\right) &= G\sinh\left(\gamma_m z'\right) + H\cosh\left(\gamma_m z'\right), \quad z' > z
\end{aligned} \quad \gamma_m \neq 0 \tag{12.50}
$$

The solution scheme becomes one of finding the coefficients A, B, C, D, E, F, G, and H by applying the adiabatic and Newtonian cooling boundary conditions, plus conditions (ix.6-2) and (ix.6-3) to Eqs. (12.49) and (12.50), so that we have four equations to get four unknowns. The algebra of finding the solution coefficients is not included here as it is very tedious reading. If the reader attempts to solve this problem, the best advice is to use Cramer's rule for finding the four coefficients as it has the advantage of imposing some degree of organization right from the beginning. The results are

$$
G_{lm}\left(z_< | z_>\right) = -\left(z_> - t - \frac{k}{h}\right) \qquad \gamma_{lm} = 0 \tag{12.51a}
$$

$$
G_{lm}\left(z_< | z_>\right) = -\frac{1}{\gamma_{lm}}\left\{
\begin{aligned}
&\sinh\left(\gamma_{lm}z_>\right)\cosh\left(\gamma_{lm}z_<\right) - \\
&\left[\frac{\cosh(\gamma_{lm}t) + \left(h/\gamma_{lm}k\right)\sinh\left(\gamma_{lm}t\right)}{\sinh(\gamma_{lm}t) + \left(h/\gamma_{lm}k\right)\cosh\left(\gamma_{lm}t\right)}\right]\cosh\left(\gamma_{lm}z_>\right)\cosh\left(\gamma_{lm}z_<\right)
\end{aligned}
\right\} \quad \gamma_{lm} \neq 0 \tag{12.52a}
$$

where we have used the Green's function convention of $z_>, z_<$ to mean the greater and lesser of z, z'. More explicitly,

$$
0 \leq z' < z
$$

$$
G_{lm}\left(z'|z\right) = \left(\frac{k}{h} + t - z\right)
$$

$$
z < z' \leq t
$$

$$
G_{lm}\left(z|z'\right) = \left(\frac{k}{h} + t - z'\right)
$$

$$
\gamma_{lm} = 0 \tag{12.51b}
$$

$$
0 \leq z' < z
$$

$$
G_{lm}\left(z'|z\right) = -\frac{1}{\gamma_{lm}}\left\{
\begin{aligned}
&\sinh\left(\gamma_{lm}z\right)\cosh\left(\gamma_{lm}z'\right) - \\
&\left[\frac{\cosh(\gamma_{lm}t) + \left(h/\gamma_{lm}k\right)\sinh\left(\gamma_{lm}t\right)}{\sinh(\gamma_{lm}t) + \left(h/\gamma_{lm}k\right)\cosh\left(\gamma_{lm}t\right)}\right]\cosh\left(\gamma_{lm}z\right)\cosh\left(\gamma_{lm}z'\right)
\end{aligned}
\right\} \quad \gamma_{lm} \neq 0
$$

$$
z < z' \leq t
$$

$$
G_{lm}\left(z|z'\right) = -\frac{1}{\gamma_{lm}}\left\{
\begin{aligned}
&\sinh\left(\gamma_{lm}z'\right)\cosh\left(\gamma_{lm}z\right) - \\
&\left[\frac{\cosh(\gamma_{lm}t) + \left(h/\gamma_{lm}k\right)\sinh\left(\gamma_{lm}t\right)}{\sinh(\gamma_{lm}t) + \left(h/\gamma_{lm}k\right)\cosh\left(\gamma_{lm}t\right)}\right]\cosh\left(\gamma_{lm}z'\right)\cosh\left(\gamma_{lm}z\right)
\end{aligned}
\right\} \quad \gamma_{lm} \neq 0
$$

$$
\tag{12.52b}
$$

Recalling that the Eqs. (12.42) for the Fourier coefficients, ϕ_{lm}, all have a direct dependence on $\delta(z)$, the integration in Eq. (12.45) is readily accomplished using the property from Eq. (vi-7). Remembering that our source is at $z = 0$ and therefore z (field point) $\geq z'$ (source point),

$$
\Psi_{00} = \frac{1}{k}\int_{z'=0}^{z'=t} G_{00}\left(z|z'\right)\phi_{00}\left(z'\right)dz'
$$

and since we *always* have $z \geq z'$, $G_{00} = (k/h) + t - z$. Integration of the preceding equation is easy:

$$\Psi_{00} = \frac{1}{k} \int_{z'=0}^{z'=t} \left(\frac{k}{h} + t - z \right) \frac{4q(x,y)\delta(z')\Delta x_s \Delta y_s}{ab} dz'$$

$$\Psi_{00} = \frac{4Q}{kab} \left(\frac{k}{h} + t - z \right) \tag{12.53}$$

The integrations to get Ψ_{10}, Ψ_{0m}, Ψ_{lm} (neither l nor m are zero in the latter) are performed in the same manner using $\Delta x = x_2 - x_1$, $\Delta y = y_2 - y_1$, $x_c = (x_1 + x_2)/2$ and $y_c = (y_1 + y_2)/2$ to arrive at

$$\Psi_{10} = \frac{8Qa}{\pi^2 bl^2 k \Delta x} \sin\left(\frac{l\pi \Delta x}{2a} \right) \cos\left(\frac{l\pi x_c}{a} \right) \left\{ \frac{\cosh\left[\dfrac{l\pi}{a}(z-t) \right] - \left(\dfrac{ha}{l\pi k} \right) \sinh\left[\dfrac{l\pi}{a}(z-t) \right]}{\sinh\left(\dfrac{l\pi t}{a} \right) + \left(\dfrac{ha}{l\pi k} \right) \cosh\left(\dfrac{l\pi t}{a} \right)} \right\} \tag{12.54}$$

$$\Psi_{0m} = \frac{8Qb}{\pi^2 am^2 k \Delta y} \sin\left(\frac{m\pi \Delta y}{2b} \right) \cos\left(\frac{m\pi y_c}{b} \right)$$

$$\times \left\{ \frac{\cosh\left[\dfrac{m\pi}{b}(z-t) \right] - \left(\dfrac{hb}{m\pi k} \right) \sinh\left[\dfrac{m\pi}{b}(z-t) \right]}{\sinh\left(\dfrac{m\pi t}{b} \right) + \left(\dfrac{hb}{m\pi k} \right) \cosh\left(\dfrac{\pi t}{b} \right)} \right\} \tag{12.55}$$

$$\Psi_{lm} = \frac{16Q}{\pi^2 lmk \Delta x \Delta y}$$

$$\times \sin\left[\frac{l\pi}{2a} \Delta x \right] \cos\left[\frac{l\pi}{a} x_c \right] \sin\left[\frac{m\pi}{2b} \Delta y \right] \cos\left[\frac{m\pi}{b} y_c \right] \tag{12.56}$$

$$\times \left\{ \frac{\cosh\left[\sqrt{\left(\dfrac{l\pi}{a}\right)^2 + \left(\dfrac{m\pi}{b}\right)^2}(z-t) \right] - \dfrac{h}{k\sqrt{\left(\dfrac{l\pi}{a}\right)^2 + \left(\dfrac{m\pi}{b}\right)^2}} \sinh\left[\sqrt{\left(\dfrac{l\pi}{a}\right)^2 + \left(\dfrac{m\pi}{b}\right)^2}(z-t) \right]}{\sqrt{\left(\dfrac{l\pi}{a}\right)^2 + \left(\dfrac{m\pi}{b}\right)^2} \left\{ \sinh\left[\sqrt{\left(\dfrac{l\pi}{a}\right)^2 + \left(\dfrac{m\pi}{b}\right)^2} t \right] + \dfrac{h}{k\sqrt{\left(\dfrac{l\pi}{a}\right)^2 + \left(\dfrac{m\pi}{b}\right)^2}} \cosh\left[\sqrt{\left(\dfrac{l\pi}{a}\right)^2 + \left(\dfrac{m\pi}{b}\right)^2} t \right] \right\}} \right\}$$

Noting that the ambient temperature is set to zero, we are now ready to write out the complete Fourier series using Eq. (12.43).

$$T(r) = \sum_{l=0}^{\infty} \sum_{m=0}^{\infty} \varepsilon_l \varepsilon_m \Psi_{lm}(z) \cos\left(\frac{l\pi x}{a} \right) \cos\left(\frac{m\pi y}{b} \right)$$

Expanding the preceding Fourier series expression,

$$T(r) = \frac{1}{4}\Psi_{00} + \frac{1}{2}\sum_{l=1}^{\infty}\Psi_{l0}\cos\left(\frac{l\pi x}{a}\right) + \frac{1}{2}\sum_{m=1}^{\infty}\Psi_{0m}\cos\left(\frac{m\pi y}{b}\right) \qquad (12.57)$$
$$+ \sum_{l=1}^{\infty}\sum_{m=1}^{\infty}\Psi_{lm}\cos\left(\frac{l\pi x}{a}\right)\cos\left(\frac{m\pi y}{b}\right)$$

Next we divide Eq. (12.57) by Q, multiply by $k\sqrt{\Delta x \Delta y}$, and insert the appropriate Fourier coefficients Ψ_{lm} from Eqs. (12.54) - (12.56), to get a dimensionless total resistance ψ from source to ambient.

Total, dimensionless resistance:

$$\psi = \frac{\sqrt{\Delta x \Delta y}}{ab}\left(t - z + \frac{k}{h}\right)$$

$$+ \frac{4}{\pi^2}\left(\frac{a}{b}\right)\left(\frac{\sqrt{\Delta x \Delta y}}{\Delta x}\right)\sum_{l=1}^{\infty}\frac{1}{l^2}\sin\left(\frac{l\pi}{2}\frac{\Delta x}{a}\right)\cos\left(l\pi \frac{x_c}{a}\right)\cos\left(l\pi \frac{x}{a}\right)$$

$$\times\left\{\frac{\cosh\left[l\pi \frac{\langle z-t\rangle}{a}\right] - \left[\frac{\langle ha/k\rangle}{l\pi}\right]\sinh\left[l\pi \frac{\langle z-t\rangle}{a}\right]}{\sinh\left(l\pi \frac{t}{a}\right) + \left[\frac{\langle ha/k\rangle}{l\pi}\right]\cosh\left(l\pi \frac{t}{a}\right)}\right\}$$

$$+ \frac{4}{\pi^2}\left(\frac{b}{a}\right)\left(\frac{\sqrt{\Delta x \Delta y}}{\Delta y}\right)\sum_{m=1}^{\infty}\frac{1}{m^2}\sin\left(\frac{m\pi}{2}\frac{\Delta y}{b}\right)\cos\left(m\pi \frac{y_c}{b}\right)\cos\left(m\pi \frac{y}{b}\right)$$

$$\times\left\{\frac{\cosh\left[m\pi \frac{\langle z-t\rangle}{b}\right] - \left[\frac{\langle hb/k\rangle}{m\pi}\right]\sinh\left[m\pi \frac{\langle z-t\rangle}{b}\right]}{\sinh\left(m\pi \frac{t}{b}\right) + \left[\frac{\langle hb/k\rangle}{m\pi}\right]\cosh\left(m\pi \frac{t}{b}\right)}\right\}$$

$$+ \frac{16}{\pi^2}\left(\frac{a}{\sqrt{\Delta x \Delta y}}\right)\sum_{l=1}^{\infty}\sum_{m=1}^{\infty}\frac{1}{lm}\sin\left(\frac{l\pi}{2}\frac{\Delta x}{a}\right)\sin\left(\frac{m\pi}{2}\frac{\Delta y}{b}\right)$$

$$\times\cos\left(l\pi \frac{x_c}{a}\right)\cos\left(m\pi \frac{y_c}{b}\right)\cos\left(l\pi \frac{x}{a}\right)\cos\left(m\pi \frac{y}{b}\right)$$

$$\times\left\{\frac{\cosh\left[\pi\sqrt{l^2+m^2\left(\frac{a}{b}\right)^2}\left(\frac{z-t}{a}\right)\right] - \left(\frac{ha/k}{\pi\sqrt{l^2+m^2\left(\frac{a}{b}\right)^2}}\right)\sinh\left[\pi\sqrt{l^2+m^2\left(\frac{a}{b}\right)^2}\left(\frac{z-t}{a}\right)\right]}{\pi\sqrt{l^2+m^2\left(\frac{a}{b}\right)^2}\left\{\sinh\left[\pi\sqrt{l^2+m^2\left(\frac{a}{b}\right)^2}\left(\frac{t}{a}\right)\right] + \left(\frac{ha/k}{\pi\sqrt{l^2+m^2\left(\frac{a}{b}\right)^2}}\right)\cosh\left[\pi\sqrt{l^2+m^2\left(\frac{a}{b}\right)^2}\left(\frac{t}{a}\right)\right]\right\}}\right\}$$

$$(12.58)$$

Clearly, there are numerous variables that the spreading resistance is dependent upon, thus it is appropriate to cast these variables in a dimensionless form. The source is assumed to be centered on the substrate so we can set these position variables, noting that ψ_U is the first term in Eq. (12.58):

$$x_c/a = (a/2)/a = 1/2, \ y_c/b = (b/2)/b = 1/2, \ y = b/2$$

$$\rho = a/b, \ \alpha = \Delta x/a, \ \beta = \Delta y/a, \ \mu = x/a, \ \xi = z/a$$

$$\tau = t/a, \ Bi = ha/k, \ Bi \cdot \tau = (ha/k)(t/a) = ht/k \tag{12.59}$$

$$\psi = Rk\sqrt{\Delta x \Delta y}, \quad \psi = \psi_U + \psi_{Sp}$$

We wish to have two formulae for the spreading problem of single-sided Newtonian cooling. The first is for contour plotting in the *xz*-plane at $y = b/2$. <u>Contouring</u> requires both the *uniform* and the *spreading* contributions.

Contouring dimensionless "resistance":

$$\psi = Rk\sqrt{\Delta x \Delta y} = \sqrt{\alpha\beta}\,\rho\left(\tau - \xi + \frac{1}{Bi}\right)$$

$$+\frac{4\rho}{\pi^2}\sqrt{\frac{\beta}{\alpha}}\sum_{l=1}^{\infty}\frac{1}{l^2}\sin\left(\frac{l\pi\alpha}{2}\right)\cos\left(\frac{l\pi}{2}\right)\cos(l\pi\mu)\left\{\frac{\cosh\left[l\pi(\xi-\tau)\right]-\left(\dfrac{Bi\tau}{l\pi\tau}\right)\sinh\left[l\pi(\xi-\tau)\right]}{\sinh(l\pi\tau)+\left(\dfrac{Bi\tau}{l\pi\tau}\right)\cosh(l\pi\tau)}\right\}$$

$$+\frac{4}{\pi^2}\left(\frac{1}{\rho}\right)\sqrt{\frac{\alpha}{\beta}}\sum_{m=1}^{\infty}\frac{1}{m^2}\sin\left(\frac{m\pi\beta\rho}{2}\right)\cos^2\left(\frac{m\pi}{2}\right)\left\{\frac{\cosh\left[m\pi(\xi-\tau)\rho\right]-\left(\dfrac{Bi\tau}{m\pi\tau\rho}\right)\sinh\left[m\pi\rho(\xi-\tau)\right]}{\sinh(m\pi\rho\tau)+\left[\dfrac{Bi\tau}{m\pi\tau\rho}\right]\cosh(m\pi\rho\tau)}\right\}$$

$$+\frac{16}{\pi^2}\frac{1}{\sqrt{\alpha\beta}}\sum_{l=1}^{\infty}\sum_{m=1}^{\infty}\frac{1}{lm}\sin\left(\frac{l\pi}{2}\alpha\right)\sin\left(\frac{m\pi}{2}\beta\rho\right)\cos\left(\frac{l\pi}{2}\right)\cos^2\left(\frac{m\pi}{2}\right)\cos(l\pi\mu)$$

$$\times\left\{\frac{\cosh\left[\pi\sqrt{l^2+m^2\rho^2}(\xi-\tau)\right]-\left(\dfrac{Bi\tau}{\pi\tau\sqrt{l^2+m^2\rho^2}}\right)\sinh\left[\pi\sqrt{l^2+m^2\rho^2}(\xi-\tau)\right]}{\pi\sqrt{l^2+m^2\rho^2}\left[\sinh\left(\pi\tau\sqrt{l^2+m^2\rho^2}\right)+\left(\dfrac{Bi\tau}{\pi\tau\sqrt{l^2+m^2\rho^2}}\right)\cosh\left(\pi\tau\sqrt{l^2+m^2\rho^2}\right)\right]}\right\} \tag{12.60}$$

The <u>maximum spreading resistance</u> is calculated at the source center and uses the same *uniform* constristion as Eq. (12.60), but $\xi = 0$, thus $z = 0 \ (\xi = 0), x = a/2 \ (\mu = 1/2) \ (y = b/2)$.

Maximum spreading resistance:

$$\psi_{Sp} = \frac{4\rho}{\pi^2}\sqrt{\frac{\beta}{\alpha}}\sum_{l=1}^{\infty}\frac{1}{l^2}\sin\left(\frac{l\pi\alpha}{2}\right)\cos^2\left(\frac{l\pi}{2}\right)\left[\frac{\cosh(l\pi\tau)+\left(\dfrac{Bi\tau}{l\pi\tau}\right)\sinh(l\pi\tau)}{\sinh(l\pi\tau)+\left(\dfrac{Bi\tau}{l\pi\tau}\right)\cosh(l\pi\tau)}\right]$$

$$+\frac{4}{\pi^2\rho}\sqrt{\frac{\alpha}{\beta}}\sum_{m=1}^{\infty}\frac{1}{m^2}\sin\left(\frac{m\pi\beta\rho}{2}\right)\cos^2\left(\frac{m\pi}{2}\right)\left[\frac{\cosh(m\pi\rho\tau)+\left(\dfrac{Bi\tau}{m\pi\rho\tau}\right)\sinh(m\pi\rho\tau)}{\sinh(m\pi\rho\tau)+\left(\dfrac{Bi\tau}{m\pi\rho\tau}\right)\cosh(m\pi\rho\tau)}\right]$$

$$\tag{12.61 cont.}$$

Maximum spreading resistance (cont.):

$$+\frac{16}{\pi^2 \sqrt{\alpha\beta}} \sum_{l=1}^{\infty} \sum_{m=1}^{\infty} \frac{1}{lm} \cos^2\left(\frac{l\pi}{2}\right) \cos^2\left(\frac{m\pi}{2}\right) \sin\left(\frac{l\pi\alpha}{2}\right) \sin\left(\frac{m\pi\beta\rho}{2}\right)$$

$$\times \left\{ \frac{\cosh\left(\pi\sqrt{l^2+m^2\rho^2}\,\tau\right) + \left(\dfrac{Bi\tau}{\pi\tau\sqrt{l^2+m^2\rho^2}}\right)\sinh\left(\pi\tau\sqrt{l^2+m^2\rho^2}\right)}{\pi\sqrt{l^2+m^2\rho^2}\left[\sinh\left(\pi\tau\sqrt{l^2+m^2\rho^2}\right) + \left(\dfrac{Bi\tau}{\pi\tau\sqrt{l^2+m^2\rho^2}}\right)\cosh\left(\pi\tau\sqrt{l^2+m^2\rho^2}\right)\right]} \right\} \quad (12.61)$$

The contour computation time can be reduced by noting that

$$\cos\left(\frac{l\pi}{2}\right) = \begin{cases} 0 & l = odd \\ -1 & l = 2,6,10...; \\ +1 & l = 4,8,12... \end{cases} \qquad \cos\left(\frac{m\pi}{2}\right) = \begin{cases} 0 & m = odd \\ -1 & m = 2,6,10... \\ +1 & m = 4,8,12... \end{cases}$$

Then set
$$l \to 2l, \; l = 1,2... \text{ then } \cos\left(\frac{l\pi}{2}\right) \to \cos(l\pi) = (-1)^l, \qquad l = 1,2,3...$$

$$m \to 2m, \; m = 1,2... \text{ then } \cos\left(\frac{m\pi}{2}\right) \to \cos(m\pi) = (-1)^m, \; m = 1,2,3...$$

The maximum spreading computation time can be similarly reduced by using

$$l \to 2l, \quad \text{then } \cos^2(l\pi/2) \to 1, \quad l = 1,2,3...$$

$$m \to 2m, \quad \text{then } \cos^2(m\pi/2) \to 1, \quad m = 1,2,3...$$

and dividing ψ_{Sp} numerators and denominators by cosh( ) to get tanh( ), which eliminates overflow in cosh( ) and sinh( ) for large arguments.

$$\mu = x/a, \xi = z/a, \rho = a/b, \alpha = \Delta x/a, \beta = \Delta y/a, \tau = t/a, Bi\tau = ht/k$$

$$\psi = \psi_U + \psi_{Sp}, \qquad \psi_{Uniform} = \rho\sqrt{\alpha\beta}\left(\tau - \xi + \frac{\tau}{Bi\tau}\right) \qquad \text{Contours} \qquad (12.62)$$

$$\psi_{Sp} = \frac{\rho}{\pi^2}\sqrt{\frac{\beta}{\alpha}} \sum_{l=1}^{\infty} \frac{(-1)^l}{l^2} \sin(l\pi\alpha)\cos(2l\pi\mu)\left\{ \frac{\cosh[2l\pi(\xi-\tau)] - \left(\dfrac{Bi\tau}{2l\pi\tau}\right)\sinh[2l\pi(\xi-\tau)]}{\sinh(2l\pi\tau) + \left(\dfrac{Bi\tau}{2l\pi\tau}\right)\cosh(2l\pi\tau)} \right\}$$

$$+ \left(\frac{1}{\pi^2}\right)\left(\frac{1}{\rho}\right)\sqrt{\frac{\alpha}{\beta}} \sum_{m=1}^{\infty} \frac{1}{m^2} \sin(m\pi\beta\rho)\left\{ \frac{\cosh[2m\pi(\xi-\tau)\rho] - \left(\dfrac{Bi\tau}{2m\pi\tau\rho}\right)\sinh[2m\pi(\xi-\tau)\rho]}{\sinh(2m\pi\tau\rho) + \left(\dfrac{Bi\tau}{2m\pi\tau\rho}\right)\cosh(2m\pi\tau\rho)} \right\}$$

$$+ \left(\frac{1}{\pi^2}\right)\left(\frac{1}{\sqrt{\alpha\beta}}\right) \sum_{l=1}^{\infty} \sum_{m=1}^{\infty} \frac{(-1)^l}{lm} \cos(2l\pi\mu)\sin(l\pi\alpha)\sin(m\pi\beta\rho)$$

$$\times \left[\frac{\cosh\left[2\pi\sqrt{l^2+m^2\rho^2}\,(\xi-\tau)\right] - \left(\dfrac{Bi\tau}{2\pi\tau\sqrt{l^2+m^2\rho^2}}\right)\sinh\left[2\pi\sqrt{l^2+m^2\rho^2}\,(\xi-\tau)\right]}{2\pi\sqrt{l^2+m^2\rho^2}\left[\sinh\left(2\pi\sqrt{l^2+m^2\rho^2}\,\tau\right) + \left(\dfrac{Bi\tau}{2\pi\tau\sqrt{l^2+m^2\rho^2}}\right)\cosh\left(2\pi\sqrt{l^2+m^2\rho^2}\,\tau\right)\right]} \right]$$

For maximum spreading resistance $\psi_{Sp} = R_{Sp} k \sqrt{\Delta x \Delta y}$:

$$\rho = a/b,\ \alpha = \Delta x/a,\ \beta = \Delta y/a,\ \tau = t/a,\ Bi\tau = ht/k$$

Max. resistance

$$\psi_{Sp} = \frac{\rho}{\pi^2} \sqrt{\frac{\beta}{\alpha}} \sum_{l=1}^{\infty} \frac{1}{l^2} \sin(l\pi\alpha) \left[\frac{1 + \left(\dfrac{Bi\tau}{2l\pi\tau}\right) \tanh(2l\pi\tau)}{\left(\dfrac{Bi\tau}{2l\pi\tau}\right) + \tanh(2l\pi\tau)} \right]$$

$$+ \left(\frac{1}{\pi^2}\right) \left(\frac{1}{\rho}\right) \sqrt{\frac{\alpha}{\beta}} \sum_{m=1}^{\infty} \frac{1}{m^2} \sin(m\pi\beta\rho) \left[\frac{1 + \left(\dfrac{Bi\tau}{2m\pi\rho\tau}\right) \tanh(2m\pi\rho\tau)}{\left(\dfrac{Bi\tau}{2m\pi\rho\tau}\right) + \tanh(2m\pi\rho\tau)} \right]$$

$$+ \left(\frac{4}{\pi^2}\right) \left(\frac{1}{\sqrt{\alpha\beta}}\right) \sum_{l=1}^{\infty} \sum_{m=1}^{\infty} \frac{1}{lm} \sin(l\pi\alpha) \sin(m\pi\beta\rho)$$

$$\times \left\{ \frac{1 + \left(\dfrac{Bi\tau}{2\pi\tau\sqrt{l^2 + m^2\rho^2}}\right) \tanh\left(2\pi\tau\sqrt{l^2 + m^2\rho^2}\right)}{2\pi\sqrt{l^2 + m^2\rho^2} \left[\left(\dfrac{Bi\tau}{2\pi\tau\sqrt{l^2 + m^2\rho^2}}\right) + \tanh\left(2\pi\tau\sqrt{l^2 + m^2\rho^2}\right) \right]} \right\} \quad (12.63)$$

When the temperature is averaged over the source dimensions $\Delta x, \Delta y$ to give an average spreading resistance,

Source averaged resistance:

$$\bar{\psi}_{Sp} = \frac{\rho}{\pi^3\alpha} \sqrt{\frac{\beta}{\alpha}} \sum_{l=1}^{\infty} \frac{1}{l^3} \sin^2(l\pi\alpha) \left[\frac{1 + \left(\dfrac{Bi\tau}{2l\pi\tau}\right) \tanh(2l\pi\tau)}{\left(\dfrac{Bi\tau}{2l\pi\tau}\right) + \tanh(2l\pi\tau)} \right]$$

$$+ \left(\frac{1}{\pi^3}\right) \left(\frac{1}{\rho^2\beta}\right) \sqrt{\frac{\alpha}{\beta}} \sum_{m=1}^{\infty} \frac{1}{m^3} \sin^2(m\pi\beta\rho) \left[\frac{1 + \left(\dfrac{Bi\tau}{2m\pi\rho\tau}\right) \tanh(2m\pi\rho\tau)}{\left(\dfrac{Bi\tau}{2m\pi\rho\tau}\right) + \tanh(2m\pi\rho\tau)} \right]$$

$$+ \left(\frac{4}{\pi^4\rho\alpha\beta}\right) \left(\frac{1}{\sqrt{\alpha\beta}}\right) \sum_{l=1}^{\infty} \sum_{m=1}^{\infty} \frac{1}{l^2 m^2} \sin^2(l\pi\alpha) \sin^2(m\pi\beta\rho)$$

$$\times \left\{ \frac{1 + \left(\dfrac{Bi\tau}{2\pi\tau\sqrt{l^2 + m^2\rho^2}}\right) \tanh\left(2\pi\tau\sqrt{l^2 + m^2\rho^2}\right)}{2\pi\sqrt{l^2 + m^2\rho^2} \left[\left(\dfrac{Bi\tau}{2\pi\tau\sqrt{l^2 + m^2\rho^2}}\right) + \tanh\left(2\pi\tau\sqrt{l^2 + m^2\rho^2}\right) \right]} \right\} \quad (12.64)$$

12.7 RECTANGULAR SOURCE ON RECTANGULAR, FINITE-MEDIA: DESIGN CURVES

If you choose to use Eq. (12.62) for contour plotting, it is reasonable to put the formula in a resistance form, but just remember that you will be plotting temperature rise per unit source dissipation. Also, don't forget to add the ψ_U contribution.

Equations (12.63) and (12.64) are not difficult to code into a math scratchpad or other high level software and, of course once coded, you have it forever and need not resort to using approximations that may or may not be accurate for your application. Professor Yovanovich once reminded a thermal conference audience of the readily available solutions accessible to all of us -- the Carslaw and Jaeger (1986) book is a perfect example. It is surprising how many folks are reluctant to travel this route; they certainly have their excuses. Perhaps the readers of this book will be sufficiently motivated to create their own versions of Eqs. (12.63) and (12.64). In the meantime, the spreading resistance for square sources is reproduced in Figures 12.11 to 12.19 for the maximum spreading resistance and Figures 12.20 to 12.28 for the source-averaged spreading resistance. Only the spreading portions, ψ_{Sp} and $\bar{\psi}_{Sp}$, are plotted. You need to add the ψ_U contribution to get the total resistance from source to ambient. The relevant independent variables for the plots are

$$\psi_{Sp} = k\sqrt{\Delta x \Delta y}\,R_{Sp}\,;\; \tau = t/a\,;\; Biot = ha/k\,;\; Biot \cdot \tau = ht/k$$

Both the maximum and average spreading resistances are safely in the *thickness-independent* region for $\tau = t/a > 0.4$. The curves where $Biot \cdot \tau \to \infty$ correspond to an isothermal substrate base. Sample calculations for air-cooled devices tend to correspond close to the $Biot \cdot \tau \to 0$ plots.

If you now jump ahead and look at Figures 12.11 to 12.28, you will note the limiting values of the spreading resistance for the thickness-independent portions of the graphs. These thickness independent resistances are plotted in Figure 12.29, which immediately follows the design curves. When the result is thickness independent, the planar dimensions are still finite until the source edge dimension is less than about a hundredth of the plate edge dimension. The maximum resistance points don't all fall nicely onto a smooth curve, particularly for $\alpha < 0.01$, due to errors incurred by the Fourier series truncation. As the dimensionless source size becomes very small, the maximum and average resistances asymptotically approach 0.56 and 0.47, respectively. It is interesting to compare these results with those from derivations earlier in this chapter as well as other researchers in Table 12.1. The tabulated data does not all have the same number of significant figures, but averages from the various investigators are very close to those from Figure 12.29.

Readers interested in more detail on the subject of thermal spreading resistance theory will profit from reading papers by Yovanovich et al. (1998) and Muzychka et al. (2001). Now we will immediately jump to an application example, the geometry for which is shown in Figure 12.30.

12.8 APPLICATION EXAMPLE: HEAT SOURCE CENTERED ON A HEAT SINK (ELLISON, 2003)

A single power transistor is centered on an aluminum heat sink as shown in Figure 12.30. The dimensions are $\Delta x = \Delta y = 0.394$ in., $a = b = 3.94$ in., $t = 0.197$ in., $L = 1.0$ in. The total heat transfer coefficient is $h = 0.0097$ W/(in.$^2 \cdot$ °C) for the finned side only, and the thermal conductivity is $k_{Al} = 5.1$ W/(in. $\cdot$ °C). The source-plane surface is adiabatic, with the exception of the source. We wish to calculate the total thermal resistance from the source to ambient.

The total convection area is calculated, neglecting the external area of the two outer fins, as

$$A_f = 2N_f Lb + ab = 2(14)(1.0\,\text{in.})(3.94\,\text{in.}) + (3.94\,\text{in.})^2 = 125.8\,\text{in.}^2$$

This problem is continued directly following the spreading resistance design curves.

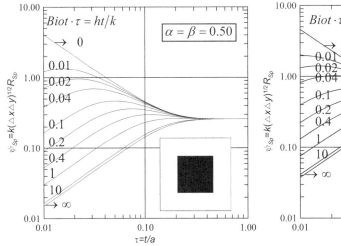

Figure 12.11. Maximum spreading resistance (From: Ellison, G.N., *IEEE Trans. CPT*, vol. 26, no. 3, 439-454, © 2003 *IEEE*. With permission).

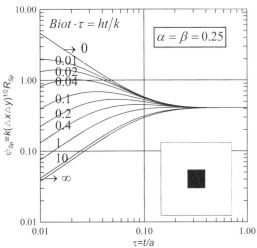

Figure 12.12. Maximum spreading resistance (From: Ellison, G.N., *IEEE Trans. CPT*, vol. 26, no. 3, 439-454, © 2003 *IEEE*. With permission).

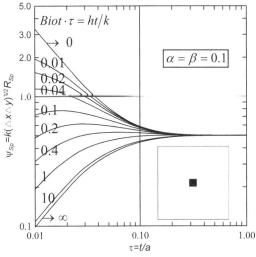

Figure 12.13. Maximum spreading resistance (From: Ellison, G.N., *IEEE Trans. CPT*, vol. 26, no. 3, 439-454, © 2003 *IEEE*. With permission).

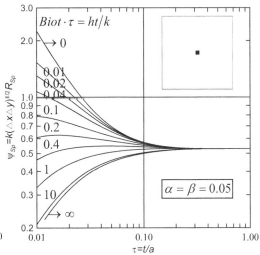

Figure 12.14. Maximum spreading resistance (From: Ellison, G.N., *IEEE Trans. CPT*, vol. 26, no. 3, 439-454, © 2003 *IEEE*. With permission).

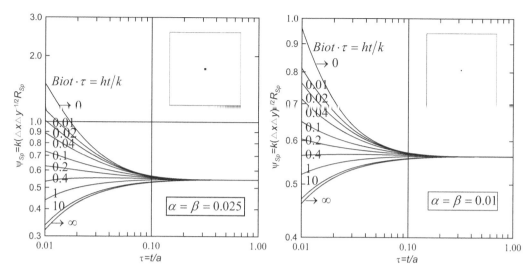

Figure 12.15. Maximum spreading resistance (From: Ellison, G.N., *IEEE Trans. CPT*, vol. 26, no. 3, 439-454, © 2003 *IEEE*. With permission).

Figure 12.16. Maximum spreading resistance (From: Ellison, G.N., *IEEE Trans. CPT*, vol. 26, no. 3, 439-454, © 2003 *IEEE*. With permission).

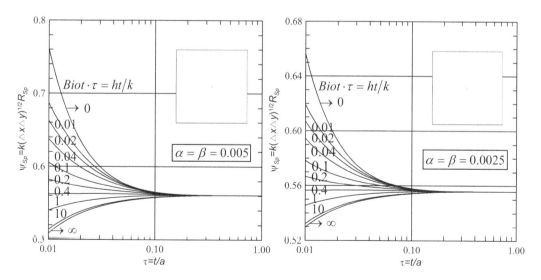

Figure 12.17. Maximum spreading resistance.

Figure 12.18. Maximum spreading resistance.

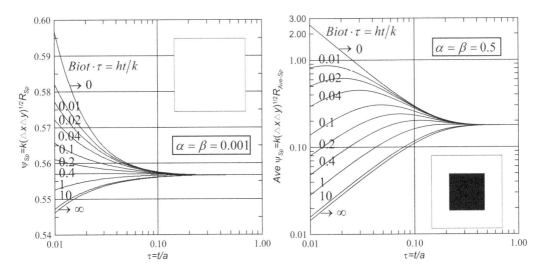

Figure 12.19. Maximum spreading resistance.   **Figure 12.20.** Average spreading resistance.

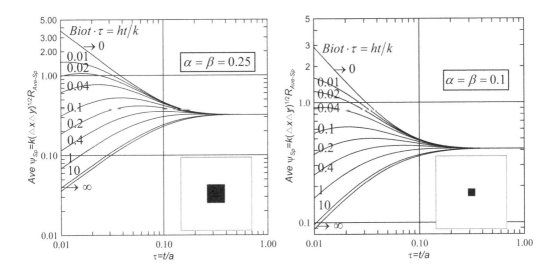

Figure 12.21. Average spreading resistance.   **Figure 12.22.** Average spreading resistance.

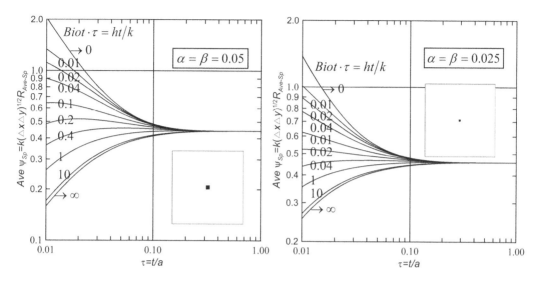

Figure 12.23. Average spreading resistance.

Figure 12.24. Average spreading resistance.

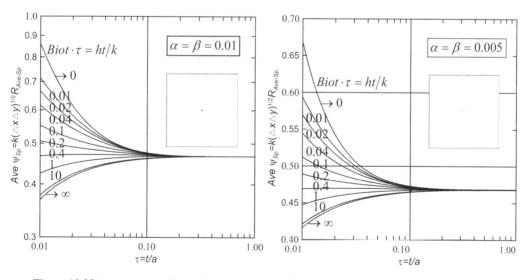

Figure 12.25. Average spreading resistance.

Figure 12.26. Average spreading resistance.

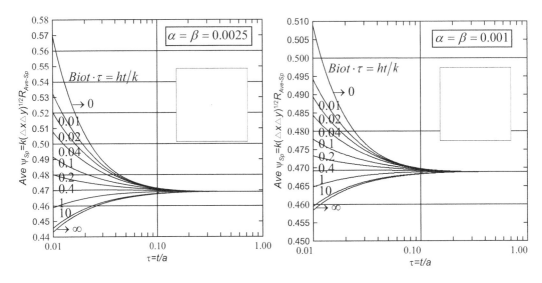

Figure 12.27. Average spreading resistance.

Figure 12.28. Average spreading resistance.

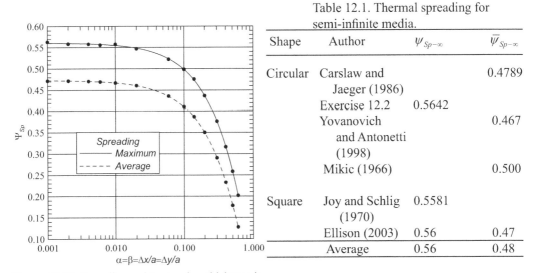

Figure 12.29. Spreading resistance when thickness-in-dependent.

Table 12.1. Thermal spreading for semi-infinite media.

Shape	Author	$\psi_{Sp-\infty}$	$\overline{\psi}_{Sp-\infty}$
Circular	Carslaw and Jaeger (1986)		0.4789
	Exercise 12.2	0.5642	
	Yovanovich and Antonetti (1998)		0.467
	Mikic (1966)		0.500
Square	Joy and Schlig (1970)	0.5581	
	Ellison (2003)	0.56	0.47
	Average	0.56	0.48

All of our spreading resistance theory is developed for a flat plate model. We need only to calculate an *effective heat transfer coefficient h_e* enhanced due to the additional surface area of the fins.

$$h_e = hA_f / ab = \left[0.0097 \, \text{W}/\left(\text{in.}^2 \cdot {}^\circ\text{C}\right) \right] \left(125.8 \, \text{in.}^2\right) / \left(3.94 \, \text{in.}\right)^2 = 0.079 \, \text{W}/\left(\text{in.}^2 \cdot {}^\circ\text{C}\right)$$

The dimensionless parameters that we need to use for the spreading resistance design curves are

$$\alpha = \beta = \Delta x / a = 0.394 \, \text{in.}/3.94 \, \text{in.} = 0.1, \tau = t/a = 0.197 \, \text{in.}/3.94 \, \text{in.} = 0.05,$$
$$Biot \cdot \tau = h_e \, t/k = (0.079)(0.197/5.1) = 3.03 \times 10^{-3}$$

Then using Figures 12.13 and 12.22, we obtain $\psi_{Sp} = 0.8$ and $\overline{\psi}_{Sp} = 0.7$, respectively. The total thermal resistances are

$$R = R_U + R_{Sp} = \left(1/hA_f\right) + \left(t/k_{Al}ab\right) + \psi_{Sp}/k_{Al}\sqrt{\Delta x \Delta y}$$

$$R = \frac{1}{(0.0097)(125.8)} + \frac{0.197}{(5.1)(3.94)^2} + \frac{0.8}{(5.1)\sqrt{(0.394)(0.394)}} = 0.82 + 0.0025 + 0.40 = 1.22 \, {}^\circ\text{C}/\text{W}$$

and

$$\overline{R} = R_U + R_{Ave-Sp} = \left(1/hA_f\right) + \left(t/k_{Al}ab\right) + \overline{\psi}_{Sp}/k_{Al}\sqrt{\Delta x \Delta y}$$

$$\overline{R} = \frac{1}{(0.0097)(125.8)} + \frac{0.197}{(5.1)(3.94)^2} + \frac{0.7}{(5.1)\sqrt{(0.394)(0.394)}} = 0.82 + 0.0025 + 0.35 = 1.17 \, {}^\circ\text{C}/\text{W}$$

Next consider the same heat sink, but with the fins replaced by a cold plate, i.e., an infinite heat sink, $Biot \cdot \tau \rightarrow \infty$. Once again we can use Figures 12.13 and 12.22 to obtain $\psi_{Sp} = 0.36$ and $\overline{\psi}_{Sp} = 0.28$, respectively. For this cold plate, we obtain

$$R = R_U + R_{Sp} = \left(t/k_{Al}ab\right) + \psi_{Sp}/k_{Al}\sqrt{\Delta x \Delta y} = \frac{0.197}{(5.1)(3.94)^2} + \frac{0.36}{(5.1)\sqrt{(0.394)(0.394)}}$$

$$R = 0.0025 + 0.18 = 0.18 \, {}^\circ\text{C}/\text{W}$$

and

$$\overline{R} = R_U + R_{Ave-Sp} = \left(t/k_{Al}ab\right) + \overline{\psi}_{Sp}/k_{Al}\sqrt{\Delta x \Delta y}$$

$$\overline{R} = \frac{0.197}{(5.1)(3.94)^2} + \frac{0.28}{(5.1)\sqrt{(0.394)(0.394)}} = 0.0025 + 0.14 = 0.14 \, {}^\circ\text{C}/\text{W}$$

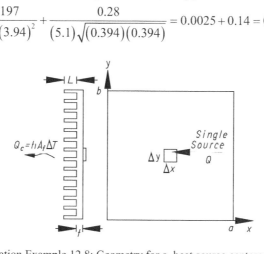

Figure 12.30. Application Example 12.8: Geometry for a heat source centered on a heat sink (From: Ellison, G.N., *IEEE Trans. CPT*, vol. 26, no. 3, 439-454, © 2003 *IEEE*. With permission).

Using the formulae from Eq. (12.62), unit-source temperature contours (temperature rise above ambient/power dissipation) and heat flux vectors are plotted in Figures 12.31 and 12.32 for the finite and infinite heat transfer coefficients, respectively. The large aspect ratio for the plate width/thickness ratio (20/1) requires that both illustrations be cropped so that adequate detail is visible in the regions of interest.

In the case of the finite h_e, isotherms computed from Eq. (12.62) are shown in Figure 12.31 (a) for the region in the vicinity of the heat source and Figure 12.31 (b) for the far left edge. Figure 12.31 (c) shows unit-source contours for an axisymmetric model computed with the professional 2-D version of FlexPDE™, a commercial finite element program from PDE Solutions, Inc. The analyses show that the isotherms tend toward a nearly vertical orientation as the distance from the source is increased. At the end region shown in Figure 12.31 (b), the isotherms bend toward the horizontal, consistent with the adiabatic edge boundary conditions. The fact that in the vicinity of the source, both the spreading theory and the *FEA* results, Figure 12.31 (b) and (c), respectively, provide isotherms of nearly identical shape, is validation that the isotherm shapes in this region are indeed correct and that Surfer™ from Golden Software, Inc. produced the correct results.

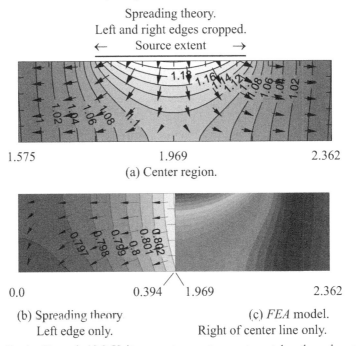

Spreading theory.
Left and right edges cropped.
← Source extent →

1.575   1.969   2.362
(a) Center region.

0.0   0.394   1.969   2.362

(b) Spreading theory   (c) *FEA* model.
Left edge only.   Right of center line only.

Figure 12.31. Application Example 12.8: Unit-source temperature contours taken through source-center plane. $Biot \cdot \tau = 3.03 \times 10^{-3}$ (From: Ellison, G.N., *IEEE Trans. CPT*, vol. 26, no. 3, 439-454, © 2003 *IEEE*. With permission).

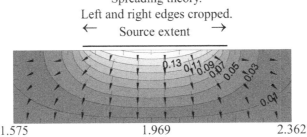

Spreading theory.
Left and right edges cropped.
← Source extent →

1.575   1.969   2.362

Figure 12.32. Application Example 12.8: Unit-source temperature contours taken through source-center plane. $Biot \cdot \tau \rightarrow \infty$. (From: Ellison, G.N., *IEEE Trans. CPT*, vol. 26, no. 3, 439-454, © 2003 *IEEE*. With permission).

The isotherms and heat flux vectors for the case of the nearly infinite h_e are illustrated in Figure 12.32 for the source region. The diminished heat spreading characteristics for the second case are clearly indicated. The isotherms outside of the truncated geometry used in the illustration indicate a small temperature gradient with the result that only one additional isotherm is hidden from view by each half of the cropped region.

12.9 ILLUSTRATION EXAMPLE: SPREADING CONTRIBUTION TO FORCED AIR COOLED HEAT SINK

This heat sink was analyzed in Section 7.9 using two different Nusselt number correlations, both of which gave similar results. The results from the correlation by Muzychka/Yovanovich are used here. The heat sink had a width $W = 4.0$ in. and length $L = 3.0$ in.1 A square device approximation is $a = \sqrt{LW} = \sqrt{(4.0)(3.0)} = 3.5$ in. Using $\tau = t/a = 0.315/3.5 = 0.09$, the largest $Biot \cdot \tau$ is calculated to be 0.016, i.e., nearly zero. Since the dimensionless source is $\alpha = \Delta x/a = 1.0/3.5 = 0.29$, Fig. 12.21 can be used to determine that $\overline{\psi}_{Sp} = 0.47$. It is easy to calculate $R_{Sp} = 0.095$ °C/W and $R_U = R_{Sink} + R_k$

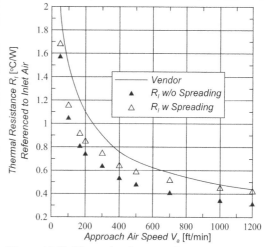

where R_{Sink} is the convective resistance result for various air speeds as calculated in Section 7.9 and plotted in Figure 7.12.

Figure 12.33 shows plots of the thermal resistance w/o (without) and w (with) spreading and also the plot of vendor data. The spreading contribution is very nearly the same value for all air speeds. The use of a one-inch-square source is pure speculation, but is the size approximating a large power transistor source that an experimentalist might use.

Figure 12.33. Thermal resistance referenced to inlet air vs. approach air speed for heat analyzed for Application Example 7.9. Average source spreading used.

EXERCISES

12.1 Derive a formula for the dimensionless maximum spreading resistance of a circular source on semi-infinite media. Hints: (1) Begin with Eq. (12.19); (2) You will need to complete an integration involving the function $J_1(u)$ for which it is unlikely that you will be able to find an identity, thus use a math scratchpad (the author used Mathcad™).

12.2 Derive the uniform source portion, G_{lm} for $l = m = 0$ of the thermal spreading resistance problem with the adiabatic source plane, *not assuming* reciprocity of the two G_{lm} arguments z', z. You are asked to (1) write the applicable *PDE*, boundary conditions, and general solution, Eqs. (ix.6-7), (ix.6-5), and (12.49), respectively; (2) derive G_{00}; and (3) show that reciprocity works for your solution. Hints: (1) Write your solution variable as $G_{00}(z'|z)$ for both $z' < z$ and $z' > z$; (2) Apply Equations (12.46), (ix.6-3), and (ix.6-2), successively.

12.3 Application Example 12.8 describes the calculation of the thermal resistance of a power transistor-shaped heat source on an aluminum heat sink where the effective heat transfer coefficient

$h_e = 0.079$ W/(in.$^2 \cdot$ °C) and the heat sink base thickness is $t = 0.179$ in. Following this same example, calculate and plot the maximum total thermal resistance $R = R_{Sp} + R_U$ for thicknesses of $t = 0.1, 0.15, 0.20, 0.30, 0.40, 0.50, 1.0, 1.5,$ and 2.0 in. What important concept does this example illustrate to you as a thermal design or packaging engineer?

12.4 Application Example 12.8 describes the calculation of the thermal resistance of a power transistor-shaped heat source on an aluminum heat sink where in one case, the heat transfer coefficient h is infinite to simulate an infinite heat sink for a heat sink base thickness $t = 0.197$ in. You are asked to calculate and compare results for the total thermal resistance from the source center to the far side of the heat sink using the theory from an *Integrated Resistance*, *Average Area Resistance*, and Section 12.6, i.e., Eqs. (12.4) and (12.6), and design curves from Section 12.7 for *average spreading*, respectively. Because the angle $\phi = 45°$ is what many people typically use, you should also use this value to see what you get for your results.

CHAPTER 13

Additional Mathematical Methods

The emphasis in the previous chapters has been on problems that are suitable for solving with an electronic calculator or some general purpose math analysis program. Now the author wishes to describe the mathematical basis of large thermal and airflow networks. These techniques require some programming effort and you should be able to apply them to practical problems when you find yourself without high-end commercial software.

This chapter also covers the most elementary aspects of finite element theory applicable to heat conduction problems. The intent is not to teach you sufficient detail to actually write an *FEA* code, but to give you an idea of what the method is all about. There are numerous books that cover the theory and coding practice in great detail.

13.1 THERMAL NETWORKS: STEADY-STATE THEORY

Figure 13.1 illustrates a nine-node model with 10 thermal resistances. Nodes 1-6 are solid material nodes and node 7 is an ambient temperature node. Resistances $R_{5,7}$ and $R_{6,7}$ represent convection elements and the remaining represent conduction. A good starting place for our math development is node 4, which may have a heat source denoted as Q_4. Assuming that T_4 is greater than any connected node temperature (if this is not actually the case, the algebra will take care of it), a steady-state heat balance may be written for this node as

$$\text{Heat in} = \text{Heat out}$$

$$Q_4 = \frac{T_4 - T_2}{R_3} + \frac{T_4 - T_3}{R_6} + \frac{T_4 - T_6}{R_5}$$

$$Q_4 = C_{4,2}\left(T_4 - T_2\right) + C_{4,3}\left(T_4 - T_3\right) + C_{4,6}\left(T_4 - T_6\right) \tag{13.1}$$

where the conductances are written with double subscripts indicating the interconnecting nodes. Solving Eq. (13.1) for the temperature at node 4 we have

$$T_4 = \frac{C_{4,2}T_2 + C_{4,3}T_3 + C_{4,6}T_6 + Q_4}{C_{4,2} + C_{4,3} + C_{4,6}} \tag{13.2}$$

Equation (13.2) is readily generalized for any *nonfixed-temperature* T_i.

$$\boxed{T_i = \left(\sum_{j \neq i} C_{ij}T_j + Q_i\right) \bigg/ \sum_{j \neq i} C_{ij}}$$

Gauss-Seidel (13.3)
solution

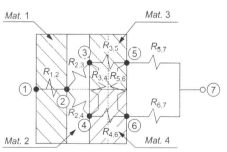

Figure 13.1(a). Thermal network model for slab-construction. Resistances are emphasized.

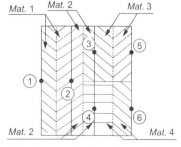

Figure 13.1(b). Thermal network model for slab-construction. Node capacitances are emphasized.

Equation (13.3) applied to a system of N nodes is commonly known as the *Gauss-Seidel* method. Clearly it is an iterative scheme. The method is applied by first guessing a set of *starting temperatures* for each nonfixed temperature node. Nonsource nodes use a $Q_i = 0$. Beginning with $i = 1$, the first node temperature is calculated using the starting temperatures for all T_j. Then using $i = 2$, the next temperature is calculated. If node 2 is connected to node 1, then the just calculated T_1 is used when $j = 1$. The other T_j use the starting temperatures. The solution proceeds for each node i where the most recently calculated values of T_j are used. When all temperatures have been calculated once, it is said that "one iteration has been completed." A second iteration begins again at node 1. The method continues for as many iterations as are necessary to get a good final solution, the criteria for which will be addressed shortly.

A *relaxed solution* is calculated using

$$T_{Ri} = T_{0i} + \beta\left(T_i - T_{0i}\right), \quad 0 < \beta < 2.0$$ Relaxation (13.4)

where T_i is the temperature calculated by Eq. (13.3) for the current iteration and T_{0i} is the temperature calculated for the *previous* iteration. The parameter β is the relaxation constant, which, as indicated, must always be less than exactly 2.0. Values of $\beta = 2.0$ lead to a rapidly divergent solution, values greater than 1.0 are said to be over-relaxed, values exactly equal to 1.0 are not relaxed, and values less than 1.0 are under-relaxed. There is no formula for predicting β for the general problem. The best method is to just experiment for a few iterations until a sufficiently optimum value is found, which is usually greater than one. A value of $\beta < 1.0$ is used to *dampen* a wildly oscillating solution. Naturally, a somewhat small, easily convergent problem solution is not worth the trouble of finding the best β, but slowly convergent, large networks may justify your time exploring the relaxation constant. Temperature dependent conductances require updating at each iteration.

The problem of determining if a math solution is good is addressed by rewriting Eq. (13.1) in the general form and placing all the T, Q, and C on the right-hand side of the energy balance.

$$r_i = \sum_{j \neq i} C_{ij}\left(T_i - T_j\right) - Q_i$$ Residual (13.5)

The *residual* r_i is zero when the mathematical solution is perfect. Of course, if you are using a relaxation method, the temperatures in Eq. (13.5) are the relaxed values. You might think that the next step is to just sum all the r_i for the nonfixed temperature nodes to obtain a *system residual*, but that might not work out well. Suppose, for example, that two nodes have very large, nearly equal residuals, but one is positive and the other is negative. Summing the residuals for these two nodes would give a small contribution to the system and suggest that the two nodal temperatures are accurate, but that would not necessarily be the case. The best procedure is shown as Eq. (13.6) where we sum the *absolute* values of the residuals, which may lead to a solution that is actually better than the system residual indicates, but you will at least know that the solution is good if the system residual is small.

Now we are left with the problem of determining what constitutes a "good residual." This quandary is resolved by comparing the system residual with the total input power for the problem, and this is exactly what we see in Eq. (13.6).

$$E.B. = \left[\sum_{i=1}|r_i| \Big/ \sum_{i=1} Q_i\right] \times 100.0$$ System energy balance (13.6)

The Gauss-Seidel solution method is easy to understand and, relative to other methods, easy to program. However, there is the occasional problem where it is difficult to obtain a converged solution. Thus it is appropriate to consider one other method. To do this, we write Eq. (13.1) in a generalized form for any node i.

$$\left(\sum_j C_{ij}\right) T_i + \sum_j \left(-C_{ij}\right) T_j = Q_i$$ Simultaneous (13.7) equations

Equation (13.7) exists for each and every nonfixed temperature node. Then we have as many Equations (13.7) as unknown nodal temperatures. If the problem is totally linear, then only one solution of the set of system equations is necessary. However, if the problem is nonlinear, i.e., there are temperature dependent conductances C_{ij}, the system of Eqs. (13.7) is solved repeatedly with the C_{ij} values updated at the beginning of each iteration. Starting values are then required for the first iteration. The relaxation method specified by Eq. (13.4) is not applicable here. If you are writing your code in a high level language such as *Basic*, *Fortran*, or *C*, you can implement an available routine in your code rather than write it yourself. If you choose to use Mathcad™, Maple™, or MATLAB™, built-in procedures are available.

13.2 ILLUSTRATIVE EXAMPLE: A SIMPLE STEADY-STATE, THERMAL NETWORK PROBLEM, GAUSS-SEIDEL AND SIMULTANEOUS EQUATION SOLUTIONS COMPARED

The simple geometry in Figure 13.1(a) is used for this example with arbitrary dimensions and material properties. The equations for both the Gauss-Seidel and simultaneous equation solution methods will be set up for solution. Materials 1, 2, 3, and 4 have the dimensions in inches for t (dimension parallel to bottom edge of page), w (dimension into plane of page), l (dimension parallel to left or right edge of page), conductivity k [W/(in. · °C)], respectively: (1) $t = 0.1, 1.0, 1.0, 1.0$ in.; (2) $w = 0.05, 1.0, 1.0, 0.02$ in.; (3) $l = 0.5, 1.0, 0.5, 10.0$ in.; (4) $k = 0.5, 1.0, 0.5, 4.0$ W/(in. · °C). It is more appropriate to calculate conductances rather than resistances (the following ij subscripts are separated by commas for clarity).

$$C_{1,2} = k_1 w_1 l_1 / t_1, \quad C_{2,3} = k_2 w_2 \left(\frac{l_2}{2}\right)\Big/t_2, \quad C_{2,4} = k_2 w_2 \left(\frac{l_2}{2}\right)\Big/t_2, \quad C_{3,5} = k_3 w_3 l_3 / t_3$$

$$C_{3,4} = \frac{1}{\left[(l_3/2)/(k_3 t_3 w_3/2)\right] + \left[(l_4/2)/(k_4 t_4 w_4/2)\right]}$$

$$C_{5,6} = \frac{1}{\left[(l_3/2)/(k_3 t_3 w_3/2)\right] + \left[(l_4/2)/(k_4 t_4 w_4/2)\right]}$$

$$C_{4,6} = k_4 w_4 l_4 / t_4, \quad C_{5,7} = h_5 w_3 l_3, \quad C_{6,7} = h_6 w_4 l_4$$

$$C_{2,1} = C_{1,2}, \quad C_{3,2} = C_{2,3}, \quad C_{4,2} = C_{2,4}, \quad C_{5,3} = C_{3,5}, \quad C_{4,3} = C_{3,4}$$

$$C_{6,5} = C_{5,6}, \quad C_{6,4} = C_{4,6}, \quad C_{7,5} = C_{5,7}, \quad C_{7,6} = C_{6,7}$$

where the preceding last three conductances use heat transfer coefficients $h_5 = 2.0$ and $h_6 = 1.0$. Also, the preceding formulae and data result in conductance values as represented in the following matrix:

$$
\mathbf{C} - \begin{pmatrix}
0 & 10 & 0 & 0 & 0 & 0 & 0 \\
10 & 0 & 0.2 & 0.2 & 0 & 0 & 0 \\
0 & 0.2 & 0 & 2.857 & 10 & 0 & 0 \\
0 & 0.2 & 2.857 & 0 & 0 & 4 & 0 \\
0 & 0 & 10 & 0 & 0 & 2.857 & 1 \\
0 & 0 & 0 & 4 & 2.857 & 0 & 0.5 \\
0 & 0 & 0 & 0 & 1 & 0.5 & 0
\end{pmatrix}
\tag{13.8}
$$

We shall use a source at node 1 of $Q = 10.0$ W. The Gauss-Seidel method with relaxation requires that we set up six equations (plus the relaxation effect) for the six nodes for which we wish to calculate temperature. The ambient node 7 is left at 20°C. We use T_N, TR_N, $T0_N$ to represent the

Gauss-Seidel temperature for the present iteration of node N, the relaxed Gauss-Seidel temperature for the present iteration of node N, and the calculated temperature for node N for the previous iteration, respectively.

$$T_1 = \frac{C_{1,2}T_2 + Q_1}{C_{1,2}}, \qquad\qquad TR_1 = T0_1 + \beta(T_1 - T0_1), TR_1 \to T_1, T_1 \to T0_1$$

$$T_2 = \frac{C_{2,1}T_1 + C_{2,3}T_3 + C_{2,4}T_4}{C_{2,1} + C_{2,3} + C_{2,4}}, \qquad TR_2 = T0_2 + \beta(T_2 - T0_2), TR_2 \to T_2, T_2 \to T0_2$$

$$T_3 = \frac{C_{3,2}T_2 + C_{3,4}T_4 + C_{3,5}T_5}{C_{3,2} + C_{3,4} + C_{3,5}}, \qquad TR_3 = T0_3 + \beta(T_3 - T0_3), TR_3 \to T_3, T_3 \to T0_3$$

$$T_4 = \frac{C_{4,2}T_2 + C_{4,3}T_3 + C_{4,6}T_6}{C_{4,2} + C_{4,3} + C_{4,6}}, \qquad TR_4 = T0_4 + \beta(T_4 - T0_4), TR_4 \to T_4, T_4 \to T0_4 \quad (13.9)$$

$$T_5 = \frac{C_{5,3}T_3 + C_{5,6}T_6 + C_{5,7}T_7}{C_{5,3} + C_{5,6} + C_{5,7}}, \qquad TR_5 = T0_5 + \beta(T_5 - T0_5), TR_5 \to T_5, T_5 \to T0_5$$

$$T_6 = \frac{C_{6,4}T_4 + C_{6,5}T_5 + C_{6,7}T_7}{C_{6,4} + C_{6,5} + C_{6,7}}, \qquad TR_6 = T0_6 + \beta(T_6 - T0_6), TR_6 \to T_6, T_6 \to T0_6$$

Each line of Eqs. (13.9) shows four successive steps: (1) calculate the Gauss-Seidel temperature T_i, (2) calculate the relaxed temperature TR_i, (3) assign the relaxed temperature TR_i to T_i, and (4) also assign T_i to $T0_i$. When calculating the Gauss-Seidel temperature T_i, the most recently calculated temperatures T_i are used on the right-hand side of the equation. A relaxed temperature T_j may not yet have been calculated and assigned, as in the situation of the very first iteration, in which case a starting temperature is used.

Now we set up equations needed for solving by the simultaneous equation method. We write each of the six equations according to Eq. (13.7).

$$C_{1,2}T_1 - C_{1,2}T_2 = Q_1$$

$$-C_{2,1}T_1 + (C_{2,1} + C_{2,3} + C_{2,4})T_2 - C_{2,3}T_3 - C_{2,4}T_4 = 0$$

$$-C_{3,2}T_2 + (C_{3,2} + C_{3,4} + C_{3,5})T_3 - C_{3,4}T_4 - C_{3,5}T_5 = 0$$

$$-C_{4,2}T_2 - C_{4,3}T_3 + (C_{4,2} + C_{4,3} + C_{4,6})T_4 - C_{4,6}T_6 = 0 \qquad (13.10)$$

$$-C_{5,3}T_3 + (C_{5,3} + C_{5,6} + C_{5,7})T_5 - C_{5,6}T_6 - C_{5,7}T_7 = 0$$

$$-C_{6,4}T_4 - C_{6,5}T_5 + (C_{6,4} + C_{6,5} + C_{6,7})T_6 - C_{6,7}T_7 = 0$$

We move the last term of the last two equations to the right-hand side because the node 7 temperature does not change.

$$C_{1,2}T_1 - C_{1,2}T_2 = Q_1$$

$$-C_{2,1}T_1 + (C_{2,1} + C_{2,3} + C_{2,4})T_2 - C_{2,3}T_3 - C_{2,4}T_4 = 0$$

$$-C_{3,2}T_2 + (C_{3,2} + C_{3,4} + C_{3,5})T_3 - C_{3,4}T_4 - C_{3,5}T_5 = 0$$

$$-C_{4,2}T_2 - C_{4,3}T_3 + (C_{4,2} + C_{4,3} + C_{4,6})T_4 - C_{4,6}T_6 = 0 \qquad (13.11)$$

$$-C_{5,3}T_3 + (C_{5,3} + C_{5,6} + C_{5,7})T_5 - C_{5,6}T_6 = C_{5,7}T_7$$

$$-C_{6,4}T_4 - C_{6,5}T_5 + (C_{6,4} + C_{6,5} + C_{6,7})T_6 = C_{6,7}T_7$$

We see the diagonal element is the sum of the absolute magnitude of the remaining row elements and the "source" vector contains not only actual sources, but also boundary conditions. In our current problem, the conductance matrix is listed in two parts due to the page width limitations.

$$\mathbf{G} = \begin{pmatrix} C_{1,2} & -C_{1,2} & 0 & 0 \\ -C_{2,1} & \left(C_{2,1}+C_{2,3}+C_{2,4}\right) & -C_{2,3} & -C_{.2,4} \\ 0 & -C_{3,2} & \left(C_{3,2}+C_{3,4}+C_{3,5}\right) & -C_{3,4} \\ 0 & -C_{4,2} & -C_{4,3} & \left(C_{4,2}+C_{4,3}+C_{4,6}\right) \\ 0 & 0 & -C_{5,3} & 0 \\ 0 & 0 & 0 & -C_{6,4} \end{pmatrix} \cdots\cdots$$

(13.12a)

Equation (13.12) continued, followed by the temperature **T** and source **S** vectors as Eq. (13.13):

$$\cdots\cdots \begin{pmatrix} 0 & 0 \\ 0 & 0 \\ -C_{3,5} & 0 \\ 0 & -C_{4,6} \\ \left(C_{5,3}+C_{5,6}+C_{5,7}\right) & -C_{5,6} \\ -C_{6,5} & \left(C_{6,4}+C_{6,5}+C_{6,7}\right) \end{pmatrix}, \quad \mathbf{T} = \begin{pmatrix} T_1 \\ T_2 \\ T_3 \\ T_4 \\ T_5 \\ T_6 \end{pmatrix}, \mathbf{S} = \begin{pmatrix} Q_1 \\ 0 \\ 0 \\ 0 \\ C_{5,7}T_7 \\ C_{6,7}T_7 \end{pmatrix}$$

(13.13)

(13.12b)

Using the conductances in Eq. (13.8) we then have

$$\mathbf{G} = \begin{pmatrix} 10 & -10 & 0 & 0 & 0 & 0 \\ -10 & 10.4 & -0.2 & -0.2 & 0 & 0 \\ 0 & -0.2 & 13.057 & -2.857 & -10 & 0 \\ 0 & -0.2 & -2.857 & 7.057 & 0 & -4 \\ 0 & 0 & -10 & 0 & 13.857 & -2.857 \\ 0 & 0 & 0 & -4 & -2.857 & 7.357 \end{pmatrix}, \quad \mathbf{S} = \begin{pmatrix} 10.0 \\ 0 \\ 0 \\ 0 \\ 20.0 \\ 10.0 \end{pmatrix}$$

(13.14)

The solution is obtained by solving Eq. (13.15) for **T**.

$$\mathbf{GT} = \mathbf{S}$$

(13.15)

Modest problems such as our current example are solved using a matrix inversion operation and larger problems are more efficiently solved by other methods. The results for the various solution methods described in this section are listed in Table 13.1.

This example, while not particularly illuminating of any significant computational concept, should be sufficient to enhance your understanding of how to set up steady-state thermal network equations for numerical analysis. The Gauss-Seidel method is one of the oldest solution techniques and is described in nearly all heat transfer textbooks. The results in Table 13.1 might lead the reader to believe that the Gauss-Seidel method, with or without relaxation, could be inadequate in large problems. Actually, in the early days of personal computers this author believed that some very simple problems were not converging. In actuality, these very same problems were later found to be easily and quickly solved by the faster computers. The single chip package described in Chapter 1 is a good example. The 900 node problem was solved with 500 iterations in less than one second to a system energy balance of less than 0.01%, according to Eq. (13.6), on a 2.8-GHz PC. Patankar (1980) states in his own symbolics that, according to Scarborough (1958), a *sufficient* condition for Gauss-Seidel convergence is given by Eq. (13.16).

Table 13.1. Illustrative Example 13.2: compared results.

Method	T_1	T_2	T_3	T_4	T_5	T_6	Eng. Bal.[1]	Eng. Bal.[2]
Gauss-Seidel, Mathcad™, 500 iterations	53.46	52.46	27.25	27.68	26.62	26.75	0.024%	0.023%
Gauss-Seidel, *TNETFA*, β =1.0	53.24	52.25	27.19	27.61	26.56	26.69	-------	0.93
TNETFA, β =1.7, 100 iterations	53.47	52.47	27.25	27.68	26.63	26.75	-------	1.23×10^{-7}
Simultaneous equations, Mathcad™	53.47	52.47	27.25	27.68	26.63	26.75	-9.2×10^{-13}	2.88×10^{-12}
Simultaneous equations, *TNETFA*	53.54	52.47	27.25	27.68	26.63	26.75	-------	1.24×10^{-12}

Note: [1] E.B. using r_i , [2] E.B. using $\left| r_i \right|$. Starting temperatures = 40.0°C.

$$T_P = \frac{\sum a_{nb} T_{nb}^* + b}{a_P} \qquad (a)$$

$$\left(\frac{\sum |a_{nb}|}{|a_P|} \right) \begin{cases} \leq 1 & \text{for all equations} \quad (b) \\ < 1 & \text{for at least one equation} \quad (c) \end{cases} \qquad (13.16)$$

We see that Eq. (13.3) would always seem to satisfy Eq. (13.16) (*b*), but never satisfy (13.16) (*c*). However, note that the preceding convergence criteria is *sufficient, but not necessary*. Thus it appears that our absolute guarantee of convergence is borderline at best. Finally, this author has not found a problem in recent years that does not converge with some effort using relaxation and restarts (from last iteration result). Very nonlinear problems may be an exception. An advantage of iterative methods such as Gauss-Seidel is that although the solution may be very inaccurate in the early to intermediate stages, the later steps may converge very quickly. The early inaccuracies and errors do not compound as round-off error because each iteration is merely a guess for the next iteration. The Gauss-Seidel method is clearly applicable to nonlinear problems. Simultaneous equation solution methods are applied to nonlinear systems by merely resolving the equations after updating the conductance matrix **G** element values.

13.3 THERMAL NETWORKS: TIME-DEPENDENT THEORY

In this section we will introduce four different formulations of time-dependent thermal networks using the network elements shown in Figure 13.1. Not shown in the illustration are symbols C^* used in this book for thermal capacitance; just use your imagination to understand that C^* are implied. Also, the node structure used in Figure 13.1 requires understanding that some nodes have a C^* for two combined materials, such as in the case of node 2 where C^* is an arithmetic sum of the capacitance for half of each of the volumes V_1 and V_2.

Our first theory is for a *forward finite difference in time*, explicit method. Referring to Figure 13.1(a), node 2, we write an energy balance for the specific node of interest:

Heat in - Heat out = Stored energy

$$Q_2 - \left[C_{2,1}\left(T_2 - T_1\right) + C_{2,3}\left(T_2 - T_3\right) + C_{2,4}\left(T_2 - T_4\right) \right] = \left(\rho c_P \Delta V\right)_2 \left(T_2^\Delta - T_2\right)/\Delta t \qquad (13.17)$$

$T_1, T_2, T_3, T_4 \equiv$ temperatures at time t

$T_2^\Delta \equiv$ temperature of node 2 at time $t + \Delta t$

$\rho \equiv$ density of node 2

$c_P \equiv$ specific heat capacity of node 2

$\Delta V \equiv$ volume of node 2

Equation (13.17) is more appropriately written as

$$Q_2 - \left[C_{2,1}\left(T_2 - T_1\right) + C_{2,3}\left(T_2 - T_3\right) + C_{2,4}\left(T_2 - T_4\right) \right] = C_2^*\left(T_2^\Delta - T_2\right)/\Delta t \qquad (13.18)$$

$C_2^* =$ thermal heat capacitance of node 2

so that C_2^* is the combined heat capacity of materials 1 and 2 that comprise node 2, i.e., $C_2^* = \left(\rho c_P\right)_1 \left(\Delta V_1/2\right) + \left(\rho c_P\right)_2 \left(\Delta V_2/2\right)$. Equation (13.18) is readily generalized for any nonfixed temperature node i.

$$Q_i - \sum_{j \neq i} C_{ij}\left(T_i - T_j\right) = \frac{C_i^*\left(T_i^\Delta - T_i\right)}{\Delta t} \qquad \begin{array}{l}\text{Forward finite}\\ \text{difference in time}\end{array} (13.19)$$

which when solved for T_i^Δ gives us

$$\boxed{T_i^\Delta = T_i\left(1 - S_i\right) + \frac{\Delta t}{C_i^*}\left(Q_i + \sum_{j \neq i} C_{ij}T_j\right)} \qquad \begin{array}{l}\text{Forward finite}\\ \text{difference in time}\end{array} (13.20)$$

The quantity C_i^* is the *capacitance* of node i and in this book, we give the name *stability constant* to S_i.

$$C_i^* = \left(\rho c_P \Delta V\right)_i, \quad S_i = \frac{\Delta t}{C_i^*}\sum_{j \neq i} C_{ij}$$

The most trivial numerical example would easily show you that if S_i is greater than 1.0, the computed temperature T_i^Δ will quickly diverge. Alternatively, if in Eq. (13.20), S_i is greater than 1.0, the temperature T_i^Δ will become cooler as T_i becomes hotter, which is nonsense. Thus we require that for a stable and sensible solution we must have S_i less than or equal to one.

$$S_i \leq 1.0$$

$$\Delta t \leq C_i^* \Big/ \sum_{j \neq i} C_{ij} \qquad \begin{array}{l}\text{Forward finite}\\ \text{difference in time}\end{array} (13.21)$$

Note also that a nodal temperature T_i^Δ at any given time step is calculated directly from temperatures at a previous time. Thus the label, *explicit forward finite difference in time*.

Holman (1990) recommends that the forward finite difference equation be written differently so that fewer round-off errors are encountered for large C_{ij}.

$$\boxed{T_i^\Delta = \frac{\Delta t}{C_i^*}\left(Q_i + \sum_{j \neq i} C_{ij}\left(T_j - T_i\right)\right) + T_i} \qquad \begin{array}{l}\text{Forward finite}\\ \text{difference (13.22)}\\ \text{in time, Holman}\end{array}$$

The stability criteria in Eq. (13.21) still applies.

The finite difference in time may also be written as a *backward difference* formulation.

Heat in - Heat out = Stored energy

$$Q_i^\Delta \sum_{j \neq i} C_{ij}\left(T_i^\Delta - T_j^\Delta\right) - \frac{\left(\rho c_P \Delta V\right)_i}{\Delta t}\left(T_i^\Delta - T_i\right) \qquad \begin{array}{l}\text{Backward}\\ \text{difference (13.23)}\\ \text{in time}\end{array}$$

where all $T_i^\Delta, T_j^\Delta, Q_i^\Delta$ are taken at time $t + \Delta t$, T_i at time t. Equation (13.23) is then solved for T_i^Δ.

$$Q_i^\Delta - T_i^\Delta \sum_{j \neq i} C_{ij} + \sum_{j \neq i} C_{ij} T_j^\Delta = \frac{(\rho c_P \Delta V)_i}{\Delta t} T_i^\Delta - \frac{(\rho c_P \Delta V)_i}{\Delta t} T_i$$

$$\boxed{T_i^\Delta = \frac{\displaystyle\sum_{j \neq i} C_{ij} T_j^\Delta + \frac{(\rho c_P \Delta V)_i}{\Delta t} T_i + Q_i^\Delta}{\displaystyle\sum_{j \neq i} C_{ij} + \frac{(\rho c_P \Delta V)_i}{\Delta t}}} \qquad \begin{array}{l} \text{Backward} \\ \text{difference} \\ \text{in time} \end{array} \quad (13.24)$$

Equation (13.24) is an implicit technique which bears a strong resemblance to the steady-state Gauss-Seidel formula, Eq. (13.3).

We return to our basic Eq. (13.23) to derive a different formulation.

$$Q_i^\Delta - \sum_{j \neq i} C_{ij} \left(T_i^\Delta - T_j^\Delta \right) = \frac{(\rho c_P \Delta V)_i}{\Delta t} \left(T_i^\Delta - T_i \right)$$

$$\boxed{\left[\frac{(\rho c_P \Delta V)_i}{\Delta t} + \sum_{j \neq i} C_{ij} \right] T_i^\Delta - \sum_{j \neq i} C_{ij} T_j^\Delta = \frac{(\rho c_P \Delta V)_i}{\Delta t} T_i + Q_i^\Delta} \qquad \begin{array}{l} \text{Backward} \\ \text{difference} \\ \text{in time} \end{array} \quad (13.25)$$

which are solvable using simultaneous equation solution methods. Note that Eq. (13.25) cannot be solved explicitly for T_i^Δ at time $t + \Delta t$ in terms of T_i and T_j at time t.

13.4 ILLUSTRATIVE EXAMPLE: A SIMPLE TIME-DEPENDENT, THERMAL NETWORK PROBLEM

We continue with the problem geometry in Figure 13.1. In particular we refer to Figure 13.1(b) in which the portions of materials 1-4 are lumped into the relevant node capacitance: Node 1 uses material 1, node 2 uses materials 1 and 2, node 3 uses materials 2 and 3, node 4 uses materials 2 and 4, node 5 uses material 3, and node 6 uses material 4. The forward finite difference and the two backward difference in time methods will be set up for solution and the results compared.

The conductances were previously calculated for the steady-state example so the capacitances remain to be calculated for nodes 1-6. We will begin with units of c_p [J/kg K], ρ [kg/m³], and ΔV [m³] with only the latter required conversion from the problem geometry so that we obtain capacitance units of C^* [J/K] or C^* [J/°C]. In those instances where the node consists of only one material, we calculate one volume and one capacitance. For those nodes where two materials are involved, we must calculate two volumes and two capacitances and add the latter. The specific heat and density values used are introduced in the calculations.

Node 1: $\quad \Delta V_1 = (t_1/2)l_1 w_1 = (0.1\,\text{in.}/2)(1.0\,\text{in.})(1.0\,\text{in.})(0.0254\,\text{m/in.})^3 = 8.19 \times 10^{-7}\,\text{m}^3$

$\quad\quad\quad C_1^* = \rho_1 c_{P1} \Delta V_1 = 4000\,\text{kg/m}^3 \left[800\,\text{J/(kg·K)} \right] (8.19 \times 10^{-7}\,\text{m}^3) = 2.62\,\text{J/K}$

Node 2: $\quad \Delta V_{2a} = (t_1/2)l_1 w_1 = (0.1\,\text{in.}/2)(1.0\,\text{in.})(1.0\,\text{in.})(0.0254\,\text{m/in.})^3 = 8.19 \times 10^{-7}\,\text{m}^3$

$\quad\quad\quad \Delta V_{2b} = (t_2/2)l_2 w_2 = (0.05\,\text{in.}/2)(1.0\,\text{in.})(1.0\,\text{in.})(0.0254\,\text{m/in.})^3 = 4.10 \times 10^{-7}\,\text{m}^3$

$\quad\quad\quad C_2^* = \rho_1 c_{P1} \Delta V_{2a} + \rho_2 c_{P2} \Delta V_{2b} = (4000\,\text{kg/m}^3) \left[800\,\text{J/(kg·K)} \right] (8.19 \times 10^{-7}\,\text{m}^3) +$

$\quad\quad\quad\quad (2000\,\text{kg/m}^3) \left[700\,J/(\text{kg·K}) \right] (4.10 \times 10^{-7}\,\text{m}^3) = 3.20\,\text{J/K}$

Node 3: $\quad \Delta V_{3a} = (t_2/2)l_3 w_3 = (0.05\,\text{in.}/2)(0.5\,\text{in.})(1.0\,\text{in.})(0.0254\,\text{m/in.})^3 = 2.05 \times 10^{-7}\,\text{m}^3$

$\quad\quad\quad \Delta V_{3b} = (t_3/2)l_3 w_3 = (0.5\,\text{in.}/2)(0.5\,\text{in.})(1.0\,\text{in.})(0.0254\,\text{m/in.})^3 = 2.05 \times 10^{-6}\,\text{m}^3$

$$C_3^* = \rho_2 c_{P2} \Delta V_{3a} + \rho_3 c_{P3} \Delta V_{3b} = \left(2000\,\text{kg/m}^3\right)\left[700\,\text{J/(kg}\cdot\text{K)}\right]\left(2.05\times10^{-7}\,\text{m}^3\right) +$$
$$\left(10^4\,\text{kg/m}^3\right)\left[400\,\text{J/(kg}\cdot\text{K)}\right]\left(2.05\times10^{-6}\,\text{m}^3\right) = 8.48\,\text{J/K}$$

Node 4:
$$\Delta V_{4a} = \Delta V_{3a} = 2.05\times10^{-7}\,\text{m}^3,\ \Delta V_{4b} = \Delta V_{3b} = 2.05\times10^{-6}\,\text{m}^3$$
$$C_4^* = \rho_2 c_{P2} \Delta V_{4a} + \rho_4 c_{P4} \Delta V_{4b}$$
$$C_4^* = \left(2000\,\text{kg/m}^3\right)\left[700\,\text{J/(kg}\cdot\text{K)}\right]\left(2.05\times10^{-7}\,\text{m}^3\right) +$$
$$\left(3000\,\text{kg/m}^3\right)\left[800\,\text{J/(kg}\cdot\text{K)}\right]\left(2.05\times10^{-6}\,\text{m}^3\right) = 5.20\,\text{J/K}$$

Node 5: $\Delta V_5 = \left(t_3\ /2\right) l_3 w_3 = \left(0.5\,\text{in.}/2\right)\left(0.5\,\text{in.}\right)\left(1.0\,\text{in.}\right)\left(0.0254\,\text{m/in.}\right)^3 = 2.05\times10^{-6}\,\text{m}^3$
$$C_5^* = \rho_3 c_{P3} \Delta V_5 = \left(10^4\,\text{kg/m}^3\right)\left[400\,\text{J/(kg}\cdot\text{K)}\right]\left(2.05\times10^{-6}\,\text{m}^3\right) = 8.19\,\text{J/K}$$

Node 6: $\Delta V_6 = \left(t_4/2\right) l_4 w_4 = \left(0.5\,\text{in.}/2\right)\left(0.5\,\text{in.}\right)\left(1.0\,\text{in.}\right)\left(0.0254\,\text{m/in.}\right)^3 = 2.05\times10^{-6}\,\text{m}^3$
$$C_6^* = \rho_4 c_{P4} \Delta V_6 = \left(3000\,\text{kg/m}^3\right)\left[800\,\text{J/(kg}\cdot\text{K)}\right]\left(2.05\times10^{-6}\,\text{m}^3\right) = 49.2\,\text{J/K}$$

The forward finite difference method requires a time step according to Eq. (13.21) so we shall calculate a maximum time step for each node.

$$\Delta t_{1\,Max} = C_1^*/C_{1,2} = 2.62/10.0 = 0.262\,\text{s}$$
$$\Delta t_{2\,Max} = C_2^*/\left(C_{1,2}+C_{2,3}+C_{2,4}\right) = 3.20/\left(10.0+0.20+0.20\right) = 0.307\,\text{s}$$
$$\Delta t_{3\,Max} = C_3^*/\left(C_{3,2}+C_{3,4}+C_{3,5}\right) = 8.48/\left(0.20+2.86+10.0\right) = 0.649\,\text{s}$$
$$\Delta t_{4\,Max} = C_4^*/\left(C_{4,2}+C_{4,3}+C_{4,6}\right) = 5.20/\left(0.2+2.86+4.00\right) = 0.737\,\text{s}$$
$$\Delta t_{5\,Max} = C_5^*/\left(C_{5,3}+C_{5,6}+C_{5,7}\right) = 8.19/\left(10.0+2.86+1.00\right) = 0.591\,\text{s}$$
$$\Delta t_{6\,Max} = C_6^*/\left(C_{6,4}+C_{6,5}+C_{6,7}\right) = 4.92/\left(4.00+2.86+0.5\right) = 0.668\,\text{s}$$

We search the preceding for the smallest time step, which is 0.262 s. We choose to set $\Delta t = 0.2\,\text{s}$.

Equations Set Up for Forward Finite Difference in Time
Each node, i.e., nodes 1-6, is set up according to Eq. (13.20).

$$S_1 = \frac{\Delta t}{C_1^*} C_{1,2},\ T_1^\Delta = T_1\left(1-S_1\right) + \frac{\Delta t}{C_1^*}\left(Q_1 + C_{1,2}T_2\right)$$
$$S_2 = \frac{\Delta t}{C_2^*}\left(C_{2,1}+C_{2,3}+C_{2,4}\right),\ T_2^\Delta = T_2\left(1-S_2\right) + \frac{\Delta t}{C_2^*}\left(C_{2,1}T_1 + C_{2,3}T_3 + C_{2,4}T_4\right)$$
$$S_3 = \frac{\Delta t}{C_3^*}\left(C_{3,2}+C_{3,4}+C_{3,5}\right),\ T_3^\Delta = T_3\left(1-S_3\right) + \frac{\Delta t}{C_3^*}\left(C_{3,2}T_2 + C_{3,4}T_4 + C_{3,5}T_5\right)$$
$$S_4 = \frac{\Delta t}{C_4^*}\left(C_{4,2}+C_{4,3}+C_{4,6}\right),\ T_4^\Delta = T_4\left(1-S_4\right) + \frac{\Delta t}{C_4^*}\left(C_{4,2}T_2 + C_{4,3}T_3 + C_{4,6}T_6\right) \quad\quad (13.26)$$
$$S_5 = \frac{\Delta t}{C_5^*}\left(C_{5,3}+C_{5,6}+C_{5,7}\right),\ T_5^\Delta = T_5\left(1-S_5\right) + \frac{\Delta t}{C_5^*}\left(C_{5,3}T_3 + C_{5,6}T_6 + C_{5,7}T_7\right)$$
$$S_6 = \frac{\Delta t}{C_6^*}\left(C_{6,4}+C_{6,5}+C_{6,7}\right),\ T_6^\Delta = T_6\left(1-S_6\right) + \frac{\Delta t}{C_6^*}\left(C_{6,4}T_4 + C_{6,5}T_5 + C_{6,7}T_7\right)$$

Equations (13.26) are stepped through for as many times as the final multiple of Δt requires to get the necessary solution. For example, if we wish to see the temperature response for a time from 0 to 200 seconds, we require *Number of Steps* = 200 s $/\Delta t$ = 200/0.2 = 1000 steps. Temperatures calculated for the first time step (T_i^Δ on left-hand side of Eqs. (13.26)) use the initial conditions (temperatures T_i and T_j on the right-hand side of Eqs. (13.26)). Temperatures calculated for the second time step (T_i^Δ on left-hand side of Eqs. (13.26)) use the temperatures results calculated for the first time step (temperatures T_i and T_j on the right-hand side of Eqs. (13.26)). The procedure is followed until solution completion.

Equations Set Up for Backward Finite Difference in Time Using Equations Resembling Gauss-Seidel. Once again each node is set up for solution, this time according to Eq. (13.24).

$$T_1^\Delta = \frac{C_{1,2}T_2^\Delta + \left(C_1^*/\Delta t\right)T_1 + Q_1}{C_{1,2} + \left(C_1^*/\Delta t\right)}, \quad T_2^\Delta = \frac{\left(C_{2,1}T_1^\Delta + C_{2,3}T_3^\Delta + C_{2,4}T_4^\Delta\right) + \left(C_2^*/\Delta t\right)T_2}{\left(C_{2,1} + C_{2,3} + C_{2,4}\right) + \left(C_2^*/\Delta t\right)}$$

$$T_3^\Delta = \frac{\left(C_{3,2}T_2^\Delta + C_{3,4}T_4^\Delta + C_{3,5}T_5^\Delta\right) + \left(C_3^*/\Delta t\right)T_3}{\left(C_{3,2} + C_{3,4} + C_{3,5}\right) + \left(C_3^*/\Delta t\right)}, \quad T_4^\Delta = \frac{\left(C_{4,2}T_2^\Delta + C_{4,3}T_3^\Delta + C_{4,6}T_6^\Delta\right) + \left(C_4^*/\Delta t\right)T_4}{\left(C_{4,2} + C_{4,3} + C_{4,6}\right) + \left(C_4^*/\Delta t\right)}$$

$$T_5^\Delta = \frac{\left(C_{5,3}T_3^\Delta + C_{5,6}T_6^\Delta + C_{5,7}T_7^\Delta\right) + \left(C_5^*/\Delta t\right)T_5}{\left(C_{5,3} + C_{5,6} + C_{5,7}\right) + \left(C_5^*/\Delta t\right)}, \quad T_6^\Delta = \frac{\left(C_{6,4}T_4^\Delta + C_{6,5}T_5^\Delta + C_{6,7}T_7^\Delta\right) + \left(C_6^*/\Delta t\right)T_6}{\left(C_{6,4} + C_{6,5} + C_{6,7}\right) + \left(C_6^*/\Delta t\right)}$$

$$(13.27)$$

Equations (13.27) are stepped through in a fashion similar to the stepping used with the forward finite time difference, but the temperatures used on the right-hand side of the solution equations are handled differently. First, all node temperatures are set to a start temperature equal to the problem initial condition. In our example then, the first line of Eqs. (13.27) uses $T_1 = 20°$, the second line uses $T_2 = 20°$, the third line uses $T_3 = 20°$, etc. However, note that in the equation for node 1 we have T_2^Δ on the right-hand side, which has not yet been calculated. Similarly for node 2 we have T_1^Δ, T_3^Δ, and T_4^Δ on the right-hand side, which have not yet been calculated, and so on for the remaining nodes 3-6. The only choice when we have a not-yet calculated temperature on the right-hand side is to use an initial condition temperature. As the stepping procedure continues, the most recently calculated temperature is used on the right-hand side. In this also, the procedure bears a resemblance to the steady-state Gauss-Seidel method. There is a condition, however; the time step needs to be sufficiently small so as to not create too large of an uncertainty in the right-hand side temperatures of Eq. (13.27).

Equations Set Up for Backward Finite Difference in Time Using Equations Solved as Simultaneous Equations. An equation is set up for each node according to Eq. (13.25).

$$\left[\left(C_1^*/\Delta t\right) + C_{1,2}\right]T_1^\Delta - C_{1,2}T_2^\Delta = \left(C_1^*/\Delta t\right)T_1 + Q_1$$

$$-C_{2,1}T_1^\Delta + \left[\left(C_2^*/\Delta t\right) + C_{2,1} + C_{2,3} + C_{2,4}\right]T_2^\Delta - C_{2,3}T_3^\Delta - C_{2,4}T_4^\Delta = \left(C_2^*/\Delta t\right)T_2$$

$$-C_{3,2}T_2^\Delta + \left[\left(C_3^*/\Delta t\right) + C_{3,2} + C_{3,4} + C_{3,5}\right]T_3^\Delta - C_{3,4}T_4^\Delta - C_{3,5}T_5^\Delta = \left(C_3^*/\Delta t\right)T_3$$

$$-C_{4,2}T_2^\Delta - C_{4,3}T_3^\Delta + \left[\left(C_4^*/\Delta t\right) + C_{4,2} + C_{4,3} + C_{4,6}\right]T_4^\Delta - C_{4,6}T_6^\Delta = \left(C_4^*/\Delta t\right)T_4 \quad (13.28)$$

$$-C_{5,3}T_3^\Delta + \left[\left(C_5^*/\Delta t\right) + C_{5,3} + C_{5,6} + C_{5,7}\right]T_5^\Delta - C_{5,6}T_6^\Delta = \left(C_5^*/\Delta t\right)T_5 + C_{5,7}T_7$$

$$-C_{6,4}T_4^\Delta - C_{6,5}T_5^\Delta + \left[\left(C_6^*/\Delta t\right) + C_{6,4} + C_{6,5} + C_{6,7}\right]T_6^\Delta = \left(C_6^*/\Delta t\right)T_6 + C_{6,7}T_7$$

Equations (13.28) are rewritten as

$$\mathbf{GT = S} \qquad (13.29)$$

where the temperature vector, source vector, and conductance matrix are given by Eqs. (13.30) and (13.31), respectively.

$$\mathbf{T} = \begin{pmatrix} T_1^{\Delta} \\ T_2^{\Delta} \\ T_3^{\Delta} \\ T_4^{\Delta} \\ T_5^{\Delta} \\ T_6^{\Delta} \end{pmatrix}, \qquad \mathbf{S} = \begin{pmatrix} \left(C_1^*/\Delta t\right)T_1 + Q_1 \\ \left(C_2^*/\Delta t\right)T_2 \\ \left(C_3^*/\Delta t\right)T_3 \\ \left(C_4^*/\Delta t\right)T_4 \\ \left(C_5^*/\Delta t\right)T_5 + C_{5,7}T_7 \\ \left(C_6^*/\Delta t\right)T_6 + C_{6,7}T_7 \end{pmatrix} \qquad (13.30)$$

$$\mathbf{G} = \begin{pmatrix} \left[\left(C_1^*/\Delta t\right) + C_{1,2}\right] & -C_{1,2} & 0 \\ -C_{2,1} & \left[\left(C_2^*/\Delta t\right) + C_{2,1} + C_{2,3} + C_{2,4}\right] & -C_{2,3} \\ 0 & -C_{3,2} & \left[\left(C_3^*/\Delta t\right) + C_{3,2} + C_{3,4} + C_{3,5}\right] \cdots \\ 0 & -C_{4,2} & -C_{4,3} \\ 0 & 0 & -C_{5,3} \\ 0 & 0 & 0 \end{pmatrix}$$

$$\begin{pmatrix} 0 & 0 & 0 \\ -C_{2,4} & 0 & 0 \\ -C_{3,4} & -C_{3,5} & 0 \\ \cdots\left[\left(C_4^*/\Delta t\right) + C_{4,2} + C_{4,3} + C_{4,6}\right] & 0 & -C_{4,6} \\ 0 & \left[\left(C_5^*/\Delta t\right) + C_{5,3} + C_{5,6} + C_{5,7}\right] & -C_{5,6} \\ -C_{6,4} & -C_{6,5} & \left[\left(C_6^*/\Delta t\right) + C_{6,4} + C_{6,5} + C_{6,7}\right] \end{pmatrix}$$

$$(13.31)$$

The reader should note that the "source vector" contains not only the true source term Q_i, but also has boundary conditions in the sixth and seventh rows. In addition, each source vector row contains a (capacitance · temperature / time step) term, but with the temperature expressed at the early time t. The temperature vector terms are node temperatures at the later time, $t + \Delta t$. The solution at the first time step $t + \Delta t$ is obtained by solving Eq. (13.29) for $\mathbf{T}$ using $\mathbf{S}$ with temperatures T_i equal to the initial conditions $T_7 = T_A = 20°C$. The solution for $\mathbf{T}$ at the second time step uses the previous solution for temperatures in $\mathbf{S}$. The solution proceeds for as many time steps as desired. The actual solution method used to obtain $\mathbf{T}$ from Eq. (13.29) may be obtained with any linear algebra method, but remembering that one time step requires the solution of the complete set of algebraic equations. Small problems consisting of a few nodes are quite easily solved by a matrix inversion technique $\mathbf{T} = \mathbf{G}^{-1}\mathbf{S}$, but large problems are probably better solved using more efficient methods.

Mathcad™ was used to compute temperatures from the three methods for the time interval $t = 0$ to 200 s. The results are nearly identical for all three methods. The answers at $t = 200$ s are listed in Table 13.2. The time-dependent results from the backward finite difference, simultaneous equations, are also plotted in Figure 13.2. Plots from the other two solution methods would be nearly identical.

Table 13.2. Illustrative Example 13.4: Node temperatures [°C] at $t = 200$ s.

Method	Eq. No.	Δt	Node 1	Node 2	Node 3	Node 4	Node 5	Node 6
Forward finite difference	(13.20)	0.2	53.450	52.450	27.245	27.674	26.618	26.742
Backward finite difference	(13.24)	0.2	53.450	52.413	27.245	27.674	26.618	26.742 (cont.)

Table 13.2. Illustrative Example 13.4: Node temperatures [°C] at $t = 200$ s.

Method	Eq. No.	Δt	Node 1	Node 2	Node 3	Node 4	Node 5	Node 6
Backward finite difference, simultaneous equations	(13.25)	5.0	53.436	52.436	27.238	27.667	26.612	26.736
Steady-state			53.47	52.47	27.25	27.68	26.63	26.75

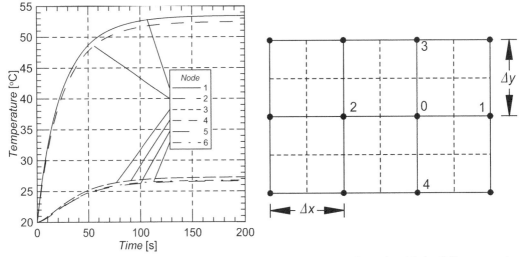

Figure 13.2. Illustrative Example 13.4: Node temperatures vs. time.

Figure 13.3. Two-dimensional finite difference mesh.

13.5 FINITE DIFFERENCE THEORY FOR CONDUCTION WITH NEWTONIAN COOLING

This short section is not intended to be a full exposition of the theory. But it is sufficient to illustrate the similarities between the finite difference and network methods. Figure 13.3 shows a portion of a two-dimensional conduction problem where the mesh size is Δx by Δy in the x and y directions, respectively. The integers 0 - 4 allow us to readily identify x and y locations and our theory will center on "node" 0.

We begin with the partial differential equation for two-dimensional isotropic heat conduction (conductivity k) with a volumetric heat source Q_V.

$$\frac{\partial^2 T}{\partial x^2} + \frac{\partial^2 T}{\partial y^2} = -\frac{Q_V}{k}$$

The "first forward" finite difference in the x-direction is

$$\left(\frac{\partial T}{\partial x}\right)_{0-1} \cong \frac{T_1 - T_0}{\Delta x}$$

The "first backward" finite difference in the x-direction is

$$\left(\frac{\partial T}{\partial x}\right)_{2-0} \cong \frac{T_0 - T_2}{\Delta x}$$

The forward and backward differences are combined in the next equation to form the "central differences" for the x and y directions.

$$\frac{\partial^2 T}{\partial x^2} \cong \frac{\left(\dfrac{\partial T}{\partial x}\right)_{0-1} - \left(\dfrac{\partial T}{\partial x}\right)_{2-0}}{\Delta x} = \frac{\left(\dfrac{T_1 - T_0}{\Delta x}\right) - \left(\dfrac{T_0 - T_2}{\Delta x}\right)}{\Delta x} = \frac{T_1 + T_2 - 2T_0}{(\Delta x)^2}$$

$$\frac{\partial^2 T}{\partial y^2} \cong \frac{\left(\dfrac{\partial T}{\partial y}\right)_{0-3} - \left(\dfrac{\partial T}{\partial y}\right)_{4-0}}{\Delta y} = \frac{\left(\dfrac{T_3 - T_0}{\Delta y}\right) - \left(\dfrac{T_0 - T_4}{\Delta y}\right)}{\Delta y} = \frac{T_3 + T_4 - 2T_0}{(\Delta y)^2}$$

$$(13.32)$$

Substituting the central differences and $Q_V = Q_0 / \Delta x \Delta y \Delta z$ into the *PDE*, we arrive at

$$\frac{T_1 + T_2 - 2T_0}{(\Delta x)^2} + \frac{T_3 + T_4 - 2T_0}{(\Delta y)^2} = -\frac{Q_0}{k \Delta x \Delta y \Delta z}$$

$$\frac{(T_1 + T_2 - 2T_0)\Delta z \Delta y}{\Delta x} + \frac{(T_3 + T_4 - 2T_0)\Delta z \Delta x}{\Delta y} = -Q_0$$

$$(13.33)$$

$$\left(\frac{k\Delta z \Delta y}{\Delta x}\right)(T_0 - T_1) + \left(\frac{k\Delta z \Delta y}{\Delta x}\right)(T_0 - T_2) + \left(\frac{k\Delta z \Delta x}{\Delta y}\right)(T_0 - T_3) + \left(\frac{k\Delta z \Delta x}{\Delta y}\right)(T_0 - T_4) = Q_0 \quad (13.34)$$

We can identify the factors of the temperature differences in Eq. (13.34) as conductances C_{ij} interconnecting the nodes i, j.

$$C_{0,1}(T_0 - T_1) + C_{0,2}(T_0 - T_2) + C_{0,3}(T_0 - T_3) + C_{0,4}(T_0 - T_4) = Q_0 \qquad (13.35)$$

which we note is identical, in principle, to the energy balance, Eq. (13.1). If node 0 is cooled by Newtonian cooling to an ambient node (such as node 1) at temperature T_A, then the source term Q_0 in Eq. (13.35) is modified by the replacement $Q_0 \rightarrow Q_0 - h\Delta y \Delta z(T_0 - T_A)$ or for the other direction, $Q_0 \rightarrow Q_0 - h\Delta x \Delta z(T_0 - T_A)$. Thus the thermal network and finite difference methods may sometimes be precisely identical, but this depends on how the network conductances are calculated. A distinct advantage of thermal networks is that the analyst has a great deal of flexibility in node placement. An example would be a thermal interface resistance where it is very straightforward to just put a resistance or conductance into the math model without concern about the "spatial meshing" for which you would have concern in a finite difference math model. Suppose that you have some kind of heat transfer (or any other physical phenomena) problem where it is not clear as to how to represent a conductance. In such an instance, you should have a partial differential equation. In this case you would apply a finite difference procedure to your *PDE*, just as we did with our heat conduction equation. The finite difference method is a very general math technique applicable to many different problem types.

13.6 PROGRAMMING THE PRESSURE/AIRFLOW NETWORK PROBLEM

Chapters 3 and 4 provide extensive detail on solving forced airflow problems for systems and components using an electronic calculator or math scratchpad. In this section some suggestions are provided for methods of solving problems using computer programming methods and languages such as *FORTRAN, BASIC, C*, or advanced math packages such as MATLAB™. The author has found the methods herein to be very useful, but you may think of your own improvements.

It is not uncommon to encounter pressure/airflow problems that contain elements where the pressure loss is neither fully laminar nor fully turbulent, i.e.,

$$\Delta h = RG^N, \quad N \neq 2$$

Perforated plates, expansions, etc., will use $N = 2$, but plate fin heat sinks with laminar airflow will use $N = 1$. The author has made measurements for flow through some grill systems where the "best" value of N is between 1.8 and 1.95.

Setup of Equations

The equations are set up for the head loss across any general element

$$\Delta h_{ij} = h_i - h_j = R_{ij} G_{ij}^{N_{ij}} \qquad \text{General head loss} \quad (13.36)$$

where i and j are used to identify the nodes between which the element with resistance R_{ij} is connected. The head h at each node as well as the airflow G_{ij} through the resistance are desired. A system of equations is used to solve for the h_i. The airflow G_{ij} through each element is calculated from Eq. (13.36). We first linearize the head loss equation.

$$\Delta h_{ij} = R_{ij} G_{ij}^{N_{ij}} = \left(R_{ij} \left| G_{ij} \right|^{N_{ij}-1} \right) G_{ij}$$

$$G_{ij} = \left(1 / R_{ij} \left| G_{ij} \right|^{N_{ij}-1} \right) \Delta h_{ij} = A_{ij} \Delta h_{ij}, \quad A_{ij} = 1 / \left(R_{ij} \left| G_{ij} \right|^{N_{ij}-1} \right) \qquad (13.37)$$

At any node i, we know that for the steady flow problem we have conservation of mass. Assuming negligible density changes (you might not want to assume this in your "code") we write the following equation for any *nonfixed pressure node*, i.e., a nonambient node:

$$\sum_j G_{ij} = S_i \qquad (13.38)$$

where S_i is the source strength (ft^3/min) at node i. At most nodes, $S_i = 0$. In fact, the only nodes at which the source strength is not zero would be those nodes which represent fans. An exhaust or blower fan would require a negative or positive source, respectively. An intermediate fan (resistance elements both before and after the fan) would require a negative source node at the fan inlet and an equal but positive source node at the fan exit. In any case, the conservation condition, Eq. (13.38), upon substituting G_{ij} from Eq. (13.37) becomes the following:

$$\sum_j A_{ij} \left(h_i - h_j \right) = S_i$$

$$\left(\sum_j A_{ij} \right) h_i - \sum_j A_{ij} h_j = S_i \quad A_{ij} = \frac{1}{R_{ij} \left| G_{ij} \right|^{N_{ij}-1}} \qquad (13.39)$$

A system of n total, nonfixed, pressure nodes, then has n such algebraic equations. The selection of an equation system solver is the analyst's choice. The Gauss-Seidel method is very slow with such equations, although it would appear appropriate because these equations must be iterated to get a solution, with the A_{ij} updated *during* each iteration. A repeatedly applied, linear equation solver requires the A_{ij} be updated at each iteration, but a technique as basic as a Gauss-Jordan reduction solver should be more than adequate for these very nonlinear equations.

Equation Solution Procedure

1. Solve the pressure/airflow equations with a first guess of the fan airflow and use as the source value S_i. A good first guess is about half of the maximum fan airflow. In this qualitative example, about $S = 16$ CFM is used for the fan curve shown in Figure 13.4.
2. Extrapolate both linear and quadratic system curves from $G = 0$ to the first guessed pressure/airflow solution.
3. Calculate the intersection of the linear and quadratic curves with the fan curve, G_l, G_q, respectively.
4. Calculate a new source weighted according to $S = 0.1\, G_l + 0.9\, G_q$.

5. Iterate the system of equations to a good solution using the fan source value from step 4.
6. Extrapolate both linear and quadratic system curves from $G = 0$ to the fan pressure/airflow solution from step 5.
7. Repeat steps 3 through 6 until the quadratic and linear extrapolations are very close, e.g.,

$$\Delta G = \left|G_q - G_l\right|\Big/\left[\left(G_q + G_l\right)\Big/2\right] < 0.1$$

Figure 13.5 illustrates the solution status at the end.

Detail Concerning Algebra of Extrapolation Through Fan Curve

The typical method of inputting fan curve data into various airflow programs, which are typically *CFD* based, is to input a series of airflow, pressure data pairs approximating the fan curve by a

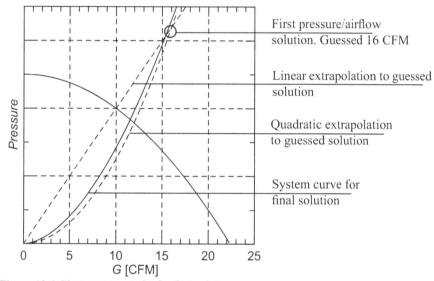

Figure 13.4. Flow system results for *first solution guess*.

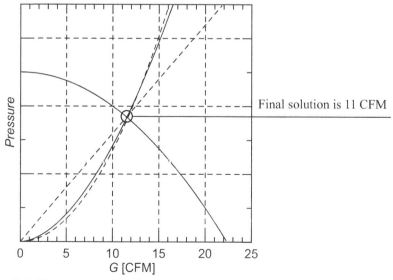

Figure 13.5. Flow system results after *final solution and extrapolation*.

series of straight line segments. A method of using each computed extrapolation is approximately as follows:

1. Each straight line segment of fan curve represented by

$$h_{Fan} = h_1 + B(G - G_1); \quad B = (h_2 - h_1)/(G_2 - G_1)$$

2. Using the appropriate G_{Fan} (guess for first iteration, then the weighted G_{Fan} for succeeding iterations) and the calculated pressure head Δh_{Fan} across the fan, calculate a quadratic system pressure $R_q = \Delta h_{Fan}/G_{Fan}^2$.

3. Using R_q , calculate two pressures, h_1 and h_2 , for the beginning and end points, respectively, of the first linear element of the fan curve. Compare these calculated h_1 and h_2 with the two pressures at the end points of the fan curve segment.

4. If h_1 and h_2 are respectively less than and greater than the two fan segment end point pressures, then this is a good straight line representation of the fan curve (in the local region) through which to extrapolate. If h_1 and h_2 are not respectively less than and greater than the two fan segment end point pressures, proceed to the next set of airflow/pressure points for the fan curve until the desired condition is met.

5. Using the resulting G_1, G_2 airflows for the "found" linear fan segment, calculate the quadratic extrapolation.

Quadratic extrapolation: $\qquad h_q = R_q G^2$

Intersection of quadratic
extrapolation and fan curve
segment:

$$h_q = h_{Fan}$$
$$R_q G_q^2 = h_1 + B(G_q - G_1)$$
$$R_q G_q^2 - BG_q + (BG_1 - h_1) = 0$$

$$\boxed{G_q = \frac{B + \sqrt{B^2 - 4R_q(BG_1 - h_1)}}{2R_q}} \quad B = \frac{h_2 - h_1}{G_2 - G_1} \quad (13.40)$$

6. Using the appropriate G_{Fan} (guess for first iteration, then the weighted G_{Fan} for succeeding iterations) and the calculated head Δh_{Fan} across the fan, calculate a linear system resistance $R_l = \Delta h_{Fan}/G$.

7. Using R_l , calculate two heads, h_1 and h_2 , for the beginning and end points, respectively, of the first linear element of the fan curve. Compare these calculated h_1 and h_2 with the two pressures at the end points of the fan curve segment.

8. If h_1 and h_2 are respectively less than and greater than the two fan segment end point pressures, then this is a good straight line representation of the fan curve (in the local region) through which to extrapolate. If h_1 and h_2 are not respectively less than and greater than the two fan segment end point pressures, proceed to the next set of airflow/pressure points for the fan curve until the desired condition is met.

9. Use the resulting G_1, G_2 airflows for the "found" linear fan segment and calculate the linear extrapolation.

Linear extrapolation: $\qquad h_l = R_l G$

Intersection of linear
extrapolation and fan
curve segment:

$$h_l = h_{Fan}$$
$$R_l G_l = h_1 + B(G_l - G_1)$$

$$\boxed{G_l = \frac{h_1 - BG_1}{R_l - B}} \quad B = \frac{h_2 - h_1}{G_2 - G_1} \qquad (13.41)$$

13.7 FINITE ELEMENT THEORY - THE CONCEPT
OF THE CALCULUS OF VARIATIONS

There are innumerable books written about the finite element theory and more than one mathematical method for developing the necessary equations. The author is partial to a variational approach for thermal problems. The calculus of variations may be new to the reader, thus a common problem will be addressed before we proceed to our true problem of interest.

Our problem (commonly referred to as *the brachistochrone problem*) is that of finding the path that a particle of mass m takes where the time is of least descent in a downward directed gravitational field. The geometry and nomenclature are shown in Figure 13.6 where the particle speed is v at any distance y in the downward direction

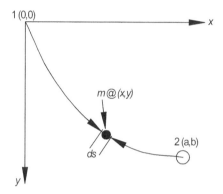

Figure 13.6. Variational solution of *time of least descent.*

We begin with the conservation of energy for the particle released at $x = y = 0$.

$$\frac{1}{2}mv^2 = mgy, \qquad v = \sqrt{2gy} \tag{13.42}$$

The time for the particle to travel from point 1, the release point, to the final point 2 is

$$\Delta t = \int_1^2 \frac{ds}{v} = \int_1^2 \frac{ds}{\sqrt{2gy}} \tag{13.43}$$

The curve element ds is written as

$$ds = \sqrt{(dx)^2 + (dy)^2} = dx\sqrt{1 + (dy/dx)^2}$$

Equation (13.43) becomes

$$\Delta t = \int_1^2 dt = \int_1^2 \frac{ds}{v} = \int_{x_1}^{x_2} \frac{\sqrt{1 + (dy/dx)^2}}{\sqrt{2gy}}\,dx = \int_{x_1}^{x_2} \sqrt{\frac{1 + (dy/dx)^2}{2gy}}\,dx \tag{13.44}$$

The problem is then to find the curve $y(x)$ such that the time Δt is a minimum. This would be a problem in variational calculus. The correct terminology is to refer to the integral, Eq. (13.44), as a *functional*. The subject of the calculus of variations is interesting and the method is powerful. The subject is well covered in many textbooks in mathematical physics and engineering. At this point we have an interest in the single subject of finite element theory. As in many other subjects, there is more than one way to approach the theory, but only one is discussed in this book.

13.8 FINITE ELEMENT THEORY - DERIVATION OF
THE EULER-LAGRANGE EQUATION

Two of many possible paths $y(x)$ are shown in Figure 13.7. The goal of variational calculus is to find that path that minimizes some functional J. y represents all possible paths and is expressed in terms of a parameter ε and a function $\eta(x)$. The latter is zero at the end points of the path. The parameter ε has some value that the path is the one that is found from the extremum of J. This discussion is begun using one dimension. If 1,2 indicate path end points, the functional is given by

$$J = \int_1^2 F(y, y', x) dx , \ y' = \frac{\partial y}{\partial x} \tag{13.45}$$

and $\qquad y(x, \varepsilon) = y(x, 0) + \varepsilon \eta(x), \ y'(x, \varepsilon) = y'(x, 0) + \varepsilon \eta'(x)$ where $y' = \frac{\partial y}{\partial x}$

The extremum in J is found using the methods of differential calculus.

$$\delta J = \frac{dJ}{d\varepsilon} d\varepsilon = 0$$

$$\delta J = \int_1^2 \left[\left(\frac{\partial F}{\partial y} \right) \frac{\partial y}{\partial \varepsilon} d\varepsilon + \left(\frac{\partial F}{\partial y'} \right) \frac{\partial y'}{\partial \varepsilon} d\varepsilon \right] dx \tag{13.46}$$

The second term in the integral is integrated by parts.

$$\int \frac{\partial F}{\partial y'} \frac{\partial y'}{\partial \varepsilon} dx = \int \frac{\partial F}{\partial y'} \frac{\partial}{\partial \varepsilon} \frac{\partial y}{\partial x} dx$$

Using $\quad u = \frac{\partial F}{\partial y'}$ and $v = \frac{\partial y}{\partial \varepsilon}$ then $du = \frac{\partial}{\partial x} \left(\frac{\partial F}{\partial y'} \right) dx$ and $dv = \frac{\partial}{\partial x} \left(\frac{\partial y}{\partial \varepsilon} \right) dx$

$$\int u dv = uv - \int v du$$

Then $\int_1^2 \frac{\partial F}{\partial y'} \frac{\partial y'}{\partial \varepsilon} dx = \int_1^2 \frac{\partial F}{\partial y'} \frac{\partial}{\partial \varepsilon} \frac{\partial y}{\partial x} dx = \left. \frac{\partial F}{\partial y'} \frac{\partial y}{\partial \varepsilon} \right|_1^2 - \int_1^2 \frac{\partial y}{\partial \varepsilon} \frac{\partial}{\partial x} \left(\frac{\partial F}{\partial y'} \right) dx$

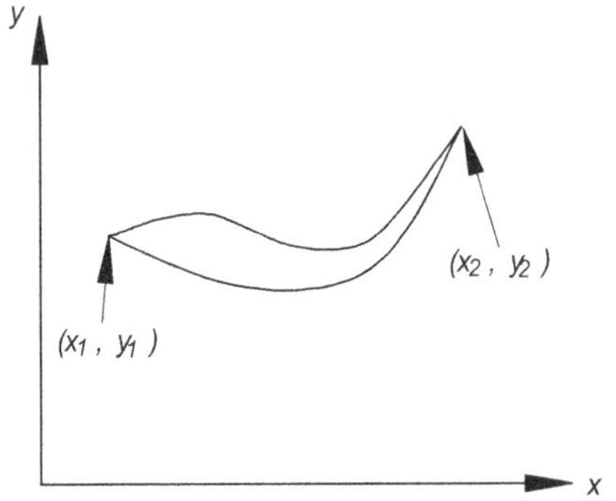

Figure 13.7. Varied paths y in one dimension x.

$$\delta J = \int_1^2 \left[\left(\frac{\partial F}{\partial y} \right) \frac{\partial y}{\partial \varepsilon} - \frac{\partial y}{\partial \varepsilon} \frac{d}{dx} \left(\frac{\partial F}{\partial y'} \right) \right] d\varepsilon dx + \frac{\partial F}{\partial y'} \frac{\partial y}{\partial \varepsilon} \bigg|_1^2 d\varepsilon$$

$$\delta J = \int_1^2 \left[\left(\frac{\partial F}{\partial y} \right) - \frac{d}{dx} \left(\frac{\partial F}{\partial y'} \right) \right] \frac{\partial y}{\partial \varepsilon} d\varepsilon dx + \frac{\partial F}{\partial y'} \frac{\partial y}{\partial \varepsilon} \bigg|_1^2 d\varepsilon$$

The variation in y is $\delta y = \frac{\partial y}{\partial \varepsilon} d\varepsilon$, then

$$\delta J = \int_1^2 \left[\frac{\partial F}{\partial y} - \frac{d}{dx} \left(\frac{\partial F}{\partial y'} \right) \right] \delta y dx + \frac{\partial F}{\partial y'} \bigg|_1^2 \delta y \tag{13.47}$$

The last term in the preceding equation is zero because $\delta y = 0$ at the end points of the path. Since the variation δy *along the path* is arbitrary, the only way to get $\delta J = 0$ is if the term in square brackets is zero.

Thus we arrive at the important result:

<u>Euler-Lagrange Equation</u>

$$\left(\frac{\partial F}{\partial y} \right) - \frac{d}{dx} \left(\frac{\partial F}{\partial y'} \right) \tag{13.48}$$

The previous paragraphs serve as a general introduction. Now the topic will be re-formulated to include a boundary condition term that is necessary for heat transfer problems such as the simple one in Section 13.9. However, we shall keep the symbolism a bit general for any field ϕ with the spatial variables x, y, and z for the generality of three-dimensions where V is a volume with surface area S. Since there are three spatial derivatives, a different notation is used where $\phi_x = \partial\phi/\partial x$, etc.

$$J = \int_V F(\phi, \phi', x, y, z) dV + \int_S f(\phi, x, y, z) dS$$

$$\delta J = \frac{\partial J}{\partial \varepsilon} d\varepsilon = \int_V \left(\frac{\partial F}{\partial \phi} \frac{\partial \phi}{\partial \varepsilon} d\varepsilon + \frac{\partial F}{\partial \phi_x} \frac{\partial \phi_x}{\partial \varepsilon} d\varepsilon + \frac{\partial F}{\partial \phi_y} \frac{\partial \phi_y}{\partial \varepsilon} d\varepsilon + \frac{\partial F}{\partial \phi_z} \frac{\partial \phi_z}{\partial \varepsilon} d\varepsilon \right) dV + \int_S \delta f(\phi, x, y, z) dS \tag{13.49}$$

But we shall use the shorthand $\delta\phi = \frac{\partial\phi}{\partial\varepsilon} d\varepsilon, \delta\phi_x = \frac{\partial\phi_x}{\partial\varepsilon} d\varepsilon, \delta\phi_y = \frac{\partial\phi_y}{\partial\varepsilon} d\varepsilon, \delta\phi_z = \frac{\partial\phi_z}{\partial\varepsilon} d\varepsilon$

$$\delta J = \int_V \left(\frac{\partial F}{\partial \phi} \delta\phi + \frac{\partial F}{\partial \phi_x} \delta\phi_x + \frac{\partial F}{\partial \phi_y} \delta\phi_y + \frac{\partial F}{\partial \phi_z} \delta\phi_z \right) dV + \int_S \delta f(\phi) dS \tag{13.50}$$

Now we need a side note derivation:

$$\delta\phi = (\phi + \varepsilon\eta) - \phi = \varepsilon\eta$$

$$\frac{\partial}{\partial x} \delta\phi = \varepsilon \frac{\partial\eta}{\partial x} = \varepsilon\eta_x$$

$$\delta\phi_x = (\phi_x + \varepsilon\eta_x) - \phi_x = \varepsilon\eta_x$$

Comparing the preceding two results, to which others can be added

by inspection $\delta\phi_x = \dfrac{\partial}{\partial x}\delta\phi,\ \delta\phi_y = \dfrac{\partial}{\partial y}\delta\phi,\ \delta\phi_z = \dfrac{\partial}{\partial z}\delta\phi$

Then $\delta J = \displaystyle\int_V \left(\dfrac{\partial F}{\partial \phi}\delta\phi + \dfrac{\partial F}{\partial \phi_x}\dfrac{\partial}{\partial x}(\delta\phi) + \dfrac{\partial F}{\partial \phi_y}\dfrac{\partial}{\partial y}(\delta\phi) + \dfrac{\partial F}{\partial \phi_z}\dfrac{\partial}{\partial z}(\delta\phi) \right) dV + \int_S \delta f(\phi)\, dS$

The second, third, and fourth terms in δJ can be integrated by parts.

For example, the first uses $\displaystyle\int u\, dv = \int d(uv) - \int v\, du = uv - \int v\, du$

where we use $u = \dfrac{\partial F}{\partial \phi_x},\ dv = \dfrac{\partial}{\partial x}(\delta\phi)\, dx,\ du = \dfrac{\partial}{\partial x}\left(\dfrac{\partial F}{\partial \phi_x} \right) dx,\ v = \delta\phi$

$\displaystyle\int_V \dfrac{\partial F}{\partial \phi_x}\dfrac{\partial}{\partial x}(\delta\phi)\, dx\, dy\, dz = \int_{S_x} \dfrac{\partial F}{\partial \phi_x}(\delta\phi)\, dy\, dz - \int_V \dfrac{\partial}{\partial x}\left(\dfrac{\partial F}{\partial \phi_x} \right) \delta\phi\, dV$, etc.

Then $\delta J = \displaystyle\int_V \left[\left(\dfrac{\partial F}{\partial \phi} - \dfrac{\partial}{\partial x}\left(\dfrac{\partial F}{\partial \phi_x} \right) - \dfrac{\partial}{\partial y}\left(\dfrac{\partial F}{\partial \phi_y} \right) - \dfrac{\partial}{\partial z}\left(\dfrac{\partial F}{\partial \phi_z} \right) \right) \right] \delta\phi\, dV$ (13.51)

$\qquad\qquad + \displaystyle\int_S \left(l_x \dfrac{\partial F}{\partial \phi_x} + l_y \dfrac{\partial F}{\partial \phi_y} + l_z \dfrac{\partial F}{\partial \phi_z} \right) \delta\phi\, dS + \int_S (\delta f)\, dS$

where the l_x, l_y, l_z are direction cosines to the bounding surfaces.

Because $\delta\phi$ is arbitrary for $\delta J = 0$, the first integral in Eq. (13.51) contains square brackets that when set to zero, constitute the Euler-Lagrange equations. The remaining two integrals are used to derive boundary condition formulae. This is deferred to the next section.

13.9 FINITE ELEMENT THEORY - APPLICATION OF THE ONE-DIMENSIONAL EULER-LAGRANGE EQUATION TO A HEAT CONDUCTION PROBLEM

Equation (13.51) applied to Figure 13.8, a one-dimensional heat conducting rod results in

$\delta J = \displaystyle\int_V \left[\left(\dfrac{\partial F}{\partial \phi} - \dfrac{\partial}{\partial x}\left(\dfrac{\partial F}{\partial \phi_x} \right) \right) \right] \delta\phi\, dV + \int_S \left(l_x \dfrac{\partial F}{\partial \phi_x} \right) \delta\phi\, dS + \int_S \delta f(\phi, x)\, dS$

$\delta J = \displaystyle\int_V \left[\left(\dfrac{\partial F}{\partial \phi} - \dfrac{\partial}{\partial x}\left(\dfrac{\partial F}{\partial \phi_x} \right) \right) \right] \delta\phi\, dV + \int_S \left(l_x \dfrac{\partial F}{\partial \phi_x} \right) \delta\phi\, dS + \int_S \dfrac{\partial f}{\partial \phi}\delta\phi\, dS$

$\delta J = \displaystyle\int_V \left[\left(\dfrac{\partial F}{\partial \phi} - \dfrac{\partial}{\partial x}\left(\dfrac{\partial F}{\partial \phi_x} \right) \right) \right] \delta\phi\, dV + \int_S \left(l_x \dfrac{\partial F}{\partial \phi_x} + \dfrac{\partial f}{\partial \phi} \right) \delta\phi\, dS$ (13.51b)

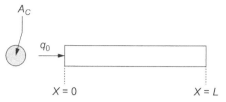

Figure 13.8. One-dimensional bar geometry.

Since $\delta\phi$ is arbitrary, the only way $\delta J = 0$ is for the integrands in Eq. (13.51) to be identically zero. Now ϕ is identified as T. The results are the Euler-Lagrange Equation and boundary conditions.

<u>1-D Euler-Lagrange</u>   <u>Boundary Conditions</u>

$$\frac{\partial F}{\partial T} - \frac{\partial}{\partial x}\left(\frac{\partial F}{\partial T_x}\right) = 0, \qquad \left[l_x\frac{\partial F}{\partial T_x} + \frac{\partial f}{\partial T}\right]_{On\ S} = 0 \qquad (13.52)$$

The main difficulty with the variational approach is finding the functions F and f for Eq. (13.52). We must often resort to a trial-and-error approach. In the case of the insulated bar with end conditions of a source at $x = 0$ and Newtonian cooling at $x = L$, it is proposed that

$$F = \frac{1}{2}\left[k\left(\frac{dT}{dx}\right)^2 - 2Q_V T\right] \text{ and } f = qT + \frac{1}{2}h\left(T - T_{Inf}\right)^2 \qquad (13.53)$$

Inserting Eqs. (13.53) into Eq. (13.52),

$$\frac{\partial F}{\partial T} = -Q_V, \ \frac{d}{dx}\left(\frac{\partial F}{\partial T_x}\right) = \frac{d}{dx}k\left(\frac{dT}{dx}\right) = k\frac{d^2T}{dx^2}$$

The Euler-Lagrange Equation becomes

$$-Q_V - k\frac{d^2T}{dx^2} = 0 \text{ or } k\frac{d^2T}{dx^2} = -Q_V \qquad (13.54)$$

$$l_x\frac{\partial F}{\partial T_x} = l_x\left(k\frac{dT}{dx}\right), \ \frac{\partial f}{\partial T} = q + h\left(T - T_\infty\right)$$

$$\left[l_x k\frac{dT}{dx} + q + h\left(T - T_\infty\right)\right]_{On\ S} = 0$$

where q is *heat out of the boundary*, consistent with Segerlind, 1976.
We see that the Euler-Lagrange equation applied to our functional gives us Eq. (13.54), the differential equation for one-dimensional heat conduction.

Consider a rod with a source at $x = 0$, insulated at $x = L$, and a convecting surface for $x = 0$ to L. Then boundary conditions in Eq. (13.52) applied to the proposed F and f result in

At $x = 0$, $l_x = -1$, $h = 0$, $q = -|q_0|$, $\qquad (13.55)$

$$-k\frac{dT}{dx} = -q, \ k\frac{dT}{dx} = -|q_0|$$

At $x = L$, $l_x = 1$, $q = 0, h = 0$

$$k\frac{dT}{dx} = 0 \qquad (13.56)$$

Then Eqs. (13.55) and (13.56) are the appropriate boundary conditions. *We can conclude that for the one-dimensional, insulated (lengthwise) bar, the integrands Eq. (13.52) correctly result in a functional J, Eq. (13.49), that gives an extremum* $\delta J = 0$.

We are ready to derive finite element method (*FEM*) equations. We will set up only enough of the problem to show the mathematical principles involved and see what conclusion(s) we might find. Thus we consider the *two element* bar in Figure 13.9. This bar is identical to that shown in Figure 13.8, but now reveals the three nodes and two elements (numbers are circled). Each of the two elements has a length *l*.

We will apply our $x = 0$ source via a boundary condition so we can immediately set $Q_V = 0$ in *J*.

$$J = A_c \int_V \left[\frac{1}{2} k \left(\frac{dT}{dx} \right)^2 \right] dx + \int_S \left[qT + \frac{1}{2} h (T - T_\infty)^2 \right] dS \qquad (13.57)$$

Interpolation equations are written for each element, and in this problem we use the lowest order equation possible, i.e., linear elements.

$$T^{(1)} = N_1^{(1)} T_1 + N_2^{(1)} T_2$$
$$N_1^{(1)} = (x_2 - x)/l, \ N_2^{(1)} = (x - x_1)/l \qquad \text{Element 1} \ (13.58)$$
$$T^{(2)} = N_2^{(2)} T_2 + N_3^{(2)} T_3$$
$$N_2^{(2)} = (x_3 - x)/l, \ N_3^{(2)} = (x - x_2)/l \qquad \text{Element 2} \ (13.59)$$

The parenthesized superscripts identify the element number and the subscripts identify the node number. T_1, T_2, and T_3 are the temperatures at each of the three nodes. You can check for yourself that substitution of the various node coordinate values in *x* does indeed return the appropriate node temperatures for T_1, T_2, and T_3. Since q_0 is *into* the surface, $q = -q_0 = -|q_0|$ consistent with Eq. (13.55). Calculating the derivatives of $T^{(1)}$ and $T^{(2)}$ and substituting into Eq. (13.57) we obtain

$$\frac{dT^{(1)}}{dx} = \frac{T_2 - T_1}{l}, \frac{dT^{(2)}}{dx} = \frac{T_3 - T_2}{l}$$
$$q = -q_0, h = 0 \text{ at } x = 0$$
$$q = 0, h \neq 0 \text{ at } x = L$$

$$J = \frac{kA_c}{2} \int_{x=0}^{x=l} \left(\frac{T_2 - T_1}{l} \right)^2 dx + \frac{kA_c}{2} \int_{x=l}^{x=2l} \left(\frac{T_3 - T_2}{l} \right)^2 dx - q_0 \int_{x=0}^{} T_1 dS$$
$$+ \frac{h}{2} \left(\frac{\pi dl}{2} \right) (T_1 - T_{Inf})^2 + \frac{h}{2} (\pi dl) (T_2 - T_{Inf})^2 + \frac{h}{2} \left(\frac{\pi dl}{2} \right) (T_3 - T_{Inf})^2$$

$$J = \frac{kA_c}{2l} (T_2 - T_1)^2 + \frac{kA_c}{2l} (T_3 - T_2)^2 - q_0 A_c T_1 + \frac{h\pi dl}{4} (T_1 - T_{Inf})^2 + \frac{h\pi dl}{2} (T_2 - T_{Inf})^2 + \frac{h\pi dl}{4} (T_3 - T_{Inf})^2$$

Define the following variables.

$$K^{(1)} = kA_c/l, \ K^{(2)} = kA_c/l, \ K_S^{(1)} = h\pi dl/2, \ K_S^{(2)} = h\pi dl, \ K_S^{(3)} = h\pi dl/2$$

$$J = \frac{K^{(1)}}{2}\left(T_2 - T_1\right)^2 + \frac{K^{(2)}}{2}\left(T_3 - T_2\right)^2 - q_0 A_c T_1$$

$$+ \frac{K_S^{(1)}}{2}\left(T_1 - T_{Inf}\right)^2 + \frac{K_S^{(2)}}{2}\left(T_2 - T_{Inf}\right)^2 + \frac{K_S^{(3)}}{2}\left(T_3 - T_{Inf}\right)^2 \qquad (13.60)$$

Once again we require that J be an extremum, but change our method of accomplishing this by minimizing J with respect to T_1, T_2, and T_3.

$$\frac{\partial J}{\partial T_1} = -K^{(1)}\left(T_2 - T_1\right) - q_o A_c + K_S^{(1)}\left(T_1 - T_{Inf}\right) = 0$$

$$\frac{\partial J}{\partial T_2} = K^{(1)}\left(T_2 - T_1\right) - K^{(2)}\left(T_3 - T_2\right) + K_S^{(2)}\left(T_2 - T_{Inf}\right) = 0$$

$$\frac{\partial J}{\partial T_3} = K^{(2)}\left(T_3 - T_2\right) + K_S^{(3)}\left(T_3 - T_{Inf}\right) = 0$$

$$\left(K^{(1)} + K_S^{(1)}\right)T_1 - K^{(1)}T_2 = q_0 A_c + K_S^{(1)}T_{Inf}$$

$$-K^{(1)}T_1 + \left(K^{(1)} + K^{(2)} + K_S^{(2)}\right)T_2 - K^{(2)}T_3 = K_S^{(2)}T_{Inf} \qquad (13.61)$$

$$-K^{(2)}T_2 + \left(K^{(2)} + K_S^{(3)}\right)T_3 = K_S^{(3)}T_{Inf}$$

We see that Eq. (13.61) is readily cast into a matrix problem where the nodal temperatures are computed. The boundary conditions and source terms are in the source vector $\mathbf{S}$. The matrix $\mathbf{K}$ is, in *FEM* terminology, called the *stiffness matrix*. We note then that the *FEM* stiffness matrix is, in thermal network terminology, the conductance matrix. This understanding can sometimes be useful when the use of a commercial *FEA* program permits the analyst to input any value of stiffness for an element.

$$\begin{pmatrix} \left(K^{(1)} + K_S^{(1)}\right) & -K^{(1)} & 0 \\ -K^{(1)} & \left(K^{(1)} + K^{(2)} + K_S^{(2)}\right) & -K^{(2)} \\ 0 & -K^{(2)} & \left(K^{(2)} + K_S^{(2)}\right) \end{pmatrix} \begin{Bmatrix} T_1 \\ T_2 \\ T_3 \end{Bmatrix} = \begin{Bmatrix} q_0 A_c + K_S^{(1)}T_{Inf} \\ K_S^{(2)}T_{Inf} \\ K_S^{(3)}T_{Inf} \end{Bmatrix} \qquad (13.62)$$

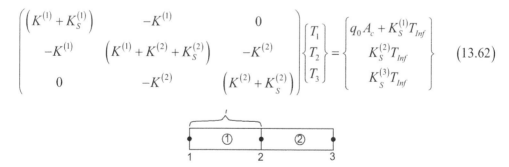

Figure 13.9. One-dimensional bar with two elements and three nodes.

APPENDIX *i*

Physical Properties of Dry Air at Atmospheric Pressure

T [°C]	k $\left[\dfrac{\text{W}}{\text{in.}\cdot\text{K}}\right]$	Pr	c_P $\left[\dfrac{\text{J}}{\text{gm}\cdot\text{K}}\right]$	ρ $\left[\dfrac{\text{gm}}{\text{in.}^3}\right]$	μ $\left[\dfrac{\text{gm}}{\text{in}\cdot\text{s}}\right]$
0	6.020×10^{-4}	0.71	1.011	0.0205	4.43×10^{-4}
20	6.375×10^{-4}	0.71	1.012	0.0191	4.63×10^{-4}
40	6.731×10^{-4}	0.71	1.014	0.0179	4.86×10^{-4}
60	7.087×10^{-4}	0.71	1.017	0.0168	5.06×10^{-4}
80	7.442×10^{-4}	0.71	1.019	0.0159	5.28×10^{-4}
100	7.798×10^{-4}	0.71	1.022	0.0150	5.50×10^{-4}
200	9.398×10^{-4}	0.71	1.035	0.0118	6.53×10^{-4}

T [°C]	ν $\left[\dfrac{\text{in.}^2}{\text{s}}\right]$	β $\left[\dfrac{1}{\text{K}}\right]$	$g\beta/\nu^2$ $\left[\dfrac{1}{\text{in.}^3\cdot\text{K}}\right]$	g/ν^2 $\left[\dfrac{1}{\text{in.}^3}\right]$
0	0.0216	3.66×10^{-3}	3.03×10^3	8.28×10^5
20	0.0243	3.41×10^{-3}	2.23×10^3	6.54×10^5
40	0.0273	3.19×10^{-3}	1.66×10^3	5.20×10^5
60	0.0301	3.00×10^{-3}	1.28×10^3	4.27×10^5
80	0.0333	2.83×10^{-3}	0.983×10^3	3.47×10^5
100	0.0366	2.68×10^{-3}	0.774×10^3	2.89×10^5
200	0.0550	2.11×10^{-3}	0.269×10^3	1.28×10^5

Gas temperature, T

Thermal conductivity, k

Prandtl number, Pr

Specific heat capacity, c_P

Density, ρ

Dynamic viscosity, μ

Kinematic viscosity, $\nu = \mu/\rho$

Thermal expansion coefficient, β

Acceleration due to gravity, g

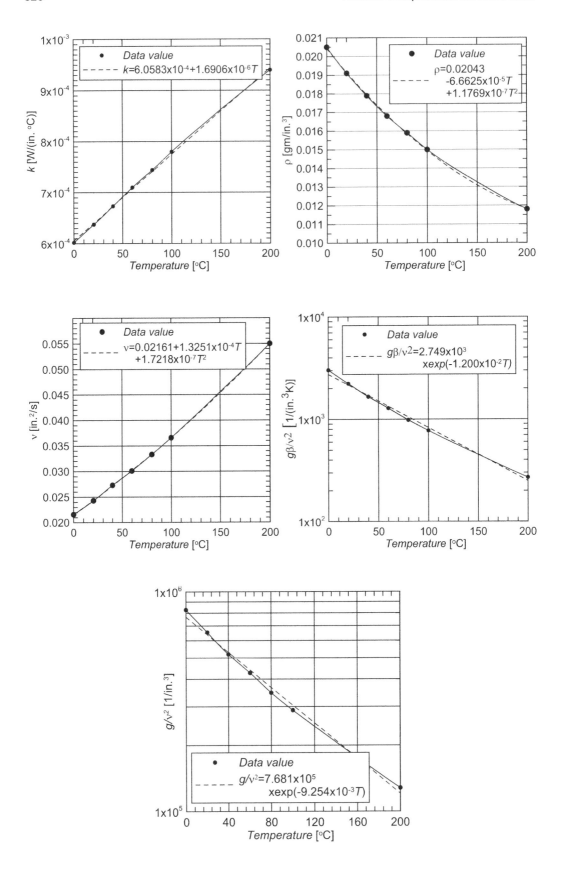

APPENDIX *ii*

Radiation Emissivity at Room Temperature
Assembled from Numerous Sources

Metals		Metals	
Material	Emissivity	Material	Emissivity
Aluminum		Magnesium	
highly polished plate	0.04-0.06	polished	0.06-0.08
polished plate	0.1	Molybdenum	
bright foil	0.04	polished	0.05-0.08
very oxidized	0.2-0	Monel	
anodized	0.7-0.8	polished	0.2
Brass		Nickel	
highly polished	0.03	electrolytic	0.05
polished	0.1	pure, polished	0.07-0.09
dull	0.2	electroplated on	0.1
oxidized	0.6	iron, not polished	
Chromium		oxidized plate	0.4
polished	0.1-0.4	Platinum	
Copper		electrolytic	0.06-0.1
highly polished	0.02	polished plate	0.05-0.1
polished	0.1	Silver	
slightly polished	0.15	polished	0.01-0.03
black anodized	0.78	Stainless Steel	
Gold		Inconel X, polished	0.2
highly polished	0.2-0.35	Inconel B, polished	0.2
polished	0.2	301, polished	0.15
Iron		310, smooth	0.4
electrolytic,	0.05-0.07	316, polished	0.24-0.3
highly polished		Steel	
wrought, highly polished	0.28	polished sheet	0.08-0.14
cast, freshly turned	0.4	mild, polished	0.3
cast, high temp.	0.6-0.8	sheet, rolled	0.7
oxidized		sheet, oxidized	0.8
cast, rough and oxidized	0.95	Tin	
new galvanized	0.23	polished sheet	0.05
and polished		Tungsten	
dirty galvanized	0.28	clean	0.03-0.08
oxide	0.96	filament	0.03
Lead		Zinc	
polished	0.05-0.07	polished	0.02-0.05
unoxidized, rough	0.4	galvanized sheet, bright	0.2
high temp. oxidized	0.6		

Nonmetals

Material	Emissivity
Alumina	0.6
Asbestos	0.95
Brick	
white refractory	0.3
red	0.9
Carbon, soot	0.95
Concrete, rough	0.9
Glass	0.9-0.95
Glass, pyrex	0.8
Mica	0.95
Paint	
all oil colors	0.90-0.95
flat black lacquer	0.95
white epoxy	0.9
aluminum lacquer	0.65
Paper, white	0.95
Plaster	0.9
Porcelain, glazed	0.9
Rubber	0.9
Thick film	
Pd/Ag	0.2 fresh - 0.4 aged
dielectric	0.74
resistor	0.7-1.0
Wood	
planed oak	0.9
beech	0.94
sawdust	0.75

Thermal Conductivity of Some Common Electronic Packaging Materials at Room Temperature: Values Are from Several Miscellaneous Sources

Material	Approximate Conductivity [W/(in. · °C)] at ≈ 300K	Material	Approximate Conductivity [W/(in. · °C)] at ≈ 300K
Diamond	20.0	Lead-tin solder	0.6
Silver	10.6	Kovar	0.5
Copper	9.6	Epoxy resin, BeO filled	0.09
Eutectic bond	7.5	Quartz	0.05
Gold	7.5	Quartz, fused	0.04
Aluminum	5.5	Silicon dioxide (SiO_2)	0.04
Beryllia	4-8	Alloy 42	0.3
Molybdenum	3.7	Glass	0.036
Gallium nitride (GaN)	3.3	Glass, borosilicate	0.026
Gallium phosphide (GaP)	2.8	Glass frit	0.024
Nickel	2.3	Conductive epoxy	0.02
Silicon, doped	4.0	Epoxy glass laminate (FR-4)	0.014
Silicon, pure	2.1		
Indium phosphide (InP)	1.7	Sylgard resin	0.01
Gallium arsenide (GaAs)	1.3	Doryl cement	0.007
Steel	1.2	Epoxy resin, unfilled	0.004
Solder (60-40)	0.9		
Lead	0.8	Silicon RTV, unfilled	0.004
Alumina (99%)	0.8		
Alumina (96%) filled	0.6	Thermal compound	0.02
Mica	0.02		

$$\text{GaAs: } k\left[W/\left(\text{in. }^{\circ}C\right)\right] = 1.45 \times 10^{3}/\left(T[K]\right)^{1.25}$$

$$\text{Si (pure) } \left[W/\left(\text{in.} \cdot {}^{\circ}C\right)\right] = 3.81/\left(T[K]/300\right)^{4/3}$$

Formula for GaAs from Wilson (2006) and Si from Lasance (1998). Consult the latter reference for more detail concerning dopant concentration.

APPENDIX *iv*

Some Properties of Bessel Functions

DEFINITIONS

Bessel's equation: $u^2 R'' + u R' + \left(u^2 - p^2\right) R = 0$

Although Bessel's equation is a differential equation of second order in the derivatives, it is often referred to as Bessel's equation of order p. This equation denotes a family of equations, with an individual member of the family for each value, real or complex, of the parameter p. Solutions in this book will not utilize any p that is not real.

When $p = 0$, the solution for R is the *Bessel function of the first kind, zeroth order* $\equiv J_0(u)$.
When $p = 1$, the solution for R is the *Bessel function of the first kind, first order* $\equiv J_1(u)$.

Solutions of Bessel's equation are typically expressed as one of two or four standardized Bessel functions:

$$J_p(u),\ Y_p(u),\ I_p(u),\ K_p(u)$$

The last two of the four solutions are used when the Bessel function is purely imaginary and we will not concern ourselves with those problems; only those whose arguments are purely real. The Bessel function J_p is continuous for all values of u, but the *Bessel function of the second kind*, Y_p is continuous at all values of u, *except at $u = 0$*, where it goes to infinity as u goes to zero. All solutions derived in this book require a finite solution at $u = 0$, and thus require rejection of Y_p as a valid solution.

A series representation for all orders p is:

$$J_p\left(u\right) = \sum_{k=0}^{\infty} \frac{\left(-1\right)^k u^{p+2k}}{2^{p+2k}\,k!\,\Gamma\left(p+k+1\right)}, \qquad \text{all real } p \neq -1, -2, -3, \cdots \qquad (iv.1)$$

and $$J_{-n}\left(u\right) = \left(-1\right)^n J_n\left(u\right), \qquad n = 1, 2, 3, \cdots \qquad (iv.2)$$

where $\Gamma(\ )$ is the Gamma function.

SOME PROPERTIES

Using Eq. $(iv.1)$, we write explicit series for the very important J_0 and J_1.

$$J_0\left(u\right) = 1 - \frac{u^2}{2^2} + \frac{u^4}{2^4\left(2!\right)^2} - \frac{u^6}{2^6\left(3!\right)^2} + \ldots \qquad (iv.3)$$

$$J_1\left(u\right) = \frac{u}{2} - \frac{u^3}{2^3\left(2!\right)} + \frac{u^5}{2^5\left(2!\right)\left(3!\right)} - \frac{u^7}{2^7\left(3!\right)\left(4!\right)} + \ldots \qquad (iv.4)$$

The plots of J_0 and J_1 in Figures $iv.1$ and $iv.2$ indicate the oscillatory nature of J_0 and J_1 about the u - axis with zeros at nearly regular intervals. It can actually be stated that J_0 and J_1 have an infinity of real zeros for any given real value of p and occur according to Eqs. $(iv.5)$ and $(iv.6)$.

Zeros of J_0: $\left(n+\dfrac{1}{2}\right)\pi < u < (n+1)\pi,\ \ n=0,1,2,3...$ $(iv.5)$

Zeros of J_1: $n\pi < u < \left(n+\dfrac{1}{2}\right)\pi,\ \ n=1,2,3,...$ $(iv.6)$

Some Useful Identities

$$\int_0^\infty J_0(\lambda r)J_1(\lambda a)d\lambda = 0,\ \ r>a \qquad (iv.7)$$

$$\int_0^\infty J_0(\lambda r)J_1(\lambda a)d\lambda = \frac{1}{2a},r=a \qquad (iv.8)$$

$$\int_0^\infty J_0(\lambda r)J_1(\lambda a)d\lambda = \frac{1}{a},\ \ r<a \qquad (iv.9)$$

$$\int_0^x J_0(y)ydy = xJ_1(x) \qquad (iv.10)$$

$$\int_0^\infty J_1^2(\lambda a)\frac{d\lambda}{\lambda^2} = \frac{4a}{3\pi} \qquad (iv.11)$$

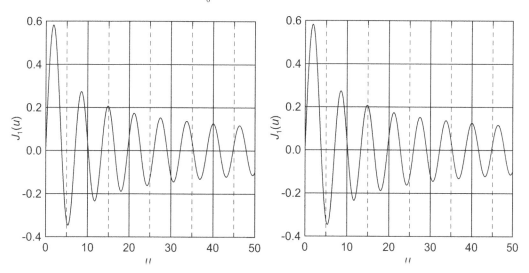

Figure *iv*.1. Bessel function of first kind, zeroth order. **Figure *iv*.2.** Bessel function of first kind, first order.

APPENDIX v

Some Properties of the Dirac Delta Function

There are many good references that you can refer to for a complete discussion of the Dirac Delta Function, e.g., Morse and Feshbach (1953) or Lea (2004). We shall only have a cursory, but sufficient for our purposes, look at this function, which has sometimes been called a *pathogenic function*.

A rigorous definition of the delta function is described by Figure $v.1$ and Eq. ($v.1$), whereas a more "relaxed" view is shown in Figure $v.2$ with defining Eq. ($v.2$).

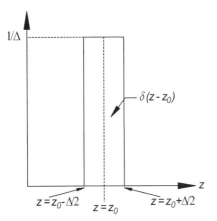

Figure $v.1$. Limit definition of Dirac delta function.   **Figure $v.2$.** Dirac delta function.

$$\delta(z-z_0) = \lim_{\Delta \to 0} \begin{cases} 0; & z < z_0 - \Delta/2 \\ 1/\Delta; & (z_0 - \Delta/2) < z < (z_0 + \Delta/2) \\ 0; & z > z_0 + \Delta/2 \end{cases} \qquad (v.1)$$

$$\delta(z-z_0) = 0, \ z \neq z_0$$

$$\int_{\text{Including } z=z_0} \delta(z-z_0)\,dz = 1 \qquad (v.2)$$

Some properties that are provable using Eq. ($v.1$):

$$\delta(z) = \delta(-z) \qquad (v.3)$$

$$z\delta'(z) = -\delta(z) \qquad (v.4)$$

$$\delta'(z) = -\delta'(-z) \qquad (v.5)$$

$$\delta(az) = \frac{1}{a}\delta(z) \qquad (v.6)$$

$$z\delta(z) = 0 \qquad (v.7)$$

$$f(z)\delta(z-z_0) = f(z_0)\delta(z-z_0) \qquad (v.8)$$

APPENDIX *vi*

Fourier Coefficients for a Rectangular Source

In Chapter 12, we saw that the source function for our rectangular geometry spreading analysis was written as follows:

$$Q_V(r) = \sum_{l=0}^{\infty} \sum_{m=0}^{\infty} \varepsilon_l \varepsilon_m \phi_{lm}(z) \cos\left(\frac{l\pi x}{a}\right) \cos\left(\frac{m\pi y}{b}\right) \qquad (vi.1)$$

$$\varepsilon_l = \begin{bmatrix} 1/2, & l = 0 \\ 1, & l \neq 0 \end{bmatrix} l = 0,1,2\ldots$$

$$\varepsilon_m = \begin{bmatrix} 1/2, & m = 0 \\ 1, & m \neq 0 \end{bmatrix} \qquad m = 0,1,2\ldots$$

where the $\varepsilon_l, \varepsilon_m$ allow a compact form to be used. In the following paragraphs, we follow standard procedures to obtain formulae for the ϕ_{lm}. We multiply both sides of Eq. ($vi.1$) by

$$\cos\left(\frac{l'\pi x}{a}\right) \cos\left(\frac{m'\pi y}{b}\right)$$

and integrate over the substrate x, y dimensions.

$$\int_{x=0}^{x=a} \int_{y=0}^{y=b} Q_V \cos\left(\frac{l'\pi x}{a}\right) \cos\left(\frac{m'\pi y}{b}\right) dxdy = \qquad (vi.2)$$

$$\int_{x=0}^{x=a} \int_{y=0}^{y=b} \sum_{l=0}^{\infty} \sum_{m=0}^{\infty} \varepsilon_l \varepsilon_m \phi_{lm}(z) \cos\left(\frac{l\pi x}{a}\right) \cos\left(\frac{m\pi y}{b}\right) \cos\left(\frac{l'\pi x}{a}\right) \cos\left(\frac{m'\pi y}{b}\right) dxdy$$

The preceding is expanded some to give

$$\int_{x=0}^{x=a} \int_{y=0}^{y=b} Q_V \cos\left(\frac{l'\pi x}{a}\right) \cos\left(\frac{m'\pi y}{b}\right) dxdy = \int_{x=0}^{x=a} \int_{y=0}^{y=b} \left(\frac{1}{2}\right)^2 \phi_{00} \cos\left(\frac{l'\pi x}{a}\right) \cos\left(\frac{m'\pi y}{b}\right) dxdy +$$

$$\int_{x=0}^{x=a} \int_{y=0}^{y=b} \sum_{l=1}^{\infty} \left(\frac{1}{2}\right) \phi_{l0} \cos\left(\frac{l'\pi x}{a}\right) \cos\left(\frac{l\pi x}{a}\right) \cos\left(\frac{m'\pi y}{b}\right) dxdy + \qquad (vi.3)$$

$$\int_{x=0}^{x=a} \int_{y=0}^{y=b} \sum_{m=1}^{\infty} \left(\frac{1}{2}\right) \phi_{0m} \cos\left(\frac{l'\pi x}{a}\right) \cos\left(\frac{m'\pi y}{b}\right) \cos\left(\frac{m\pi y}{b}\right) dxdy +$$

$$\int_{x=0}^{x=a} \int_{y=0}^{y=b} \sum_{l=1}^{\infty} \sum_{m=1}^{\infty} \phi_{lm} \cos\left(\frac{l'\pi x}{a}\right) \cos\left(\frac{l\pi x}{a}\right) \cos\left(\frac{m'\pi y}{b}\right) \cos\left(\frac{m\pi y}{b}\right) dxdy$$

The integrations are straightforward.

$$\int\limits_{x=0}^{x=a}\cos\left(\frac{l'\pi x}{a}\right)dx = \left(\frac{a}{l'\pi}\right)\int\limits_{u=0}^{u=l'\pi}\cos u\,du = \left(\frac{a}{l'\pi}\right)\sin u\Big|_{u=0}^{u=l'\pi} = \left(\frac{a}{l'\pi}\right)\sin\left(l'\pi\right)$$

$$\int\limits_{x=0}^{x=a}\cos\left(\frac{l'\pi x}{a}\right)dx = \left\{\begin{matrix} a, l' = 0 \\ 0, l' \neq 0 \end{matrix}\right\} = a\delta_{l'0}$$

where we are using the short-hand of the Kronecker delta, not to be confused with the Dirac delta function:

$$\delta_{ij} = \left\{\begin{matrix} 1, i = j \\ 0, i \neq j \end{matrix}\right.$$

Similarly:

$$\int\limits_{y=0}^{y=b}\cos\left(\frac{m'\pi y}{b}\right)dy = \left\{\begin{matrix} b, m' = 0 \\ 0, m' \neq 0 \end{matrix}\right\} = b\delta_{0m'}$$

The next integral is

$$\int\limits_{x=0}^{x=a}\cos\left(\frac{l'\pi x}{a}\right)\cos\left(\frac{l\pi x}{a}\right)dx = \frac{a}{\pi}\int\limits_{x=0}^{u=\pi}\cos\left(l'u\right)\cos\left(lu\right)du = \frac{a}{\pi}\left\{\frac{\sin\left[(l'-l)u\right]}{2(l'-l)} + \frac{\sin\left[(l'+l)u\right]}{2(l'+l)}\right\}\Big|_0^\pi$$

$$= \frac{a}{\pi}\left\{\begin{matrix} \dfrac{\pi}{2}, l' = l \\ 0, l' \neq l \end{matrix}\right\}$$

$$\int\limits_{x=0}^{x=a}\cos\left(\frac{l'\pi x}{a}\right)\cos\left(\frac{l\pi x}{a}\right)dx - \frac{a}{2}\delta_{l'l}$$

Similarly:

$$\int\limits_{y=0}^{y=b}\cos\left(\frac{m'\pi y}{b}\right)\cos\left(\frac{m\pi y}{b}\right)dy = \frac{b}{2}\delta_{mm'}$$

Eq. (vi.3), after incorporating the completed integrals, becomes

$$\int\limits_{x=0}^{x=a}\int\limits_{y=0}^{y=b} Q_V \cos\left(\frac{l'\pi x}{a}\right)\cos\left(\frac{m'\pi y}{b}\right)dxdy = \frac{ab}{4}\phi_{00}\delta_{l'0}\delta_{0m'} + \frac{ab}{4}\sum\limits_{l=1}^{\infty}\phi_{l0}\delta_{l'l}\delta_{0m'} + \frac{ab}{4}\sum\limits_{m=1}^{\infty}\phi_{0m}\delta_{l'0}\delta_{mm'}\quad (vi.4)$$

$$+\frac{ab}{4}\sum\limits_{l=1}^{\infty}\sum\limits_{m=1}^{\infty}\phi_{lm}\delta_{l'l}\delta_{m'm}$$

Careful study of Eq. (*vi*.4) and considering the meaning of the Kronecker deltas give us

$$\phi_{00} = \frac{4}{ab} \int\limits_{x=0}^{x=a} \int\limits_{y=0}^{y=b} Q_V \, dxdy$$

$$\phi_{l0} = \frac{4}{ab} \int\limits_{x=0}^{x=a} \int\limits_{y=0}^{y=b} Q_V \cos\left(\frac{l\pi x}{a}\right) dxdy \qquad\qquad (vi.5)$$

$$\phi_{0m} = \frac{4}{ab} \int\limits_{x=0}^{x=a} \int\limits_{y=0}^{y=b} Q_V \cos\left(\frac{m\pi y}{b}\right) dxdy$$

$$\phi_{lm} = \frac{4}{ab} \int\limits_{x=0}^{x=a} \int\limits_{y=0}^{y=b} Q_V \cos\left(\frac{l\pi x}{a}\right)\cos\left(\frac{m\pi y}{b}\right) dxdy$$

We now turn our attention to any single, planar, source Q_V at $z = z_0$:

$$Q_V = q(x,y)\delta(z - z_0) \qquad\qquad (vi.6)$$

Then with source beginning x-y coordinates x_1, y_1 and edge dimensions $\Delta x = x_2 - x_1, \Delta y = y_2 - y_1$, the total source dissipation (W):

$$Q = \int_V Q_V \, dxdydz = \int_V q(x,y)\delta(z-z_0) \, dxdydz = \int_z \delta(z-z_0) \, dz \iint_{\Delta x\,\Delta y} q(x,y) \, dxdy$$

and if we set the source at the $z = z_0 = 0$ plane and include a *model assumption* that $q(x,y)$ is uniform, we write

$$Q = \int_z \delta(z) \, dz \iint_{\Delta x\,\Delta y} q(x,y) \, dxdy = q\Delta x\Delta y$$

noting that we have used one of the properties of the Dirac delta function. This also tells us that we identify q as a flux density with dimensions of *heat dissipation* (W) / *unit area*. Substituting Eq. (*vi*.6) into Eq. (*vi*.5), we can carry out source Fourier coefficients a little further.

$$\phi_{00} = \frac{4}{ab} \int\limits_{x=x_1}^{x=x_2} \int\limits_{y=y_1}^{y=y_2} q(x,y)\delta(z) \, dxdy$$

$$\phi_{l0} = \frac{4}{ab} \int\limits_{x=x_1}^{x=x_2} \int\limits_{y=y_1}^{y=y_2} q(x,y)\delta(z)\cos\left(\frac{l\pi x}{a}\right) dxdy = \frac{4q(x,y)\delta(z)\Delta y}{ab} \int\limits_{x_1}^{x_2} \cos\left(\frac{l\pi x}{a}\right) dx$$

$$\phi_{0m} = \frac{4}{ab} \int\limits_{x=x_1}^{x=x_2} \int\limits_{y=y_1}^{y=y_2} q(x,y)\delta(z)\cos\left(\frac{m\pi y}{b}\right) dxdy = \frac{4q(x,y)\delta(z)\Delta x}{ab} \int\limits_{y_1}^{y_2} \cos\left(\frac{m\pi y}{b}\right) dy$$

$$\phi_{lm} = \frac{4}{ab} \int\limits_{x=x_1}^{x=x_2} \int\limits_{y=y_1}^{y=y_2} Q_V \cos\left(\frac{l\pi x}{a}\right)\cos\left(\frac{m\pi y}{b}\right) dxdy$$

$$\phi_{lm} = \frac{4q(x,y)\delta(z)}{ab} \int\limits_{x=x_1}^{x=x_2} \int\limits_{y=y_1}^{y=y_2} \cos\left(\frac{l\pi x}{a}\right)\cos\left(\frac{m\pi y}{b}\right) dxdy$$

Performing the integrations and simplifying a bit, we arrive at the results for a source centered at $\bar{x}_s = (x_1 + x_2)/2$, $\bar{y}_s = (y_1 + y_2)/2$ with dimensions of $\Delta x_s = x_2 - x_1$, $\Delta y_s = y_2 - y_1$:

$$
\begin{aligned}
\phi_{00} &= \frac{4q(x,y)\delta(z)\Delta x_s \Delta y_s}{ab} \\[2mm]
\phi_{l0} &= \frac{8q(x,y)\delta(z)\Delta y_s}{l\pi b} \sin\left(\frac{l\pi\Delta x_s}{2a}\right)\cos\left(\frac{l\pi\bar{x}_s}{a}\right) \\[2mm]
\phi_{0m} &= \frac{8q(x,y)\delta(z)\Delta x_s}{m\pi a} \sin\left(\frac{m\pi\Delta y_s}{2b}\right)\cos\left(\frac{m\pi\bar{y}_s}{b}\right) \\[2mm]
\phi_{lm} &= = \frac{16q(x,y)\delta(z)}{lm\pi^2} \sin\left(\frac{l\pi\Delta x_s}{2a}\right)\sin\left(\frac{m\pi\Delta y_s}{2b}\right)\cos\left(\frac{l\pi\bar{x}_s}{a}\right)\cos\left(\frac{m\pi\bar{y}_s}{b}\right)
\end{aligned}
$$

$$(vi.7)$$

APPENDIX *vii*

Derivation of the Green's Function Properties for the Spreading Problem of a Rectangular Source and Substrate - Method A

In this section, we are going to use the first of two methods to derive the Green's function for Eq. (*vii*.1) with the boundary conditions expressed in Eq. (*vii*.2).

$$\frac{d^2 \Psi_{lm}}{dz^2} - \gamma_{lm}^2 \Psi_{lm} = -\frac{1}{k}\phi_{lm} \qquad (vii.1)$$

$$\gamma_{lm}^2 = \alpha_l^2 + \beta_m^2, \ \alpha_l^2 = \left(l\pi/a\right)^2, \ \beta_m^2 = \left(m\pi/b\right)^2$$

$$k\frac{\partial \Psi_{lm}}{\partial z} - h_1 \Psi_{lm} = 0 \text{ at } z = 0; \ k\frac{\partial \Psi_{lm}}{\partial z} + h_2 \Psi_{lm} = 0 \text{ at } z = t \qquad (vii.2)$$

Note that these boundary conditions are slightly different than those given by Eq. (12.46), in particular, a Newtonian cooling boundary condition is shown for both $z = 0$ and $z = t$. At the conclusion of this section, we merely let $h_1 \to 0$ and $h_2 = h$ for Newtonian cooling only at $z = t$. At this point, you no doubt wonder what the Green's function really is. The following paragraphs are intended to clarify this issue.

We begin the analysis by a straightforward multiplication of Eq. (*vii*.1) by the Green's function G_{lm} and integrate over the total thickness t (we have anticipated the probable dependence on the integers l, m). Note that we change the name of the independent variable from z to z' (you will see the reason for this small change as we proceed).

$$\int_{z'=0}^{z'=t} G_{lm}\left(z|z'\right)\left[\frac{d^2\Psi_{lm}}{dz'^2} - \gamma^2\Psi_{lm}\right]dz' = -\frac{1}{k}\int_{z'=0}^{z'=t}\phi_{lm}G_{lm}\left(z|z'\right)dz' \qquad (vii.3)$$

You should look carefully at Figure *ix*.1 because we will be very careful in the integration. We will integrate the *left side* up to, but just short of $z' = z$, and then begin again at $z' = z + \varepsilon$, and then let $\varepsilon \to 0$. There is no point in breaking up the integration of the right-hand side because we will assume G_{lm} is continuous at $z' = z$.

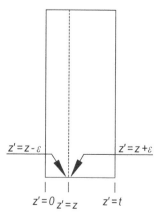

Figure *vii*.1. Illustration of care taken in integration over $z' = 0$ to $z' = t$. z' is the variable of integration and $z' = z$ is the field point (where you wish to calculate the temperature).

The integration of Eq. (*vii*.3) is

$$
\int_0^t G_{lm}\left(z\middle|z'\right)\left[\frac{d^2\Psi_{lm}}{dz'^2}-\gamma_{lm}^2\Psi\right]dz'=\int_{z'=0}^{z'=z-\varepsilon}G_{lm}\left(z\middle|z'\right)\frac{d^2\Psi_{lm}}{dz'^2}dz'+\int_{z'=z+\varepsilon}^{z'=t}G_{lm}\left(z\middle|z'\right)\frac{d^2\Psi_{lm}}{dz'^2}dz'
$$

$$
-\gamma_{lm}^2\int_{z'=0}^{z'=z-\varepsilon}G_{lm}\left(z\middle|z'\right)\Psi_{lm}dz'-\gamma_{lm}^2\int_{z'=z+\varepsilon}^{z'=t}G_{lm}\left(z\middle|z'\right)\Psi_{lm}dz' \qquad (vii.4)
$$

$$
=-\frac{1}{k}\int_{z'=0}^{z'=t}\phi_{lm}G_{lm}\left(z\middle|z'\right)dz'
$$

where we have carefully excluded integration near the field point $z'=z$ in the integration of Eq. (*vii*.4). The right side needs no such consideration. First note from your elementary calculus that an integration by parts is performed according to

$$
d\left(uv\right)=vdu+udv,\ udv=d\left(uv\right)-vdu
$$

$$
\int udv=\int d\left(uv\right)-\int vdu,\ \int udv=uv-\int vdu
$$

which we can use to our advantage with the first two terms on the right side of Eq. (*vii*.4):

$$
\int_{z'=0}^{z'=z-\varepsilon}G_{lm}\left(z\middle|z'\right)\frac{d^2\Psi_{lm}}{dz'^2}dz'+\int_{z'=z+\varepsilon}^{z'=t}G_{lm}\left(z\middle|z'\right)\frac{d^2\Psi_{lm}}{dz'^2}dz'=\left[G_{lm}\left(z\middle|z'\right)\frac{d\Psi_{lm}}{dz'}\right]\Bigg|_{z'=0}^{z'=z-\varepsilon}
$$

$$
-\int_{z'=0}^{z'=z-\varepsilon}\left[\frac{dG_{lm}\left(z\middle|z'\right)}{dz'}\right]\left[\frac{d\Psi_{lm}}{dz'}\right]dz'+\left[G_{lm}\left(z\middle|z'\right)\frac{d\Psi_{lm}}{dz'}\right]\Bigg|_{z'=z+\varepsilon}^{z'=t}-\int_{z'=z+\varepsilon}^{z'=t}\left[\frac{dG_{lm}\left(z\middle|z'\right)}{dz'}\right]\left[\frac{d\Psi_{lm}}{dz'}\right]dz'
$$

$$
=\left[G_{lm}\left(z\middle|z'\right)\frac{d\Psi_{lm}}{dz'}\right]\Bigg|_{z'=0}^{z'=z-\varepsilon}-\left\{\left[\frac{dG_{lm}\left(z\middle|z'\right)}{dz'}\right]\Psi_{lm}\right\}\Bigg|_{z'=0}^{z'=z-\varepsilon}+\int_{z'=0}^{z'=z-\varepsilon}\left[\frac{d^2G_{lm}\left(z\middle|z'\right)}{dz'^2}\right]\Psi_{lm}dz'
$$

$$
+\left[G_{lm}\left(z\middle|z'\right)\frac{d\Psi_{lm}}{dz'}\right]\Bigg|_{z'=z+\varepsilon}^{z'=t}-\left\{\left[\frac{dG_{lm}\left(z\middle|z'\right)}{dz'}\right]\Psi_{lm}\right\}\Bigg|_{z'=z+\varepsilon}^{z'=t}+\int_{z'=z+\varepsilon}^{z'=t}\left[\frac{d^2G_{lm}\left(z\middle|z'\right)}{dz'^2}\right]\Psi_{lm}dz'
$$

$$(vii.5)$$

Combining Eqs. (*vii*.4) and (*vii*.5),

$$
\left[G_{lm}\left(z\middle|z'\right)\frac{d\Psi_{lm}}{dz'}\right]\Bigg|_{z'=0}^{z'=z-\varepsilon}-\left[\frac{dG_{lm}\left(z\middle|z'\right)}{dz'}\Psi_{lm}\right]\Bigg|_{z'=0}^{z'=z-\varepsilon}+\int_{z'=0}^{z'=z-\varepsilon}\left[\frac{d^2G_{lm}\left(z\middle|z'\right)}{dz'^2}\right]\Psi_{lm}dz'
$$

$$
+\left[G_{lm}\left(z\middle|z'\right)\frac{d\Psi_{lm}}{dz'}\right]\Bigg|_{z'=z+\varepsilon}^{t}-\left[\frac{dG_{lm}\left(z\middle|z'\right)}{dz'}\Psi_{lm}\right]\Bigg|_{z'=z+\varepsilon}^{t}+\int_{z'=z+\varepsilon}^{t}\left[\frac{d^2G_{lm}\left(z\middle|z'\right)}{dz'^2}\right]\Psi_{lm}dz'
$$

$$
-\gamma_{lm}^2\int_{z'=0}^{z'=z-\varepsilon}G_{lm}\left(z\middle|z'\right)\Psi_{lm}dz'-\gamma_{lm}^2\int_{z'=z+\varepsilon}^{z'=t}G_{lm}\left(z\middle|z'\right)\Psi_{lm}dz'=-\frac{1}{k}\int_{z'=0}^{z'=t}\phi_{lm}G_{lm}\left(z\middle|z'\right)dz'
$$

and rearranging some terms, we obtain,

$$
\left[G_{lm}\left(z|z'\right)\frac{d\Psi_{lm}}{dz'} - \Psi_{lm}\frac{dG_{lm}\left(z|z'\right)}{dz'} \right]_{z'=0}^{z'=z-\varepsilon} + \int_{z'=0}^{z'=z-\varepsilon}\left[\frac{d^2 G_{lm}\left(z|z'\right)}{dz'^2} - \gamma_{lm}^2 G_{lm}\left(z|z'\right)dz' \right]\Psi_{lm}dz'
$$

$$
+\left[G_{lm}\left(z|z'\right)\frac{d\Psi_{lm}}{dz'} - \Psi_{lm}\frac{dG_{lm}\left(z|z'\right)}{dz'} \right]_{z'=z+\varepsilon}^{z'=t} + \int_{z'=z+\varepsilon}^{z'=t}\left[\frac{d^2 G_{lm}\left(z|z'\right)}{dz'^2} - \gamma_{lm}^2 G_{lm}\left(z|z'\right) \right]\Psi_{lm}dz'
$$

$$
= -\frac{1}{k}\int_{z'=0}^{z'=t}\phi_{lm}G_{lm}\left(z|z'\right)dz'
$$

which when arranged according to the *just before - just after* values of $z' \rightarrow z, 0, t$ gives us the next form.

$$
\left[\Psi_{lm}\frac{dG_{lm}\left(z|z'\right)}{dz'}\Big|_{z'=z+\varepsilon} - \Psi_{lm}\frac{dG_{lm}\left(z|z'\right)}{dz'}\Big|_{z'=z-\varepsilon} \right] + \left[\begin{array}{l} G_{lm}\left(z|z'\right)\dfrac{d\Psi_{lm}}{dz'}\Big|_{z'=z-\varepsilon} \\ -G_{lm}\left(z|z'\right)\dfrac{d\Psi_{lm}}{dz'}\Big|_{z'=z+\varepsilon} \end{array} \right]
$$

$$
+\left[\Psi_{lm}\frac{dG_{lm}\left(z|z'\right)}{dz'} - G_{lm}\left(z|z'\right)\frac{d\Psi_{lm}}{dz'} \right]_{z'=0} + \left[G_{lm}\left(z|z'\right)\frac{d\Psi_{lm}}{dz'} - \Psi_{lm}\frac{dG_{lm}\left(z|z'\right)}{dz'} \right]_{z'=t}
$$

$$
+\int_{z'=0}^{z'=z-\varepsilon}\left[\frac{d^2 G_{lm}\left(z|z'\right)}{dz'^2} - \gamma_{lm}^2 G_{lm}\left(z|z'\right)dz' \right]\Psi_{lm}dz' + \int_{z'=z+\varepsilon}^{z'=t}\left[\frac{d^2 G_{lm}\left(z|z'\right)}{dz'^2} - \gamma_{lm}^2 G_{lm}\left(z|z'\right) \right]\Psi_{lm}dz'
$$

$$
= -\frac{1}{k}\int_{z'=0}^{z'=t}\phi_{lm}G_{lm}\left(z|z'\right)dz' \tag{$vii.6$}
$$

Now we are in a position to write down the *properties of the Green's function for this problem*.

Equation (*vii*.6), line 1, first [...] term:

$$
\boxed{\Psi_{lm} = \text{continuous at } z' = z} \tag{$vii.6\text{-}1$}
$$

$$
\boxed{\left[\left(dG_{lm}\left(z|z'\right)/dz'\right)\Big|_{z'=z+\varepsilon} - \left(dG_{lm}\left(z|z'\right)/dz'\right)\Big|_{z'=z-\varepsilon} \right] = -1 \text{ at } z' = z} \tag{$vii.6\text{-}2$}
$$

where (*vii*.6-2) is always referred to as the *jump condition*.

Equation (*vii*.6), line 1, second [...] term contributes zero if

$$
\boxed{G_{lm}\left(z|z'\right) \text{ continuous at } z' = z} \tag{$vii.6\text{-}3$}
$$

$$
\boxed{d\Psi_{lm}/dz' \text{ continuous at } z' = z} \tag{$vii.6\text{-}4$}
$$

where Eq. (*vii*.6-4), if multiplied by k, is seen to be *flux density continuity*.

Equation (vii.6), line 2, first [...] term: this line contributes zero if the Newtonian cooling boundary condition is applied to both the temperature and the Green's function, i.e.,

$$\boxed{k\,dG_{lm}\left(z|z'\right)/dz' - h_1 G_{lm}\left(z|z'\right) = 0,\ k\,d\Psi_{lm}/dz' - h_1\Psi_{lm} = 0 \text{ at } z' = 0} \qquad (vii.6\text{-}5)$$

Equation (vii.6), line 2, second [...] term: this line contributes zero if the Newtonian cooling boundary condition is applied to both the temperature and the Green's function, i.e.,

$$\boxed{k\,dG_{lm}\left(z|z'\right)/dz' + h_2 G_{lm}\left(z|z'\right) = 0,\ k\,d\Psi_{lm}/dz' + h_2\Psi_{lm} = 0 \text{ at } z' = t} \qquad (vii.6\text{-}6)$$

Equation (vii.6), line 3: this line contributes zero if both [...] terms under the integrals are zero, i.e.,

$$\boxed{\frac{d^2 G_{lm}\left(z|z'\right)}{dz'^2} - \gamma_{lm}^2 G_{lm}\left(z|z'\right) = 0,\ z' \neq z} \qquad (vii.6\text{-}7)$$

Finally then, from Eq. (vii.6) we are left with

$$\boxed{\Psi_{lm}\left(z\right) = \frac{1}{k}\int_{z'=0}^{z'=t} \phi_{lm} G_{lm}\left(z|z'\right)dz'} \qquad (vii.6\text{-}8)$$

The problem is not nearly as complex as it first appears. Succinctly stated, our Green's function is the solution of the homogeneous, one-dimensional differential equation (vii.6-7), satisfying the same boundary conditions at $z = 0$, t as the temperature must satisfy. In addition, G and its first derivative must satisfy continuity and the jump condition, respectively, at $z' = z$. The continuity condition on G at $z' = z$, Eq. (vii.6-3), validates the statement before and following Eq. (vii.4), which is that the right side of same needs no such consideration. If our problem does not have Newtonian cooling, but instead has an adiabatic boundary at $z = 0$, then we merely set $h_1 = 0$ and $h_2 = h$.

Derivation of the Green's Function Properties for the Spreading Problem of a Rectangular Source and Substrate - Method B

In this section, we are going to use the second of two methods to derive the Green's function for Eq. (*viii*.1) with the boundary conditions expressed in Eq. (*viii*.2).

$$\frac{d^2\Psi_{lm}}{dz^2} - \gamma_{lm}^2 \Psi_{lm} = -\frac{1}{k}\phi_{lm} \qquad (viii.1)$$

$$\gamma_{lm}^2 = \alpha_l^2 + \beta_m^2, \ \alpha_l^2 = (l\pi/a)^2, \ \beta_m^2 = (m\pi/b)^2$$

$$k\frac{\partial\Psi_{lm}}{\partial z} - h_1\Psi_{lm} = 0 \text{ at } z = 0$$

$$k\frac{\partial\Psi_{lm}}{\partial z} + h_2\Psi_{lm} = 0 \text{ at } z = t \qquad (viii2)$$

As in Appendix *vii*, these boundary conditions are slightly different than those given by Eq. (12.46), in that a Newtonian cooling boundary condition is shown for both $z = 0$ and $z = t$. At the conclusion of this section, we merely let $h_1 \to 0$ and $h_2 = h$. We use the following derivation to define what the Green's function actually is. Additionally, the method used here follows more conventional Green's function analysis.

First we change the independent variable from z to z' and try a differential equation similar to Eq. (*viii*.1).

$$\frac{d^2 G_{lm}}{dz'^2} - \gamma_{lm}^2 G_{lm} = -\delta(z'-z) \qquad (viii.3)$$

We multiply Eq. (*viii*.1) by G_{lm}, and Eq. (*viii*.3) by Ψ_{lm}, then subtract the second result from the first.

$$G_{lm}\left(\frac{d^2\Psi_{lm}(z')}{dz'^2} - \gamma_{lm}^2\Psi_{lm}(z')\right) - \Psi_{lm}(z')\left(\frac{d^2 G_{lm}}{dz'^2} - \gamma_{lm}^2 G_{lm}\right) = -\frac{1}{k}\phi_{lm}G_{lm} + \delta(z'-z)\Psi_{lm}(z')$$

$$\left(\Psi_{lm}(z')\frac{d^2 G_{lm}}{dz'^2} - G_{lm}\frac{d^2\Psi_{lm}(z')}{dz^2}\right) = -\delta(z'-z)\Psi_{lm}(z') + \frac{1}{k}\phi_{lm}G_{lm}$$

$$\frac{d}{dz'}\left(\Psi_{lm}(z')\frac{dG_{lm}}{dz'} - G_{lm}\frac{d\Psi_{lm}(z')}{dz'}\right) = -\delta(z'-z)\Psi_{lm}(z') + \frac{1}{k}\phi_{lm}G_{lm} \qquad (viii.4)$$

Equation (*viii*.4) is integrated over $z' = 0$ to $z' = t$:

$$\int_{z'=0}^{z'=t}\frac{d}{dz'}\left(\Psi_{lm}(z')\frac{dG_{lm}}{dz'} - G_{lm}\frac{d\Psi_{lm}(z')}{dz'}\right)dz' = -\int_{z'=0}^{z'=t}\delta(z'-z)\Psi_{lm}(z')dz' + \frac{1}{k}\int_{z'=0}^{z'=t}\phi_{lm}G_{lm}dz' \ (viii.5)$$

Using property (*v*.8) for the first right-hand-side term of Eq. (*viii*.5), and requiring the continuity of G_{lm} at $z' = z$ (the latter is thus one of the derived properties of G_{lm}),

$$\Psi_{lm}(z) = -\int_{z'=0}^{z'=t} \frac{d}{dz'}\left(\Psi_{lm}(z')\frac{dG_{lm}}{dz'} - G_{lm}\frac{d\Psi_{lm}(z')}{dz'}\right)dz' + \frac{1}{k}\int_{z'=0}^{z'=t}\phi_{lm}G_{lm}dz' \qquad (viii.6)$$

Integration of the first term of the right-hand side of Eq. (*viii*.6) is immediate.

$$\Psi_{lm}(z) = -\left(\Psi_{lm}(z')\frac{dG_{lm}}{dz'} - G_{lm}\frac{d\Psi_{lm}(z')}{dz'}\right)\Bigg|_{z'=0}^{z'=t} + \frac{1}{k}\int_{z'=0}^{z'=t}\phi_{lm}G_{lm}dz'$$

$$\Psi_{lm}(z) = -\left(\Psi_{lm}(z')\frac{dG_{lm}}{dz'} - G_{lm}\frac{d\Psi_{lm}(z')}{dz'}\right)\Bigg|_{z'=t} + \left(\Psi_{lm}(z')\frac{dG_{lm}}{dz'} - G_{lm}\frac{d\Psi_{lm}(z')}{dz'}\right)\Bigg|_{z'=0} \qquad (viii.7)$$

$$+ \frac{1}{k}\int_{z'=0}^{z'=t}\phi_{lm}G_{lm}dz'$$

Our next property, the condition that the boundary conditions are identical for both G_{lm} and Ψ_{lm} sets the first two terms on the right-hand side of Eq. (*viii*.7) identically equal to zero so that we have

$$\Psi_{lm} = \frac{1}{k}\int_{z'=0}^{z'=t}\phi_{lm}G_{lm}dz' \qquad (viii.8)$$

One additional calculation remains: The proof of the "jump condition" for the derivative of G_{lm} at $z' = z$. Beginning with Eq. (*viii*.3), we integrate over $z' = 0$ to $z' = z - \varepsilon$ and from $z' = z + \varepsilon$ to $z' = t$ according to Figure *viii*.1.

$$\int_{z'=z-\varepsilon}^{z'=z+\varepsilon}\left(\frac{d^2G_{lm}}{dz'^2}\right)dz' - \gamma_{lm}^2\int_{z'=z-\varepsilon}^{z'=z+\varepsilon}G_{lm}dz' = -\int_{z'=z-\varepsilon}^{z'=z+\varepsilon}\delta(z'-z)dz'$$

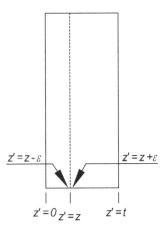

Figure *viii*.1. Illustration of care taken in integration over $z' = 0$ to $z' = t$. z' is the variable of integration and $z' = z$ is the field point (where you wish to calculate the temperature).

$$\frac{dG_{lm}}{dz'}\bigg|_{z'=z+\varepsilon} - \frac{dG_{lm}}{dz'}\bigg|_{z'=z-\varepsilon} - \gamma_{lm}^2 \int_{z'=z-\varepsilon}^{z'=z+\varepsilon} G_{lm}\,dz' = -1 \tag{viii.9}$$

Since $G_{lm} = continuous$ at $z' = z$, we write more explicitly,

$$\lim_{\varepsilon \to 0}\left[\frac{dG_{lm}}{dz'}\bigg|_{z'=z+\varepsilon} - \frac{dG_{lm}}{dz'}\bigg|_{z'=z-\varepsilon} - \gamma_{lm}^2 \int_{z'-z-\varepsilon}^{z'=z+\varepsilon} G_{lm}\,dz' \right] =$$

$$\lim_{\varepsilon \to 0}\left[\frac{dG_{lm}}{dz'}\bigg|_{z'=z+\varepsilon} - \frac{dG_{lm}}{dz'}\bigg|_{z'=z-\varepsilon} - \gamma_{lm}^2 G_{lm}\left(z'|z\right)2\varepsilon \right]$$

$$\lim_{\varepsilon \to 0}\left[\frac{dG_{lm}}{dz'}\bigg|_{z'=z+\varepsilon} - \frac{dG_{lm}}{dz'}\bigg|_{z'=z-\varepsilon} - \gamma_{lm}^2 G_{lm}\left(z'|z\right)2\varepsilon \right] \to \frac{dG_{lm}}{dz'}\bigg|_{z'=z^+} - \frac{dG_{lm}}{dz'}\bigg|_{z'=z^-}$$

$$\frac{dG_{lm}}{dz'}\bigg|_{z'=z^+} - \frac{dG_{lm}}{dz'}\bigg|_{z'=z^-} = -1 \tag{viii.10}$$

where the z^+ and z^- mean just to the right-hand and left-hand side of z, respectively. Thus we are left with the jump condition and the concluding properties of G_{lm} are exactly those of Appendix *vii*, Eqs. (*vii*.6-1) to (*vii*.6-8).

APPENDIX *ix*

Proof of Reciprocity for the Steady-State Green's Function

In Chapter 12, we derived the Green's function as part of the solution for a specific thermal spreading resistance problem. The notation $G_{lm}(z'|z)$ or $G_{lm}(z|z')$ was used without elaboration. This notation is usually found in any work regarding Green's functions and is intended to reflect the property of the Green's function *reciprocity* (in z, z'), which we shall prove here. We begin by writing two differential equations for the Green's function using the delta function source discussed in Appendix *viii*, with the second argument replaced with z_1 and z_2.

$$\frac{d^2 G_{lm}(z'|z_1)}{dz'^2} - \gamma_{lm}^2 G_{lm}(z'|z_1) = -\delta(z'-z_1) \tag{ix.1}$$

$$\frac{d^2 G_{lm}(z'|z_2)}{dz'^2} - \gamma_{lm}^2 G_{lm}(z'|z_2) = -\delta(z'-z_2) \tag{ix.2}$$

Multiplying Eqs. (*ix*.1) and (*ix*.2) by $G_{lm}(z'|z_2)$ and $G_{lm}(z'|z_1)$, respectively, and subtract.

$$G_{lm}(z'|z_2)\frac{d^2 G_{lm}(z'|z_1)}{dz'^2} - \gamma_{lm}^2 G_{lm}(z'|z_1)G_{lm}(z'|z_2)$$

$$-G_{lm}(z'|z_1)\frac{d^2 G_{lm}(z'|z_2)}{dz'^2} + \gamma_{lm}^2 G_{lm}(z'|z_2)G_{lm}(z'|z_1) =$$

$$-\delta(z'-z_1)G_{lm}(z'|z_2) + \delta(z'-z_2)G_{lm}(z'|z_1)$$

$$G_{lm}(z'|z_2)\frac{d^2 G_{lm}(z'|z_1)}{dz'^2} - G_{lm}(z'|z_1)\frac{d^2 G_{lm}(z'|z_2)}{dz'^2} = -\delta(z'-z_1)G_{lm}(z'|z_2) + \delta(z'-z_2)G_{lm}(z'|z_1)$$

$$\frac{d}{dz'}\left[G_{lm}(z'|z_2)\frac{dG_{lm}(z'|z_1)}{dz'} - G_{lm}(z'|z_1)\frac{dG_{lm}(z'|z_2)}{dz'} \right] =$$
$$-\delta(z'-z_1)G_{lm}(z'|z_2) + \delta(z'-z_2)G_{lm}(z'|z_1) \tag{ix.3}$$

Now we integrate Eq. (*ix*.3) from $z' = 0$ to $z' = t$.

$$\int_{z'=0}^{z'=t} \frac{d}{dz'}\left[G_{lm}(z'|z_2)\frac{dG_{lm}(z'|z_1)}{dz'} - G_{lm}(z'|z_1)\frac{dG_{lm}(z'|z_2)}{dz'} \right] dz' =$$

$$-\int_{z'=0}^{z'=t} \delta(z'-z_1)G_{lm}(z'|z_2)dz' + \int_{z'=0}^{z'=t} \delta(z'-z_2)G_{lm}(z'|z_1)dz'$$

$$\left[G_{lm}(z'|z_2)\frac{dG_{lm}(z'|z_1)}{dz'} - G_{lm}(z'|z_1)\frac{dG_{lm}(z'|z_2)}{dz'} \right]_{z'=0}^{z'=t} = -G_{lm}(z_1|z_2) + G_{lm}(z_2|z_1) \tag{ix.4}$$

Placing the integration limits into Eq. (*ix*.4),

$$
G_{lm}\left(t\middle|z_2\right)\frac{dG_{lm}\left(t\middle|z_1\right)}{dz'}-G_{lm}\left(t\middle|z_1\right)\frac{dG_{lm}\left(t\middle|z_2\right)}{dz'}
$$

$$
-G_{lm}\left(0\middle|z_2\right)\frac{dG_{lm}\left(0\middle|z_1\right)}{dz'}+G_{lm}\left(0\middle|z_1\right)\frac{dG_{lm}\left(0\middle|z_2\right)}{dz'}=-G_{lm}\left(z_1\middle|z_2\right)+G_{lm}\left(z_2\middle|z_1\right)
$$

(*ix*.5)

Now we write the boundary conditions for the Green's function for arguments of z_1 and z_2.

$$
\frac{dG_{lm}\left(t\middle|z_1\right)}{dz'}=-\frac{h_2}{k}G_{lm}\left(t\middle|z_1\right),\qquad\frac{dG_{lm}\left(t\middle|z_2\right)}{dz'}=-\frac{h_2}{k}G_{lm}\left(t\middle|z_2\right)
$$

$$
\frac{dG_{lm}\left(0\middle|z_1\right)}{dz'}=\frac{h_1}{k}G_{lm}\left(0\middle|z_1\right),\qquad\frac{dG_{lm}\left(0\middle|z_2\right)}{dz'}=\frac{h_1}{k}G_{lm}\left(0\middle|z_2\right)
$$

and when placed in Eq. (*ix*.5) gives us

$$
-\left(\frac{h_2}{k}\right)G_{lm}\left(t\middle|z_2\right)G_{lm}\left(t\middle|z_1\right)+\left(\frac{h_2}{k}\right)G_{lm}\left(t\middle|z_1\right)G_{lm}\left(t\middle|z_2\right)
$$

$$
-\left(\frac{h_1}{k}\right)G_{lm}\left(0\middle|z_2\right)G_{lm}\left(0\middle|z_1\right)+\left(\frac{h_1}{k}\right)G_{lm}\left(0\middle|z_1\right)G_{lm}\left(0\middle|z_2\right)
$$

$$
=-G_{lm}\left(z_1\middle|z_2\right)+G_{lm}\left(z_2\middle|z_1\right)
$$

$$
G_{lm}\left(z_1\middle|z_2\right)=G_{lm}\left(z_2\middle|z_1\right)\qquad\qquad\text{Reciprocity (}ix.6)
$$

thus proving the *reciprocity* property for the two arguments of the Green's function. Even if you studied Chapter 12 very carefully, you might not have been aware of this property because in that chapter, the field point z was always equal to or greater than the source location at $z'=0$. However, the same problem could have been solved for a source at a $z'>0$. For the person doing the mathematics, this means that he/she only need derive a solution for $z'<z$ and then using Eq. (*ix*.6), you can write down the solution for $z<z'$.

Proof of Reciprocity for the Three-Dimensional, Time-Dependent Green's Function

The previous paragraphs proved the reciprocity of the z-direction portion of the 3-D steady-state Green's function solution for the spreading problem in Chapter 12. Only a spatial component was required. In this section, the proof is extended to the 3-D time-dependent problem.

The time-dependent PDEs with a source in isotropic media are given by Eqs. (12.24), (12.25). The time-dependent heat conduction equation is

$$
\nabla^2 T\left(\underline{r},t\right)+\frac{1}{k}Q_V\left(\underline{r},t\right)=\frac{1}{\alpha}\frac{\partial T\left(\underline{r},t\right)}{\partial t},\ \alpha=k/(\rho c)\qquad(12.24)
$$

and the time-dependent auxiliary equation is then

$$
\nabla^2 G\left(\underline{r},t\middle|\underline{r}',t'\right)+\frac{1}{\alpha}\delta\left(\underline{r}-\underline{r}'\right)\delta\left(t-t'\right)=\frac{1}{\alpha}\frac{\partial G\left(\underline{r},t\middle|\underline{r}',t'\right)}{\partial t},\ \alpha=k/(\rho c)\quad(12.25)
$$

Equations (12.24) and (12.25) are re-written.

$$\nabla^2 T\left(\underline{r},t\right) - \frac{1}{\alpha}\frac{\partial T\left(\underline{r},t\right)}{\partial t} = -\frac{1}{k}Q_V\left(\underline{r},t\right) \qquad (ix.7)$$

$$\nabla^2 G\left(\underline{r},t\big|\underline{r}',t'\right) - \frac{1}{\alpha}\frac{\partial G\left(\underline{r},t\big|\underline{r}',t'\right)}{\partial t} = -\frac{1}{\alpha}\delta\left(\underline{r}-\underline{r}'\right)\delta\left(t-t'\right) \qquad (ix.8)$$

Equation (ix.8) is re-written for

(1) a source time $t' = t_0$, source point $\underline{r}' = \underline{r}_0$ and field time t, field point $\underline{r}$

$$\nabla^2 G\left(\underline{r},t\big|\underline{r}_0,t_0\right) - \frac{1}{\alpha}\frac{\partial G\left(\underline{r},t\big|\underline{r}_0,t_0\right)}{\partial t} = -\frac{1}{\alpha}\delta\left(\underline{r}-\underline{r}_0\right)\delta\left(t-t_0\right) \qquad (ix.9)$$

(2) a source time $t' = -t_1$, source point $\underline{r}' = \underline{r}_1$ and field time $-t$, field point $\underline{r}$.

$$\nabla^2 G\left(\underline{r},-t\big|\underline{r}_1,-t_1\right) + \frac{1}{\alpha}\frac{\partial G\left(\underline{r},-t\big|\underline{r}_1,-t_1\right)}{\partial t} = -\frac{1}{\alpha}\delta(\underline{r}-\underline{r}_1)\delta\left(t-t_1\right) \qquad (ix.10)$$

Equation (ix.10) was derived from Eq. (ix.8) using

$$\frac{\partial}{\partial(-t)} = -\frac{\partial}{\partial t} \text{ and } \delta\left(-t+t_1\right) = \delta\left[-\left(t-t_1\right)\right] = \delta\left(t-t_1\right)$$

Now we multiply Eq. (ix.9) by $G\left(\underline{r},-t\big|\underline{r}_1,-t_1\right)$ and multiply Eq. (xi.10) by $G\left(\underline{r},t\big|\underline{r}_0,t_0\right)$ and subtract the results to get

$$\left[G\left(\underline{r},-t\big|\underline{r}_1,-t_1\right)\nabla^2 G\left(\underline{r},t\big|\underline{r}_0,t_0\right) - G\left(\underline{r},t\big|\underline{r}_0,t_0\right)\nabla^2 G\left(\underline{r},-t\big|\underline{r}_1,-t_1\right)\right]$$

$$-\left[\frac{1}{\alpha}G\left(\underline{r},-t\big|\underline{r}_1,-t_1\right)\frac{\partial G\left(\underline{r},t\big|\underline{r}_0,t_0\right)}{\partial t} + \frac{1}{\alpha}G\left(\underline{r},t\big|\underline{r}_0,t_0\right)\frac{\partial G\left(\underline{r},-t\big|\underline{r}_1,-t_1\right)}{\partial t}\right] \qquad (ix.11)$$

$$= -\frac{1}{\alpha}G\left(\underline{r},-t\big|\underline{r}_1,-t_1\right)\delta\left(\underline{r}-\underline{r}_0\right)\delta\left(t-t_0\right) + \frac{1}{\alpha}G\left(\underline{r},t\big|\underline{r}_0,t_0\right)\delta\left(\underline{r}-\underline{r}_1\right)\delta\left(t-t_1\right)$$

Integrate the preceding, Eq. (ix.11) over the space of interest and over $t = -\infty$ to $t = t_0^+$, where t_0^+ is greater than t_1.

Integration of the first term on the left side of Eq. (ix.11) is accomplished using the three-dimensional form of Green's theorem:

$$\int_V \left[G\left(\underline{r},-t\big|\underline{r}_r,-t_1\right)\nabla^2 G\left(\underline{r},t\big|\underline{r}_0 t_0\right) - G\left(\underline{r},t\big|\underline{r}_0 t_0\right)\nabla^2 G\left(\underline{r},-t\big|\underline{r}_r,-t_1\right)\right]dV =$$

$$\qquad (ix.12)$$

$$\int_A \left[G\left(\underline{r},-t\big|\underline{r}_r,-t_1\right)\overline{\nabla}G\left(\underline{r},t\big|\underline{r}_0 t_0\right) - G\left(\underline{r},t\big|\underline{r}_0 t_0\right)\overline{\nabla}G\left(\underline{r},-t\big|\underline{r}_r,-t_1\right)\right]\cdot d\underline{A}$$

where A is the closed surface and V is the entire volume enclosed by A.

Green's function methods are readily applied to problems with various kinds of boundary conditions. A noteworthy issue is that irrespective of the actual problem boundary condition, i.e., Dirichlet, Neumann, or Robin, homogeneous or inhomogenous, the auxiliary equation uses a homogeneous boundary condition, otherwise, identical to the actual problem. The problems in this book use the homogeneous Robin boundary condition so that the Green's function boundary conditions are

$$-k\bar{\nabla}G\left(\underline{r},t|\underline{r}_0,t_0\right)=hG\left(\underline{r},t|\underline{r}_0,t_0\right); \quad -k\bar{\nabla}G\left(\underline{r},-t|\underline{r}_1,-t_1\right)=hG\left(\underline{r},-t|\underline{r}_1,-t_1\right) \qquad (ix.13)$$

Substitution of the first of Eq. ($ix.13$) into the first term of the right-hand side of Eq. ($ix.12$) and the second of Eq. ($ix.13$) into the second term of the right-hand side of Eq. ($ix.12$) reduces that side to zero. Then the integration of Eq. ($ix.11$) becomes

$$-\int_V dV\int_{t=0}^{t=t_0^+}\left[G\left(\underline{r},-t|\underline{r}_1,-t_1\right)\frac{\partial G\left(\underline{r},t|\underline{r}_0,t_0\right)}{\partial t}+G\left(\underline{r},t|\underline{r}_0,t_0\right)\frac{\partial G\left(\underline{r},-t|\underline{r}_1,-t_1\right)}{\partial t}\right]dt$$

$$(ix.14)$$

$$=-\int_V\int_{t=0}^{t=t_0^+}G\left(\underline{r},-t|\underline{r}_1,-t_1\right)\delta\left(\underline{r}-\underline{r}_0\right)\delta\left(t-t_0\right)dtdV+\int_V\int_{t=0}^{t=t_0^+}G\left(\underline{r},t|\underline{r}_0,t_0\right)\delta\left(\underline{r}-\underline{r}_1\right)\delta\left(t-t_1\right)dtdV$$

The two left–hand–side terms of Eq. ($ix.14$) can be combined and integrated over t as

At the upper limit t_0^+, the first factor is $G\left(\underline{r},-t|\underline{r}_1,-t_1\right)\Big|^{t=t_0^+}=0$ *by causality*, since $-t_0^+<-t_1$ and there can be no response at $-t_0^+$ before the cause at $-t_1$.

At the lower limit $t=0$, the second factor $G\left(\underline{r},t|\underline{r}_0,t_0\right)\Big|_{t=0}=0$ *by causality*, since $t_0>0$ and there can be no response at $t=0$ before the cause at t_0.

Thus the left-hand-side of the following, Eq. ($ix.14$), is zero.

- -

$$-\int_V dV\int_{t=0}^{t=t_0^+}\left[G\left(\underline{r},-t|\underline{r}_1,-t_1\right)\frac{\partial G\left(\underline{r},t|\underline{r}_0,t_0\right)}{\partial t}+G\left(\underline{r},t|\underline{r}_0,t_0\right)\frac{\partial G\left(\underline{r},-t|\underline{r}_1,-t_1\right)}{\partial t}\right]dt$$

$$=-\int_V\int_{t=0}^{t=t_0^+}G\left(\underline{r},-t|\underline{r}_1,-t_1\right)\delta\left(\underline{r}-\underline{r}_0\right)\delta\left(t-t_0\right)dtdV+\int_V\int_{t=0}^{t=t_0^+}G\left(\underline{r},t|\underline{r}_0,t_0\right)\delta\left(\underline{r}-\underline{r}_1\right)\delta\left(t-t_1\right)dtdV$$

- -

The right-hand side of Eq. ($ix.14$) is readily integrated using the properties of the Dirac delta function to give

$$G\left(\underline{r}_0,-t_0|\underline{r}_1,-t_1\right)=G\left(\underline{r}_1,t_1|\underline{r},t_0\right)$$

If we presume $\underline{r}_1$, t_1 to be any arbitrary $\underline{r}$, t, then we have the desired general result.

$$\boxed{G\left(\underline{r}_0,-t_0|\underline{r},-t\right)=G\left(\underline{r},t|\underline{r}_0,t_0\right)} \qquad (ix.15)$$

APPENDIX *x*

Finned Surface to Flat Plate *h* Conversion

This book explores finned heat sinks in quite a lot of detail, but no attention was paid to the effect of heat sources of a size that is less than the sink base width and length. Sometimes the problem of a small heat source is solved only by conduction codes, particularly when more than one heat source is required. Whether the problem is one of single or multiple heat sources, if a finned surface is involved, we can at least reduce the fin problem to that of a flat plate, which simplifies the analysis considerably. If the heat sink has only a single, centered source, then you can probably even use the spreading theory and design curves provided in this book. In this section you will learn how to convert the thermal resistance of a set of fins to a single effective heat transfer coefficient that you can use in a spreading theory or conduction code. If you are quite experienced in thermal analysis you probably won't even read these paragraphs, but otherwise it is worth your brief attention.

Suppose that the finned surface has a thermal resistance, $R_{Sink\ Fins}$, from the heat sink base to ambient. You wish to calculate the heat transfer coefficient for an unfinned, flat plate that has the same resistance as the finned plate:

$$R_{Plate} = R_{Sink\ Fins}$$
$$R_{Plate} = 1/(h_e ab)$$

where the plate has dimensions of length a, width b, and a heat transfer coefficient h_e for the flat plate surface. The heat sink might have the same dimensions as the plate, but not necessarily. Set the right-hand sides of the above two equations equal and solve for h_e.

$$h_e = 1/(R_{Sink\ Fins}\, ab)$$

The heat transfer coefficient h_e may be used as the required heat transfer coefficient in the *Biot* number required by spreading resistance curves or as a boundary condition in a conduction code.

APPENDIX *xi*

Some Conversion Factors

There are many quality texts and reference books on the subjects of heat transfer and fluid flow. The reader is advised to consult any of these. However, in recognition of the fact that this book uses a slightly odd mix of units, a short list is offered for conversion to and from the common quantities frequently used.

Table *xi*.1. Conversion constants.

Quantity	From	×	To
Density, ρ	kg/m³	6.243×10^{-2}	lb_m/ft^3
Mass Flow rate, $\dot{m}$	kg/s	1.936×10^{-2}	m³/min
Volumetric Flow rate, G	m³/s	2.119×10^3	ft³/min
	ft³/min	35.317	m³/min
Pressure, p	N/m²	4.019×10^{-3}	in. H_2O
	N/m²	9.80	mm H_2O
	in. H_2O	25.4	mm H_2O
	lb_f/ft^2	5.198	in. H_2O
Thermal conductivity, k	W/(m K)	0.0254	W/(in. K)

Note: for dry air at 20°C at standard temperature pressure and using $\dot{m} = \rho G$,

$$\dot{m}\left[\text{kg/s}\right] = \left(1.164\,\text{kg/m}^3\right) G\left[\text{ft}^3/\text{min}\right]\left(\frac{\text{m}^3/\text{s}}{2.119 \times 10^3\,\text{ft}^3/\text{min}}\right) = 5.481 \times 10^{-4}\,G\left[\text{ft}^3/\text{min}\right]$$

APPENDIX *xii*

Altitude Effects for Fan-Driven Airflow and Forced Convection Cooled Enclosures

Earlier chapters concerned with forced air assumed sea level operation. When fans are operated at altitude, the effect of a diminished air density must be considered. Several articles are readily found from a search of the Internet. Most of the articles that consider altitude effects necessarily begin with a listing of what are known as the "fan laws," also referred to as the "affinity laws." The various mechanical aspects of fans and blowers are outside of the realm of heat transfer in electronics; nevertheless, the following paragraphs provide an introduction to the subject and are sufficient for many design purposes. In the event that you have concerns about a particular fan operation at altitude, you should contact an applications engineer at a fan manufacturer, in which case you will be well prepared following this study. Only those effects for what the author considers ground-based electronic equipment are considered here. Readers interested in a high-altitude environment as encountered with military aircraft should look elsewhere for those considerations. Therefore, altitudes up to about 10,000 ft are considered herein.

XII.1 DERIVATION OF THE FAN LAWS

A perusal of several articles and books listing the fan laws show somewhat different results. Thus a derivation of these laws is necessary, an advantage of which is an enumeration of assumptions used. The reader is also better prepared to question a fan manufacturer's claims. A common assertion by some authors is that the Buckingham-π theorem is used to derive the fan laws. This theorem indicates that a set of n_G dimensionless groups is equal to the number of physical quantities n_P minus the number of primary dimensions n_D. In the case of fans, the procedure should be to use $n_P = 3$ and $n_D = 3$, resulting in $n_G = 0$. Clearly the Buckingham-π theorem is of no value in this instance. A simple dimensional analysis is appropriate.

It was shown in Chapter 3 that the important fan characteristics are shown as graphs of fan static pressure h_{fs} vs. airflow G. The other important property is the power Q, e.g., the *shaft power* (delivered to the shaft) or the *hydraulic power* (power imparted to the air by the fan). For purposes of the dimensional analysis, which type of power is used need not be specified. However, in a later calculation, the *hydraulic power* will be used. Thus three variables, h_{fs}, G, and Q are assumed. The three primary dimensions are length L, time t, and mass M.

Fan Power
First we derive the dimensions of the fan power Q. Using the dimensions of the force F to push the air through the fan we write this force as the product of mass m and acceleration a. The pressure Δp is the force F per unit area A. Then

$$[F] = [m][a] = M\left(\frac{L}{t^2}\right) = \frac{ML}{t^2}, \ [\Delta p] = \frac{[F]}{[A]} = \left(\frac{ML}{t^2}\right)\Big/L^2 = \frac{M}{Lt^2}$$

$$[Q] = \frac{[\text{Energy}]}{[\text{time}]} = \frac{[\text{Work}]}{[\text{time}]} = \frac{[\text{Force}][\text{Distance}]}{[\text{time}]} = \left(\frac{ML}{t^2}\right)(L)\Big/t = \frac{ML^2}{t^3}$$

Now we calculate a simple derivation of Q based on the fan pressure Δp and the volumetric flow rate G. Use the definition of

Power = Rate of doing work

$$Q = dE/dt = d\left(mV^2/2\right)\Big/dt = mV\left(dV/dt\right) = mVa = VF$$

The result will be shown to actually be one of the fan laws.

$$Q = (VA)(F/A) = G\Delta p$$

Fan Laws Derivations

Now we can proceed with three-dimensional analyses. Assume that G, Δp, and Q are functions only of the shaft speed N [1/t], a "size" or fan diameter D, and the air density ρ. First we look for a form of G.

$$G = f_1(D, N, \rho) = C_1 D^a N^b \rho^c$$

C_1 = a constant of proportionality

a, b, c = numerical constants to be determined

Writing the above with the correct dimensions,

$$L^3/t = C_1 L^a t^{-b} M^c L^{-3c}$$
$$L^3 t^{-1} = C_1 L^{a-3c} t^{-b} M^c$$

Set the exponents equal for like quantities on the left-and right-hand sides.

$3 = a - 3c$		$a = 3 + 3c$	$a = 3$
$-1 = -b$	or	$b = 1$	$\rightarrow$ $b = 1$
$0 = c$		$c = 0$	$c = 0$

$$\boxed{G = C_1 D^3 N} \qquad (xii.1)$$

Next we look for a form of Δp.

$$\Delta p = f_2(D, N, \rho) = C_2 D^a N^b \rho^c$$

C_2 = a constant of proportionality

a, b, c = numerical constants to be determined

Writing the above with the correct dimensions,

$$M/Lt^2 = C_2 L^a t^{-b} M^c L^{-3c}$$
$$ML^{-1}t^{-2} = C_2 L^{a-3c} t^{-b} M^c$$

Set the exponents equal for like quantities on the left-and right-hand sides.

$-1 = a - 3c$		$a = 3c - 1$	$a = 2$
$-2 = -b$	or	$b = 2$	$\rightarrow$ $b = 2$
$1 = c$		$c = 1$	$c = 1$

$$\boxed{\Delta p = C_2 D^2 N^2 \rho} \qquad (xii.2)$$

Try the equation for the fan power, Q.

$$Q - f_3(D, N, \rho) = C_3 D^a N^b \rho^c$$

C_3 = a constant of proportionality

a, b, c = numerical constants to be determined

Writing the above with the correct dimensions,

$$ML^2/t^3 = C_3 L^a t^{-b} M^c L^{-3c}$$
$$ML^2 t^{-3} = C_3 L^{a-3c} t^{-b} M^c$$

Set the exponents equal for like quantities on the left-and right-hand sides.

$$2 = a - 3c \qquad\qquad a = 2 + 3c \qquad\qquad a = 5$$
$$-3 = -b \qquad\text{or}\qquad b = 3 \qquad\rightarrow\qquad b = 3$$
$$1 = c \qquad\qquad c = 1 \qquad\qquad c = 1$$

$$\boxed{Q = C_3 D^5 N^3 \rho} \qquad\qquad (xii.3)$$

In summary, there are three fan laws.

$$(1) \qquad G = C_1 D^3 N$$
$$(2) \qquad \Delta p = C_2 D^2 N^2 \rho$$
$$(3) \qquad Q = C_3 D^5 N^3 \rho$$

Note that the product of the right-hand sides of the first and second equations above, i.e., Q, is identical to the earlier result when we defined $Q = dE/dt$.

The reader who makes the effort to learn more about fans will find significant discrepancies in the "literature" when comparing various stated "*fan laws.*" This is what led to the need for a derivation herein. Some of those errors are no doubt typographical. But there seems to be at least one exception wherein A.J. Steonoff in "Helik, Leu, *Centrifugal & Rotary Pumps*, CRC Press, 1999," proposed that empirical data suggests (1) above better suited as

$$G \propto C_1 DN, (2) \text{ as above, but } (3)\ Q \propto C_3 D^3 N^3$$

which is consistent with $Q = G \Delta p$. Most electronics cooling practitioners are not experts in fan and pump aerodynamics, but also seem to use some form of the fan laws with a dependence of G, Δp, and Q on D, N, and ρ as shown in (1), (2), and (3) above. The fan laws should only be used to predict property variations for a specific fan, i.e., a fan with identical aerodynamic, flow, and geometric properties. Additionally, it should be clear that calculations made using the fan laws are only approximate. Nevertheless, this author has had some success in applying these principles.

XII.2 AIR PRESSURE AT AN ELEVATED ALTITUDE

The next step is to develop an equation that describes the effect of altitude on the air density ρ. We use the US Standard Atmosphere model, which is described in many texts and online documents. Atmospheric pressures from sea level to twenty thousand feet are listed in Table *xii*.1

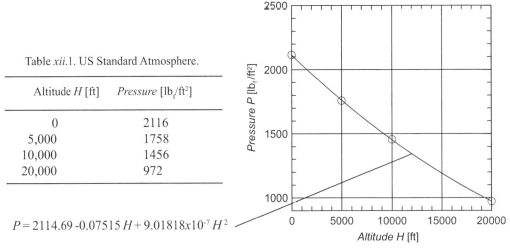

Table *xii*.1. US Standard Atmosphere.

Altitude H [ft]	Pressure [lb$_f$/ft^2]
0	2116
5,000	1758
10,000	1456
20,000	972

$$P = 2114.69 - 0.07515\, H + 9.01818x10^{-7}\, H^2$$

Figure *xii*.1. Atmospheric pressure vs. altitude.

XII.3 THE IDEAL GAS LAW AND AN AIR DENSITY MODEL AT ALTITUDE

A review of the ideal gas law is in order. This law states the following:

$$PV = nRT'$$

$$P = \text{Pressure}, \text{lb}_f/\text{ft}^2, \text{Pa}, \text{N}/\text{m}^2$$

$$n = \text{number of moles}$$

$$R = \text{universal gas constant} = 8314.5 \text{ J}/(\text{kg mole} \cdot \text{K})$$

$$T' = \text{temperature}, \text{K}$$

In addition we define

$$m = \text{mass of gas}$$

$$W_M = \text{molecular weight or molecular mass or mass/mole}$$

$$= 28.97 \text{ kg}/\text{kg} \cdot \text{mole for air}$$

Then $\quad n = m/W_M$

The ideal gas law becomes

$$PV = \frac{m}{W_M} RT', P = (m/V)(RT'/W_M) = \rho(RT'/W_M)$$

$$\rho = \frac{PW_M}{RT'}$$

It is customary to use the symbol for sigma to define the ratio σ of air density ρ at some altitude compared to the air density ρ_0 at sea level. Then

$$\sigma = \frac{\rho}{\rho_0} = \left(\frac{P}{P_0}\right)\left(\frac{T_0'}{T'}\right)$$

If we can assume that we are operating our equipment in nearly the same ambient temperature, e.g., room temperature, as the fan was tested, the temperature ratio is identically equal to one. Any units may be used for the pressure as long as they are consistent. Using a sea level reference pressure P_0 of 2114.69 lb$_f$/ft^3, the equation for σ is obtained from the curve fit associated with Fig. *xii*.2.

$$\sigma = 1 - 3.554 \times 10^{-5} H + 4.265 \times 10^{-10} H^2 \qquad (xii.4)$$

Figure *xii*.2. Sigma $\sigma = \rho/\rho_0$ vs. altitude H.

It is interesting to note that the air density at the approximate elevation of Denver, Colorado, is diminished to about eighty percent of the sea level value.

Specific heat, thermal conductivity, and the Prandtl number for air are nearly constant with altitude, i.e., pressure, thus the density is the major factor to be considered (see, for example, Electronics Cooling, vol. 5, no.3, Sept. 1999, pp. 44-45 and Steinberg, Dave, Cooling Techniques fo Electronic Equipment, Wiley-Interscience, 1980, pg. 129).

XII.4 FAN AND SYSTEM INTERACTION AT ALTITUDE

From the second fan law, Eq. (xii.2), we surmise that the fan static pressure head h_{fs} is linearly proportional to the air density.

The effect on the various elements in an airflow-pressure circuit must also be considered. In Chapter 3, the basic head loss, *assuming a turbulent flow*, was written as

$$\Delta h = RG^2 = RA^2V^2 \qquad (xii.5)$$

It was also noted that a head loss may be written as the product of one velocity head h_V and a loss constant K.

$$\Delta h = Kh_V = K\left(\frac{1}{2}\rho V^2\right) \qquad (xii.6)$$

Then
$$RA^2V^2 = K\left(\rho V^2/2\right)$$
$$R = K\rho/2A^2 \propto \rho \qquad (xii.7)$$

i.e., the head loss for a system or element in turbulent flow varies linearly with the air density.

Since the fan and system loss curves change identically with density for a system loss $H_L \propto G^2$, the operating point airflow G does not change with density or altitude. The fan and system curves need not be replotted.

Some resistances, e.g., a channel contraction or expansion, have loss coefficients that are a known function of a Reynold's number. In this case the Reynold's number computed for sea level must be adjusted for altitude.

$$Re = \frac{VD\rho}{\mu} = \left(\frac{VD\rho_0}{\mu}\right)\left(\frac{\rho}{\rho_0}\right)$$
$$Re = Re_0\sigma \qquad (xii.8)$$

If a system contains one or more laminar flow elements, then the fan and system curves must be adjusted from the sea level value. The second fan law suggests

$$h_{fs} = h_{fs0}\sigma \qquad (xii.9)$$

for a fan curve h_{fs0} at sea level.

Equation (xii.7) shows that each quadratic element Δh, must be adjusted similarly.

$$\Delta h = \Delta h_0 \sigma \qquad (xii.10)$$

Laminar flow elements with a loss Δh_l must also be corrected. For example, Eq. (4.1), the resistance for duct flow has a friction term that may be written as

$$R_{\bar{f}} = \frac{1.29 \times 10^{-3}}{N_c^2 A_c^2} K_{\bar{f}} \qquad (xii.11)$$

where
$$K_{\bar{f}} = 4\bar{f}\left(\frac{L}{D}\right) \qquad (xii.12)$$

The coefficient of $K_{\bar{f}}$ in Eq. (xii.11) is one velocity head at sea level.

As in Chapters 3 and 4, the quadratic dependence of Δh on G is used as in Eq. (*xii*.5) so that Eq. (*xii*.11) may be written as

$$R_{\bar{f}} = \frac{1.29 \times 10^{-3}}{N_c^2 A_c^2} \sigma K_{\bar{f}} \qquad (xii.13)$$

and the loss constant is a function of the Reynold's number.

$$K_{\bar{f}} = K_{\bar{f}}(Re), \qquad Re = Re_0 \sigma$$

Re must therefore be corrected for altitude and the solution may require iteration.

Many fans are constant speed and this shall be assumed here. In this situation a given volumetric air flow will, at some elevated altitude, produce an increased air temperature rise due to the lower density. The most appropriate equation for air temperature rise is taken from Chapter 2.

$$\Delta T = Q / (\rho G C_P) \qquad (2.9)$$

In the present context, Eq. (2.9) is re-written.

$$\Delta T = \frac{Q}{\rho G C_P} \left(\frac{\rho_0}{\rho_0} \right) = \frac{Q}{\rho_0 G C_P} \left(\frac{\rho_0}{\rho} \right)$$

$$\boxed{\Delta T = \frac{Q}{\rho_0 G C_P} \left(\frac{1}{\sigma} \right)} \qquad (xii.14)$$

XII.5 APPLICATION EXAMPLE - FAN COOLED HEAT SINK

The finned heat sink in Figure *xii*.3 is similar to the one analyzed in Section 4.4. The associated fan curve is shown in Figure *xii*.4. The system airflow is calculated for both sea level and also for an altitude of 10,000 ft. The device is intended to be used in a laboratory (i.e., not in an aircraft).

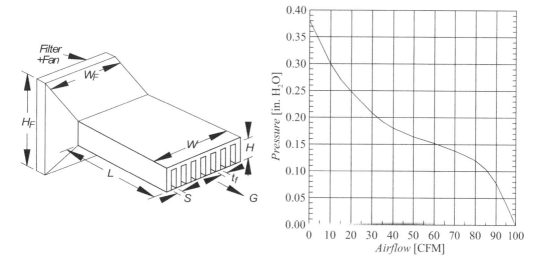

Figure *xii*.3. Fan cooled heat sink with seven fin channels.

Figure *xii*.4. Fan curve for Application Example *xii*.5.

The airflow path in the sink system is first through an air filter, through the fan, into a contraction from the fan to the heat sink, and then through the seven heat sink channels. A series airflow resistance network is readily drawn for this problem.

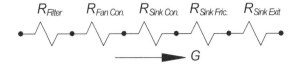

Figure *xii*.5. Airflow circuit for Application Example *XII*.5.

The fan and filter dimensions are $W_F = H_F = 4.06$ in. The heat sink dimensions are $W = 4.06$ in., $L = 6.0$ in., $H = 1.0$ in., $t_f = 0.1$ in., $S = 0.466$ in., and the number of fins $N_f = 8$. A vendor-supplied filter airflow resistance is given as $R_{Filter} = 4.0 \times 10^{-3} / (H_F W_F)^2$ [in. H_2O/CFM^2]. The first calculation is for a total airflow of 40 CFM and an altitude of 0 ft., i.e., $\sigma = \rho/\rho_0 = 1.0$. An array of uniformly distributed power transistors on the heat sink dissipates a total of $Q = 200$ W.

$$R_{Filter} = \frac{4.0 \times 10^{-3}}{(H_F W_F)^2} \sigma = \frac{4.0 \times 10^{-3}}{\left[(4.06\,\text{in.})^2 \right]^2}(1.0) = 1.47 \times 10^{-5}\ \text{in.}\ H_2O/\text{CFM}^2$$

$$R_{Fan\,Con.} = \frac{0.5 \times 10^{-3}}{(HW)^2}\left[1 - \frac{HW}{H_F W_F} \right]^{3/4} \sigma == \frac{0.5 \times 10^{-3}}{\left[(1.0)(4.06) \right]^2}\left[1 - \frac{(1.0)(4.06)}{(4.06)(4.06)} \right]^{3/4}(1.0)$$

$$R_{Fan\,Con.} = 2.45 \times 10^{-5}\ \text{in.}\ H_2O/\text{CFM}^2$$

The resistances for the heat sink contraction at inlet, friction in the channels, and the exit expansion require calculation of the hydraulic diameter D_H and Reynold's number Re_{DH}.

$$D_H = 2SH/(S+H) = 2(0.466)(1.0)/(0.466+1.0) = 0.635\,\text{in.}$$

$$V_f = \left[\frac{G}{(N_f - 1)} \right] \bigg/ \left(\frac{SH}{144} \right) = \left(\frac{40}{8-1} \right) \bigg/ \left[\frac{(0.466)(1.0)}{144} \right] = 1767\ \text{ft/min.}$$

$$Re_{DH} = \frac{V_f D_H}{5\nu}\sigma = \frac{(1767)(0.635)}{5(0.023)}(1.0) = 9764$$

where Re_{DH} is in mixed units.

The heat sink inlet and exit resistances require the calculation of the free area ratio. In order to not confuse the free area ratio with the air density ratio σ, the free area ratio is subscripted here with an *s* denoting *sink*.

$$\sigma_s = \frac{(N_f - 1)SH}{WH} = \frac{(8-1)(0.466)(1.0)}{(4.06)(1.0)} = 0.80$$

Using the calculated $Re_{DH} = 9764$ and $\sigma_s = 0.80$ and Figures 4.2, 4.3, the inlet and exit loss factors are estimated to be $K_c = 0.205$ and $K_E = 0$. The contraction exit resistances use the first and second parts of Eq. (4.1), respectively.

$$R_{Sink\,Con.} = \frac{1.29 \times 10^{-3}}{\left[(N_f - 1)SH \right]^2}K_c\sigma = \frac{1.29 \times 10^{-3}}{\left[(8-1)(0.466)(1.0) \right]^2}(0.205)(1.0)$$

$$R_{Sink\,Con.} = 2.49 \times 10^{-5}\ \text{in.}\ H_2O/\text{CFM}^2$$

$$R_{Sink\,Exit} = \frac{1.29 \times 10^{-3}}{\left[(N_f - 1)SH \right]^2}K_E\sigma = 0$$

The heat sink frictional resistance $R_{Sink\,Fric.}$ uses the third part of Eq. (4.1).

$$R_{SinkFric.} = \frac{1.29 \times 10^{-3}}{\left[(N_f - 1)SH\right]^2}\left(\frac{4\overline{f}L}{D_H}\right)\sigma = \frac{1.29 \times 10^{-3}}{\left[(8-1)(0.466)(1.0)\right]^2}\left(\frac{4\overline{f}(6.0)}{0.635}\right)(1.0) = 4.58 \times 10^{-3}\,\overline{f}$$

The friction factor $\overline{f}$ is determined from Figure 4.4 using the calculation

$$\frac{L}{D_H\,Re_{DH}} = \frac{6.0}{(0.635)(9764)} = 9.67 \times 10^{-4}$$

$$\overline{f}Re_{DH} = 50,\ \overline{f} - 50/Re_{DH} = 50/9764 = 5.12 \times 10^{-3}$$

Then $R_{Sink\,Fric.} = 4.58 \times 10^{-3}\,\overline{f} = \left(4.58 \times 10^{-3}\right)\left(5.12 \times 10^{-3}\right) = 2.35 \times 10^{-5}$

According to Figure *xii*.5, the total airflow resistance is given by

$$R_{af} = R_{Filter} + R_{Fan\,Con.} + R_{Sink\,Con.} + R_{Sink\,Fric.} + R_{Sink\,Exit}$$

and substituting the resistance values for a total airflow of $G = 40$ CFM,

$$R_{af} = 1.47 \times 10^{-5} + 2.45 \times 10^{-5} + 2.49 \times 10^{-5} + 2.35 \times 10^{-5} + 0 = 8.76 \times 10^{-5}\ \text{in.}\,H_2O/CFM^2$$

$$\text{and } H_L = \left(8.76 \times 10^{-5}\right)(40)^2 = R_{af}G^2 = 0.140\ \text{in.}\,H_2O$$

The preceding calculations are repeated for $G = 40$ CFM, but for an altitude of 10,000 ft. By referring to Figure *xii*.2, it is determined that $\sigma = 0.687$. The various resistances must be recalculated.

$$R_{Filter} = \frac{4.0 \times 10^{-3}}{(H_F W_F)^2}\sigma = \frac{4.0 \times 10^{-3}}{\left[(4.06\,\text{in.})^2\right]^2}(0.687) = 1.01 \times 10^{-5}\ \text{in.}\,H_2O/CFM^2$$

$$R_{Fan\,Con.} = \frac{0.5 \times 10^{-3}}{(HW)^2}\left[1 - \frac{HW}{H_F W_F}\right]^{3/4}\sigma == \frac{0.5 \times 10^{-3}}{\left[(1.0)(4.06)\right]^2}\left[1 - \frac{(1.0)(4.06)}{(4.06)(4.06)}\right]^{3/4}(0.687)$$

$$R_{Fan\,Con.} = 1.69 \times 10^{-5}\ \text{in.}\,H_2O/CFM^2$$

$$Re_{DH} = \frac{V_f D_H}{5\nu}\sigma = \frac{(1767)(0.635)}{5(0.023)}(0.687) = 6710$$

Noting that from Figure 4.2, $K_c = 0.21$, $R_{Sink\,Con.} = \dfrac{1.29 \times 10^{-3}}{\left[(N_f - 1)SH\right]^2}K_c\sigma$

$$R_{Sink\,Con.} - \frac{1.29 \times 10^{-3}}{\left[(8-1)(0.466)(1.0)\right]^2}(0.21)(0.687) - 1.75 \times 10^{-5}\ \text{in.}\,H_2O/CFM^2$$

From Figure 4.3, $K_E = -0.01$,

$$R_{Sink\,Exit} = \frac{1.29 \times 10^{-3}}{\left[(N_f - 1)SH\right]^2}K_E\sigma = \frac{1.29 \times 10^{-3}}{\left[(8-1)(0.466)(1.0)\right]^2}(-0.01)(0.687)$$

$$R_{Sink\,Exit} = -8.34 \times 10^{-7}\ \text{in.}\,H_2O/CFM^2$$

$$R_{SinkFric.} = \frac{1.29 \times 10^{-3}}{\left[\left(N_f - 1\right) SH\right]^2} \left(\frac{4\overline{f}L}{D_H}\right) \sigma = \frac{1.29 \times 10^{-3}}{\left[(8-1)(0.466)(1.0)\right]^2} \left(\frac{4\overline{f}(6.0)}{0.635}\right)(0.687) = 3.15 \times 10^{-3}\,\overline{f}$$

$$\frac{L}{D_H Re_{DH}} = \frac{6.0}{(0.635)(6710)} = 1.41 \times 10^{-3} \text{ and } \overline{f}Re_{DH} = 42,\ \overline{f} = 42/Re_{DH} = 42/6710 = 6.26 \times 10^{-3}$$

$$R_{SinkFric.} = 3.15 \times 10^{-3}\,\overline{f} = 3.15 \times 10^{-3}\left(6.26 \times 10^{-3}\right) = 1.97 \times 10^{-5} \text{ in. H}_2\text{O}/\text{CFM}^2$$

$$R_{af} = R_{Filter} + R_{Fan\ Con.} + R_{Sink\ Con.} + R_{Sink\ Fric.} + R_{Sink\ Exit}$$

$$R_{af} = 1.01 \times 10^{-5} + 1.69 \times 10^{-5} + 1.75 \times 10^{-5} + 1.97 \times 10^{-5} - 8.34 \times 10^{-7} = 6.34 \times 10^{-5} \text{ in. H}_2\text{O}/\text{CFM}^2$$

$$\text{and } H_L = R_{af}G^2 = \left(6.34 \times 10^{-5}\right)(40)^2 = 0.101 \text{ in. H}_2\text{O}$$

It is interesting to note that the ratio for H_L at 10,000 ft. and 0 ft. is $0.101/0.14 = 0.721$, which is not the value of $\sigma = 0.687$. This slight difference is no doubt due to the non-linearities caused by Re_{DH} in some of the resistances. Results for other volumetric airflows with intermediate variable values are listed in Table *xii*.2.

The head loss H_L from Table *xii*.2 and the fan static pressure are plotted in Figure *xii*.6. The fan curve for an altitude of 10,000 ft. is adjusted according to the second of the three fan laws derived earlier. The estimated operating points are $G = 45$ and 43 CFM for sea level and 10,000 ft., respectively. It should be noted that according to Eq. (3.12), the "fan curve" is actually $h_{fs} + (h_{Vd} - h_{Vi})$, but in this example, the fan inlet and exit areas are identical so that $h_{Vd} = h_{Vi}$.

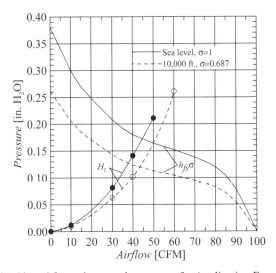

Figure *xii*.6. Plotted fan and system loss curves for Application Example *XII*.5.

The effect of altitude on forced convective heat transfer is also accounted for via the Reynold's number density dependence. In the case of the current problem, the constant wall temperature formula, Eq. (7.13) is used.

$$Nu_{DH} = r_{Nu}\left[3.66 + \frac{0.104\left(Re_{DH}Pr\Big/\dfrac{L}{D_H}\right)}{1 + 0.016\left(Re_{DH}Pr\Big/\dfrac{L}{D_H}\right)^{0.8}}\right] \qquad (xii.15)$$

Table *xii*.2. Head loss calculations. Resistances [in. H_2O/CFM^2], head loss [in. H_2O].

G	Re_{DH}	K_c	$R_{Sink\ Con.}$	$L/Re_{DH}D_H$	$\bar{f}\,Re_{DH}$	$\bar{f}$	$R_{Sink\ Fric.}$	$K_{Sink\ Exit}$	$R_{Sink\ Exit}$	R_{af}	H_L
					Sea level, $\sigma = 1$						
10	2441	0.22	2.67×10^{-5}	3.87×10^{-3}	28	0.0110	5.26×10^{-5}	0	0	1.19×10^{-4}	0.0119
30	7323	0.21	2.55×10^{-5}	1.29×10^{-3}	42	0.00574	2.63×10^{-5}	-0.01	-1.21×10^{-6}	9.00×10^{-5}	0.0811
40	9764	0.205	2.49×10^{-5}	9.67×10^{-4}	50	0.00512	2.35×10^{-5}	0	0	8.76×10^{-5}	0.140
50	12200	0.20	2.43×10^{-5}	7.74×10^{-4}	55	0.00450	2.07×10^{-5}	0	0	8.44×10^{-5}	0.211
					Altitude = 10,000, $\sigma = 0.687$						
10	1677	0.30	2.50×10^{-5}	5.63×10^{-3}	28	0.017	5.26×10^{-5}	-0.1	-8.30×10^{-6}	9.64×10^{-5}	0.0096
30	5032	0.22	1.84×10^{-5}	1.88×10^{-3}	40	0.00795	2.50×10^{-5}	-0.02	-1.67×10^{-6}	6.89×10^{-5}	0.0620
40	6710	0.21	1.75×10^{-5}	1.41×10^{-3}	42	0.00626	1.97×10^{-5}	-0.01	-8.34×10^{-7}	6.34×10^{-5}	0.101
60	10060	0.20	1.69×10^{-5}	9.38×10^{-4}	52	0.00517	1.63×10^{-5}	0	0	6.01×10^{-5}	0.216

$$Nu_{DH} = \left(\frac{Nu_\varepsilon}{Nu_{Circ.}}\right)\left[3.66 + \frac{0.104\left(Re_{DH}Pr / \dfrac{L}{D_H}\right)}{1 + 0.016\left[Re_{DH}Pr / \dfrac{L}{D_H}\right]^{0.8}}\right]$$

where

$$\boxed{V_f = \frac{G/(N_f - 1)}{SH/144}, \quad Re_{DH} = \frac{V_f D_H}{5v}\sigma \text{ in mixed units}}$$

Using the fin height to space ratio $H/S = 1.0/0.466 = 2.15$, the Nusselt number ratio is found from Figure 7.6 and Table 7.1 as $r_{Nu} = 3.4/3.66 = 0.929$.

The first calculation is the heat transfer solution for sea level where the total airflow was found to be $G = 45$ CFM.

$$V_f = \frac{G/(N_f - 1)}{SH/144} = \frac{45/(8-1)}{(0.466)(1.0)/144} = 1988\,\text{ft/min.}$$

$$Re_{DH} = \frac{V_f D_H}{5v}\sigma = \frac{(1987)(0.636)}{5(0.023)}(1.0) = 1.098 \times 10^4$$

$$Nu_{DH} = \left(\frac{Nu_\varepsilon}{Nu_{Circ.}}\right)\left[3.66 + \frac{0.104\left(\dfrac{Re_{DH}Pr}{\dfrac{L}{D_H}}\right)}{1 + 0.016\left(\dfrac{Re_{DH}Pr}{\dfrac{L}{D_H}}\right)^{0.8}}\right]$$

$$Nu_{DH} = (0.929)\left\{3.66 + \frac{0.104\left[\dfrac{(1.098\times10^4)(0.71)}{6.0/0.636}\right]}{1 + 0.016\left[\dfrac{(1.098\times10^4)(0.71)}{6.0/0.636}\right]^{0.8}}\right\} = 21.3$$

$$h = \frac{k_{Air}}{D_H}Nu_{DH} = \frac{6.6\times10^{-4}}{0.636}(21.3) = 0.022$$

Although it might be expected that the fin efficiency is nearly unity, it should be calculated. According to Eq. (11.22),

$$R_k = \frac{L}{k_{Al}Lt_f} = \frac{6.0}{(5.0)(6.0)(0.1)} = 0.333, \quad R_S = \frac{1}{2hLH} = \frac{1}{2(0.022)(6.0)(1.0)} = 3.76$$

$$\eta = \sqrt{\frac{R_S}{R_k}}\tanh\sqrt{\frac{R_k}{R_S}} = \sqrt{\frac{3.76}{0.333}}\tanh\sqrt{\frac{0.333}{3.76}} = 0.97$$

Equation (7.8) is used to calculate the heat sink resistance referenced to the inlet air.

$$A_S = (N_f - 1)(2H + S)L = (8-1)[2(1.0) + 0.466] = 103.56 \text{ in.}^2,$$

$$A_{C-Total} = (N_f - 1)HS = (8-1)(1.0)(0.466) = 3.26 \text{ in.}^2$$

$$R_C = \frac{1}{2\eta h A_S} = \frac{1}{2(0.97)(0.022)(103.56)} = 0.224 \,^0C/W$$

$$\dot{m}C_P = \frac{V_f A_{C-Total}}{262}\sigma = \frac{(1988)(3.26)}{262}(1.0) = 24.7, \ \beta = \frac{1}{R_C \dot{m}C_P} = \frac{1}{(0.224)(24.7)} = 0.18$$

$$R_I = \frac{R_C \beta}{1 - e^{-\beta}} = \frac{(0.224)(0.18)}{1 - e^{-0.18}} = 0.245 \,^0C/W$$

The average heat sink temperature rise above the inlet air is then

$$\Delta T_{Sink} = R_I Q = (0.245)(200) = 49.0 \,^0C/W$$

The total air temperature rise from inlet to exit, neglecting a fan contribution is calculated from Eq. (*xii*.14).

$$\Delta T_{Air} = \frac{Q}{\rho_0 G C_P}\left(\frac{1}{\sigma}\right) = \frac{1.76Q}{G}\left(\frac{1}{\sigma}\right) = \frac{1.76(200)}{45}\left(\frac{1}{1.0}\right) = 7.82 \,^0C$$

The heat transfier for the 10,000 ft altitude is similarly calculated using $\sigma = 0.687$.

$$V_f = \frac{G/(N_f - 1)}{SH/144} = \frac{43/(8-1)}{(0.466)(1.0)/144} = 1898 \text{ ft/min.}$$

$$Re_{DH} = \frac{V_f D_H}{5\nu}\sigma = \frac{(1898)(0.636)}{5(0.023)}(0.687) = 7209$$

$$Nu_{DH} = \left(\frac{Nu_\varepsilon}{Nu_{Circ.}}\right)\left[3.66 + \frac{0.104\left(\dfrac{Re_{DH}Pr}{\dfrac{L}{D_H}}\right)}{1 + 0.016\left(\dfrac{Re_{DH}Pr}{\dfrac{L}{D_H}}\right)^{0.8}}\right]$$

$$Nu_{DH} = (0.929)\left\{3.66 + \frac{0.104\left[\dfrac{(7209)(0.71)}{6.0/0.636}\right]}{1 + 0.016\left[\dfrac{(7209)(0.71)}{6.0/0.636}\right]^{0.8}}\right\} = 18.53$$

$$h = \frac{k_{Air}}{D_H}Nu_{DH} = \frac{6.6 \times 10^{-4}}{0.636}(18.53) = 0.019$$

$$A_S = (N_f - 1)(2H + S)L = (8-1)[2(1.0) + 0.466] = 103.56 \text{ in.}^2,$$

$$A_{C-Total} = (N_f - 1)HS = (8-1)(1.0)(0.466) = 3.26 \text{ in.}^2$$

The fin efficiency should be little changed from the previous case, so we shall use the prior value.

$$R_C = \frac{1}{2\eta h A_S} = \frac{1}{2(0.97)(0.019)(103.56)} = 0.257\,^0\text{C}/\text{W}$$

$$\dot{m}C_P = \frac{V_f A_{C-Total}}{262}\sigma = \frac{(1898)(3.26)}{262}(0.687) = 16.24,\ \beta = \frac{1}{R_C \dot{m}C_P} = \frac{1}{(0.257)(16.24)} = 0.239$$

$$R_I - \frac{R_C\beta}{1-e^{-\beta}} = \frac{(0.257)(0.239)}{1-e^{-0.239}} = 0.289\,^0\text{C}/\text{W}$$

$$\Delta T_{Sink} = R_I Q = (0.289)(200) = 57.9\,^0\text{C}/\text{W}$$

$$\Delta T_{Air} = \frac{Q}{\rho_0 GC_P}\left(\frac{1}{\sigma}\right) = \frac{1.76Q}{G}\left(\frac{1}{\sigma}\right) = \frac{1.76(200)}{43}\left(\frac{1}{0.687}\right) = 11.9\,^0\text{C}$$

The calculations show that the change from sea level to 10,000 ft results in a heat sink temperature increase of 9 ^{0}C and an air temperature rise increase of 4 ^{0}C.

APPENDIX *xiii*

Altitude Effects for Buoyancy Driven Air and Natural Convection Cooled Enclosures

This is a continuation of Appendix *xii*, but is devoted to non-fan cooled enclosures. Clearly there is no need to consider the fan laws, but altitude effects must still be considered for the buoyancy effects on both the airflow and the heat transfer from surfaces. Considerable care is devoted to the re-derivation of the buoyance draft so that the altitude effect is correctly developed and understood.

XIII.1 DERIVATION OF AIR TEMPERATURE RISE FORMULA

From Chapter 2 the air density dependence on temperature formula from the Ideal Gas Law is used.

ρ_0 = density of air at sea level and temperature T, $^0\mathrm{C}$

ρ_R = density of air at sea level and reference temperature T_R, $^0\mathrm{C}$

ρ_A = density of air at the ambient temperature T and altitude

$$\rho_0 = \rho_R \left(\frac{T_R + 273.17}{T + 273.17} \right)$$

$$\sigma = \rho_A / \rho_0$$

$$\rho_A = \sigma\rho_0 = \sigma\rho_R \left(\frac{T_R + 273.17}{T + 273.17} \right) \tag{xiii.1}$$

Using Equation (2.9).

$$\Delta T = Q/\dot{m}C_P = Q/\rho_A G C_P$$

$$\Delta T = \frac{Q(T + 273.16)}{\sigma\rho_R C_P (T_R + 273.16) G} \tag{xiii.2}$$

and substituting some constants,

$$T_R = 0\,^0\mathrm{C}, \ \rho_R = 0.021\,\mathrm{gm/in.}^3, C_P = 1.01\,\mathrm{J/(gm\cdot K)}, Q[\mathrm{J/s}], G\left[\mathrm{in.}^3/\mathrm{s}\right]$$

$$G\left[\mathrm{in.}^3/\mathrm{s}\right] = G\left[\mathrm{ft}^3/\mathrm{min}\right](\mathrm{min}/60\,\mathrm{s})(12\,\mathrm{in./ft})^3$$

$$\Delta T\left[^0\mathrm{C}\right] = \frac{Q[\mathrm{J/s}](T + 273.16)\,\mathrm{K}}{\sigma\left(0.021\,\mathrm{gm/in.}^3\right)\left(1.01\dfrac{\mathrm{J}}{\mathrm{gm\cdot K}}\right)(273.16\,\mathrm{K})G\left[\dfrac{\mathrm{ft}^3}{\mathrm{min}}\right]\left(\dfrac{\mathrm{min}}{60\,\mathrm{s}}\right)\left(\dfrac{12\,\mathrm{in.}}{\mathrm{ft}}\right)^3}$$

$$\boxed{\Delta T\left[^0\mathrm{C}\right] = \frac{C\left(T\left[^0\mathrm{C}\right] + 273.16\right)Q[\mathrm{W}]}{\sigma G\left[\mathrm{ft}^3/\mathrm{min}\right]}}, \ C = 5.99 \times 10^{-3} \tag{xiii.3}$$

XIII.2 DERIVATION OF AIR TEMPERATURE RISE USING AVERAGE BULK TEMPERATURE FOR PROPERTIES

As in Chapter 2, we choose to identify the temperature T in Eq. (*xiii*.3) as the average of the inlet ambient and enclosure air temperatures.

$$T = \bar{T}_B = \frac{T_A + T_E}{2} \tag{xiii.4}$$

Inserting Eq. ($xiii$.4) into Eq. ($xiii$.3) and noting $T_E = T_A + \Delta T$,

$$\Delta T = \frac{C\left(\bar{T}_B + 273.16\right)Q}{\sigma G} = \frac{C\left[\left(\dfrac{T_A + T_E}{2}\right) + 273.16\right]Q}{\sigma G} = \frac{C\left[\left(\dfrac{2T_A + \Delta T}{2}\right) + 273.16\right]Q}{\sigma G}$$

$$\Delta T = \left(\frac{CQ}{\sigma G}\right)\left[\left(\frac{\Delta T}{2}\right) + T_A + 273.16\right] = \left(\frac{CQ}{\sigma G}\right)\left(\frac{\Delta T}{2}\right) + \left(\frac{CQ}{\sigma G}\right)\left(T_A + 273.16\right)$$

$$\Delta T\left[1 - \left(\frac{CQ}{2\sigma G}\right)\right] = \left(\frac{CQ}{\sigma G}\right)\left(T_A + 273.16\right)$$

$$\Delta T = \left(\frac{CQ}{\sigma G}\right)\left(T_A + 273.16\right)\frac{1}{\left[1 - \left(\dfrac{CQ}{2\sigma G}\right)\right]} = \left(\frac{2\sigma G}{CQ}\right)\left(\frac{CQ}{\sigma G}\right)\left(T_A + 273.16\right)\frac{1}{\left[\left(\dfrac{2\sigma G}{CQ}\right) - 1\right]}$$

$$\boxed{\Delta T = \frac{2\left(T_A + 273.16\right)}{\left[\left(\dfrac{2\sigma G}{CQ}\right) - 1\right]}, \quad C = 5.99 \times 10^{-3}} \qquad (xiii.5)$$

where Eq. ($xiii$.5) is, with the exception of σ, identical to Eq. (2.14).

XIII.3 HEAD LOSS DUE TO BUOYANCY PRESSURE

In a manner nearly identical to the method used in Chapter 5, the infinitesimal interior pressure difference dp_B over a height dz is

$$dp_B = g\left(\rho_A - \rho\right)dz$$

where $\rho_A \equiv$ external (ambient) density at inlet

$\rho \equiv$ internal density at height z

At an ambient temperature T_A the internal density at temperature T is

$$\rho = \rho_A \frac{\left(T_A + 273.16\right)}{\left(T + 273.16\right)}$$

Then

$$dp_B = g\left[\rho_A - \rho_A \frac{\left(T_A + 273.16\right)}{\left(T + 273.16\right)}\right]dz = g\rho_A\left[1 - \frac{\left(T_A + 273.16\right)}{\left(T + 273.16\right)}\right]dz$$

into which the linear air temperature profile from Chapter 5 is substituted.

$$T = \left(\Delta T/d_H\right)\left(z - z_0\right) + T_A$$

so that dp_B becomes

$$dp_B = g\rho_A\left[1 - \frac{\left(T_A + 273.16\right)}{\left(\Delta T/d_H\right)\left(z - z_0\right) + T_A + 273.16}\right]dz$$

$$dp_B = g\rho_A dz - g\rho_A\left(T_A + 273.16\right)\left[\frac{1}{\left(\Delta T/d_H\right)\left(z - z_0\right) + T_A + 273.16}\right]$$

The preceding is integrated from $z = z_0$ to $z = z_0 + d_H$

$$\Delta p_B = g\rho_A d_H - g\rho_A\left(T_A + 273.16\right) \int_{z_0}^{z_0+d_H} \left[\frac{dz}{\left(\Delta T/d_H\right)\left(z-z_0\right)+T_A+273.16}\right]$$

$$\Delta p_B = g\rho_A d_H\left[1-\left(\frac{T_A+273.16}{\Delta T}\right)\ln\left(1+\frac{\Delta T}{T_A+2731.6}\right)\right]$$

Making the same approximation to the preceding as was done in Chapter 5,

$$\Delta p_B = g\rho_A d_H\left[\frac{\left(\Delta T/2\right)}{\left(\Delta T/2\right)+T_A+273.16}\right]$$

But $\sigma = \rho_A/\rho_0$, $\rho_A = \sigma\rho_0$, where $\rho_0 = $ air density at sea level and

$$\Delta p_B = g\rho_0 d_H\sigma\left[\frac{\left(\Delta T/2\right)}{\left(\Delta T/2\right)+T_A+273.16}\right]$$

Using

$$\rho_0 = 1.2\,\text{kg/m}^3, \; g = 9.8\;\text{m/s}^2, \; d_H\,[\text{m}], \; T\left[{}^0\text{C}\right], \; \Delta T\left[{}^0\text{C}\right]$$

$$\Delta h_B\left[\text{in. H}_2\text{O}\right] = \Delta p_B\left[\text{Pa, N/m}^2\right]\left(4.019\times10^{-3}\right)$$

the buoyancy driven head becomes

$$\Delta h_B = 0.00120 d_H\sigma\left[\frac{\left(\Delta T/2\right)}{\left(\Delta T/2\right)+T_A+273.16}\right] = 0.00120 d_H\sigma\left[\frac{1}{1+\dfrac{T_A+273.16}{\left(\Delta T/2\right)}}\right]$$

$$\left(xiii.6\right)$$

Noting that Eq. (*xiii*.5) may be rewritten slightly differently as,

$$\frac{\Delta T}{2} = \frac{\left(T_A+273.16\right)}{\left[\left(\dfrac{2\sigma G}{CQ}\right)-1\right]},$$

$$\Delta h_B = 0.00120 d_H\sigma\left[\frac{1}{1+\dfrac{T_A+273.16}{\left(\Delta T/2\right)}}\right] = 0.00120 d_H\sigma\left[\frac{1}{1+\left(\dfrac{2\sigma G}{CQ}\right)-1}\right] = 0.00120 d_H\sigma\left(\frac{CQ}{2\sigma G}\right)$$

$$\boxed{\Delta h_B = 0.00120 d_H\left(\frac{CQ}{2G}\right)}$$

$$\left(xiii.7\right)$$

The σ term has cancelled!

XIII.4 MATCHING SYSTEM LOSS TO BUOYANCY PRESSURE HEAD

In the instance of first-order buoyancy head driven systems, it would be unusual for the total system resistance to be anything other than of a turbulent form. We therefore use a total system resistance that is calculated for a sea level circuit and explicitly use σ as a factor of R_{Sys} in the total head loss H_L.

$$H_L = \sigma R_{Sys} G^2$$

Matching the system head loss to the buoyancy pressure,

$$\sigma R_{Sys} G^2 = 0.00120 d_H \left(\frac{5.99 \times 10^{-3}}{2G} \right) Q$$

$$\boxed{G = 1.53 \times 10^{-2} \left(\frac{d_H Q_d}{\sigma R_{Sys}} \right)^{1/3}}$$

$$(xiii.8)$$

The well mixed air temperature rise formula is listed again.

$$\Delta T = \frac{Q}{\rho_0 G C_P} \left(\frac{1}{\sigma} \right)$$

$$\boxed{\Delta T = \frac{1.76Q}{G} \left(\frac{1}{\sigma} \right)}$$

$$(xiii.9)$$

where a 20 ^{0}C inlet temperature is assumed.

XIII.5 APPLICATION EXAMPLE: BUOYANCY - DRAFT COOLED ENCLOSURE FROM SECTION 5.3

One can use the total sea level resistance $R_{Sys} = 7.11 \times 10^{-5}$ in. H$_2$O/CFM^2 calculated in Section 5.3 where the dissipation height is $d_H = 10$ in. and a total heat dissipation is $Q = 20$ W. If an altitude of 10,000 ft is assumed, then from Figure $xiv.2$, $\sigma = 0.687$.

The air draft is readily calculated using Eq. ($xiii.8$).

$$G = 1.53 \times 10^{-2} \left(\frac{d_H Q_d}{\sigma R_{Sys}} \right)^{1/3} = 1.53 \times 10^{-2} \left[\frac{(10.0)(20.0)}{(0.687)(7.11 \times 10^{-5})} \right]^{1/3} = 2.45 \text{ CFM}$$

The well-mixed air temperature rise at the exit is calculated using Eq. ($xiii.9$).

$$\Delta T = \frac{1.76Q}{G} \left(\frac{1}{\sigma} \right) = \frac{1.76(20)}{2.45} \left(\frac{1}{0.687} \right) = 20.9 \,^0\text{C}$$

The draft at 10,000 ft is greater than the sea level result of 2.16 CFM, but the new ΔT is 4 ^{0}C greater (28%) than the calculated 16.3 ^{0}C for sea level.

XIII.6 APPLICATION EXAMPLE: VERTICAL CONVECTING PLATE FROM SECTION 8.4

The classical vertical and horizontal flat plate correlations for laminar flow, natural convection follow Eq. (8.1) where $n = 0.25$. Therefore

$$Nu_P = C \left(Gr_P Pr \right)^{1/4}, \quad \text{Gr} = \frac{g \rho^2}{\mu^2} \beta \left(T_S - T_A \right) P^3$$

Then

$$Nu_P \propto \rho^{1/2}$$

whereby one may write

$$Nu_P = C \sqrt{\sigma} \left(Gr_P Pr \right)_0^{1/4}$$

the subscript 0 implying "calculated for sea level."

The simplified laminar flow heat transfer coefficients in Table 8.7 are each to be multiplied by $\sqrt{\sigma}$, e.g., for a vertical plate.

$$h = 0.0024\sqrt{\sigma}\left(\Delta T/H\right)^{1/4}$$

The heat transfer formula for this two-sided 6.0 in. x 9.0 in. plate is

$$Q = h_c A_S \Delta T = 0.0024\sqrt{\sigma}\left(\Delta T/H\right)^{1/4} A_S \Delta T = \frac{0.0024\sqrt{\sigma} A_S \left(\Delta T\right)^{1.25}}{H^{0.25}}$$

and

$$\Delta T = \left(\frac{Q}{0.0024\sqrt{\sigma} A_S / H^{0.25}}\right)^{1/1.25}$$

Inserting the various values into the preceding,

$$\Delta T = \left[\frac{8.0}{\left(0.0024\sqrt{0.687}\left(108\right)/\left(6.0\right)^{0.25}\right)}\right]^{1/1.25} = 25.9\,^0C$$

$$h = 0.0024\sqrt{\sigma}\left(\Delta T/H\right)^{1/4} = 0.0024\sqrt{0.687}\left(25.9/6.0\right)^{1/4} = 0.0029\,W/\left(in.^2 \cdot \,^0C\right)$$

thus ΔT is modestly increased by 4 ^0C from 22.2 ^0C to 25.9 ^0C.

XIII.7 COMMENTS CONCERNING OTHER HEAT TRANSFER COEFFICIENTS AT AN ELEVATED ALTITUDE

It should be clear that all heat transfer coefficients must be corrected for altitude by appropriately placing σ into a formula. One merely identifies the location of the air density symbol and insert σ as a factor.

In Chapter 8, the quantity γ was introduced as a calculation convenience.

$$\gamma = \left(g\beta/v^2\right)Pr = \left(g\beta\rho^2/\mu^2\right)Pr$$

In particular, the channel Rayleigh numbers are dependent on ρ so that, e.g., Eq. (8.11) may be re-written as

$$Ra_b = \sigma^2\left(\frac{\gamma_0}{\beta}\right)\beta\left(\frac{b^4}{L}\right)\left(T_W - T_I\right) \qquad\qquad (xiii.10)$$

where the Rayleigh number dependencies γ_0 and γ_0/β are calculated exactly according to Eq. (8.3) and Figure 8.3 at sea level conditions, and Ra_b, Eq. (xiii.10), contains σ^2.

The vertical plate-fin heat sinks in Chapter 9 are modified in the Van de Pol and Tierney Eq.(9.1) using a re-written modified channel Rayleigh number

$$Ra_r^* = \left(\frac{r}{H}\right)\sigma^2 Gr_0 Pr \qquad\qquad (xiii.11)$$

where Gr_0 is calculated for sea level. The author has not re–calculated the impact of σ on the various design plots of Figure 9.2. Therefore the reader is advised to use Eq. (9.1) directly, rather than the design plots. On the other hand, rough estimates using h_c/h_H could be adequate as long as h_H is corrected for altitude. Bear in mind that your calculations are of first–order accuracy anyway.

Bibliography

Adam, J. 1998. Pressure drop coefficients for thin perforated plates. *Electronics Cooling*, 4:40-41.

Agonafer, D. and Moffat, R.J. 1989. Numerical modeling of forced convection heat transfer for modules mounted on circuit boards. *Numerical Simulation of Convection in Electronic Equipment Cooling, ASME HTD,* 121:1-5.

AMCA, Air Movement and Control Association. http://www.amca.org. Accessed on Aug. 22, 2010.

American Society of Heating, Refrigerating, and Air Conditioning Engineers. Standard 51-75.

Anderson, A.M. and Moffat, R.J. 1992. The adiabatic heat transfer coefficient for different flow conditions. *J. Electron. Packag.*, 12:14-21.

Anderson, A. M. and Moffat, R.J. 1990. A heat transfer correlation for arbitrary geometries in electronic equipment, in *Proc. ASME Winter Annual Meeting*, Dallas, TX.

Andrews, J.A. 1988. Package thermal resistance model: dependency on equipment design. *IEEE Trans. CHMT*, 11:528-537.

Bar-Cohen, A., Elperin, T., and Eliasi, R. 1989. R_{th} characterization of chip packages - justification, limitations, and future. *IEEE Trans. CHMT*, 12:724-731.

Bar-Cohen, A. and Rohsenow, W.M. 1984. Thermally optimum spacing of vertical natural convection cooled, parallel plates. *ASME J. of Heat Transfer*, 106:116-123.

Barzelay, M.E., Kin Nee Tong, and Holloway, G. May 1955. *National Advisory Committee for Aeronautics*, Tech. Note 3295.

Beck, J.V., Cole, K.D., Haji-Sheikh, A., and Litkouhi, B. 2010. *Heat conduction using Green's functions*. 2nd ed. CRC Press, Taylor and Francis Group. Boca Raton.

Beck, J.V., Cole, K.D., Haji-Sheikh, A., and Litkouhi, B. 1992. *Heat conduction using Green's functions*. Hemisphere Publishing Corp.

Bilitsky, A. 1986. The Effect of Geometry on Heat Transfer by Free Convection from a Fin Array. Master's thesis. Dept. of Mech. Eng., Ben-Gurion University of the Negev, Beer Sheva, Israel.

Blazej, D. 2003. Thermal interface materials. *Electronics Cooling*. 9(4): 14-20.

Boelter, L.M.K., Young, G., and Iversen, H.W. July 1948. *NACA TN* 1451 (now NASA), Washington, D.C.

Brebbia, C.A. and Dominguez, J. 1989. *Boundary elements, an introductory course*. Computational Mechanics Pubs, Boston. Co-published with McGraw-Hill Book Co., New York.

Burnett, D.S. 1987. *Finite element analysis*. Addison-Wesley Publishing Company.

Carslaw, H.S. and Jaeger, J.C. 1986. *Conduction of heat in solids*. Clarendon Press, Oxford.

Childs, G.E., Ericks, L.J., and Powell, R.L. 1973. *Thermal conductivity of solids at room temperature and below*. U.S. Dept. of Commerce, p. 507, B.Z.S.-2.

Church, S.W., and Chu, H.S. 1975. Correlating equations for laminar and turbulent free convection from a vertical plate. *Int. J. Heat and Mass Transfer*, 18:1323.

Copeland, D. 1992. Effects of channel height and planar spacing on air cooling of electronic components. *J. Elect. Packag.*, 114: 420-424.

David, R.F. 1977. Computerized thermal analysis of hybrid circuits. *Trans. Elect. Comp. Conf.*, Arlington, VA.

Dean, D.J. 1985. *Thermal design of electronic circuit boards and packages*. Electrochemical Pubs. Limited, 8 Barns St, Ayr, Scotland.

Dittus, F.W. and Boelter, L.M.K. 1930. *Univ. Calif. Berkeley Publ. Eng.*, 2:433.

Dunkle, R.V. 1954. Thermal radiation tables and applications. *Trans. ASME*, 65: 549-552.

Ellison, G.N. 2003. Maximum thermal spreading resistance for rectangular sources and plates with nonunity aspect ratios. *IEEE Trans. CPT*, 26: 439-454.

Ellison, G.N. 1996. Thermal analysis of circuit boards and microelectronic components using an analytical solution to the heat conduction equation, an invited lecture presented at the 12th Annual Semiconductor Thermal Measurement and Management Symposium, Austin, Texas, March 5-7.

Ellison, G.N., Hershberg, E.L., and Patelzick, D.L. 1994. The crycooler implications of flexible, multi-channel I/O cables for low temperature superconducting microelectronics. *Advances in Cryogenic Engineering*, 39: 1167-1176, Spring Publishing Co.

Ellison, G.N. 1993. Methodologies for thermal analysis of electronic components and systems, in *Advances in thermal modeling of electronic components and systems*, Eds. A. Bar-Cohen and A.D. Kraus, vol. III, ASME Press, New York and IEEE Press, New York.

Ellison, G.N. 1992. Extensions of a closed form method for subtrate thermal analyzers to include thermal resistances from source to-substrate and source-to-ambient. *IEEE Trans. Comp., Hybrids, Mfg. Tech.*, 15: 658-666.

Ellison, G.N. 1990. TAMS: A thermal analyzer for multilayer structures, *Electrosoft*, 1: 85-97.

Ellison, G.N. and Patelzick, D.L. 1986. The thermal design of a forced air cooled power supply. *Proced. 6th Intl. Elect. Packag. Conf.*, pp. 829-838.

Ellison, G.N. 1984a. *Thermal computations for electronic equipment*. Van Nostrand Reinhold. Reprint edition by Robert E. Kreiger Pub. Co., Malabar, FL, 1989.

Ellison, G.N. 1984b. A review of a thermal analysis computer program for micro-electronic devices, *Thermal Mgt. Concepts in Microelec. Packaging*, ISHM Tech. Monograph Series 6984-003.

Ellison, G.N. 1979. Generalized computations of the gray body shape factor for thermal radiation from a rectangular U-channel. *IEEE Trans. Comp., Hybrids, Mfg. Tech.*, CHMT-2: 517-522.

Ellison, G.N. 1976. The thermal design of an LSI single chip package. *IEEE Trans. Parts, Hybrids, Packag.*, PHP-12:371-378.

Ellison, G. N. 1973. The effect of some composite structures on the thermal resistance of substrates and integrated circuit chips, *IEEE Trans. Electron Dev.*, ED-20:233-238.

Estes, R. 1987. *Thermal design considerations for COB applications*. Paper presented at the 7th Ann. Intl. Elec. Packag. Conf.

Faghri, M. 1996. Entrance design correlations for circuit boards in forced-air cooling, *Air cooling technology for electronic equipment*, Sung Jin Kim and Sang Woo Lee, Eds., CRC Press, Boca Raton, FL, pp. 47-80.

Farrell, O. J. and Bertram, R. 1963. *Solved problems - gamma and beta functions, Legendre polynomials, Bessel functions*. The MacMillan Company, New York, NY.

Fermi, E. 1937. *Thermodynamics*. Dover Publications, Inc. New York, NY.

Fried, E. and Idelchick, I.E. 1988. *Flow resistance, a design guide for engineers*. Hemisphere Publishing Corp., Washington, D.C.

Graebner, J.E. and Azar, K. 1995. Thermal conductivity of printed wiring boards. *Electronics Cooling*, 1:27.

Hay, D. 1964. Cooling card mounted solid-state component circuits. McLean Engineering Division of Zero Corporation, Trenton, NJ 08691.

Hein, V.L. and Lenzi, V.K. 1969. Thermal analysis of substrates and integrated circuits. Paper presented at the 1969 Electronic Components Conf.

Holman, J.P. 1990. *Heat transfer*, 7th Ed. McGraw-Hill Book Co, New York, NY.

Holman, J.P. 1974. *Thermodynamics*. McGraw-Hill Publishing Co., New York, NY.

Howell, J.R. 1982. *A catalog of radiation configuration factors*. McGraw-Hill Book Co., New York, NY.

Hultberg, J.A. and O'Brien, P.F. 1971. *TAS (Thermal Analyzer System)*. Jet Propulsion Laboratory, California Institute of Technology, Pasadena, CA, Tech. Rep. 32-1416.

Jonsson, H. and Moshfegh, B. 2001. Modeling of the thermal and hydraulic performance of plate fin, strip fin, and pin fin heat sinks-influence of flow bypass, *IEEE Trans. CPT*, CPT-24:142-149.

Joy, R.C. and Schlig, E.S. Thermal properties of very fast transistors. *IEEE Trans. Electron Devices*, vol. ED-17, no. 8, pp. 586-594, Aug. 1970.

Kang, S.S. 1994. The thermal wake function for rectangular electronic modules. *J. Electron. Pkg.* 116: 55-59.

Kays, W.M. and Crawford, M.E. 1980. *Convective heat and mass transfer*, 2nd. Ed., McGraw-Hill Book Co., New York, NY.

Kays, W.M. and London, A.L. 1964. *Compact heat exchangers*, McGraw-Hill Book Co., New York, NY.

Kays, W.M. 1955. Numerical solutions for laminar-flow heat transfer in circular tubes, *Trans. ASME* 58:1265-1274.

Kennedy, D.P. 1960. Spreading resistance in cylindrical semicoductor devices. *J. Applied Physics*, 31:1490-1497.

Khan, W.A., Culham, J.R., and Yovanovich, M.M. 2005. Modeling of cylindrical pin-fin heat sinks for electronic packaging. Paper presented at the 21st *IEEE Semi-Therm Symposium*.

Kraus, A.D. and Bar-Cohen, A. 1995. *Design and analysis of heat sinks*, John Wiley & Sons, Inc., New York, NY.

Kraus, A.D. and Bar-Cohen, A. 1983. *Thermal analysis and control of electronic equipment*, Mc-Graw-Hill, New York, NY.

Kreith, F. and Bohn, M.S. 2001. *Principles of heat transfer*, 6th Ed., Brooks/Cole, New York, NY.

Langhaar, H.L. 1942. Steady flow in the transition length of a straight tube, *J. Appl. Mech., Trans. ASME* 64:A-55.

Lasance, C. 2006. Thermal conductivity of III-V semiconductors. *Electronics Cooling* 12:4.

Lasance, C., Rosten, H.I., and Parry, J.D. 1997. The world of thermal characterization according to DELPHI, part I: background to DELPHI. *IEEE Trans. Comp., Hybrids, Mfg. Tech.,* CHMT-20: 384-391.

Lasance, C., Rosten, H.I., and Parry, J.D. 1997. Experimental and numerical methods, part I: background to DELPHI. *IEEE Trans. Comp., Hybrids, Mfg. Tech.,* CHMT-20: 392-398.

Lasance, C. 1998. The thermal conductivity of silicon. *Electronics Cooling* 4:12.

Lea, S.M. 2004. *Mathematics for physicists*, Brooks/Cole- Thomson Learning, Belmont, CA.

Lee, S., Song, V.A.S., and Moran, K.P.1995. Constriction/spreading resistance model for electronics packaging. *Proc.* 4th *ASME/JSME Thermal Eng. Joint Conf.* 4:199-206.

Lee, S. 1995. Optimum design and selection of heat sinks. Proceedings of the 11th Annual IEEE Semiconductor Thermal Measurement and Management Symposium, San Jose, CA, pp. 48-54.

Linton, R.A. and Agonafer, D. 1989. Thermal model of a PC, numerical simulation of convection in electronic equipment cooling. ASME Pub HTD121: 69-72.

Mahajan, R. 2004. Thermal interface materials. *Electronics Cooling* 10:10-16.

Marotta, E., Mazzuca, S., and Norley, J. 2002. Thermal joint conductance. *Electronics Cooling* 8:16-22.

McAdams, W.H. 1954. *Heat transmission*, 3rd Ed., McGraw-Hill Book Company, New York.

Mikic, B.B. 1966. *Thermal contact resistance*. Sc.D. Thesis, Dept. of Mech. Eng., MIT, Cambridge, MA.

Moffat, R.J. and Ortega, A. 1988. Direct air cooling of electronic components. In *Advances in thermal modeling of electronic components and systems*, Ed. A. Bar-Cohen and A.D. Kraus, vol. 1, chap. 3, p. 240, Hemisphere Publishing Co., New York, NY.

Morse, P.M. and Feshbach, H.F. 1953. *Methods of theoretical physics*, Part I, McGraw-Hill Book Company, New York, NY.

Muzychka, Y.S., Yovanovich, M.M., and Culham, J.R. 2001. Application of thermal spreading resistance in compound systems, in *Proc. 39th AIAA Aerosp. Sci. Meeting Exhibit*, Reno, NV, Jan. 8-11.

Muzychka, Y.S. and Yovanovich, M.M. 1998. Modeling friction factors in non-circular ducts for developing laminar flow. Presented at the *2nd AIAA Theoretical Fluid Mechanics Meeting*, Albuqerque, NM.

Muzychka, Y.S. and Yovanovich, M.M. 1998. Modeling Nusselt numbers for thermally developing laminar flow in non-circular ducts. Paper presented at the *AIAA/ASME Joint Thermophysics Heat Transfer Conf.*, paper no. 98-2586.

Palisoc, A.L. and Lee, C.C. 1988. Thermal design of integrated circuit devices. Paper presented at the 4th Ann. IEEE semiconductor thermal measurement and management symposium.

Patankar, S. V. 1980. *Numerical heat transfer and fluid flow*. Hemisphere Publishing Corp., New York, NY.

PDE Solutions, Inc., Spokane Valley, WA. http://www.pdesolutions.com.

Pinto, E.J. and Mikic, B.B. 1986. Temperature prediction on substrates and integrated circuit chips. Paper presented at the 1986 *AIAA/ASME Thermophysics and Heat Trans Conf.*

Present, R.D. 1958. *Kinetic theory of gases*, McGraw-Hill Book Co. New York, NY.

Raithby, G.D. and Hollands, K.G.T. 1985. Natural convection. *Handbook of heat transfer fundamentals,* Eds. Rohsenow, W.M., Hartnett, J.P. and Granic, E.M., McGraw-Hill, chap. 6, pp. 34-36.

Ramakrishna, K. 1982. Thermal analysis of composite multilayer structures with multiple heat sources. Paper presented at the 2nd *Intl. Elect. Packag. Conf.*

Rantala, J. 2004. Surface flatness, *Electronics Cooling* 10:12.

Rea, S.N. and West, S.E. 1976. Thermal radiation from finned heat sinks. *IEEE Trans. on Parts, Hybrids, and Packaging* PHP-12:115-117.

Reddy, J.N. 1984. *An introduction to the finite element method*. McGraw-Hill, New York, NY.

Rhee, J. and Bhatt, A.D. 2007. Spatial and temporal resolution of conjugate conduction-convection thermal resistance. *IEEE Trans. Comp. and Pkg. Tech.* 30:673-682.

Rosten, H., Lasance, C. and Parry, J. 1997. The world of thermal characterisation according to DELPHI-Part I: Background to DELPHI. *IEEE Trans. CHMT* 20:384-391.

Rosenberg, H.M. 1963. *Low temperature solid state physics*. Oxford University Press, London.

Saums, D.L. 2007. Developments with metallic thermal interface materials. *Electronics Cooling* 13:26-30.

Scarborough, J.B. 1958. *Numerical mathematical analysis*, 4th Ed. Johns Hopkins Press, Baltimore.

Schlichting, H. 1968. *Boundary layer theory*. 6th Ed. McGraw-Hill Publishing Co., New York, NY.

Schnipke, R J. 1989. A fluid flow and heat transfer analysis for evaluating the effectiveness of an IC package heat sink. Paper presented at the 5th Ann. IEEE semiconductor thermal measurement and management symposium.

Segerlind, L.J. 1976. *Applied finite element analysis.* John Wiley & Sons, Inc, New York, NY.

Simons, R.E. and Schmidt, R.R. 1997. A simple method to estimate heat sink air flow bypass. *Electronics Cooling* 2: 36-37.

Sparrow, E.M., Niethammer, J.E., and Chaboki, A. 1982. Heat transfer and pressure drop charcateristics of arrays of rectangular modules encountered in electronic equipment. *Int. J. Heat Mass Transfer,* 25(7): 961-973.

Sparrow, E.M., Niethammer, J.E., and Chaboki, A. 1982. Heat transfer and pressure drop characteristics of arrays of rectangular modules encountered in electronic equipment. *Int. J. Heat Mass Transfer* 25: 961-973.

Spoor, J. 1974. *Heat sink application handbook*, AHAM, Inc., Azusa, CA.

Steinberg, D.S. 1991. *Cooling techniques for electronic equipment.* Wiley-Interscience Pubs, New York, NY.

Teertstra, P., Culham, J.R., and Yovanovich, M.M. 1996. Comprehensive review of natural and mixed convection heat transfer models for circuit board arrays. *Proceedings of the International Electronics Packaging Symposium* pp. 156-171.

Teertstra, P., Yovanovich, M.M., Culham, J.R. 1997. Pressure loss modeling for surface mounted cuboid-shaped packages in channel flow. *IEEE Transactions on Components, Packaging, and Manufacturing Technology*-Part A, CPMT-20:463-469.

Touloukian, Y.S., Powell, R.W., Ho, C.Y., and Klemens, P.G., editors. *Thermophysical Properties of Matter, TPRC Data Series.* 1970. Undoped germanium 1:131; undoped silicon,1:339.

Van de Pol, D.W. and Tierney, J.K. 1974. Free convection heat transfer from verical fin-arrays. *IEEE Trans. on Parts, Hybrids, and Packaging* PHP-10:542-543.

Watson, G.N. 1922. *A treatise on the theory of Bessel functions*. Cambridge University Press, London.

Wills, M. 1983. Thermal analysis of air-cooled PCBs. *Electronic Production*, May, 11-18, pp. 11-18.

Wilson, J. 2006. Thermal conductivity of III-V semiconductors. *Electronics Cooling* 12:4.

Wirtz, R. 1996. Forced air cooling of low profile package arrays. *Air cooling technology for electronic equipment*, Eds Sung Jin Kim and Sang Woo Lee. CRC Press, pp. 81-102.

Wirtz, R.A. and Colban, D.M. 1995. Comparison of the cooling performance of staggered and in-line arrays of electronic packages. *Proc. ASME/JSME Thermophys. Conf.*, Maui, Hawaii 4:215-221.

Wirtz, R.A., Chen, W., and Zhou, R. 1994. Effect of flow bypass on the performance of longitudinal fin heat sinks. *J. Electronic Pkg.* 116:206-211.

Wirtz, R.A. and Dykshoorn, P. 1984. Heat transfer from arrays of flat packs in a channel flow. *Proc. 4th IEPS Conf.*, Baltimore, MD, pp. 318-326.

Wirtz, R.A. and Mathur, A. 1994. Convective heat transfer distribution on the surface of an electrronic package. *J. Electronic Pkg.* 116:49-54.

Yovanovich, M.M. 2005. Four decades of research on thermal contact, gap, and joint resistance in microelectronics. *IEEE Trans. CPMT*, 28:182-206.

Yovanovich, M.M. and Antonetti, V.W. 1998. Application of thermal contact resistance theory to electronic packages. Eds. A. Bar-Cohen and A.D. Kraus. In *Advances in thermal modeling of electronic components and systems*, Ed. A. Bar-Cohen and A.D. Kraus, vol. 1, ch. 2, Hemisphere Publishing Co., New York.

Yovanovich, M.M. and Teerstra, P. 1998. Laminar forced convection from isothermal rectangular plates from small to large Reynold's numbers. American Institute of Aeronautics and Astronautics, paper AIAA-98-2675.

Yovanovich, M.M., Myzychka, Y.S., and Culham, J.R. 1998. Spreading resistance of isoflux rectangles on compound flux channels, in *Proc. 36th AIAA Aerospace Sci. Meeting Exhibit*, Reno, NV, Jan. 12-15, 1998.

Yovanovich, M.M., Culham, J.R., and Teertstra, P. 1997. Calculating interface resistance. *Electronics Cooling* 3:24-29.

Index

Printed and bound by CPI Group (UK) Ltd, Croydon, CR0 4YY

17/10/2024

01775672-0005